Composite Materials Science and Engineering

Composite Materials Science and Engineering

Editor:

Harinirina Randrianarisoa, Ph.D.

ArclerPress

www.arclerpress.com

Composite Materials Science and Engineering

Editor: Harinirina Randrianarisoa, Ph.D.

708 3rd Avenue, 6th Floor
New York
NY 10017
United States of America
www.arclerpress.com

ISBN: 978-1-68094-403-7

Library of Congress Control Number: 2016951051

About the Editor

Harinirina Randrianarisoa, Ph.D.

Hari Randrianarisoa obtained is Ph.D from Institute of Electronics, Microelectronics and Nanotechnology (IEMN) of the University of Science and Technology – Lille (France). His thesis focused for designing and manufacturing of thermal micro sensors. His interests span from on renewable energy especially solar energy. He has a strong knowledge in technology transfer field. He is currently working as a Business advisor for economic development organization in Québec City and as a consultant Scientist advisor.

Contents

List of Contributors

James N. Boland
CSIRO Exploration and Mining, Kenmore QLD 4069, Australia

Xing S. Li
CSIRO Exploration and Mining, Kenmore QLD 4069, Australia

P. Del Linz
School of Civil and Environmental Engineering, Nanyang Technological University, 50 Nanyang Avenue, 639798 Singapore, Singapore

X. Liang
Ocean College, Zhejiang University, Room 302, Ocean Engineering Building, Zhoushan Campus, Zhejiang University, Dinghai, Zhoushan 316021, China

P.A. Hooper
Department of Mechanical Engineering, Imperial College London, South Kensington Campus, London SW7 2AZ, United Kingdom

L.Z. Wang
College of Civil Engineering and Architecture, Zhejiang University, Yuhangtang Road 866#, Hangzhou 310058, China

J.P. Dear
Department of Mechanical Engineering, Imperial College London, South Kensington Campus, London SW7 2AZ, United Kingdom

Gergely Czél
MTA–BME Research Group for Composite Science and Technology, Budapest University of Technology and Economics, Műegyetem rkp. 3., H-1111 Budapest, Hungary
Advanced Composites Center for Innovation and Science, University of Bristol, Queen's Building, BS8 1TR Bristol, United Kingdom

Meisam Jalalvand
Advanced Composites Center for Innovation and Science, University of Bristol, Queen's Building, BS8 1TR Bristol, United Kingdom

Michael R. Wisnom
Advanced Composites Center for Innovation and Science, University of Bristol, Queen's Building, BS8 1TR Bristol, United Kingdom

Marco Corradi
Mechanical and Construction Engineering Department, Northumbria University, Wynne-Jones Building, NE1 8ST Newcastle upon Tyne, United Kingdom
Department of Engineering, University of Perugia, Via Duranti, 93 06125 Perugia, Italy

Antonio Borri
Department of Engineering, University of Perugia, Via Duranti, 93 06125 Perugia, Italy

Giulio Castori
Department of Engineering, University of Perugia, Via Duranti, 93 06125 Perugia, Italy

Emanuela Speranzini
Department of Engineering, University of Perugia, Via Duranti, 93 06125 Perugia, Italy

M.R. Wisnom
Advanced Composites Centre for Innovation and Science, University of Bristol, Queen's Building, BS8 1TR Bristol, United Kingdom

G. Czél
Advanced Composites Centre for Innovation and Science, University of Bristol, Queen's Building, BS8 1TR Bristol, United Kingdom
MTA–BME Research Group for Composite Science and Technology, Budapest University of Technology and Economics, Műegyetem rkp. 3, H-1111 Budapest, Hungary

Y. Swolfs
Department of Materials Engineering, KU Leuven, Kasteelpark Arenberg 44 bus 2450, B-3001 Leuven, Belgium

M. Jalalvand
Advanced Composites Centre for Innovation and Science, University of Bristol, Queen's Building, BS8 1TR Bristol, United Kingdom

L. Gorbatikh
Department of Materials Engineering, KU Leuven, Kasteelpark Arenberg 44 bus 2450, B-3001 Leuven, Belgium

I. Verpoest
Department of Materials Engineering, KU Leuven, Kasteelpark Arenberg 44 bus 2450, B-3001 Leuven, Belgium

F. Gommer
Composites Research Group, Faculty of Engineering, University of Nottingham, University Park, Nottingham NG7 2RD, UK
Department of Aeronautics, Imperial College London, South Kensington, London SW7 2AZ, UK

A. Endruweit
Composites Research Group, Faculty of Engineering, University of Nottingham, University Park, Nottingham NG7 2RD, UK

A.C. Long
Composites Research Group, Faculty of Engineering, University of Nottingham, University Park, Nottingham NG7 2RD, UK

Dmitry S. Ivanov
Advanced Composites Centre for Innovation and Science, University of Bristol, Queen's Building, University Walk, Bristol BS8 1TR, United Kingdom

Yann M. Le Cahain
Advanced Composites Centre for Innovation and Science, University of Bristol, Queen's Building, University Walk, Bristol BS8 1TR, United Kingdom

Surush Arafati
Advanced Composites Centre for Innovation and Science, University of Bristol, Queen's Building, University Walk, Bristol BS8 1TR, United Kingdom

Alexandre Dattin
Advanced Composites Centre for Innovation and Science, University of Bristol, Queen's Building, University Walk, Bristol BS8 1TR, United Kingdom

Sergey G. Ivanov
Institute of Polymer Mechanics, University of Latvia, 23 Aizkraukles Str., Riga LV-1006, Latvia

Andrey Aniskevich
Institute of Polymer Mechanics, University of Latvia, 23 Aizkraukles Str., Riga LV-1006, Latvia

Mingzhong Zhang
Advanced and Innovative Materials (AIM) Group, Department of Civil, Environmental and Geomatic Engineering, University College London, London WC1E 6BT, UK
School of Mechanical, Aerospace and Civil Engineering, The University of Manchester, Manchester M13 9PL, UK

Andrey P. Jivkov
School of Mechanical, Aerospace and Civil Engineering, The University of Manchester, Manchester M13 9PL, UK

O.J. Nixon-Pearson
University of Bristol, Advanced Composites Centre for Innovation and Science, Queens Building, University Walk, Bristol BS8 1TR, UK

S.R. Hallett
University of Bristol, Advanced Composites Centre for Innovation and Science, Queens Building, University Walk, Bristol BS8 1TR, UK

Runzhou Huang
College of Materials Science and Engineering, Nanjing Forestry University, Nanjing 210037, China

Xinwu Xu
College of Materials Science and Engineering, Nanjing Forestry University, Nanjing 210037, China

Sunyoung Lee
Korea Forest Research Institute, Seoul 130-712, Korea

Yang Zhang
College of Materials Science and Engineering, Nanjing Forestry University, Nanjing 210037, China

Birm-June Kim
Department of Forest Products and Biotechnology, Kookmin University, Seoul 136-702, Korea

Qinglin Wu
College of Materials Science and Engineering, Nanjing Forestry University, Nanjing 210037, China
School of Renewable Natural Resources, Louisiana State University Agricultural Center, Baton Rouge, LA 70803, USA

Yong X. Gan
Department of Mechanical, Industrial and Manufacturing Engineering, College of Engineering, University of Toledo, 2801 W Bancroft Street, Toledo, OH 43606, USA
Department of Mechanical Engineering, Albert Nerken School of Engineering, The Cooper Union for the Advancement of Science and Art, 41 Cooper Square, New York, NY 10003, USA

Daniel Solomon
Department of Mechanical Engineering, Albert Nerken School of Engineering, The Cooper Union for the Advancement of Science and Art, 41 Cooper Square, New York, NY 10003, USA

Michael Reinbolt
Department of Mechanical, Industrial and Manufacturing Engineering, College of Engineering, University of Toledo, 2801 W Bancroft Street, Toledo, OH 43606, USA

René Salgado-Delgado
Division de Estudios de Posgrado e Investigación del Instituto Tecnológico de Zacatepec, Calzada Tecnológico No. 27, Col. Centro, 62780 Zacatepec, MOR, Mexico

Alfredo Olarte-Paredes
Division de Estudios de Posgrado e Investigación del Instituto Tecnológico de Zacatepec, Calzada Tecnológico No. 27, Col. Centro, 62780 Zacatepec, MOR, Mexico

Areli Marlen Salgado-Delgado
Division de Estudios de Posgrado e Investigación del Instituto Tecnológico de Zacatepec, Calzada Tecnológico No. 27, Col. Centro, 62780 Zacatepec, MOR, Mexico

Zully Vargas-Galarza
Division de Estudios de Posgrado e Investigación del Instituto Tecnológico de Zacatepec, Calzada Tecnológico No. 27, Col. Centro, 62780 Zacatepec, MOR, Mexico

Teresa Lopez-Lara
Division de Estudios de Posgrado, Facultad de Ingeniería, Universidad Autónoma de Querétaro, Cerro de las Campanas S/N, Col. de lãs Campanas, 76010 Querétaro, QRO, Mexico

Juan Bosco Hernández-Zaragoza
Division de Estudios de Posgrado, Facultad de Ingeniería, Universidad Autónoma de Querétaro, Cerro de las Campanas S/N, Col. de lãs Campanas, 76010 Querétaro, QRO, Mexico

Israel Rico-Rodríguez
Division de Estudios de Posgrado e Investigación del Instituto Tecnológico de Zacatepec, Calzada Tecnológico No. 27, Col. Centro, 62780 Zacatepec, MOR, Mexico

Gonzalo Martínez-Barrera
Laboratorio de Investigación y Desarrollo de Materiales Avanzados, Facultad de Química, Universidad Autónoma del Estado de México, Km 12 Toluca-Atlacomulco, 50200 San Cayetano, MEX, Mexico

Chia-Liang Sun
Department of Chemical and Materials Engineering, Chang Gung University, Guishan, Taoyuan 333, Taiwan
Biosensor Group, Biomedical Engineering Research Center, Chang Gung University, Guishan, Taoyuan 333, Taiwan

Jheng-Sin Su
Department of Chemical and Materials Engineering, Chang Gung University, Guishan, Taoyuan 333, Taiwan

Shun-Yi Lai
Department of Chemical and Materials Engineering, Chang Gung University, Guishan, Taoyuan 333, Taiwan

Yu-Jen Lu
Department of Neurosurgery, Chang Gung Memorial Hospital, No. 5, Fu-Shing Road, Guishan, Taoyuan 333, Taiwan

Laura Cristina Rusu
Department of Technology of Materials and Devices in Dental Medicine, "Victor Babeş" University of Medicine and Pharmacy Timişoara, Piaţa Eftimie Murgu, No. 2, 300041 Timioara, Romania

Ioan Avram Nedelcu
Faculty of Applied Chemistry and Material Science, Politehnica University of Bucharest, 1-7 Polizu Street, 011061 Bucharest, Romania

Mădălina Georgiana Albu
Collagen Department, Leather and Footwear Research Institute, 93 Ion Minulescu, 031215 Bucharest, Romania

Maria Sonmez
Collagen Department, Leather and Footwear Research Institute, 93 Ion Minulescu, 031215 Bucharest, Romania

Georgeta Voicu
Faculty of Applied Chemistry and Material Science, Politehnica University of Bucharest, 1-7 Polizu Street, 011061 Bucharest, Romania

Marius Radulescu
Faculty of Applied Chemistry and Material Science, Politehnica University of Bucharest, 1-7 Polizu Street, 011061 Bucharest, Romania

Denisa Ficai
Faculty of Applied Chemistry and Material Science, Politehnica University of Bucharest, 1-7 Polizu Street, 011061 Bucharest, Romania

Anton Ficai
Faculty of Applied Chemistry and Material Science, Politehnica University of Bucharest, 1-7 Polizu Street, 011061 Bucharest, Romania

Meda-Lavinia Negrutiu
Department of Technology of Materials and Devices in Dental Medicine, "Victor Babeş" University of Medicine and Pharmacy Timişoara, Piaţa Eftimie Murgu, No. 2, 300041 Timi□oara, Romania

Cosmin Sinescu
Department of Technology of Materials and Devices in Dental Medicine, "Victor Babeş" University of Medicine and Pharmacy Timişoara, Piaţa Eftimie Murgu, No. 2, 300041 Timi□oara, Romania

Magdalena Streckova
Institute of Materials Research, Slovak Academy of Sciences, Watsonova 47, 040 01 Košice, Slovakia

Radovan Bures
Institute of Materials Research, Slovak Academy of Sciences, Watsonova 47, 040 01 Košice, Slovakia

Maria Faberova
Institute of Materials Research, Slovak Academy of Sciences, Watsonova 47, 040 01 Košice, Slovakia

Pavel Kurek
Institute of Materials Research, Slovak Academy of Sciences, Watsonova 47, 040 01 Košice, Slovakia

Pavla Roupcova
Institute of Physics of Materials ASCR v.v.i., Zizkova 22, 616 62 Brno, Czech Republic

Hynek Hadraba
CEITEC IPM, Institute of Physics of Materials ASCR, Zizkova 513/22, 616 62 Brno, Czech Republic

Vladimir Girman
Institute of Physics, Faculty of Science, Pavol Jozef Šafárik University, Park Angelinum 9, 040 01 Košice, Slovakia

Jozef Strecka
Institute of Physics, Faculty of Science, Pavol Jozef Šafárik University, Park Angelinum 9, 040 01 Košice, Slovakia

E.P. Bowyer
Department of Aeronautical and Automotive Engineering, Loughborough University, Loughborough, Leicestershire LE11 3TU, UK

V.V. Krylov
Department of Aeronautical and Automotive Engineering, Loughborough University, Loughborough, Leicestershire LE11 3TU, UK

Stefano Del Rosso
Imperial College London, Exhibition Road, SW7 2AZ London, UK

Lorenzo Iannucci
Imperial College London, Exhibition Road, SW7 2AZ London, UK

Paul T. Curtis
Imperial College London, Exhibition Road, SW7 2AZ London, UK
Defence Science and Technology Laboratory, Porton Down, Salisbury SP4 0JQ, UK

M.P.F. Sutcliffe
Cambridge University Engineering Department, Trumpington Street, Cambridge CB2 1PZ, UK

S.L. Lemanski
Cambridge University Engineering Department, Trumpington Street, Cambridge CB2 1PZ, UK

A.E. Scott
School of Engineering Science, University of Southampton, Southampton, Hampshire SO17 1BJ, UK

Farhan Ansari
Department of Fiber and Polymer Technology, KTH Royal Institute of Technology, SE-100 44 Stockholm, Sweden

Mikael Skrifvars
Swedish Centre for Resource Recovery, University of Borås, S-50190 Borås, Sweden

Lars Berglund
Department of Fiber and Polymer Technology, KTH Royal Institute of Technology, SE-100 44 Stockholm, Sweden
Wallenberg Wood Science Center (WWSC), KTH Royal Institute of Technology, SE-100 44 Stockholm, Sweden

Joseph Mills-Brown
Advanced Composites Centre for Innovation and Science (ACCIS), University of Bristol, Queen's Building, Bristol BS8 1TR, United Kingdom

Kevin Potter
Advanced Composites Centre for Innovation and Science (ACCIS), University of Bristol, Queen's Building, Bristol BS8 1TR, United Kingdom

Steve Foster
McLaren Racing Limited, McLaren Technology Centre, Chertsey Road, Woking, Surrey GU21 4YH, United Kingdom

Tom Batho
McLaren Racing Limited, McLaren Technology Centre, Chertsey Road, Woking, Surrey GU21 4YH, United Kingdom

Tanvir Manzur
Department of CE, BUET, Dhaka 1000, Bangladesh

Nur Yazdani
Department of CE, University of Texas at Arlington, Arlington, TX 76019, USA

Md. Abul Bashar Emon
Department of CE, BUET, Dhaka 1000, Bangladesh

S. Gholizadeh
Universiti Putra Malaysia, 43400 UPM Serdang, Selangor Darul Ehsan, Malaysia

Cristina Dueso
Newcastle University, Merz Court, NE1 7RU Newcastle upon Tyne, United Kingdom

Claire Thompson
Newcastle University, Merz Court, NE1 7RU Newcastle upon Tyne, United Kingdom

Ian Metcalfe
Newcastle University, Merz Court, NE1 7RU Newcastle upon Tyne, United Kingdom

Preface

Composites are one of the most widely used materials because of their adaptability to different situations and the relative ease of combination with other materials to serve specific purposes and exhibit desirable properties. In surface transportation, reinforced plastics are the kind of composites used because of their huge size. They provide ample scope and receptiveness to design changes, materials and processes. The strength-weight ratio is higher than other materials. Their stiffness and cost effectiveness offered, apart from easy availability of raw materials, make them the obvious choice for applications in surface transportation. Composite materials have a wide range of properties and uses, making them the most common building and manufacturing materials in use today. They are created by combining two different components in order to create a new material with different properties than those which were started with. Concrete, fabric blends, paint, particle board, plastics, glass and even paper are all examples of composite materials. Metallic alloys are a different type of composite, requiring metal ores to be combined in extreme conditions. These metal composites are often used in items such as utensils, ornaments, trinkets, statues and as building materials. In 1940 fiberglass was invented and the modern era of composite materials began. The basic premise of a modern composite material is some sort of fiber material for the structure, and some sort of plastic or glue to hold the shape and provide strength. These fiber-reinforced plastic composites are used in manufacturing boat and kayak hulls, swimming pool liners, turbine blades, sporting goods and many other possible applications. Composite materials distinguish themselves from other plastics products of synthesis by characteristics which allow them properties of inalterability and low weight, to substitute for metal pieces. Since a few years, the market of composite materials doesn't stop growing. In the present day scenario, use of advanced composite materials has been extended to a large number of segments.

The details of the research results and the new development in fields are highlighted in this book. The subject discussed the advances of specific idea and technologies. With regards to composite materials research, the specifics advances discussed in this book include the experimentation, investigation and the characterization of the behavior mechanical, physical property for new products. Advances of the use of the composite materials to produce a new material like nano composite and the applications are also discussed.

Editor
Harinirina Randrianarisoa, Ph.D.

INTRODUCTION

A composite material is a combination of two or more materials that results in better properties than those of the individual components used alone. In contrast to metallic alloys, each material retains its separate chemical, physical, and mechanical properties. The two constituents are a reinforcement and a matrix. The main advantages of composite materials are their high strength and stiffness, combined with low density, when compared with bulk materials, allowing for a weight reduction in the finished part. The reinforcing phase provides the strength and stiffness. In most cases, the reinforcement is harder, stronger, and stiffer than the matrix. The reinforcement is usually a fiber or a particulate. Particulate composites have dimensions that are approximately equal in all directions. They may be spherical, platelets, or any other regular or irregular geometry. Particulate composites tend to be much weaker and less stiff than continuous fiber composites, but they are usually much less expensive. Particulate reinforced composites usually contain less reinforcement (up to 40 to 50 volume percent) due to processing difficulties and brittleness.

Composite materials have a wide range of properties and uses, making them the most common building and manufacturing materials in use today. They are created by combining two different components in order to create a new material with different properties than those which were started with. Concrete, fabric blends, paint, particle board, plastics, glass and even paper are all examples of composite materials. Metallic alloys are a different type of composite, requiring metal ores to be combined in extreme conditions. These metal composites are often used in items such as utensils, ornaments, trinkets, statues and as building materials. In 1940 fiberglass was invented and the modern era of composite materials began. The basic premise of a modern composite material is some sort of fiber material for the structure, and some sort of plastic or glue to hold the shape and provide strength. These fiber-reinforced plastic composites are used in manufacturing boat and kayak hulls, swimming pool liners, turbine blades, sporting goods and many other possible applications.

Composites are made up of individual materials referred to as constituent materials. There are two main categories of constituent materials: matrix and reinforcement. At least one portion of each type is required. The matrix material surrounds and supports the reinforcement materials by maintaining their relative positions. The reinforcements impart their special mechanical and physical properties to enhance the matrix properties. A synergism produces material properties unavailable from the individual constituent materials, while the wide variety of matrix and strengthening materials allows the designer of the product or structure to choose an optimum combination.

Engineered composite materials must be formed to shape. The matrix material can be introduced to the reinforcement before or after the reinforcement material is placed into the mold cavity or onto the mold surface. The matrix material experiences a melding event, after which the part shape is essentially set. Depending upon the nature of the matrix material, this melding event can occur in various ways such as chemical polymerization or solidification from the melted state.

A variety of molding methods can be used according to the end-item design requirements. The principal factors impacting the methodology are the natures of the chosen matrix and reinforcement materials. Another important factor is the gross quantity of material to be produced. Large quantities can be used to justify high capital expenditures for rapid and automated manufacturing technology. Small production quantities are accommodated with lower capital expenditures but higher labor and tooling costs at a correspondingly slower rate.

Many commercially produced composites use a polymer matrix material often called a resin solution. There are many different polymers available depending upon the starting raw ingredients. There are several broad categories, each with numerous variations. The most common are known as polyester, vinyl ester, epoxy, phenolic, polyimide, polyamide, polypropylene, and others. The reinforcement materials are often fibers but also commonly ground minerals. The various methods described below have been developed to reduce the resin content of the final product, or the fiber content is increased. As a rule of thumb, lay up results in a product containing 60% resin and 40% fiber, whereas vacuum infusion gives a final product with 40% resin and 60% fiber content. The strength of the product is greatly dependent on this ratio. Martin Hubbe and Lucian a Lucia consider wood to be a natural composite of cellulose fibers in a matrix of lignin.

What is a composite?

A composite is a structural material that consists of two or more combined constituents that are combined at a macroscopic level and are not soluble in each other. One constituent is called the reinforcing phase and the one in which it is embedded is called the matrix. The reinforcing phase material may be in the form of fibers, particles, or flakes. The matrix phase materials are generally continuous. Examples of composite systems include concrete reinforced with steel and epoxy reinforced with graphite fibers, etc.

Composite materials are extending the horizons of designers in all branches of engineering, and yet the degree to which this is happening can easily pass unperceived. The eye, after all, does not see beyond the glossy exterior or the race performance of a GRP1 yacht, nor does it sense the complexity of the structure of a composite helicopter rotor blade or of a modern CFRP2 tennis racket. Nevertheless, this family of synthesized materials offers the possibility of exciting new solutions to difficult engineering problems. In composites, materials are combined in such a way as to enable us to make better use of their virtues while minimizing to some extent the effects of their deficiencies. This process of optimization can release a designer from the constraints associated with the selection and manufacture of conventional materials. He can make use of tougher and lighter materials, with properties that can be tailored to suit particular design requirements. And because of the ease with which complex shapes can be manufactured, the complete rethinking of an established design in terms of composites can often lead to both cheaper and better solutions. The 'composites' concept is not a human invention. Wood is a natural composite material consisting of one species of polymer — cellulose fibers with good strength and stiffness — in a resinous matrix of another polymer, the polysaccharide lignin. Nature makes a much better job of design and manufacture than we do, although Man was able to recognize that the way of overcoming two major disadvantages of natural wood — that of size (a tree has a limited transverse dimension), and that of anisotropy (properties are markedly different in the axial and radial directions) — was to make the composite material that we call plywood. Bone, teeth and mollusk shells are other natural composites, combining hard ceramic reinforcing phases in natural organic polymer matrices. Man was aware, even from the earliest times, of the concept that combining materials could be advantageous, and the down-to-earth procedures of wattle-and-daub (mud and straw) and 'pied' (heather incorporated in hard-rammed earth) building construction, still in use today, pre-date the use of reinforced concrete by the Romans which foreshadowed the pre-tensioned and post-tensioned reinforced concretes of our own era. But it is only in the last half century that the science and technology of composite materials have developed to provide the engineer with a novel class of materials and the necessary tools to enable him to use them advantageously.

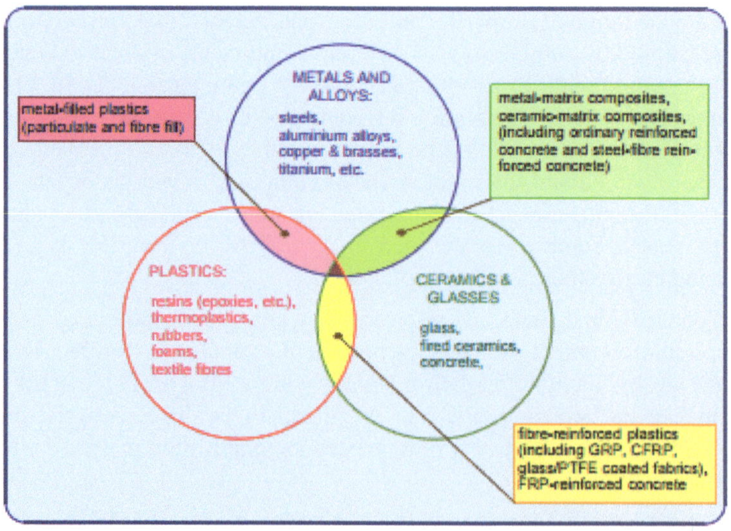

Figure 1 Relationships between classes of engineering materials, showing the evolution of composites.

The simple term 'composites' gives little indication of the vast range of individual combinations that are included in this class of materials. We have mentioned some of the more familiar ones, but the diagram Figure 1 gives a clearer idea of the scope for ingenuity which is available to the Materials Scientist and his customer, the Design Engineer. First, within each group of materials — metallic, ceramic and polymeric — there are already certain familiar materials which can be described as composites. Many members of the commonest and largest group of engineering materials, the family of steels, consist of combinations of particles of hard ceramic compounds in a softer metallic matrix. These particles are sometimes plate-like, sometimes needle-shaped, and sometimes spherical or polygonal. Polymers, too, are often two-phased, consisting of a matrix of one polymer with distributions of harder or softer particles contained within it; wood is a perfect example of this, as we have seen. And concrete is a classic example of a ceramic/ceramic composite, with particles of sand and aggregate of graded sizes in a matrix of hydrated Portland cement. These materials have been well known for many years, and Materials Scientists have learned to control their properties by controlling their microstructures; that is to say, the quantity, the form, and the distribution of what we might refer to as the 'reinforcing phase'. The idea of mixing components across the materials class boundaries is a natural extension of this idea. Making additions of hard, or fire-resistant, or simply cheap, ceramic powders to plastics to make filled polymers; and making additions of very hard, or abrasive, or thermally stable ceramic particles to metals to make the class of materials known as 'cermet's' to produce machine tool tips capable of cutting hard metals at high speeds or high temperatures; are only two examples of important developments in our exploitation of these materials. But even more significant is the extension of this principle to incorporate filamentary metals, ceramics and polymers into the bulk forms of any of these three classes of materials to make fiber composites — reinforced plastics, like CFRP and GRP, metal-matrix composites (MMCs) like silicon-carbide-fiber-reinforced aluminum, and ceramic-matrix composites (CMCs) like carbonfibre-reinforced glass.

Examples of naturally found composites

It include wood, where the lignin matrix is reinforced with cellulose fibers and bones in which the bone-salt plates made of calcium and phosphate ions reinforce soft collagen.

What are advanced composites?

Advanced composites are composite materials that are traditionally used in the aerospace industries. These composites have high performance reinforcements of a thin diameter in a matrix material such as epoxy and aluminum. Examples are graphite/epoxy, Kevlar®†/epoxy, and boron/ aluminum composites. These materials have now found applications in commercial industries as well.

What is the world market of composites?

The world market for composites is only 10 × 109 US dollars as compared to more than 450 × 109 US dollars for steel. The annual growth of composites is at a steady rate of 10%. Presently, composite shipments are about 3 × 109 lb annually.

Conventional Materials and Their Limitations

It is difficult to draw up a table of materials characteristics in order to assess the relative strengths and weaknesses of metals, plastics and ceramics because each of these terms covers whole families of materials within which the range of properties is often as broad as the differences between the three classes. A comparison in general terms, however, can identify some of the more obvious advantages and disadvantages of the different types of material. At a simplistic level, then:

- Plastics are of low density. They have good short-term chemical resistance but they lack thermal stability and have only moderate resistance to environmental degradation (especially that caused by the photo-chemical effects of sunlight). They have poor mechanical properties, but are easily fabricated and joined.

- Ceramics may be of low density (although some are very dense). They have great thermal stability and are resistant to most forms of attack (abrasion, wear, corrosion). Although intrinsically very rigid and strong because of their chemical bonding, they are all brittle and can be formed and shaped only with difficulty.

- Metals are mostly of medium to high density — only magnesium, aluminum and beryllium can compete with plastics in this respect. Many have good thermal stability and may be made corrosion resistant by alloying. They have useful mechanical properties and high toughness, and they are moderately easy to shape and join. It is largely a consequence of their ductility and resistance to cracking that metals, as a class, became (and remain) the preferred engineering materials.

Strong Fibers

The engineer who uses materials for structural or load-bearing purposes is quickly aware of an important feature of engineering solids, which is that they are never as strong as we would expect them to be from our knowledge of the strengths of the chemical bonds which hold them together. The reason for this is that all materials contain defects of various kinds which can never be entirely eliminated in practical manufacturing operations. For example, the strength of bulk glass and other ceramics is determined not by their strong covalent or ionic bonds, but by the many tiny pores or sharp cracks that exist either on the surface or in the interior. The most highly polished and dense bulk ceramic will rarely have a strength that exceeds one thousandth of the theoretically predicted strength. Similarly, metals contain faults in the stacking of atoms in their crystalline arrays, and the most damaging of these imperfections — dislocations — cause most bulk metal samples to deform plastically at loads which are, again, perhaps a thousandth or less of the theoretically calculated shear strength. The strength of any sample of a glass or ceramic is actually determined by the size of the largest defect, or crack, which it happens to contain. Roughly, the strength is proportional to the inverse square root of the length of the largest flaw, a relationship developed in a thermodynamic argument by Griffith in the 1920s. The relationship in its simplest form is:

$$\sigma_{max} = \sqrt{\frac{2E\gamma_F}{\pi a}} \qquad (1)$$

where σ_{max} is the strength of the material, E is its elastic stiffness (Young's modulus), γ_F is the work required to fracture the sample (a slight modification of Griffith's original idea that it represented the surface free energy of the material), and a is the flaw size. Since materials of this kind inevitably contain cracks and other defects with a wide range of sizes, one of the unfortunate consequences of the Griffith model is that the measured strengths of a batch of supposedly identical samples of a brittle material will be very variable simply because of the spread of flaw sizes. If the flaw sizes can be reduced by careful control of the manufacturing process, the general level of strength of the material will be raised and its variability reduced. Glass fibers, for example, are manufactured by drawing molten glass very rapidly down to form fine filaments of the order of only ten microns (10μm) in diameter. The resulting freshly formed surface is free of macroscopic defects and the fiber itself is too fine to contain any defects of the size that are found in bulk glass. Careful measurements of the strengths of freshly drawn glass fibers shows them to be up to 5GPa by comparison with the very modest 100MPa or so of bulk glass. The fresh fiber needs to be protected from abrasion by contact with hard objects and from corrosion by moisture in the atmosphere, both of which would result in the reintroduction of surface flaws and loss of the high strength achieved by drawing. Similarly, the strengths of polymeric filaments, such as the polyacrylonitrile fibers used to make 'acrylic' textiles, are limited by the weak chemical bonding between the molecular chains which make up the filament and by the defects caused by the manufacturing process. By subjecting such filaments to carefully controlled stretching, oxidation and carbonization processes, however, the polymer can be converted into fibers which are chemically almost entirely carbon, and with a crystal structure that is approximately that of graphite, in which the proportion of strong C—C bonds that lie in the direction of the fiber axis is very great. The load-bearing ability of these filaments, which are about 7μm in diameter, is high, and the sizes of the residual defects are very small. As a result, each filament, finer than a human hair, is capable of supporting a load of 10-20g without breaking, which is equivalent to strengths of up to 5GPa. And so it is with the many other types of fibres, organic and inorganic, that are used in composite materials. The finer the filament that can be made from a given solid, the stronger it will be.

The theoretical strength of a given type of solid is determined by the strengths of the atomic or molecular bonds that hold the solid together. And although the practical strengths of solids are determined by the defects which they contain, it is nonetheless necessary to seek materials with the strongest chemical bonds if we are to have the best chance of exploiting the principle of composite materials construction. An appreciation of the nature and magnitude of chemical bonding leads us to scan the Periodic Table for elements in which there is a high density of covalent or mixed covalent/ionic bonds, and we naturally alight on the area containing light elements with directional bonds. The elements most likely to provide what we seek are those like carbon, boron and silicon. In the 1950s, when ideas about composite materials were first being formulated, the plain elements would not have been considered to be particularly promising as high-performance structural materials, but the invention of carbon fibers at the Royal Aircraft Establishment, Farnborough, in the UK, and the simultaneous development of high-strength boron fibers at Texaco in the USA, resulted in a very rapid expansion of research and development programs to exploit these new filamentary materials in engineering composites.

Glass fibers

Glass fibers are manufactured by drawing molten glass into very fine threads and then immediately protecting them from contact with the atmosphere or with hard surfaces in order to preserve the defectfree structure that is created by the drawing process. Glass fibers are as strong as any of the newer inorganic fibers but they lack rigidity on account of their molecular structure. The properties of glasses can be modified to a limited extent by changing the chemical composition of the glass, but the only glass used to any great extent in composite materials is ordinary borosilicate glass, known as E-glass. The largest volume usage of composite materials involves E-glass as the reinforcement. S-glass (called Rglass in France) has somewhat better properties than E-glass, including higher thermal stability, but its higher cost has limited the extent of its use. Wallenberger and Brown (1994) have recently described the properties of experimental calcium aluminate glass fibers with stiffness's as high as 180GPa.

Carbon fibers By oxidizing and paralyzing a highly drawn textile fiber such as polyacrylonitrile (PAN), preventing it from shrinking in the early stages of the degradation process, and subsequently hot-stretching it, it is possible to convert it to a carbon filament with an elastic modulus that approaches the value we would predict from a consideration of the crystal structure of graphite, although the final strength is usually well below the theoretical strength of the carbon-carbon chain. The influence of strength limiting defects is considerable, and clean-room methods of production can result in substantial increases in the tensile strength of commercial materials. Prior to sale, fibres are usually surface-treated by chemical or electrolytic oxidation methods in order to improve the quality of adhesion between the fiber and the matrix in a composite. Depending on processing conditions, a wide range of mechanical properties (controlled by structural variation) can be obtained, and fibres can therefore be chosen from this range so as to give the desired composite properties.

Silicon carbide

Continuous SiC monofilaments were first produced by pyro lytic decomposition of gaseous silences onto fine filaments of carbon or tungsten. These are thick fibres, of the order of 100μm in diameter, which continue to be of major interest to manufacturers of metal- and ceramic-matrix composites. The alternative production method, analogous to that for carbon described earlier, is based on the controlled thermal degradation of a polymer precursor. This process typically takes a precursor such as some mixture of dimethyl dichlorosilane and diphenyl dichlorosilane and converts it in an autoclave to a polycarbosilane from which a continuous fiber is made by melt-spinning. The fiber is then converted by pyrolysis at 1300°C into a fiber consisting mainly of β-SiC of about 15μm diameter. The characteristic commercial fiber of this type is that known as 'Nicalon' which is marketed by the Nippon Carbon Company. It has a rough surface, making for good fiber/matrix adhesion, but is somewhat reactive towards oxygen. It is well wetted by molten metals and is reasonably stable as a reinforcement for MMCs based on aluminum and copper although it lacks long-term thermal stability.

Alumina and alumina/silica compounds

The earliest work involving Al_2O_3 as a reinforcement for composite materials concerned the use of tiny filamentary crystals, of the order of 50μm long and a few microns in diameter, which could be grown from the vapor phase in a highly perfect state and which, in consequence, had great strength and stiffness. These crystals, known as 'whiskers', proved to be very expensive and difficult to handle in processing operations, and their potential could not be satisfactorily realized. The DuPont Company in the USA subsequently developed an improved alumina fiber in the form of a polycrystalline yarn known as FP fiber. These filaments, which were manufactured by a sol-gel process, had a modulus comparable with those of boron and carbon, and a strength of the order of 1.8GPa. Like SiC, this fiber also had a rough surface and was potentially well-suited for the reinforcement of metals such as aluminum and magnesium on account of its chemical inertness, high-temperature stability, and its ability to form a good bond with matrix alloys. As far as can be ascertained at present, however, FP has never been an economic prospect for its manufacturer, and alternative forms of alumina, usually containing other constituents, like Sumitomo's Altex fiber.

Organic Fibres

Bulk polymers have elastic moduli no greater than 100MPa, but if the polymer is spun into fibres and cold drawn so as to develop a high degree of molecular orientation, substantial improvements in both strength and rigidity can be achieved. In particular, polyolefin fibres with an exceptionally highly oriented, extended molecular chain structure can be achieved by super drawing in the solid state. In this process, both the crystalline and non-crystalline phases of the initially isotropic polymer are stretched out and aligned and there is an increase in crystal continuity. Such fibres have high strengths, and their elastic moduli are similar to those of glass and aluminum. Apart from their excellent mechanical properties such fibres have the important advantage over inorganic fibres that they are not brittle.

CHAPTER 1

MICROSTRUCTURAL CHARACTERISATION OF COMPOSITE MATERIALS

James N. Boland and Xing S. Li

Keywords: Diamond composite; superhard materials; wear rate; cutting brittle material; Raman; XRD; cathodoluminescence

INTRODUCTION

Diamond is one of the allotropic forms of carbon [1,2] and is renowned for its outstanding physical, chemical, electrical and mechanical properties. It is also prized for its brilliance, which has led to being exploited for both its scientific applications and its unique, socially accepted form in the jewelry market. On the industrial front it is remarkable to note that in the early usage of diamonds for mineral exploration applications, large carat-sized pieces of diamond were bonded into the crown of drill bits [3].

Diamond composite tools should be considered as a multi-component system requiring the successful performance of many subsystems. These subsystems include: (a) the diamond grains themselves, (b) the binder phase or phases used in the sintering operation to form the composite and (c) the method of bonding the composite element into the body of the tool. As cutting elements, these composites are subjected to severe abrasive/erosive wear regimes leading to the generation of high temperatures. So not only is the wear resistance of the composite critical to its operational performance but equally important is its thermal stability, thermal conductivity, impact resistance, thermal fatigue limit and fracture toughness. These properties determine the suitability of diamond composite materials for cutting, drilling and sawing operations in the mining and civil construction industries.

A wide-ranging study has been undertaken by the SMART*CUT team of researchers at CSIRO's division of Earth Science and Resource Engineering and the Flagship Minerals Down Under. Over the past 15 years, the wear Behavior of hard and superhard materials has been researched with particular emphasis on the cutting of hard and abrasive rocks. Initially, the research concentrated on the abrasive wear Behavior of thermally stable diamond composite (TSDC) cutting elements with some minor studies of the performance of cubic boron nitride (cBN) and cemented tungsten carbide. However, with the increasing importance of tools manufactured with a polycrystalline diamond composite (PCD) coating on a WC substrate, the results of wear tests on these tools are also included in this review. These studies are restricted to tools manufactured using the high pressure, high temperature (HPHT) processes, although it is acknowledged that chemical vapor deposition (CVD) coatings of diamond on WC and other superhard carbide materials play a major part of the new-age diamond tooling industry. This review covers the latest research results pertaining to materials Characterization techniques that can be profitably exploited in the study of the microstructural features in a wide range of commercially produced diamond composites. Further, the results of a simple, low cost abrasive wear testing method for superhard diamond composites is presented, leading to an empirical measurement of the wear rate of these composites. Based on these results, the wear rate expressed as the weight loss of the cutting element per unit volume of material cut is proposed as a quality factor for assessing the Behavior of these diamond composite cutting elements. Using the newly developed solid TSDC and diamond composite coated WC cutting elements, the micromechanics of the cutting of brittle materials such as stone and rocks has necessitated the development of a more general phenomenological model of the macro chipping processes.

Citation: James N. Boland and Xing S. Li. "Microstructural Characterisation of Composite Materials", Materials 2010, 3(2), 1390-1419; DOI: 10.3390/ma3021390

DIAMONDS

A great deal of research has already been undertaken to understand the properties and defect structures of both natural and synthetic diamond grains with excellent coverage given in the following references [4,5,6,7]. What emerges from these fundamental studies is that the properties of both natural and synthetic diamonds can vary significantly and the choice of the most appropriate source material for tool manufacture depends on the application to which diamonds will be put. Nearly all the research referenced in this review paper will relate to synthetic diamonds with only an occasional referral to naturally occurring diamonds.

There are three main diamond manufacturing process: (1) high pressure, high temperature (HPHT) synthesis, usually operating in the diamond stability field [2,8], (2) chemical vapor deposition (CVD) [9,10] and (3) detonation synthesis [11]. Three important characteristics of diamond grit/grain used in composite manufacture are: (1) grain size, (2) grain shape and (3) strength [13]. Other physical properties that have important implications for the performance of diamond composites are: thermal stability, thermal fatigue strength, fracture toughness and impact strength. As these properties have been reviewed elsewhere [4,5,6] they will only be referred to indirectly whenever they relate to the wear Behavior of diamond composites.

DEVELOPMENT OF DIAMOND COMPOSITES

Diamond composites have a wide range of compositions and applications. They have been variously referred to as PCD (polycrystalline diamond), PDC (polycrystalline diamond compact/cutter) and TSP (thermally stable polycrystalline diamond composite, sometimes represented as TSDC). Although not commonly referenced as diamond composite material, diamond impregnated metal matrix composites (MMC) are an integral part of the industrial diamond industry. The common parameters to be considered for this wide range of diamond composites are: the quality of the diamond itself, the strength of the bonding between the diamond grain and the matrix, the diamond concentration in the composite and the wear rates of the diamond and matrix as a function of the abrasivity of the material being drilled, machined and sawn.

The quality of individual diamond grains used in the production of these composites will depend on the source material. Synthetic diamonds, the most commonly used variety, are produced under HPHT conditions and their quality will depend on the solvent-catalyst system used in production of the diamonds. The usual catalysts are Ni, Fe, Co, Mn and Ti [8]. These elements can be incorporated into the diamond structure to produce a wide range of point defects that have been extensively studied [5]. While such point defects are not expected to significantly influence the mechanical and wear properties of diamond, if their concentration is too high, they can aggregate into fine-scale nanoprecipitates. The coefficient of thermal expansion of these nanoprecipitates will be different from that of diamond itself so thermally induced residual stresses will be generated in these diamond grains by thermal cycling events during cutting. These stresses may well be sufficient to initiate microcracks which will have a detrimental effect on the strength, wear resistance and fatigue properties of the diamond grains. One outstanding exception to the apparent neutral role of point defects has recently been reported with boron-doped diamond crystals [15]. Thermal stability, compressive strength and impact toughness all showed significant increases in the doped crystals compared with the normal diamonds. The solvent/catalyst was based on Fe-Ni alloy and the HPHT conditions were given as 1300–1400 °C and 5.4–5.7 GPa, which is in the diamond stability field of the carbon phase diagram.

In addition to their being several manufacturing processes, the form and shape of the product grains from these manufacturing processes can vary widely. In the case of PCD and PDC, the actual diamond composite is normally molded onto a substrate during sintering and, in most applications, the substrate is cemented tungsten carbide, referred to as WC in this review. The major advantages of this product format are (a) the binder phase in both components is cobalt and (b) in this hybrid form, the PCD element can be easily brazed onto any shaped tool. The obvious disadvantage is that such tooling components are limited to low-temperature conditions as the cobalt is a catalyst for the back transformation of diamond to graphite [16,17].

TSDC components are produced by either using silicon carbide as binder or leaching the unwanted cobalt component from the surface layers of the standard PCD. The earliest known product in the category was de Beer's (now Element 6) Syndax 3 [18]. It was applied by Asahi Diamond in its hybrid core bits [12]. Despite the higher thermal stability of this material (stable to 1200 °C) early production samples were reported to be unsuitable because of their lower strength and wear resistance than PCD. Nevertheless the fracture toughness was twice that of diamond single crystals [14]. Because residual silicon was observed in these samples, it can be concluded that the reactive bonding-sintering process did not proceed to completion.

A new composite material was patented by A.E. Ringwood based on research in the mid 1980s [19]. This was known as ADC (advanced diamond composite). It was originally produced from natural diamond grit at pressures of about 2–2.5 GPa and at temperature ~1500 °C which is outside the diamond stability field and in the graphite phase field, Figure 1 [20].

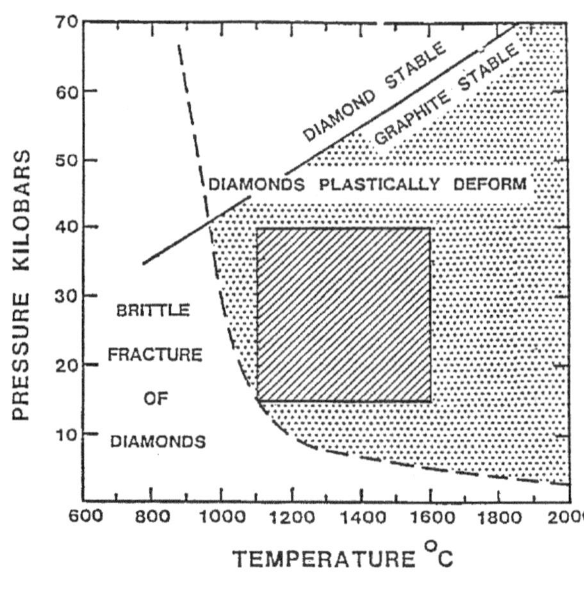

Figure 1. P-T phase diagram for carbon relevant to diamond composite manufacture. Note: 10kb = 1 GPa (Copy from US Patent: 4,948,388, 1989).

Under these HPHT conditions the silicon metal melted and infiltrated the interstices of the granular structure of diamond grit. The silicon reacted with the diamond to form the refractory ceramic SiC which enabled strong bonding to occur between the diamond grains. To achieve optimum performance of the composite, the reactive sintering process should proceed to completion leaving no residual silicon metal in the microstructure. Another essential stage in the process is to rapidly quench the sintered product in order to avoid any back transformation of the diamond phase to graphite. The outstanding property exhibited by this type of composite is that it produces one of the earliest examples of thermally stable diamond composite–TSDC. Further, the microstructure indicated a high proportion of diamond-to-diamond grain boundary bonding. Such ceramic-type grain boundary bonding (equivalent to metallurgical bonds in metallic systems) produces a high contiguity value which is known to result in good mechanical properties in carbide systems [21,22]. The importance of microstructural features in diamond composites is given greater coverage in the next section.

The use of TSDC appears to be limited in the industry despite its potential to enable a much wider range of applications in the mining and manufacturing industries. The major issue to be addressed in the application of TSDC is the difficulty of bonding these cutting elements to a suitable substrate. Active brazing compounds, especially those containing Ti, are recommended to ensure strong bonding. The brazing has to be performed either at high vacuum or in an inert environment [23,24]. A simpler, less expensive bonding process has recently been patented, overcoming this difficulty in the usage of TSDC cutting elements in rock cutting picks and saw blades [25].

The major diamond composite type used in mining and manufacturing operations is PCD/PDC. An excellent history of the development of these diamond composites is given by Scott in [26]. PCD cutters have been shown to perform well in both civil construction and mining/exploration industries [27,28,29].

The physical properties of these composites are markedly dependent on both grain size and grain shape. In ceramic-type materials, the relationship between strength and grain size is complex. With PCD, being ceramic-like, it is not surprising to observe that the rupture strength is inversely proportional to $d^{1/2}$ over the size range 2–30 microns [30]; for grain sizes greater than 30 microns, the strength drops off rapidly. The relationship between fracture toughness and grain size is less clearly defined but the normal trend, within the same grain size range, is an inverse relationship.

A major innovative step was taken with the patenting of nanostructured diamond-SiC composites [31,32]. This composite, based on at least one of the phases being in a nano grain size format, has produced excellent fracture toughness which both contradicts the usual inverse relationship between hardness and toughness and results in a product with toughness close to that of WC [13]. In this nanocomposite, the SiC phase was reported as being in the nano size range. Nanodiamonds are now being produced via a number of processes such as detonation, CVD and HPHT and they offer an opportunity for further research into the outstanding properties of nanocomposites.

MICROSTRUCTURE OF DIAMOND COMPOSITES

From the outset, diamond composite materials present some challenges in undertaking detailed studies of their microstructures. Because they are superhard, sample preparation needs careful attention. Recently, a series of papers have been published detailing a successful approach to this issue [35]. Clearly, the better the surface preparation the more clearly defined the microstructural features become. However, not all surfaces need to be polished and this situation pertains particularly to the examination of worn surfaces in which it is essential to study the spatial distribution and near-neighbor relationships of all the phases exposed on that surface.

Observation Techniques

A detailed study of the microstructures of diamond composites is essential to a proper understanding of their physical properties. It is important to note that this study has been undertaken on commercially available diamond composite products using optical and scanning electron microscopy (SEM equipped with energy dispersive system - EDS), electron probe microanalytical analysis (EPMA equipped for cathodoluminescence (CL), wavelength (WDS) and energy dispersive systems) with additional observations being made using micro-focus X-ray shadow imaging, X-ray diffraction (XRD) and Raman spectroscopy (RS). The surface preparation of samples for SEM imaging was varied. In some instances polished surfaces were studies but equally important for understanding the micromechanics of fracture and wear, are studies of as-fractured or as-worn surfaces.

Results

Thermally stable diamond composites (TSDC) have a wide range of uses apart from cutting elements. However, only those composites recommended for cutting have been studies in any detail and this review will concentrate on composites used mainly in cutting brittle materials such as rocks. The dominant phase in TSDC is diamond with a matrix of silicon carbide. Unless otherwise indicated, the specific TSDC samples obtained from the various manufacturers referred to in this review have a nominal 80 percent diamond and 20 percent binder matrix (by volume). Being thermally stable, the matrix is supposed to be SiC produced during the reactive bonding process at HPHT conditions in the diamond stability field—seeFigure 1. While variations to this compositional mix will become obvious in the results presented in this review, the responsibility for these variations rests with the suppliers and such variations can be interpreted as issues of quality assurance which is outside the control of researchers.

As shown in Figure 2 (left), the fracture surface as viewed in the secondary electron image mode (SEI) indicates that the diamond grains have bonded to the matrix with no evident pluck-out of the diamond grains. With a moderate polish, the surface details are better viewed in the backscattered electron image mode (BEI) with the contrast arising from differences in the mean atomic number of the phases. The TSDC sample in Figure 2 (right) has a bi-modal grain size distribution but the polish is not sufficiently relief-free, with the surface roughness masking the complex phase distribution in the matrix.

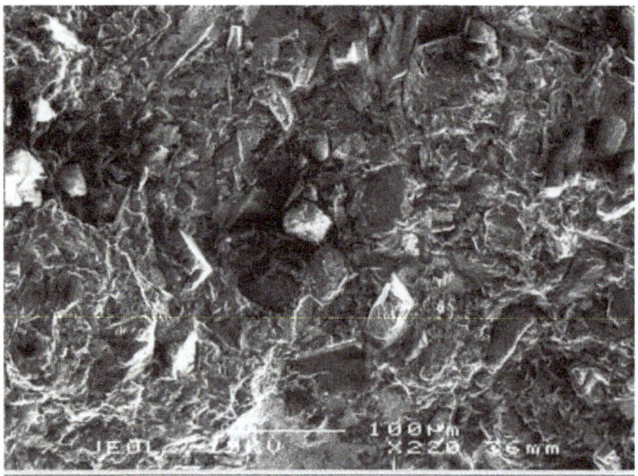

Figure 2. SEM images of TSDC cutting elements; (left) SEI of fracture surface with no evident pluck-out of diamond grains; (right) BEI of a polished surface with surface relief masking detailed phase information in the image.

Using better polished surfaces, more specific microstructural and phase details can be observed. The following SEM-based microstructural images in Figure 3, Figure 4, Figure 5 and Figure 6 are from the same commercially produced TSDC sample. As the contrast in the BEI mode of imaging is related to variations in the mean atomic number of the phases, it is clear that there is a plethora of phases in the matrix interstitial areas. The determination of a particular phase may be simple as in the case of the single carbon peak in the EDS spectrum at point A in Figure 3 which could result from either graphite or diamond. Similarly, the single Si peak at the marked spot in Figure 4 indicates residual silicon metal in the TSDC. However, it has not been possible to identify the other minor phase constituents with the more complex elemental composition identified in the EDS spectra shown in Figure 3 (point B) and Figure 5. The presence of residual, unreacted silicon metal and/or graphite in TSDC samples has disastrous consequences for the wear Behavior of such materials as discussed in Section 5.

While this chemical information is useful in deciphering the possible phases at specific locations, the distribution of these phases can be analysed using X-ray mapping. However, a much more powerful approach to this issue is provided by combining the signals from several detectors attached to the SEM or probe. This patching together of detector signals has been referred to as hyperspectral or holistic mapping [36,37]. Examples of hyperspectral mapping are shown in Figure 6. The BEI image in Figure 6(a) shows similar image contrast effects to that presented in Figure 3, Figure 4 and Figure 5. By mapping signals from the various detectors via a color code (or color table), a modified SEM image is produced. It is now possible to identify distinct phases and their distribution in the sample as shown in Figure 6(b). For example, in this color rendered image the CL active phases—diamond and SiC—and the characteristic X-ray photons for specific elements such as Ni, Si (metal) and a Si-Ti alloy have been color coded, making it possible to identify the presence of residual silicon and a Si-Ti alloy as well as the reactively produced SiC phase. Additionally, small patches of Ni metal have been identified in this image.

From previous studies the zero-phonon lines in diamond have been identified [5]. Within the wavelength range 350–800 nm CL spectra have been recorded for the wide range of TSDC samples used in this study. Three dominant bands have been noted: band A with a peak at ~447 nm, band B with a peak at ~517 nm and a broad band C ranging from 576–626 nm—seeFigure 6(c). Again, by mapping these peaks via the specified color table, an image of the distribution of these CL active regions in the diamond can be identified—see Figure 6(d). This particular sample of TSDC is dominated by bands A and B and no evidence for band C. Band B has been previously identified with Type Ib synthetic diamond in which pure Co was the solvent-catalyst [5].

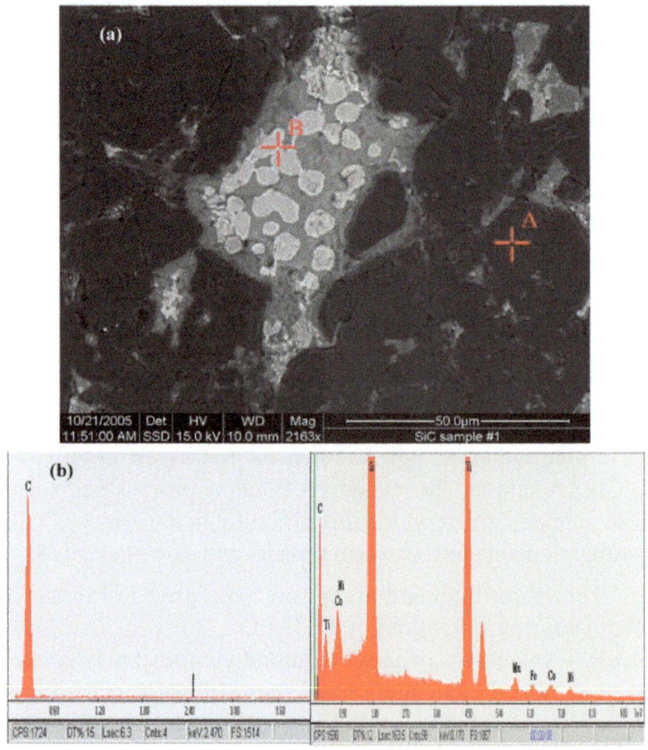

Figure 3. (a) SEM-BEI image of a TSDC sample showing the dominant diamond phase (A) and the complex, multi-phase matrix material. (b). EDS spectra from points A (left) and B (right). The single carbon peak at A clearly indicates the diamond phase; the multi-elemental spectrum analysed at point B clearly highlights the complexity of the matrix material in this commercially produced TSDC sample.

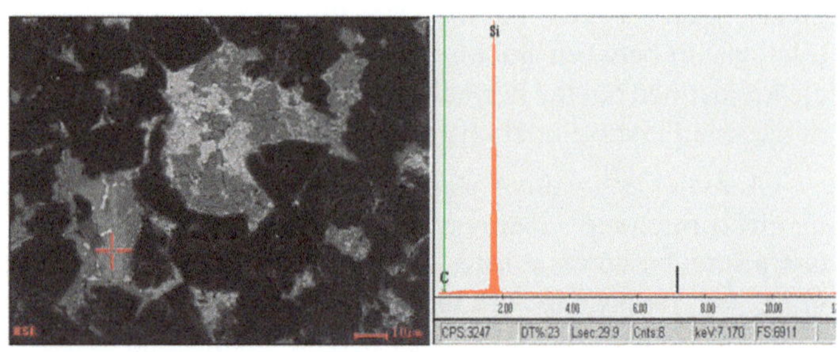

Figure 4. SEM-BEI image (left) from a TSDC sample with the corresponding EDS spot analysis of the light-grey phase indicating residual, unalloyed Si metal in the matrix.

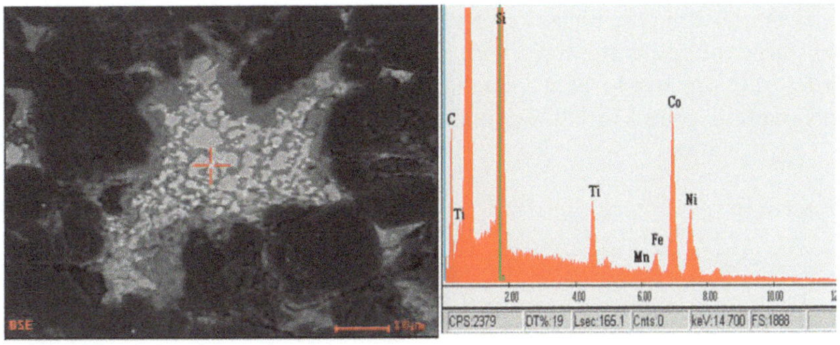

Figure 5. SEM-BEI image (left) from TSDC sample with the EDS spectrum (right) from the designed spot.

Another valuable optical technique used to study diamond composite materials is Raman spectroscopy. This technique is capable of examining the finer details in the microstructure of the composite material by focusing a 2 μm laser beam onto any selected microstructural feature. It is capable of readily distinguishing between the allotropic forms of carbon. For example, diamond has a unique Raman peak in the spectrum at 1332 cm^{-1} and a typical spectrum from a diamond grain in a good quality (equivalent to high wear resistance) TSDC sample is shown in Figure 7. The spectrum from graphitic material is different, with broad peaks centred around 1350 and 1550 cm^{-1} and no peak at 1332 cm^{-1}. These graphite peaks are referred to as the D and G modes (or bands) of disordered carbon [39].

(a)

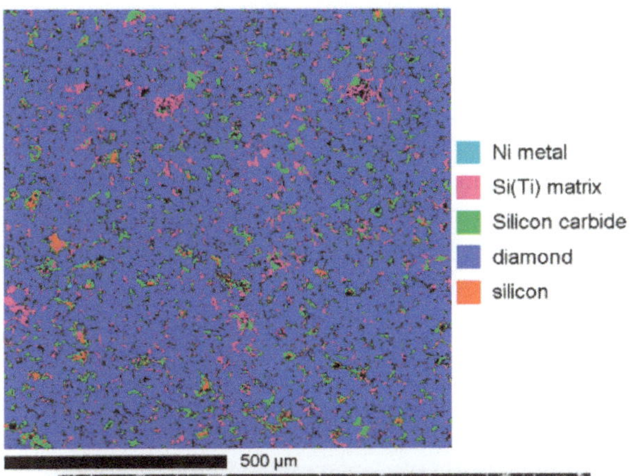

(b)

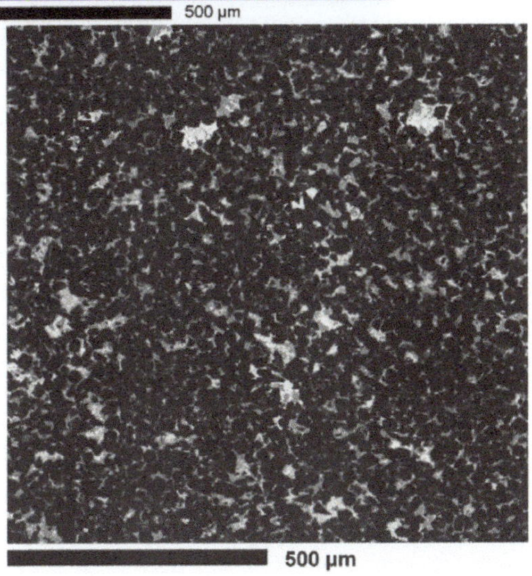

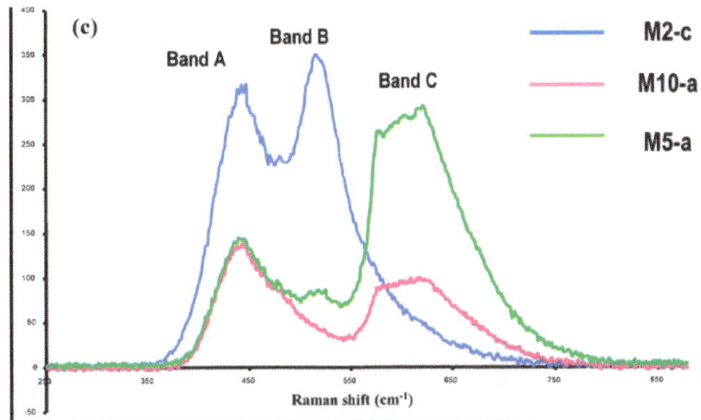

(d)

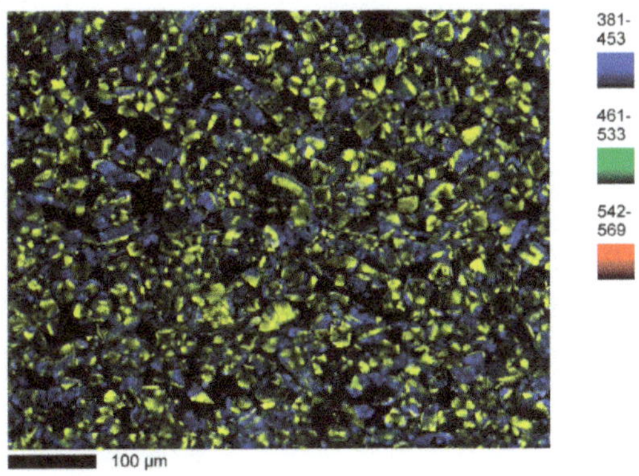

Figure 6. (a). SEM-BEI image of a TSDC sample. The dark grains are diamond and the various shades of lighter contrast in the matrix regions indicate the presence of several phases. (b). Hyperspectral image of the same field-of-view as in (a) in which the specific elements have been color-rendered according to the attached color table—see text for details. (c). Cathodoluminescence from TSDC samples listed in Table 1showing room temperature CL spectra and the three prominent bands A, B and C. Note: M10-a and M3-f as shown in Table 1 are equivalent samples with respect to their wear resistance. (d). Mapping of the cathodoluminescent signals from a TSDC sample with the color code showing the spectral range in nm—see text for details.

The optical micrograph in Figure 8 shows two different grain types, as indicated by their different reflectivities, within a poor quality (low wear resistance) TSDC sample. The larger (~25 μm), euhedral grains with the high reflectivity are the diamond phase but the smaller (~10 μm) grains exhibit a totally different Raman spectrum as shown in Figure 8. The broad peak centered around 1520 cm^{-1} is typical of graphitic material and is most likely associated with the D and G bands of disordered carbon. Further, there is no evidence of the diamond peak at 1332 cm^{-1}. Raman signals may also be used to produce hyperspectral images or chemical images of a sample in a manner not too dissimilar from the images created in EPMA—see Figure 6(c). Such microspectroscopy with a spatial resolution of the order of 2 μm would be a valuable aid in detecting diamond and graphitic phases in TSDC cutting elements.

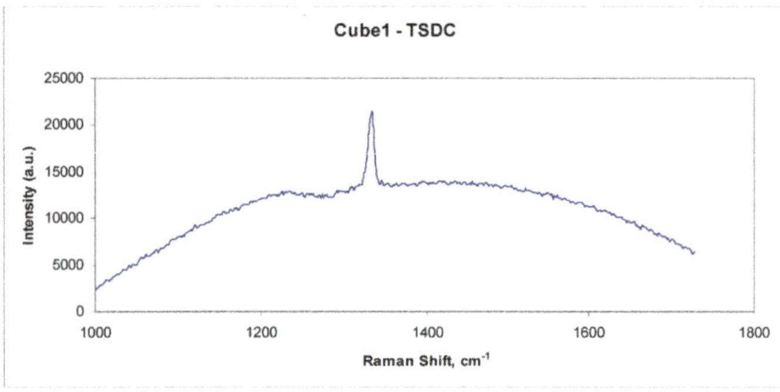

Figure 7. Raman spectrum from a good quality (low wear rate or high wear resistance) TSDC sample with no evidence of graphitic material.

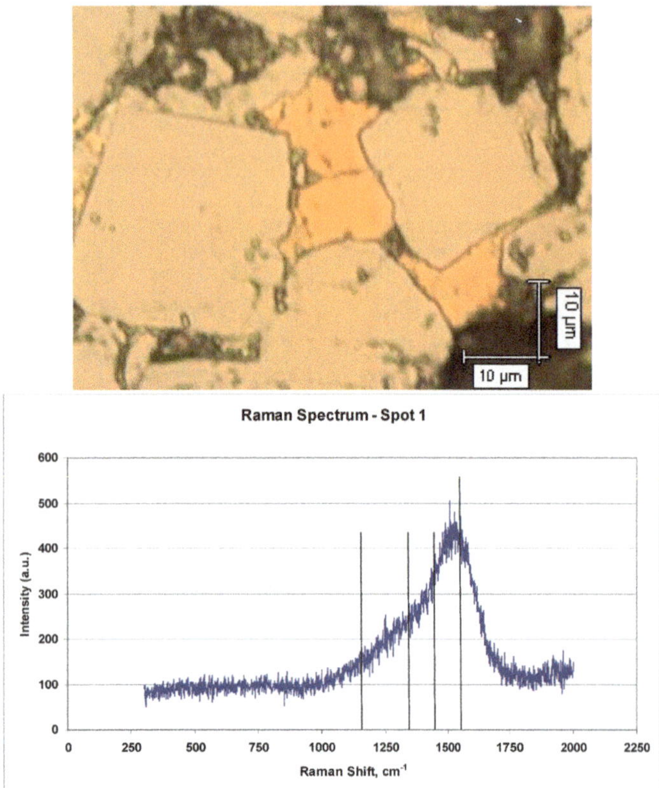

Figure 8. Optical micrograph (top) of poor quality (low wear rate or high wear resistance) TSDC material showing large (~25 μm), diamond grains with smaller (~10 μm) golden colored grains with the distinctive graphitic Raman spectrum (bottom).

Finally, standard X-ray diffraction or XRD can be used to determine the phases in TSDC samples but it has limitations due to the low detectability limits of the minor phases. A major advance in this technique has been achieved with the use of micro-XRD systems, especially on beamlines at synchrotron sources. With the high brilliance and small spot sizes available in this technique several square millimetres of sample can be quickly scanned to determine the phases in materials. A more macroscopic approach to XRD is given in the next section.

WEAR BEHAVIOR OF DIAMOND COMPOSITE TOOLING

Diamond composite along with cubic boron nitride form the group of superhard, abrasive resistant materials that have great potential to revolutionise the cutting, drilling and sawing operations in the mining and civil construction industries. Because of their extreme wear resistance, standard wear testing procedures [42,43] are inappropriate for such materials. For example, using the pin-on-disk wear test for a TSDC cutting element, there was no measurable weight loss of the sample and no detectable wear flat produced. The SEM-SEI image (Figure 9) of the tip of the sample shows not observable wear of the diamond composite material.

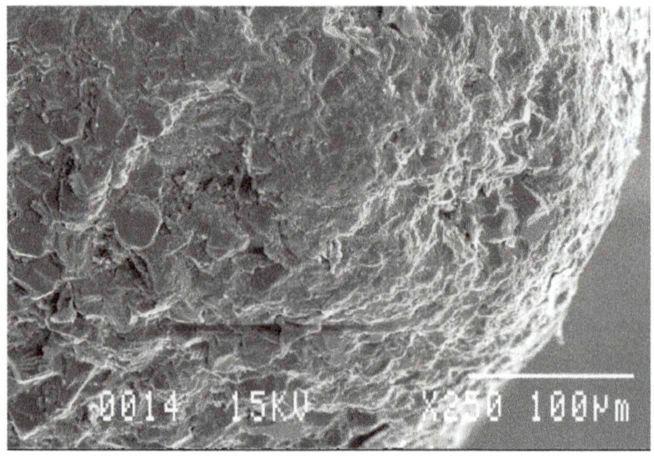

Figure 9. SEM-SEI image of the wear surface of a TSDC cutting element subjected to the standard pin-on-disk abrasive wear test.

One recent approach to this issue of low abrasive wear was tested using single point PCD cutting tools with WC (13% Co) and Duralcan with 20% SiC as the workpieces [44]. The emphasis in the testing was low cost and rapid turnaround time and was especially geared to machining operations of these engineering materials.

A more abrasive wear test procedure has been devised by CSIRO that spans the range from hard to superhard tool materials [45]. In these tests, a vitrified bonded alumina grinding wheel of standard manufacture is used as the workpiece. The cutting elements in these tests were solid, moulded TSDC samples and diamond composite coated tools on a WC substrate. In the latter case, the diamond composite was a PCD type and hence not thermally stable. The alumina grinding wheel was chosen in preference to the more abrasive SiC grinding wheel because of the requirement to cover WC cutting tools, diamond composite as well as cubic boron nitride tools. The properties of the alumina grinding wheel and test conditions were:

- Density: 2.18 g/cc
- Compressive strength: 121 MPa
- Tensile strength: 20 MPa
- Cerchar Abrasivity Index (CAI): 3.54
- Porosity: ~42 vol %
- Al_2O_3 grain size: ~400 microns
- Wheel dimensions: 350mm OD × 51mm width

- Rotational speed: 480 rpm
- Depth of cut: ~0.3mm
- Feed rate: ~1.2mm
- Testing conditions: Dry cutting

The wheel was "machined" across its face (Figure 10) for a set number of traverses which, for the TSDC cutting elements, was 100. Depending on the wheel diameter this could amount to cumulative cutting distances of over 4000m. However, for poor performing tools the number of cuts had to be limited either because the tool had worn excessively as in the case for WC (Figure 11) or because the diamond composite was of poor quality and the tool wore through to the underlying carbide substrate (Figure 12).

Figure 10. Photos of the CSIRO's abrasive wear testing rig for diamond composite cutting elements with an alumina grinding wheel as the counter piece for the test. Note: the high temperature region in the vicinity of the cutting tip (left) and tool holder (right).

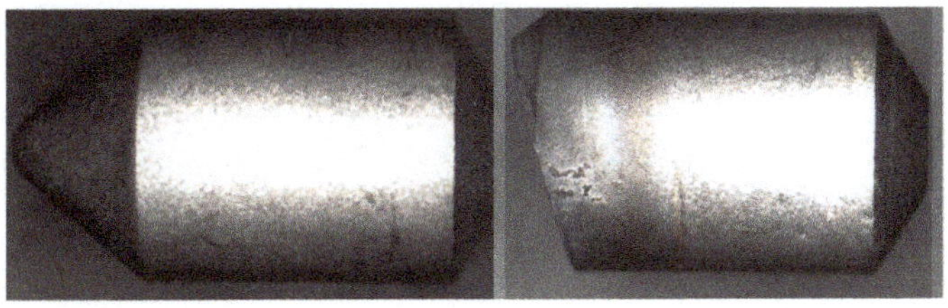

Figure 11. Cemented tungsten carbide cutting element (WC) before (left) and after only 20 cuts on the abrasive wear testing rig (right).

Two levels of tests have been used in CSIRO's wear testing studies. Level one is that specified in an earlier study [45] in which cumulative wear rates, dimensional wear coefficients (*k*) and specific energies may be calculated while level two consists in measuring the weight loss of the cutting element to the nearest 0.0001 g and the volume or weight loss of the work piece during the test. These simple data yield the empirical wear rate - loss of weight of the cutting element (grams) per unit volume cut from work piece (m^3) or the empirical wear resistance (viewed as a figure-of-merit) which is the weight of material cut per unit weight loss of cutting tool—either measure may be used as a quality factor for assessing the performance of diamond composite cutting elements. Any cutting element can be assessed by this latter method, once the tool holder has been modified to accept the shape of the element. The results of a wide ranging study of diamond composite cutting elements from various manufacturers are presented in Table 1, Table 2 and Table 3. (Note: no individual manufacturer has been identified in these tests as it is not the purpose of the study to rank the quality of the manufacturers, only the quality of their products. The microstructural investigations have been specifically directed at determining the possible causes for the poor wear performance of the various grades of diamond composites used.)

Table 1. Wear of ballistic [1] shaped TSDC cutting elements.

Element ID	**Wear Rate (g/m³)**	**Wear Resistance (×10³) (wt material cut/wt loss tool)**	**Depth of Cut (mm)**	**Length of Cut (m)**
M1-a	12.0	183.6	0.27	4181
M2-a	53.0	41.5	0.29	4197
M2-b	256.1	8.6	0.18	883
M2-c	6.9	318.8	0.28	4211
M3-a	4.8	458.6	0.27	3478
M3-b	111.4	19.7	0.24	4265
M3-c	3102.5	0.7	0.05	365
M3-d	5.1	432.0	0.26	3508
M3-e	4.3	516.9	0.27	4222
M3-f	3.6	607.0	0.29	4193
M4-a	19.7	111.7	0.26	4215
M5-a	18.1	121.4	0.38	3152
M6-a	8.4	261.1	0.27	2197
M6-b	2.9	749.2	0.29	3854
M6-c	13.4	164.4	0.29	4193
M7-a	232.3	9.5	0.15	404
M7-b	1960.6	1.1	0.08	375

M7-c	4765.2	0.5	0.08	410
M7-d	6865.8	0.3	0.10	905

Note: variations in the depth of cut arise from the relative wear rates of the cutting element, the poor quality tools are not able to maintain the set depth of cut.1 the precise shape of the ballistic cutting element is subject to "commercial-in-confidence" agreement.

Table 2. Wear of cylindrical cutting elements.

Element ID	Wear Rate (g/m^3)	Wear Resistance ($\times 10^3$) (wt material cut/wt loss tool)	Depth of Cut (mm)	Length of Cut (m)
M2-d[1]	2.7	818.9	0.28	4206
M7-e[2]	6.3	351.0	0.29	3482
M8-a[2]	10.3	213.7	0.29	3511
M9-a[3]	14.2	155.3	0.26	4227

Note: 1. solid TSDC; 2. PCD layer on WC substrate; 3. solid TSDC with 98% diamond.

Table 3. Wear on ballistic[1] shaped cutting elements - PCD coating on WC.

Element ID	Wear Rate (g/m^3)	Wear Resistance ($\times 10^3$) (wt material cut/wt loss tool)	Depth of Cut (mm)	Length of Cut (m)
M3-g	4.9	448.3	0.30	3946
M5-b	3.1	716.8	0.29	2798

1 The precise shape of the ballistic cutting element is subject to "commercial-in-confidence" agreement.

Even within the category of diamond composite cutting elements, the wear resistance across several manufacturers can vary from 0.3 to 818.9 $\times$ 10^3. For comparative purposes, the wear resistance of mining-grade WC was observed to vary between 0.01 and 0.12 $\times$ 10^3. For the purposes of assessing the relative wear Behavior of hard and superhard cutting elements, CSIRO's wear testing procedure is a simple and low cost operation.

Microstructural observations

Solid TSDC

The CSIRO abrasive test methodology is not registered as a standard assessment procedure hence the results are only indicative of the relative wear Behavior of this class of superhard material. However, it is clear from the results shown inTable 1, Table 2 and Table 3 that the wear rates of solid TSDC samples cover an extremely wide range of values. It has yet to be determined what is an acceptable wear rate for this class of material but, based on the wear results, one may tentatively assign the qualitative terms such as "good wear resistance", "low wear rate" or "good quality" to TSDC elements with wear rates less than 10 g/m^3 (or equivalent wear resistances greater than ~ 200 $\times$ 10^3), while samples with wear rates greater than this value as "poor quality". Certainly, more research and cross-laboratory studies are required to determine more precisely the quality of these superhard materials and to relate wear test parameters to cutting performance of tools in field trials.

In an idealised microstructure, a TSDC element is expected to contain SiC as the binder phase, produced during the HPHT reactive bonding process. As the microstructural studies presented in Section 4 indicate, such an idealised mix of phases has not been observed. Certainly these phases are present in most samples but there are other phases such as residual silicon metal, graphite and a range of chemically distinct but as yet unidentified phases that occur in the matrix of these TSDC samples. The presence of such contaminant phases is the most likely cause of the variable wear resistance. For example, in TSDC sample M2-b (Table 1) with an unacceptably high wear rate of 256.1 g/m^3, an SEM study of wear surface revealed that there were significant amounts of unreacted silicon metal in the microstructure, indicating that the reactive bonding process at the HPHT conditions of manufacture was incomplete—Figure 12.

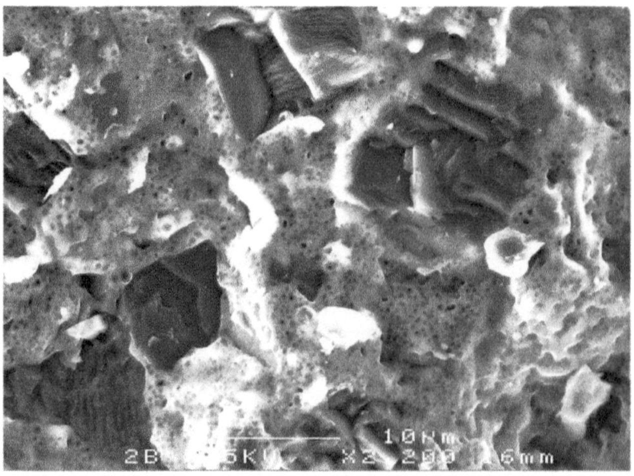

Figure 12. SEM-SEI image of a poor quality TSDC cutting element (see M2-b, Table 1). The diamond grains are clearly delineated with cleavage fracture features while the matrix is unreacted silicon metal with the porous microstructure. The bonding of the diamond to the unreacted Si appears to be

weak as there is evidenced by the diamond grains having been plucked out during the wear test.

A detailed study of two solid TSDC samples (designed as samples 1 and 5) with markedly contrasting wear Behavior was undertaken using GADDS™ (General Area Detector Diffraction System) with Cu K_{α} radiation ($\lambda = 1.5418$ Å). The major advantages of this system are: (1) no special surface preparation is required; (2) it is possible to maintain spatial integrity and hence examine the homogeneity of the phase distribution across the worn surface; (3) reasonably large-sized samples can be mounted in the specimen holder. The results from a study of the worn surfaces in these samples are shown in Figure 13, Figure 14, Figure 15 and Figure 16.

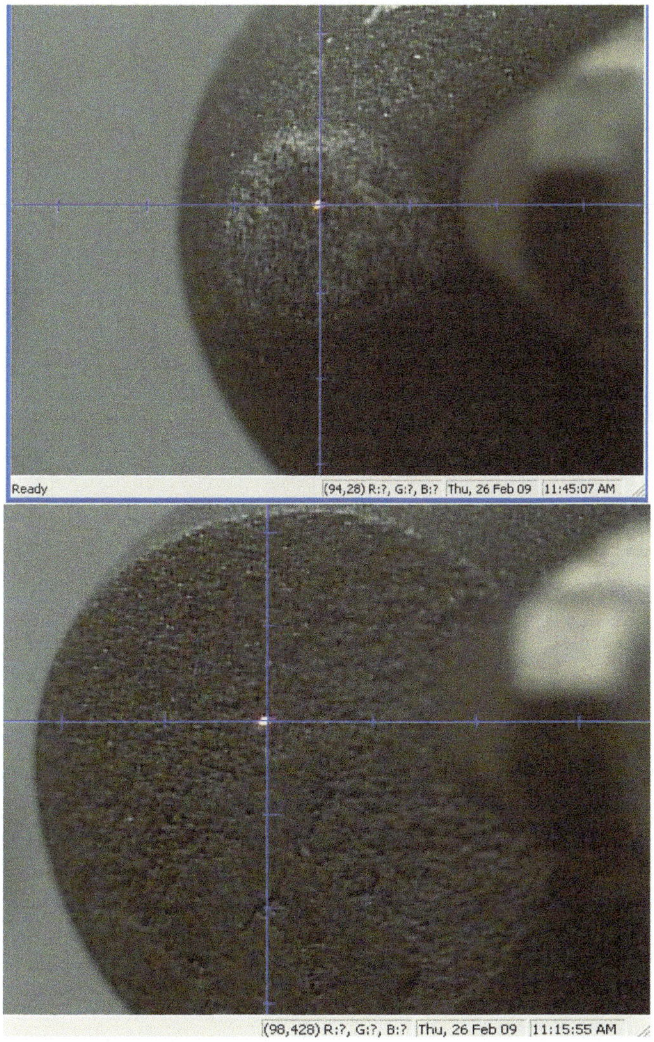

Figure 13. Optical images of both samples as mounted in the sample holder of GADDS™. Upper image is of the high quality TSDC (*i.e.*, high wear resistance, sample 1) with a small wear flat (upper) and the lower image of poor quality TSDC (*i.e.*, low wear resistance, sample 5) showing the large wear flat. The bright laser spot on each of the wear surfaces indicates the location of the x-ray beam used in the analyses (Magnifications: upper image ×40, lower image ×30).

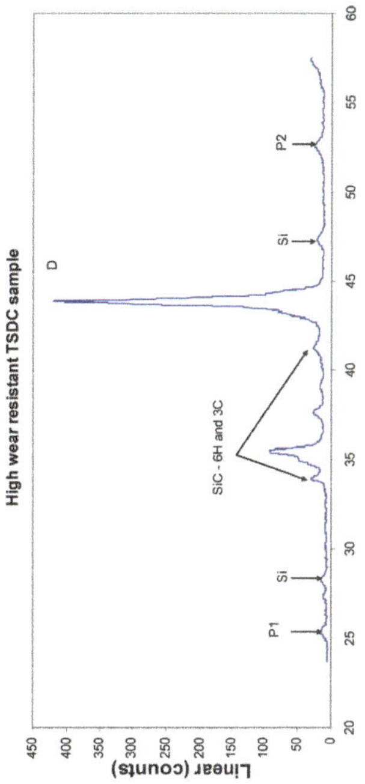

Figure 14. X-ray diffraction spectrum of sample 1 (good quality TSDC with high resistance). The automated peak search revealed diamond, silicon metal, the 6H and 3C polytypes of SiC with two unknown peaks, P1 and P2.

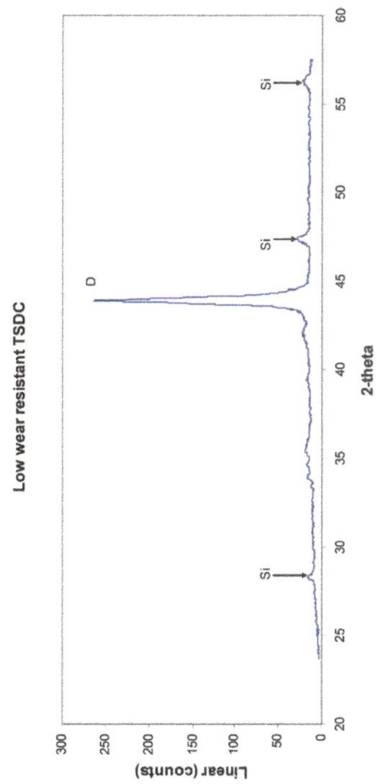

Figure 15. X-ray diffraction spectrum of sample 5 (poor quality TSDC) with diamond and Si metal positively identified. There are several smaller peaks between 33° and 43° that can tentatively assigned to the SiC polytypes—refer to XRD trace above - but appear to be in much smaller quantitatives than in sample 1.

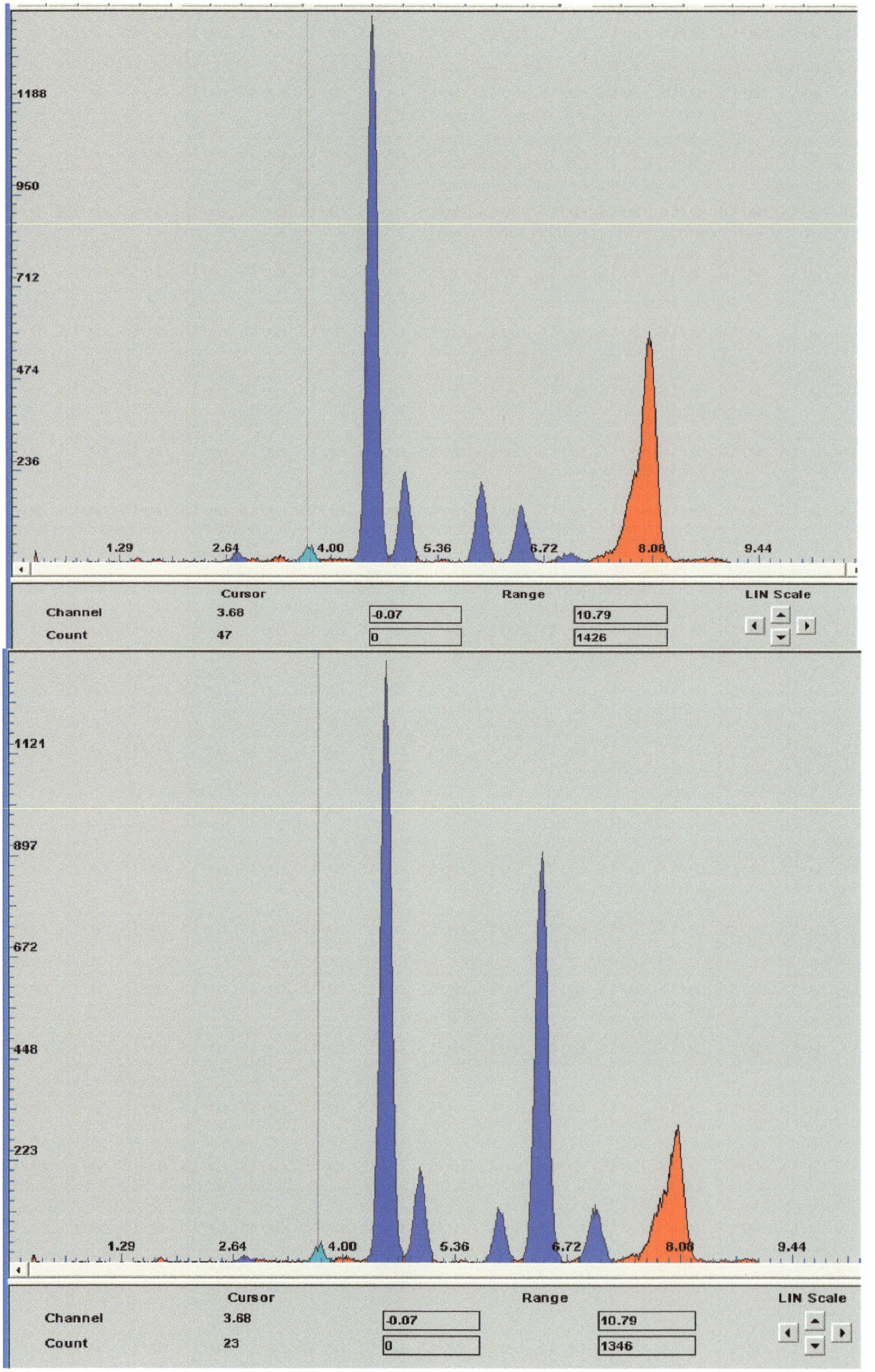

Figure 16. EDS spectra observed in the GADDS™ analyses from Sample 1 (upper spectrum) and sample 5 (lower spectrum).

The phases and the sample chemistries vary greatly between these samples. Sample 1 shows x-ray diffraction evidence of hexagonal and cubic SiC polytypes as well as metallic silicon in addition to the strong diamond peak. There are two as yet unidentified peaks (P1 and P2) at 3.507 Å and 1.736 Å. The metallic Si is evidence of residual, unreacted Si as previously discovered in the SEM-EDS observations shown in Figure 4. Sample 5 has both the diamond and silicon metal with trace amounts of SiC as well as three unidentified peaks at 3.154, 1.917 and 1.635 Å. Perhaps the more significant data are shown in the EDS spectra of the two samples. Both samples have strong Ti peaks with the poor wear-resistant sample 5 containing significant Fe and Co contents (note that the broad peak around 8 keV is produced by the Cu Kα radiation source while there is poor sensitivity to X-ray photons at the low energy end of the spectra). The diamond concentration of these samples has yet to be determined.

PCD Coating on WC Substrate

PDC/PCD-coated WC elements are used extensively in drill bits in the mining, exploration and civil construction industries [27,28,29]. Although the composite coatings may have varying thicknesses, the samples examined in this study have thickness of approximately 0.5 mm and are sintered onto a WC substrate under HPHT conditions within the diamond stability phase field—refer Figure 1. The coatings are limited to the upper section of the cutting element which may have either a conic shape with variable included angle or a proprietary ballistic shape. At present, the diamond composite used in coating the cutting elements is not classified as thermally stable, having a complex mix of phases, including cobalt, which is also the dominant binder phase used in the WC substrate. As with the solid TSDC cutting elements, these PCD-coated cutting elements exhibited an extremely wide range of wear resistance.

The macrographs shown in Figure 17 are examples of the variability in quality of these coated samples. The upper sample was classified as good quality with a wear rate of ~6 g/m^3 after 3399 m of abrasive wear on the corundum wheel—this sample has similar wear parameters to sample M2-c in Table 1. The small wear flat was totally contained within the diamond composite coating.

The lower sample in Figure 17 had such poor wear resistance that the diamond coating had worn through to the underlying WC substrate after only 152 m of abrasive wear on the corundum wheel—this corresponds to only 4 cuts across the face of the workpiece compared with 100 cuts in the good quality sample. In such cases, any measurement of weight loss of the sample would be heavily biased by the removal of some of the high density WC substrate. To run a valid test in which only the diamond composite coating is abraded away, two cuts or ~75 m of abrasive wear was the maximum distance. Even then, the wear rate was measured at 7800 g/m^3—equivalent to M7-d in Table 1.

Both samples were examined in the SEM (Figure 18 and 19) and an image analysis showed the only obvious difference between them is the volume percentage diamond (Table 4). Further investigation is underway to positively identify other significant differences between these two samples using EDS, XRD and Raman techniques.

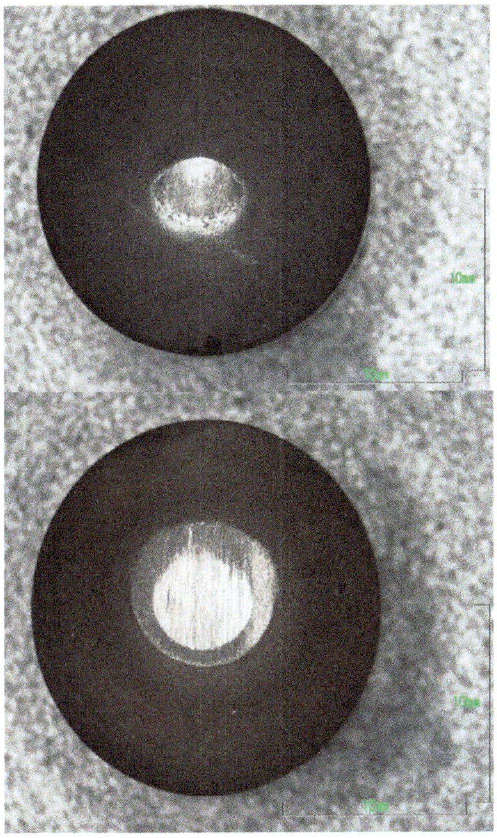

Figure 17. Top views of diamond coated cutting elements sintered onto a WC substrate: upper image showing the wear flat generated on a high quality sample after 100 cuts (approximate cutting distance of 3399 m and wear rate of 6 g/m^3); lower image showing the excessive wear produced on a poor quality sample in which the diamond coating abraded through to expose the WC substrate after 4 cuts (~152 m) (scale markers: 10 mm).

Table 4. Average grain size and diamond content in PCD-WC cutting element.

ID	Average diamond grain size, microns	Volume % diamond
Poor quality	13.6	39
Good quality	15.1	57

Figure 18. SEM-BEI image of the top worn surface of poor quality

PCD-coated WC cutting element shown in Figure 17 (lower image).

Figure 19. SEM-BEI image of top worn surface of good quality PCD-coated WC cutting element shown in Figure 17 (upper image).

X-ray Imaging

A very recent study has been initiated in which the quality of the diamond composite coating on WC is being assessed using X-ray imaging. The first set of images revealed, in somewhat surprising detail, the macroscopic defects generated in this form of coating (Figure 20). This is a projected image, generated from a point-focus X-ray source (70 keV, 80µA), of a PCD-coated WC sample sectioned longitudinally. Clearly, the interface between the coating and the WC substrate is rough with frequent intrusions of the WC substrate into cracks that were created at some stage in the manufacturing process. These cracks run almost to the outer surface of the sample. Such poor quality coatings would account for the high wear rates sometimes observed in coated samples.

Subsurface integrity of diamond coatings would be difficult to assess from optical and scanning microscopy. An X-ray micro-tomographic study could also be used on whole samples to generate 3D images of the coating, provided a sufficiently intense beam were available, such as a synchrotron source. Future research work will be undertaken to develop this technique.

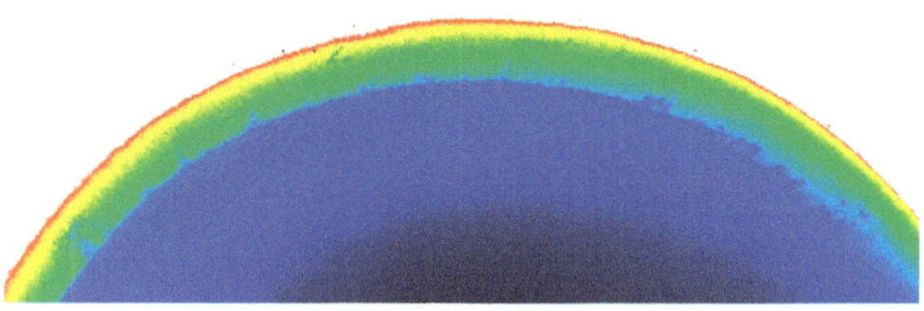

Figure 20. X-ray projection image of a PCD coating on WC substrate (blue region). Note the extent of intrusion of the WC substrate into the coating as well as the small cracks confined with the coating itself (scale: coating thickness ~454 µm.)

Cutting Mechanisms and Wear of Diamond Composite Cutting Elements

Diamond composite tooling can be employed in cutting a wide variety of materials. The most significant parameters to monitor are the wear rates of the individual phases in the composite itself as these determine the useful lifetime of the tool. However, wear rates can also be influenced by the retention rate of the diamond cutting elements which in some operations is the individual diamond grains (referred to as grit by some tool manufacturers). To strengthen the bonding of the diamond grains in the matrix, some manufacturers pre-coat the diamond grit with carbide-forming metals such at Ti. This treatment has been reported to increase the life of the cutting tool, especially for diamond impregnated MMC tools [46].

In order to understand the wear Behavior of diamond composite tools it is important to examine in detail both the chip formation process and the wear mechanisms. In the cutting of brittle materials, and especially rocks and stone, the dominant consideration has generally been the interaction between an individual diamond grain and the material surface. Most descriptive models envisage an individual diamond grain, embedded in the matrix of the tool, cutting into the work piece. With the diamond grain size ranging for 10 to 50 micron, the exposed portion of the active cutting grain is approximately half that value resulting in chip sizes in the micron to sub-micron size range [47,52]. Such a model is illustrated in Figure 21for a diamond impregnated MMC segment cutting rock or stone. It is not intended to analysis this model in detail but it is clear that the chipping process is microscopic with submicroscopic fracture and fragmentation occurring in the work piece.

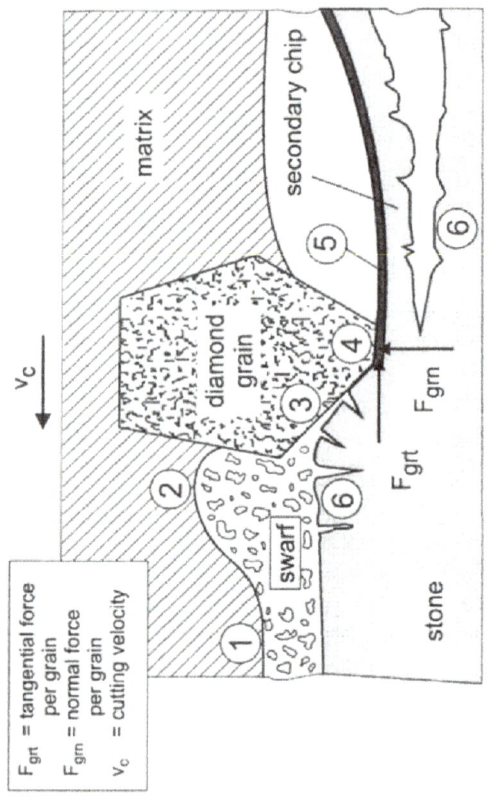

Figure 21. Microscopic model of the chip forming processes based on the interaction between an individual diamond grain and the work piece. This model is specific to brittle materials such a stone (Reproduced with permission of Trans Tech Publications – [52]).

The major factors influencing cutting performance are:

- Quality of the diamond composite relative to abrasive wear mechanisms;
- The physical and chemical properties of the work

piece—the latter aspect is vitally important in metal cutting although this area has purposefully been omitted from this review;

- Reaction forces on the tool;
- Stress distribution in both the tool and the work piece;
- Temperature at the tool—workpiece interface resulting from the frictional forces;
- Cutting speed and depth of cut;
- Removal of swath in the zone in front of the diamond grain.

A more innovative approach to the cutting operation in brittle materials, especially in the areas of rock and mineral excavation for mining and civil construction, is to use diamond composite cutting elements that will enable the depth of cut to be increased by at least two orders of magnitude. This approach has been adequately demonstrated in linear cutting operations [45] as well as in rotary cutting using a cutter drum. In the research studies presented in this review, the cutting element has been, in the first instance, TSDC. More recently this work has been extended to include cutting elements consisting of a diamond composite coating, generally PCD, on a cemented WC substrate. As discussed in Section 5.2.1, the PCD coating must be rated as non-thermally stable.

Phenomenological model of cutting brittle materials

The cutting operation can be arbitrarily partitioned into two forces acting on the work piece: the normal force required to maintain depth-of-cut and the cutting force. The effect of the normal force is equivalent to the action of an indenter with its accompanying deformation, especially the accompanying cracking and fracture under the indenter [48,49,50]. The chip formation mechanism consists in the propagation of a series of cracks initiated in the deformed volume surrounding the cutting element (Figure 22). Using high speed photography, it has been observed that the first detectable movement of macro chips corresponds to a tensile opening mode of fracture (mode I—see [48]), followed by a sliding of the separated chip across the fracture plane. Using an optical pyrometer, the temperature on the rock surface immediately behind the cutting tool can rise to ~1350 °C—these observations were made using cemented WC cutting elements [51]. Using this phenomenological model of chip formation, the more important aspects of the wear Behavior of diamond composite tools can be highlighted. Clearly, micro-mechanisms are the fundamental processes for both rock fracture and wear of the diamond composite. However, instead of microchip formation at the front surface of individual diamond grains with the concomitant gauging out of the binder phase directly ahead of the grain [47,52], somewhat less severe abrasive/erosive mechanisms occur on the rake and flank surfaces of the diamond composite cutting element. Further, the normal force will be distributed over the larger flank surface of the diamond composite cutting element. This force distribution contrasts greatly with the resultant forces expected on individual grains as envisaged in the micro chipping process for diamond impregnated MMC cutting elements. A more detailed analysis of this cutting model is the subject of ongoing research.

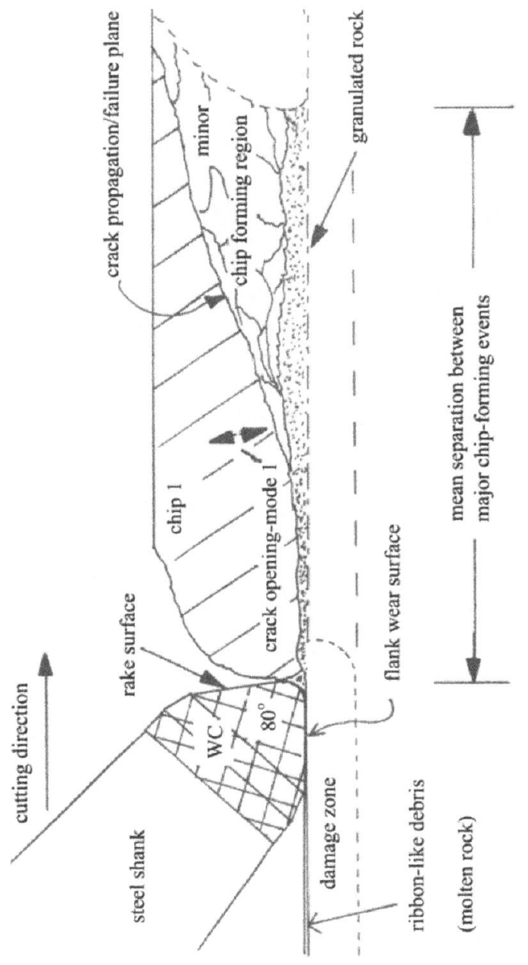

Figure 22. Phenomenological model for macro and micro-chip formation in brittle materials such as rock.

CONCLUSION

The wear Behavior of commercially produced diamond composite tools is highly variable. Since one of the major operational areas for diamond composites is in cutting, drilling and sawing for the mining, exploration and civil construction industries, there is a need for a simple and cost effective abrasive wear test for these superhard tool materials. As a first assessment procedure, an appropriate wear test of the type developed at CSIRO is recommended.

In order to determine the source of any poor wear performance of diamond composite materials, it is also essential to study their microstructural development. . It has been shown in this review that optical microscopy, scanning electron microscopy and X-ray microscopy provide important data for assessing the homogeneity and phase integrity of diamond composites. These methods are supplemented by XRD, EDS and Raman spectroscopy. It appears that, for TSDC cutting elements, the most significant processes or microstructural features responsible of high wear rates can be any of the following:

- Incomplete reaction of the molten silicon to SiC in the HPHT reaction bonding process;
- Failure to quench in the HPHT phases, leading to back transformation of diamond to graphite—this reaction may be catalysed by the presence of contaminants such as Co, Ni or Fe used in the initial production of

the synthetic diamonds;

- Residual graphitic material in the initial diamond charge used in the HPHT sintering process.

For the PCD-coated WC tools, the more likely wear mechanisms are:

- Delamination of the PCD coating due to weak bonding between the WC substrate and the coating;
- Premature abrasive wear accompanied by cracking at the macrostructural defects in the coating.

Based on these observations, there is clearly a need to improve the quality of TSDC and PCD-coated WC cutting elements by better quality control of the reactive sintering process. All poor quality components have both residual silicon metal and graphitic material in their microstructures. The presence of the former component indicates incomplete reactive sintering under the HPHT conditions, while the source of the graphite has yet to be determined. It is essential that the only remaining allotropic form of carbon in the cutting element is diamond.

Although not specifically covered in this study, x-ray photoelectron spectroscopy (XPS) in combination with Raman spectroscopy will also assist in deciphering the type of carbon bonds (sp^2 in graphitic material and sp3 in diamond), and hence the crystalline structure of carbon in these microstructures. Finally, cathodoluminescence has been shown to elucidate the phase distribution in a range of diamond composites as well as providing a means of assessing the quality of the diamond phase itself.

ACKNOWLEDGEMENTS

The authors acknowledge the assistance of Peter Clark and Craig Harbers at CSIRO Rock Cutting Laboratory, QCAT Brisbane for performing the wear testing on all diamond composite samples. We also thank Roger Rassool, Vivien Lee and Stefanie Elbracht of the Physics Department at Melbourne University for performing the x-ray imaging. Preliminary Raman spectroscopy studies on a sample of TSDC by Scott Furman at CSIRO Material Science and Engineering, Clayton Victoria are gratefully acknowledged along with thanks to David Hay for access to GADDS™.

REFERENCES

1. Falcao, E.H.L.; Wudl, F. Carbon Allotropes: Beyond Graphite and Diamond. *J. Chem. Technol. Biotechnol.* 2007, *82*, 523–531.
2. Bundy, F.P.; Bassett, W.A.; Weathers, M.S.; Hemley, R.J.; Mao, H.K.; Goncharov, A.F. The Pressure-Temperature Phase and Transformation Diagram for Carbon; Updated through 1994. *Carbon* 1996, *34*, 141–153.
3. Heinz, W.F. *Diamond Drilling Handbook*, 3rd ed.; W.F. Heinz publisher: Republic of South Africa, 1994.
4. Field, J.E., Ed.; *The Properties of Diamond*; Academic Press: London, UK, 1978.
5. Field, J.E., Ed.; *The Properties of Natural and Synthetic Diamond*; Academic Press: London, UK, 1992.
6. Wilks, J.; Wilks, E. *Properties and Applications of Diamond*; Butterworth-Heinemann Press: Oxford, UK, 1991.
7. Mainwood, A.; Newton, M.E.; Stoneham, M. Science's Gem: Diamond Science 2009. *J. Phys.: Condens. Matter* 2009, *21*, 360301.
8. Burns, R.C.; Davies, G.J. Growth of Synthetic Diamond. In *The Properties of Natural and Synthetic Diamond*; Field, J.E., Ed.; Academic Press: London, UK, 1992; pp. 395–422.
9. Chattopadhyay, R. *Surface Wear: Analysis, Treatment and Prevention*; ASM International: Materials Park, OH, USA, 2001.
10. Sato, Y.; Kamo, M. Synthesis of Diamond from Vapor Phase. In *The Properties of Natural and Synthetic Diamond*; Field, J.E., Ed.; Academic Press: London, UK, 1992; pp. 423–470.
11. Tatsii, V.; Bochko, A.V.; Oleinik, G.S. Structure and Properties of Dalan Detonation Diamonds. *Combust. Explosion Shock Waves* 2009, *45*, 95–103.
12. Wilks, J.; Wilks, E. *Properties and Applications of Diamond*; Butterworth-Heinemann Press: Oxford, UK, 1991; pp. 337–379.
13. Konstanty, J. *Powder Metallurgy Diamond Tools*; Elsevier: Amsterdam, The Netherlands, 2005.
14. Wilks, J.; Wilks, E. *Properties and Applications of Diamond*; Butterworth-Heinemann Press: Oxford, UK, 1991; p. 12.
15. Li, H.S.; Qi, Y.X.; Gong, J.H.; Wang, M.; Li, M.S. High-Pressure Synthesis and Characterization of Thermal-Stable Boron-doped Diamond Single Crystals. *Int. J. Refract. Met. Hard Mater.* 2009, *27*, 564–570.
16. Sexton, T.N.; Cooley, C.H. Polycrystalline Diamond Thrust Bearings for Down-Hole Oil and Gas Drilling Tools. *Wear*2009, *267*, 1041–1045.
17. Westraadt, J.E.; Dubrovinskaia, N. Thermally Stable Polycrystalline Diamond Sintered with Calcium Carbonate.*Diamond Related Materials* 2007, *16*, 1929–1935.
18. Tomlinson, P.N. Application of Diamond Grits and Composites. In *The Properties of Natural and Synthetic Diamond*; Field, J.E., Ed.; Academic Press: London, UK, 1992; pp. 637–665.
19. Ringwood, A.E. Diamond Compacts and Process for Making Same. US Pat. 4,948,388, 1990.
20. Irifune, T.; Utsumi, W.; Yagi, T. Use of a New Diamond Composite for Multianvil High Pressure Apparatus. *Proc. Jpn. Acad. B* 1992, *68*, 161–166.
21. Upadhyaya, G.S. *Cemented Tungsten Carbides, Production, Properties, and Testing*; Noyes Publications: Westwood, NJ, USA, 1998.
22. German, R.M. *Sintering Theory and Practice*; John Wiley & Sons, Inc.: New York, NY, USA, 1996.
23. Sung, J.C.; Sung, M. The Brazing of Diamond. *Int. J. Refract. Met. Hard Mater* 2009, *27*, 382–393.
24. Sun, F.; Feng, J.; Li, D. Bonding of CVD Diamond Thick Films using an Ag-Cu-Ti Brazing Alloy. 2001, *115*, 333–337.

25. Boland, J.N. Cutting Tool and Method of Using Same. US Pat. 6,868,848, 2005.
26. Scott, D.E. The History and Impact of Synthetic Diamond Cutters and Diamond Enhanced Inserts on the Oil and Gas Industry. *Industr. Diamond Rev.* 2006, *1*, 48–55.
27. Moseley, S.G.; Bohn, K.-P.; Goedickemeier, M. Core Drilling in Reinforced Concrete using Polycrystalline Diamond (PCD) Cutters: Wear and Fracture Mechanisms. *Int. J. Refract. Met. Hard Mater.* 2009, *27*, 394–402.
28. Belnap, D.; Griffo, A. Homogeneous and Structured PCD/WC-Co Materials for Drilling. *Diamond Related Materials*2004, *13*, 1914–1922.
29. Fang, Z.; Griffo, A.; White, B.; Lockwood, G.; Belnap, D.; Hilmas, G.; Bitler, J. Fracture Resistant Superhard Materials with Functionally Designed Microstructure. *Int. J. Refract. Met. Hard Mater.* 2001, *19*, 453–459.
30. Field, J.E. Strength, Fracture and Erosion Properties of Diamond. In *The Properties of Natural and Synthetic Diamond*; Field, J.E., Ed.; Academic Press: London, UK, 1992; pp. 473–513.
31. Qian, J. Diamond-Silicon Carbide Composite. US Pat. 7,060,641 B2, 2006.
32. Qian, J.; Voronin, G.; Zerda, T.W.; He, D.; Zhao, Y. High-Pressure, High-Temperature Sintering of Diamond-SiC Composites by Ball-Milled Diamond-Si Mixtures. *J. Mater. Res.* 2002, *17*, 2153–2160.
33. Upadhyaya, G.S. *Cemented Tungsten Carbides, Production, Properties, and Testing*; Noyes Publications: Westwood, NJ, USA, 1998; p. 212.
34. Zhao, S.; Song, X.; Wei, C.; Zhang, L.; Liu, X.; Zhang, J. Effects of WC Particle Size on Densification and Properties of Spark Plasma Sintered WC-Co Cermet. *Int. J. Refract. Met. Hard Mater.* 2009, *27*, 1014–1018.
35. Chen, Y.; Nguyen, L.C.; Zhang, L.C. Polishing of Polycrystalline Diamond by the Technique of Dynamic Friction—Part 5: Quantitative Analysis of Material Removal. *Int. J. Mach. Tools Manufact.* 2009, *49*, 515–520.
36. MacRae, C.M.; Wilson, N.C.; Johnson, S.A.; Phillips, P.L.; Otsuki, M. Hyperspectral Mapping—Combining Cathodoluminescence and X-ray Collection in an Electron Microprobe. *Microsc. Res. Tech.* 2005, *67*, 271–277.
37. MacRae, C.M.; Wilson, N.C.; Otsuki, M. Holistic Mapping. In AMAS VI-The Sixth Biennial Symposium in conjunction with SM III-The Third Scanned Probe Microscopy Conference, Melbourne, Australia, 23 May 2001; pp. 68–69.
38. Burns, R.C.; Davies, G.J. Growth of Synthetic Diamond. In *The Properties of Natural and Synthetic Diamond*; Field, J.E., Ed.; Academic Press: London, UK, 1992; p. 400.
39. Ferrari, A.C.; Robertson, J. Origin of the 1150-cm^{-1} Raman Mode in Nanocrystaalline Diamond. *Phys. Rev. B* 2001, *63*, 121405.
40. Kromka, A.; Breza, J.; Kadlecikova, M.; Janik, J.; Balon, F. Identification of Carbon Phases and Analysis of Diamond/Substrate Interfaces by Raman Spectroscopy. *Carbon* 2005, *43*, 425–429.
41. Prawer, S.; Nemanich, R.J. Raman Spectroscopy of Diamond and Doped Diamond. *Phil. Trans. R. Soc. Lond. A* 2004,*362*, 2537–2565.
42. Neale, M.J.; Gee, M. *Guide to Wear Problems and Testing for Industry*; William Andrew Publishing: New York, NY, USA, 2001.
43. Hutchings, I.M. *Tribology Friction and Wear of Engineering Materials*; Butterworth-Heinemann: Oxford, UK, 1992.
44. Slutz, D.; Clark, R. A New Low Cost Accelerated PCD Abrasion Test Method. *Finer Points* 2009, *Summer*, 8–12.
45. Li, X.S.; Boland, J.N. The Wear Characteristics of Superhard Composite Materials in Abrasive Cutting Operations.*Wear* 2005, *259*, 1128–1136.
46. Wang, Y.H.; Zang, J.B.; Wang, M.Z.; Guan, Y.; Zheng, Y.Z. Properties and Applications of Ti-Coated Diamond Grits. *J. Mater. Process. Technol.* 2002, *129*, 369–372.
47. Xu, X.P.; Zhang, G.Q.; Huang, H. Face Grinding of Granite with Brazed Diamond Grit. *Key Eng. Mater.* 2004, *259–260*, 26–32.
48. Lawn, B. *Fracture of Brittle Solids*, 2nd ed.; Cambridge University Press: Cambridge, UK, 1993.
49. Boland, J.N.; Clark, P.; Harbers, C.; Li, S.; Alehossein, H. Proceeding European Conference on Hard Materials and Diamond Tooling, EURO PM, Lausanne, Switzerland, 7–9 October 2002; pp. 103–108.
50. Alehossein, H.; Li, X.S.; Boland, J.N. Towards Improving Rock Cutting Tools using Thermally-Stable Diamond Composites. *Adv. Mater. Res.* 2009, *76–78*, 585–590.
51. Boland, J.N.; Li, X.S.; MacRae, C.; Wilson, N. Wear of Diamond Composites in Hard Rock Cutting. In Euro PM2006, Ghent, Belgium, 23–25 October 2006; Volume 3 Diamond Tooling, pp. 75–80.
52. Tönshoff, H.K.; Denkena, B.; Apmann, H.H.; Asche, J. Diamond Tools in Stone and Civil Engineering Industry—Cutting Principles, Wear and Applications. *Key Eng. Mater.* 2003, *250*, 103–109.
53.

CHAPTER 2

AN ANALYTICAL SOLUTION FOR PRE-CRACK BEHAVIOR OF LAMINATED GLASS UNDER BLAST LOADING

P. Del Linz, X. Liang, P.A. Hooper, L.Z. Wang, and J.P. Dear

Keywords: Laminated glass; Blast; Large deflections; Nonlinear

INTRODUCTION

The façade system of buildings is of high importance for designers attempting to improve the security of structures against accidental or malicious explosions. The outer layer of the affected buildings needs to prevent blast waves from penetrating the building interior, protecting the occupants and internal fittings. Windows often represent particularly weak areas in this context. Traditional annealed glass is not well suited to protect from blast shock waves, as it tends to fracture early, projecting fragments inside and outside the building and allowing further blast pressure waves to penetrate the building envelope. Laminated glass has been shown to significantly increase the protection of structures [2]. This composite material is formed by two or more layers of glass interposed with layers of Polyvinyl Butyral (PVB), a polymer membrane with good optical properties. When loaded with the pressure wave, the glass layers will tend to fracture. Following this, the PVB membrane is able to stretch, absorbing a significant quantity of incident energy. Additionally, the glass fragments will generally remain attached to the interlayer, preventing them from being flown into the building space.

However, this composite material is difficult to account for in design, as its behavior is complex. Additionally, the impulsive nature of the loading and the large deflections caused invalidate several simple analytical techniques. To obviate this, single degree of freedom approximations are often used, employing several empirical constants to obtain the required design data [2] and [3]. To obtain such constants and to study the effect of such loadings on window panes, blast experimental programs have been carried out in the past [4]. Hooper et al. [1] performed tests where Digital Image Correlation (DIC) techniques were also used to collect deformation data throughout the extent of the glazing pane. These data was used subsequently by Del Linz et al. [5] to calculate the reaction forces acting on the supports and therefore provide further data to determine empirical constants to be used in design. Several other authors [6], [7], [8], [9] and [10]also developed finite element analysis (FEA) models to perform analyses of the system. Whilst the results obtained are often realistic, the models can become very complex, requiring significant expertise and computing power to be utilized.

Wei and Dharani [11] developed an analytical model to predict the large deformations of the glazing before the failure of the glass layers, assuming a first order cosine wave to represent the deflected shape. They used this model to reproduce results from several experiments, which generally utilized relatively small charges. The method was then also extended to calculate likely crack densities in different areas of the pane [12]. Whilst their results agreed well with the experimental data for low explosive quantities, the solution proposed was not able to account for the behavior shown by the DIC results for heavier blast loadings. In these, the measured deflected shape of the window indicated that a single term solution would not be able to represent the deflections accurately, as higher order cause the deflected shape to differ from a simple sinusoidal deflection in the early phase of the loading [1].

Therefore, in this paper, the analytical solution was expanded to account for these effects. Both the deflection and the Airy's stress function were considered as summations of

Citation: P. Del Linz, X. Liang, P.A. Hooper, L.Z. Wang, J.P. Dear, An analytical solution for pre-crack behavior of laminated glass under blast loading, Composite Structures, Volume 144, 1 June 2016, Pages 156-164, ISSN 0263-8223, http://dx.doi.org/10.1016/j.compstruct.2016.02.058.

several terms, and a general solution for any number of terms was derived. This was employed to produce the required differential equations, which were solved numerically. The deflected shapes were then compared with the experimental results. Additionally, the results were used to calculate the stresses acting on the window pane, and from these reaction forces and likely crack densities were found. These again were compared with experimental results available, with the aim of validating the proposed method.

METHOD

Three of the blast tests (Tests 1–3) performed by Hooper on laminated glass were considered for this analysis [1]. One of these tests (Test 1) used 12.8 kg of C4 explosive, (15 kg TNT equivalent), whilst two more tests (Test 2 and Test 3) employed 25.6 kg of C4 explosive, (30 kg TNT equivalent). The charge weights and explosive stand-offs are summarized in Table 1.

All the glazing samples were 1.5 × 1.2 m in planar dimensions and were composed of two 3 mm thick layers of annealed glass and a 1.52 mm PVB layer. Data was collected using pressure gauges, strain gauges and 3D DIC. This last technique allowed the measurements of full field 3D deflections throughout the panes. The experimental set up is shown in Fig. 1, whilst typical DIC results are shown in Fig. 2.

Table 1. Blast tests explosive quantities and stand offs.

Test	C4 charge weight (kg)	TNT equivalent charge weight (kg)	Stand-off (m)
1	12.8	15	13
2	25.6	30	16
3	25.6	30	14

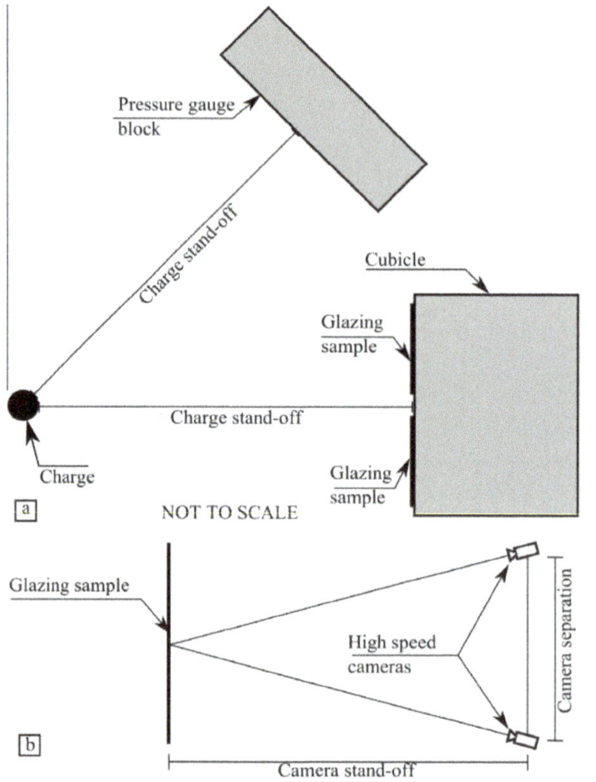

Figure 1. Typical blast test set up, showing the charge, glazing sample and pressure gauges (a) and the DIC equipment in the cubicle (b). Adapted from [1]).

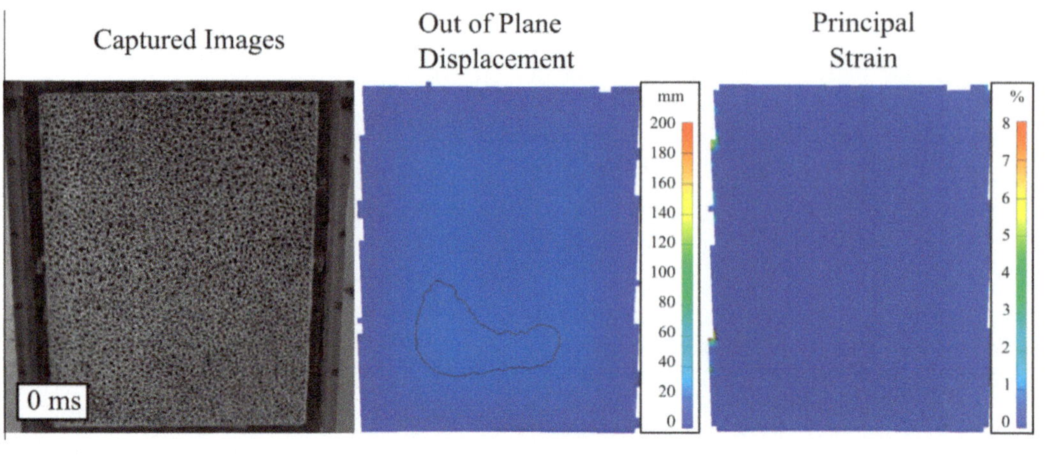

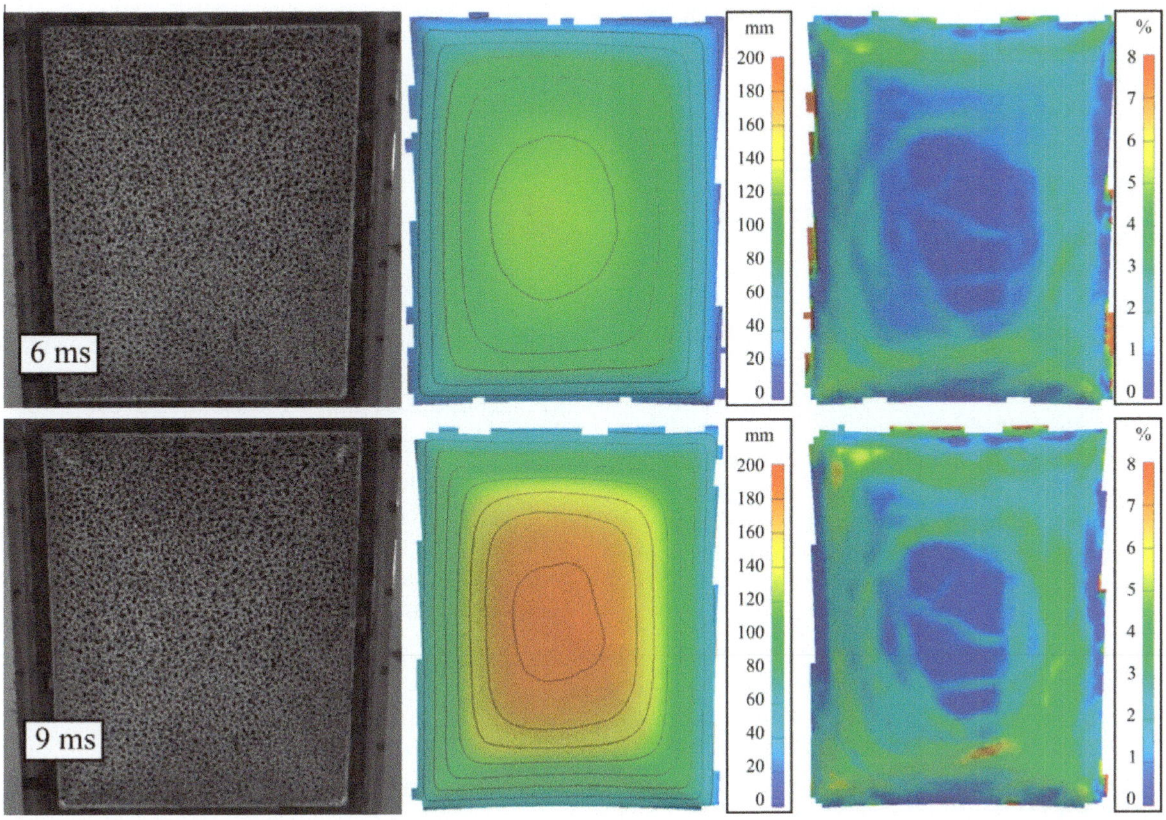

Figure 2. Typical DIC results at 3 time points (adapted from [1]). The strain measurements after the glass cracking highlight the position of cracks crossing the middle of the glass pan.

These data collected by Hooper et al. [1] were employed to validate an analytical solution of the plate deformation developed using the Von Karaman theory for large deformations, a commonly used method to represent these phenomena [11], [13] and [14]. The plate was assumed to be thin, therefore through thickness effects were ignored in the analysis. Using this theory, however, required some assumptions both regarding the material properties of the panel and the loading. These will be considered first.

MATERIAL PROPERTIES

Laminated glass is composed of two or more layers of glass interposed with layers of Polyvinyl Butyral, a rubbery polymer. Whilst any type of glass could be used, in this analysis annealed glass was assumed, as this was the variety used in the blast experiments.

Different levels of composite deformation theory or action between the composite components were also presumed. However, as in previous studies full composite action was considered a reasonable assumption previous to the glass failure point [1] and [12]. Therefore, the rule of mixtures could be used to estimate the stiffness and Poisson's ratio of the system. The formulae used were:

$$E = \frac{2E_g h_g + E_p h_p}{h_g + h_p} \qquad (1)$$

$$\upsilon = \frac{2\nu_g h_g + \nu_p h_p}{h_g + h_p} \qquad (2)$$

where E_g and E_p were the Young's modulus of the glass and PVB respectively, ν_g and ν_p were the Poisson's ratios and h_g and h_p were the thicknesses of each layer. The mass per unit area of the section could be found through:

$$M=2\rho_g h_g+\rho_p h_p \qquad (3)$$

where ρ_g and ρ_p were the density of glass and PVB respectively.

Therefore, to apply the formulae above, it was necessary to assume appropriate material properties for each material. Whilst the glass layers would be normally considered as acting in a linear elastic manner previous to failure, the PVB membrane material was significantly more complex to model. Authors performed several experimental studies on this material, which showed significant non linearity and rate dependency in its behavior[15] and [16]. Including a material model to accurately represent this complexity in the analytical structural model would have proven impractical. However, before the glass failure, the strains of the PVB membrane were small and the stresses significantly lesser than those in the glass layers, due to the much higher stiffness of the outer material. Because of this, previous authors normally considered the PVB as linear elastic for the purpose of pre-crack analysis [5]. The same assumption was made here, and the relevant material properties were obtained from the work of Hooper et al. [1]. The material properties assumed for the analysis are summarized in Table 2.

Table 2. Material properties of the glass and PVB layers.

Material	Density (kg/m³)	Young's modulus (MPa)	Poisson's ratio
Glass	2530	72×10^9	0.22
PVB	1100	0.53×10^9	0.485

Part of the analysis also covered the analysis of the glass fragment density in the windows area and the calculation of reaction forces at the edges. To decide at which time point the crack density was to be calculated and the deflection and reaction calculations stopped, a failure condition for the glass plies had to be assumed. Glass failure is difficult to predict accurately, as it strongly depends on the presence of local random imperfections. Previous authors have used tensile stress levels of 80–100 MPa as the cracking limit for glass subject to explosions [1] and [12]. In this case an overall limit of 100 MPa was assumed. The resulting failure times were then compared with the experimental results, where the failure of the window could be determine within an error of ±1 ms from the DIC data. Once the failure time was decided it was used as described above, as it was also supposed that, as seen in the experiments, most of the crack surfaces would have been formed in less than 1 ms from the initial crack formation.

BLAST PROPERTIES

The pressure time history of blast pressure waves is generally composed of an initial near instantaneous rise, followed by an approximately exponential decay which eventually lowers the pressure to below atmospheric. Henceforth, a negative phase takes place, where the pressure on obstacles is reversed. In this analysis, the blast pressures were included using the blast curve utilized in several previous studies [11],[13], [14] and [17]. The equation representing the blast wave was [18]:

$$p(t) = p_0\left(1 - \frac{t}{t_d}\right)e^{-\alpha t/t_d} \tag{4}$$

where p_0 is the initial pressure peak above the atmospheric pressure (overpressure), tis the time, t_d is the time duration of the positive phase and α is a decay constant.

As the reflected pressure information was not directly available for all the blast tests considered, for this analysis the computational fluid dynamics results also used by Hooper et al. [1] where used. Specifically, the peak pressure p_0 and impulse data were used to fit the parameters of the blast wave equation for each case. The final constants used are shown in Table 3.

Table 3. The blast curve constants used in the analytical calculations and the blast impulse for each case considered.

Test	p_o (Pa)	*Td* (s)	α	Impulse (Pa s)
1	92,000	0.0106	1.90	284
2	99,000	0.010	0.88	344
3	127,000	0.00693	1.1096	413

ANALYTICAL METHOD

The analysis method used needed to allow for the large deflections which took place during the loading. This meant that membrane actions were likely to be present in the deformed system even before the glass failure and could not be ignored in the solution. The von Karman large deflection theory could take this into consideration and therefore was used in this study. The basic equations describing the motion were [11]:

$$\nabla^4 F = E(w_{,xy}^2 - w_{,yy}w_{,yy}) \tag{5}$$

$$D\nabla^4 w + Mw_{,tt} - h(w_{,xx}F_{,yy} + w_{,yy}F_{,xx} - w_{,xy}F_{,xy}) = p_0\left(1 - \frac{t}{t_d}\right)e^{-\alpha t/t_d} \tag{6}$$

where h was the total thickness of the sample, w was the out of plane deflection, F was the Airy's stress function and M was the mass of the section.

In this solution, it was assumed that the glass would be simply supported along all four edges, without allowing any movement in the out-of-plane or in-plane direction. Therefore, assuming that the coordinates 0,0 were at the center of the panel, a was the panel length and b was the panel width, the conditions reduced to:

$$\begin{aligned} w = 0, \quad \frac{\partial^2 w}{\partial x^2} = 0 \text{ at } x = \pm a/2 \\ w = 0, \quad \frac{\partial^2 w}{\partial y^2} = 0 \text{ at } y = \pm b/2 \end{aligned} \tag{7}$$

Additionally, an initial condition of zero deflections and rotations throughout the panel att = 0 was utilized in the solution.

Dharani and other authors [11], [13], [14] and [17] assumed a deflected shape given by a combination of cosine waves. However, most of the authors, except for Türkmen and Mecitoglu, assumed a single term for their series, arguing this would be sufficient to represent the pre-cracking deformations of the pane. Whilst this assumption was appropriate for lower intensity blasts, the DIC results showed that such a deflection shape was not suitable for more intense loading situations. In these cases the curvature was not uniformly distributed throughout the panel, being instead concentrated in bands along the edges, as can be seen in Fig. 2. This was especially the case in the earlier stages of the deformation. To model this appropriately, it was decided to utilize more terms in the series. The equation was therefore assumed to be a summation:

$$w = h \times \sum_m \sum_n \phi_{mn} \times \cos\left(\frac{\pi m x}{a}\right) \times \cos\left(\frac{\pi n y}{b}\right) \tag{8}$$

where $\varphi_{mn}(t)$ were time dependent functions to be determined and m and n were odd positive integers. Eq. (8) could then be substituted in Eq. (5) giving:

$$\nabla^4 F = -\frac{E\pi^4 h^2}{a^2 b^2} \sum_m \sum_n \frac{1}{2}\phi_{mn}^2 m^2 n^2 \left(\cos\left(\frac{2\pi m x}{a}\right) + \cos\left(\frac{2\pi n y}{b}\right)\right) \tag{9}$$

Therefore a summation for F of a form similar to the function used by previous authors[11] could be assumed:

$$F = Eh^2 \sum_m \sum_n \phi_{mn}^2 \times \left(f_{mn} \times \cos\left(\frac{2\pi m x}{a}\right) + k_{mn} \times \cos\left(\frac{2\pi n y}{b}\right)\right) \tag{10}$$

where f_{mn} and k_{mn} were coefficients to be determined. Differentiating Eq. (10) and substituting into Eq. (9) produced:

$$\sum_m \sum_n \left(f_{mn}\frac{32m^4}{a^4} \times \cos\left(\frac{2\pi m x}{a}\right) + k_{mn}\frac{32n^4}{b^4} \times \cos\left(\frac{2\pi n y}{b}\right)\right) = \sum_m \sum_n -\frac{m^2 n^2}{a^2 b^2}\cos\left(\frac{2\pi m x}{a}\right) - \frac{m^2 n^2}{a^2 b^2}\cos\left(\frac{2\pi n y}{b}\right) \tag{11}$$

This could be manipulated to give:

$$f_{mn} = -\frac{n^2a^2}{32m^2b^2} \tag{12}$$

$$k_{mn} = -\frac{m^2b^2}{32n^2a^2} \tag{13}$$

$$F = \frac{Eh^2}{32}\sum_m\sum_n \phi_{mn}^2 \times \left(-\frac{n^2a^2}{m^2b^2}\times\cos\left(\frac{2\pi mx}{a}\right) - \frac{m^2b^2}{n^2a^2}\times\cos\left(\frac{2\pi ny}{b}\right)\right) \tag{14}$$

Eq. (6) could be changed into:

$$L(w,F) = D\nabla^4 w + Mw_{,tt} - h(w_{,xx}F_{,yy} + w_{,yy}F_{,xx} - w_{,xy}F_{,xy}) - p_0\left(1-\frac{t}{t_d}\right)e^{-\alpha t/t_d} \tag{15}$$

The Galerkin method could then be applied to Eq. (15):

$$\int_{-a/2}^{a/2}\int_{-b/2}^{b/2} L(w,F)w\,dxdy = 0 \tag{16}$$

Eqs. (14) and (8) could be substituted in Eqs. (15) and (16). When proceeding with the derivation of the differential equation, it was possible to use the orthogonality of the cosine terms in the following situation:

$$\int_{-a/2}^{a/2} \cos\left(\frac{\pi mx}{a}\right)\times\cos\left(\frac{\pi nx}{a}\right) = \frac{a}{\pi}\left(\frac{\sin\left(\frac{1}{2}\pi(n-m)\right)}{n-m} + \frac{\sin\left(\frac{1}{2}\pi(n+m)\right)}{n+m}\right) \tag{17}$$

which, for odd positive n and m, was equal to 0 unless n = m.

Using this result and following some simplifications, the following system of differential equations was obtained:

$$\sum_m\sum_n\phi''_{mn} + \frac{D\pi^4}{M}\left(\frac{1}{a^4}\sum_m\sum_n\phi_{mn}m^4 + \frac{1}{b^4}\sum_m\sum_n\phi_{mn}n^4\right) + \frac{E\pi^4h^3}{16M}\left(\begin{array}{l}\frac{1}{a^4}\sum_m\sum_n\sum_i\sum_j\sum_r\sum_s\phi_{ij}^2\phi_{rs}m^2i^2\left(\begin{array}{l}\delta(s-n-2j)\\+\delta(s+n-2j)+\delta(s-n+2j)\end{array}\right)\delta(r-m)\\+\frac{1}{b^4}\sum_m\sum_n\sum_i\sum_j\sum_r\sum_s\phi_{ij}^2\phi_{rn}n^2J^2\left(\begin{array}{l}\delta(r-m-2i)\\+\delta(r+m-2i)+\delta(r-m+2i)\end{array}\right)\delta(s-n)\end{array}\right)$$
$$-\frac{16p_0}{Mh\pi^2}\left(1-\frac{t}{t_d}\right)e^{-\alpha t/t_d}\sum_m\sum_n\frac{1}{mn}\sin\left(\frac{\pi m}{2}\right)\sin\left(\frac{\pi n}{2}\right)=0 \tag{18}$$

where δ represents the Dirac function. Eq. (18) was solved using the numerical differential equation solver capabilities of Matlab [19]. It was decided to use m and n up to 7 as this combination produced results which did not change significantly when further increasing the number of terms.

REACTIONS

The out of plane reactions imposed on the window edge are of prime importance to the design of the window system. Whilst most of the blast energy is normally assumed to be absorbed in the post-crack stage of the deformation, previous results suggested that significant reactions could be imposed in the pre-crack stage [5]. The experimental data used to obtain the reaction forces though presented significant noise, preventing an accurate calculation of the forces involved. Therefore it was considered useful to calculate these directly from the results of this analytical analysis.

It was assumed that both the bending and the membrane mechanisms produced significant forces in the pre-crack phase and needed to be considered in the analysis. A diagram showing the relevant forces is shown in Fig. 3.

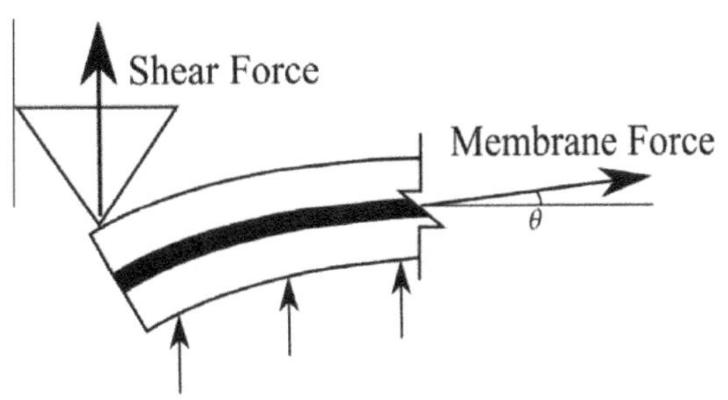

Figure 3. Diagram showing the forces acting on the glass edge (adapted from [5]).

The bending stresses were given by:

$$\sigma_{bx} = -\frac{Ez}{1-\nu^2}\left(\frac{\partial^2 w}{\partial x^2} + \nu\frac{\partial^2 w}{\partial y^2}\right) \tag{19}$$

$$\sigma_{by} = -\frac{Ez}{1-\nu^2}\left(\frac{\partial^2 w}{\partial y^2} + \nu\frac{\partial^2 w}{\partial x^2}\right) \tag{20}$$

$$\tau_{bxy} = -\frac{Ez}{1+\nu}\times\frac{\partial^2 w}{\partial x\partial y} \tag{21}$$

where σ_{bx}, σ_{by} and τ_{bxy} are respectively the stresses in the x and y direction and planar shear. z is the position in the plate thickness and was assumed to be 0 at the center. Eqs. (19), (20) and (21) represent an approximation, as the stresses will not be constant at the interface between the glass and the PVB, however the level of errors introduced was considered acceptable for this analysis.

The membrane stresses are instead given by:

$$\sigma_{mx} = \frac{\partial^2 F}{\partial y^2} \tag{22}$$

$$\sigma_{my} = \frac{\partial^2 F}{\partial x^2} \tag{23}$$

where σ_{mx}, σ_{my} were the membrane stresses in the x and y directions.

Finding the contribution to the reactions of the membrane forces required firstly calculating the angle of the glass at the edges, as it was assumed that these reactions would be acting parallel to the glass surface. This angle was calculated at all locations using the deflection data. The results were then used to calculate the out-of-plane component of the membrane stress. The stresses at each point were then integrated to find the reaction force. An area given by the glass thickness and the calculation points spacing was used for each datum.

Reactions due to the bending mechanism were taken to be equal to the shear force acting at the edge locations. In turns, this shear force was assumed to be equal to the differential of the bending moment. Therefore, to calculate this component of the reactions, firstly the bending moment was found from the bending stresses assuming a symmetrical distribution across the sample thickness. This bending moment was then differentiated in the x and y directions to find the shear forces at all points. The values at the edges were considered to be the out of plane reactions due to bending.

The membrane and bending reactions found in this manner where then integrated and compared with the experimental values.

CRACK SPACING

In all the cases considered the glass fractured during the loading. As the location of these cracks could help to assess the behavior and residual capacity of the system after the failure of the glass, the stress results available were used to calculate the likely crack density on the window area. This was compared with descriptions of the experimental outcomes.

To achieve this, a method inspired by that used by Wei and Dharani [12] was applied. The energy balance of the glazing system was given by:

$$U=U_0-U_a+U_\gamma \qquad (24)$$

where U is the total energy, U_0 is the strain energy, U_a is the drop in energy due to a crack being formed and U_γ is the surface energy required to create a crack. When a crack is forming Eq. (24) reduced to:

$$U_a=U_\gamma \qquad (25)$$

The surface energy required to form a crack was given by:

$$U_\gamma=2lh_g\gamma_s \qquad (26)$$

where l is the length of crack, h_g is the glass thickness and γ_s is 3.9 J/m² as given in[20].

The strain energy available in a facet was:

$$U_a=\frac{1}{2}\int_{x=0}^{x=\Delta x}\int_{y=0}^{y=\Delta y}\int_{z=0}^{z=h/2}\left(\sigma_x^2+\sigma_y^2+2(1+\upsilon_g)\tau_{xy}^2\right)dzdydx \qquad (27)$$

where Δx and Δy are the dimension of each facet. If these two energies are equated, the value of l can be found. This represented the average length of cracks present in one of the calculation facets and was used as a measure of the cracking density.

As the crack density was not measured during the blast tests, the results of this analysis could only be compared qualitatively with the description and images of typical results.

RESULTS

The formulae derived above were used to calculate the deflection of the window pane. A quarter of the area was used for the calculations as, due to symmetry, the results would have been the same in all other quadrants. The calculated deflections were compared with the experimental DIC results to ensure that both the shape and magnitude of the reactions were comparable. Typical results along a cut located as shown in Fig. 4 are shown in Figs. 5, 6 and 7. The results for an equivalent cut taken in the opposite direction were similar. The analytical results managed to predict the general behavior of the glass, generally showing a similar deflected shape as seen in the experimental results.

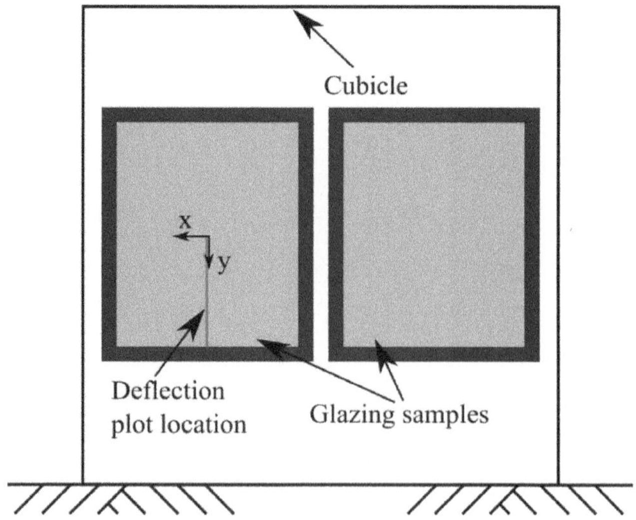

Figure 4. Location on the glazing samples of the deflection plots.

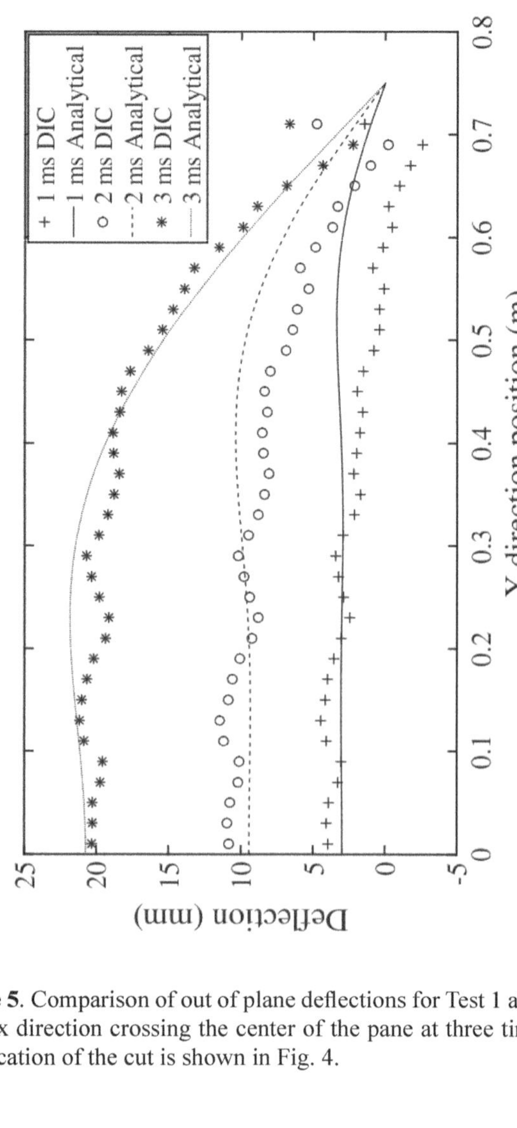

Figure 5. Comparison of out of plane deflections for Test 1 along cuts in the x direction crossing the center of the pane at three time steps. The location of the cut is shown in Fig. 4.

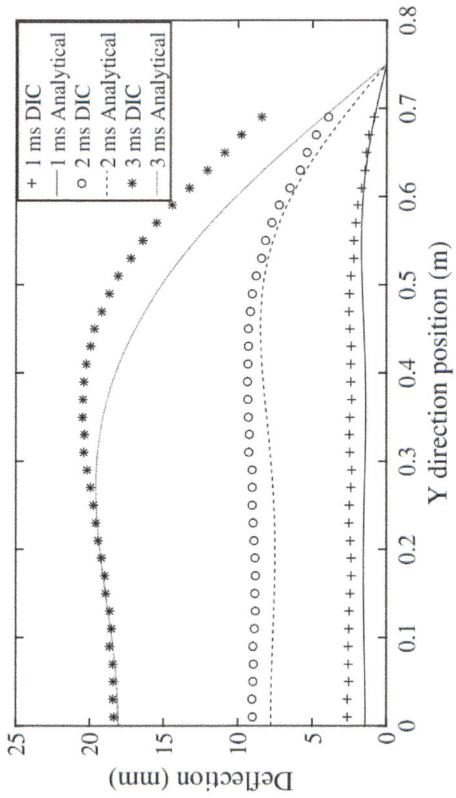

Figure 6. Comparison of out of plane deflections for Test 2 along cuts in the x and direction crossing the center of the pane at three time steps. The location of the cut is shown in Fig. 4.

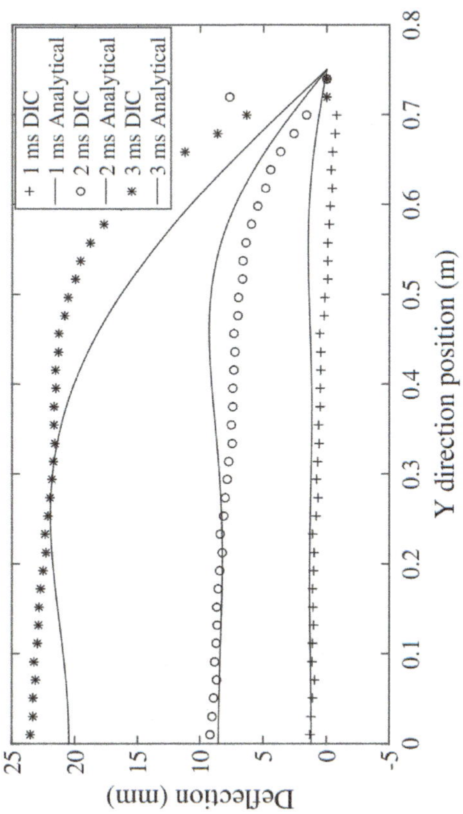

Figure 7. Comparison of out of plane deflections for Test 3 along cuts in the x direction crossing the center of the pane at three time steps. The location of the cut is shown in Fig. 4.

Some differences could be seen at the edges of the window panes. These were due to the idealisation of the boundary conditions. In reality the supporting steel members showed some deflections, though this was not measured directly. It can though be seen in the experimental results that the deflections do not always start from 0 at the glazing edges. As this was not included in the analytical model, the edge locations and central deflections are often smaller than found in the experiments. These differences are though limited to approximately 15% in the worst locations away from the edges. Additionally, DIC data was easily lost near the supports and part of the experimental data had to be interpolated, as described by Del Linz et al. [5]. This would have also introduced some irregularities.

The results were then used to calculate the stresses across the window pane. As quarter symmetry was used, the results plots show the data for an area as shown in Fig. 8. Fig. 9shows typical results for Test 2 at 2.25 ms. The results at this time peaked at approximately 100 MPa, suggesting the window was likely to fail at approximately this moment. Whilst this was approximately 0.5 ms earlier than assumed from the experiments, such a difference is compatible with the uncertainty in the assessment of the experimental results. The failure time for the other blasts varied, being affected by the explosion energy. In Test 1 the failure time was higher, approximately 2.5 ms, whilst in test 3 due to higher blast intensity the failure took place at 1.75 ms.

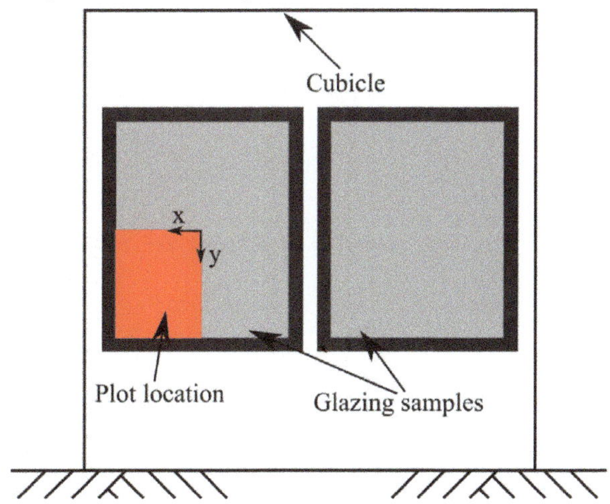

Figure 8. Location on the glazing samples of the stress, bending moment and fragment densities plots.

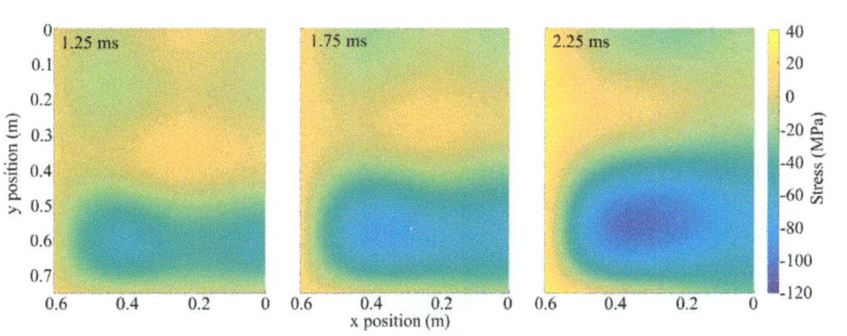

Figure 9. Stresses at the top surface across the window pane for Test 2 at 2.25 ms. The area covered by the plot is shown in Fig. 8.

Following this, the reaction forces were considered. The bending moments across the slab were firstly calculated. Typical results are shown in Fig. 10. As could be expected from the deflected shape and stress calculations, the peak moments were located in a band approximately 1/3 of the distance from the supports to the center of the pane. This was also the area with the highest curvature in the deflections.

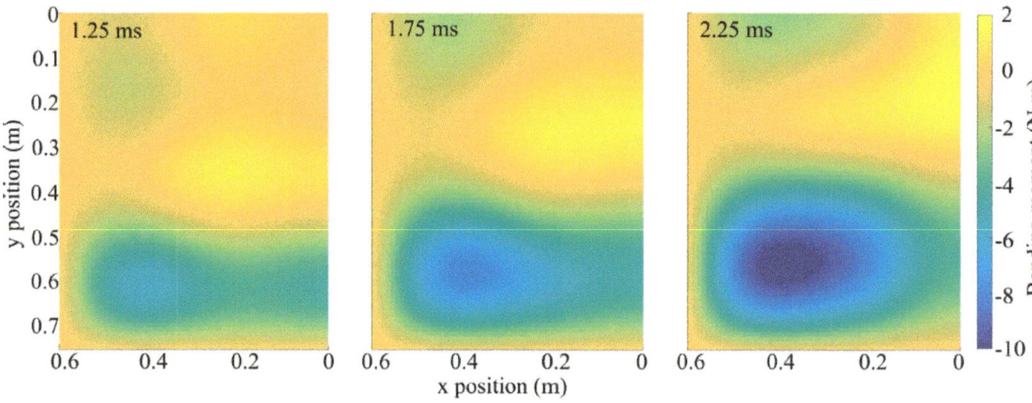

Figure 10. x direction bending moment across the window pane for test 2 at 2.25 ms. The area covered by the plot is shown in Fig. 8.

This data was differentiated and the shear forces were found along each corner. These were then added with the membrane forces, providing an overall estimate of the out-of-plane reactions. The results for the three tests are shown in Fig. 11. The analytical results showed a similar magnitude as those seen in the experimental results, though with some deviations. Test 2 showed higher reactions in the analytical case, whilst in Tests 1 and 3 the analytical and experimental data showed a better agreement. The differences though could be due to the paucity of experimental time steps which could be used, only three except for test 1.

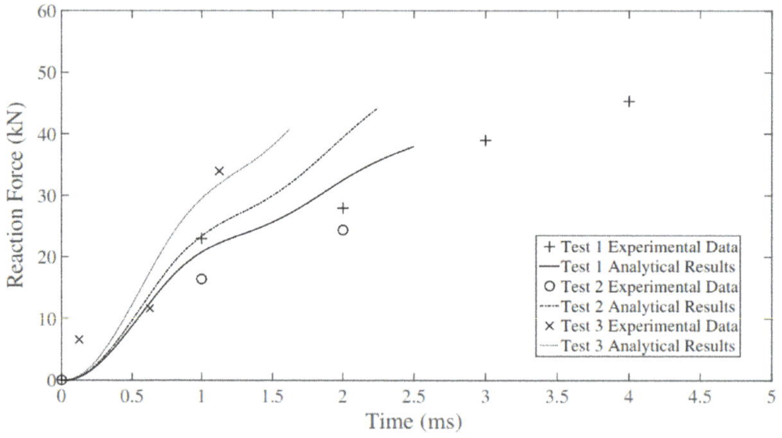

Figure 11. Out of plane reaction forces for the Tests 1–3. The experimental results are shown for comparison.

The stress data was also used to calculate likely crack densities in the panes. Typical results are shown in Fig. 12, together with an image from a typical experiment. The data showed a much higher density of cracks near the corner, followed by bands approximately 1/3 of the way between the edges and the center of the window, with very few cracks present in the center of the pane. This was similar to the previously observed stress concentrations, and was comparable to the distribution suggested by experimental observations [21].

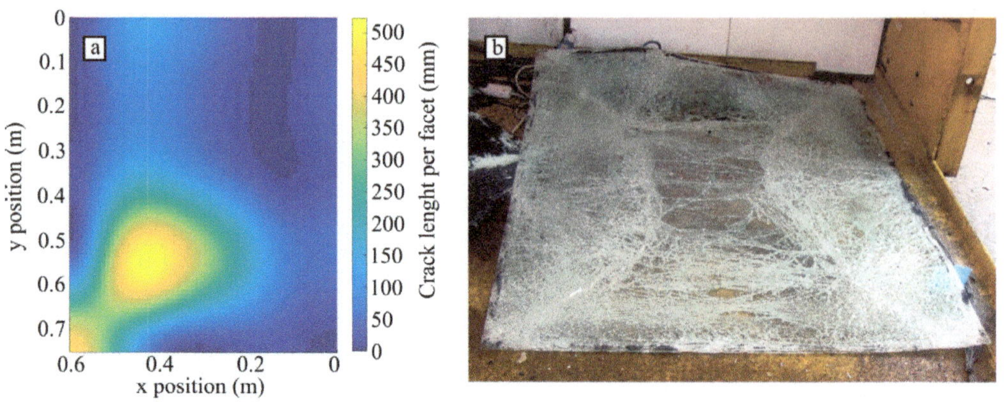

Figure 12. Test 2 calculated crack density at 2.25 ms in the area shown in Fig. 8(a) and an image from a typical test (b)[21]. The areas of higher crack density are similar.

DISCUSSION

The results of the analysis showed that considering more than one term in the deflection function was essential to capture the pane behavior in the pre-crack phase for the large blasts considered. Whilst lower blast intensities are likely to excite simpler deflection modes which could be adequately represented by the first term in the cosine series, the more extreme excitation of the structure in these heavier loading cases caused the glass to deform in a non-uniform manner. A consequence was that the highest bending moments were generated away from the glass center. This affected the stress distribution throughout the pane, causing significant differences both in the reaction forces and cracking pattern, both aspects of direct interest to glass façade designers.

The calculated deflected shapes agreed with the experimental records. Whilst the absolute quality varied between the tests, the general behavior of the pane was captured in all cases, with some showing very accurate predictions. Improvements could be made by refining the boundary conditions modelling. It is likely that in reality the edges would show a small degree of rigidity, which is not represented here. The results showed that a fully pinned model represented the results adequately, however introducing some moment resistance might improve the predictions. More importantly though, it would be beneficial to introduce the ability of the supports to move in the out-of-plane direction. Whilst the effect was relatively limited here, the movement of the supports already introduced some observable errors in the estimates. Also, the set up used in the experiment employed heavy steel sections as supports, thereby limiting their displacements. In practical glazing applications much lighter aluminium sections would be likely to be used, making their deflection much larger. This phenomenon might also become more relevant should the analysis be taken further to the post-crack phase. A further source of errors is the failure of the glass plies. Whilst for ease of comparison and due to uncertainties in the failure time estimations the deflection results were shown for the first 3 ms, it should be remembered that the glass failure is likely to have happened earlier than this point, by more than a millisecond in Test 3. Whilst the cracking event will not immediately affect the deflection shapes, some deviations will be introduced, causing some of the discrepancies.

The failure time estimates are comparable with the experimental results. The estimated failure times of 2.5, 2.25 and 1.75 ms for Tests 1, 2 and 3 respectively are comparable with the experimental times, which were approximately 3, 4 and 2 ms. Whilst the calculated failures took place sooner, especially in the case of Test 2, there was some uncertainty on the times obtained from the experimental data, as they had to be derived from indirect observations. Most importantly, the extra flexibility of the supports in the experiments would have reduced the stresses in the glass, delaying its ultimate failure.

The reactions also showed the same trend as seen in the experimental cases, though the calculated values differed in some cases significantly from the experimental estimates. This was due to both imprecisions in the analytical solutions, which would be amplified during the reaction calculation process, and the high levels of noise in the experimental data used for the reaction measurements. Del Linz et al. [5] stated that the pre-crack reaction results were not likely to be precise. This, together with the simplifications in the supports assumptions, could justify the differences seen in this analysis. In general though, the reactions were shown to be of a similar magnitude to the post-crack reactions calculated from the experiments. This highlighted that these forces need to be considered during the design process of the glazing supports.

The crack spacing results were qualitatively similar to the experimental results. Hooper[21] stated that more dense cracks were observed in bands parallel to the supports, with even denser cracks joining these bands to the corners. A similar situation can be seen inFig. 12, lending credibility to the analysis.

CONCLUSION

An analytical method to calculate the deflections and stresses in laminated glass subjected to blast pressure waves was extended by including further terms of the deflection function to produce improved results for larger explosions. The results produced showed that the method was able to predict the deflections magnitude and shape for the panes considered. The results were in all cases representative of the experimental data and in some cases the differences between the simulation and experimental results were minimal. Therefore the results showed the importance of considering the higher order terms in the deflection function when performing analyses for these cases.

The results were then used to find the reactions forces acting on the window supports. Both the bending and the shear forces were considered for this. The results showed a pattern similar to that seen in experiments, though with some differences in magnitude. This could be due both to imprecisions in the analytical model and to the high degree of uncertainty in the experimental data used for the measurements.

An estimate of cracking density was also found and compared with experimental records. The patterns found were similar to those described, showing the cracks to be concentrated in similar areas.

The results produced can effectively be used to account more precisely for the pre-crack phase of the blast event on a glass façade. The ability to calculate reaction forces will enable much more accurate designs, increasing their safety and reducing their cost. The analytical approach developed is a very valuable contribution to engineering companies engaged in risk assessment and glass façade design where simpler-to-apply analytically based models are preferable over full-scale FE analysis which can be time consuming for design assessments.

ACKNOWLEDGMENTS

The authors acknowledge the Engineering and Physical Sciences Research Council (EPSRC) and Arup Resilience, Security and Risk for financially supporting Dr Paolo Del Linz during his PhD and Dr Paul A. Hooper during his PhD. The authors also thank the Center for the Protection of National Infrastructure (CPNI) for providing access to the GL Group test facilities at RAF Spadeadam. The authors also acknowledge the National Natural Science Foundation of China (No. 51325901, 51209185) and the Fundamental Research Funds for the Central Universities for the support of Dr Liang and Professor Wang.

REFERENCES

1. Hooper PA, Sukhram RAM, Blackman BRK, Dear JP. On the blast resistance of laminated glass. Int J Solids Struct 2012;49:899–918.
2. Cormie D, Mays G, Smith P. Blast effects on buildings. 2nd ed. ICE Publishing; 2009.
3. Fischer K, Häring I. SDOF response model parameters from dynamic blast loading experiments. Eng Struct 2009;31:1677–86.
4. Stephens RAC. Determination of the resistance of laminated glass subjected to blast loading using high speed video: Cranfield University; 2008.
5. Del Linz P, Hooper PA, Arora H, Smith D, Pascoe L, Cormie D, et al. Reaction forces of laminated glass windows subject to blast loads. Compos Struct 2015;131:193–206.
6. Larcher M. Simulation of laminated glass loaded by air blast waves; 2009. p. 1553–9.
7. Zhang X, Hao H, Ma G. Parametric study of laminated glass window response to blast loads. Eng Struct 2013;56:1707–17.
8. Hidallana-Gamage HD, Thambiratnam DP, Perera NJ. Failure analysis of laminated glass panels subjected to blast loads. Eng Fail Anal 2014;36:14–29.
9. Amadio C, Bedon C. Blast analysis of laminated glass curtain walls equipped by viscoelastic dissipative devices. Buildings 2012;2:359–83.
10. Chen S, Zang M, Wang D, Zheng Z, Zhao C. Finite element modelling of impact damage in polyvinyl butyral laminated glass. Compos Struct 2016;138:1–11.
11. Wei J, Dharani LR. Response of laminated architectural glazing subjected to blast loading. Int J Impact Eng 2006;32:2032–47.
12. Wei J, Dharani LR. Fracture mechanics of laminated glass subjected to blast loading. Theoret Appl Fract Mech 2005;44:157–67.
13. Birman V, Bert CW. Behavior of laminated plates subjected to conventional blast. Int J Impact Eng 1987;6:145–55.
14. Teng TL, Liang CC, Liao CC. Transient dynamic large-deflection analysis of panel structure under blast loading. JSME Int. J. Ser. A – Mech. Mater. Eng. 1996;39:591–7.
15. Hooper PA, Blackman BRK, Dear JP. The mechanical behavior of poly(vinyl butyral) at different strain magnitudes and strain rates. J Mater Sci 2012;47:3564–76.
16. Iwasaki R, Sato C, Latailladeand JL, Viot P. Experimental study on the interface fracture toughness of PVB (polyvinyl butyral)/glass at high strain rates. Int J Crashworthiness 2007;12:293–8.
17. Turkmen HS, Mecitoglu Z. Nonlinear structural response of laminated composite plates subjected to blast loading. AIAA J. 1999;37:1639–47.
18. Kinney GF, Graham KJ. Explosive shocks in air. Berlin and New York: SpringerVerlag; 1985. 282 p. 1985;1.
19. MathWorks. MATLAB – Documentation; 2012.
20. Wiederhorn SM. Fracture surface energy of glass. J Am Ceram Soc 1969;52:99–105.
21. Hooper P. Blast performance of silicone-bonded laminated glass. Ph.D. Thesis. London: Imperial College; 2011.

CHAPTER 3

DESIGN AND CHARACTERISATION OF ADVANCED HYBRID COMPOSITES

Gergely Czél, Meisam Jalalvand, and Michael R. Wisnom

Keywords: Pseudo-ductility; Thin-ply composites; Fragmentation; Delamination; Damage and fracture mechanics; Mechanical testing

INTRODUCTION

High performance fiber reinforced composites offer outstanding strength and stiffness together with low density, therefore they are traditionally considered for advanced lightweight applications such as aero-structures, spacecraft, motorsports, high specification sports equipment, etc. However, their relatively high material and manufacturing cost and their usually brittle, catastrophic failure without sufficient warning restricts their spread towards many high volume applications including mass produced automotive and construction, where unexpected failure and poor residual integrity cannot be tolerated. The unfavorable failure characteristic is usually compensated for by conservative design limits, which hinders component manufacturers from fully exploiting the excellent mechanical properties of composites. High performance ductile or pseudo-ductile composites delivering safe failure mechanisms similar to metals' yielding and strain hardening with detectable warning and a wide margin before final failure are therefore of significant interest.

Adding ductility as a new feature to composite materials is particularly challenging because both traditional constituents of high performance long fiber reinforced thermoset polymer matrix composites (i.e. fibers such as glass, carbon and thermosetting resins such as epoxy, unsaturated polyester) are intrinsically brittle [1]. Modified matrix systems including hybrid resins [2] can only provide a small increase in properties such as impact and fatigue resistance because the mechanical response of composites is usually fiber dominated.

Development of new ductile fibers is challenging although nanotube fiber spinning seems promising [3] but it is a long process to verify and commercialize a new fiber. Ductile steel fibers have also been investigated as reinforcement for composites especially after low diameter filaments of high performance annealed stainless steel emerged recently. Excellent ductility with tensile failure strains well beyond 10% was reported [4], [5],[6] and [7]. The trade-off for high failure strains and ductile failure in these composites seems to be their high density and moderate performance compared to carbon/epoxy.

An alternative approach is to modify the architecture of traditional laminated composites e.g. by generating additional strain through the realignment of off-axis fibers [8],[9] and [10] or out of plane waviness [11], which can generate extra strain before failure by allowing reorientation of the fibers. Interface modification on the fiber [12] and on the ply level [13] as well as designed discontinuities [14] and [15] are also suitable for delaying fracture and generating non-linearity through controlled damage before final failure. Hybridization of commercially available unidirectional (UD) plies is another approach to maintain high initial modulus and potentially introduce a gradual failure, although it usually results in unfavorable major load drops when the lower strain fibers break. Our intention is to address this issue and fully exploit the benefits of hybridization for progressive, pseudo-ductile failure.

Intensive research on hybrid composites typically containing glass and carbon fibers was initiated early in the 1970s and focused on improving the low strain and brittle failure of early carbon composites. A few review papers summarized

Citation: Gergely Czél, Meisam Jalalvand, Michael R. Wisnom, Design and characterisation of advanced hybrid composites , http://dx.doi.org/10.1016/j.compstruct.2016.02.010.

the structure and properties of the hybrids studied over the first decades [16], [17], [18], [19] and [20]. Good potential for improving the brittle failure of single fiber type composites was reported with two major approaches: (i) interlayer or ply-by-ply hybrids [21], [22], [23] and [24] and (ii) intimately mixed or intermingled hybrids [25] and [26]. A more recent review by Swolfs et al. [27] pointed out the potential for adding pseudo-ductility to high performance composites as an emerging field for hybrid composites. Intermingling of continuous fibers proved to be challenging and only moderate degree of dispersion has been reported using productive continuous technologies such as tow spreading [28]. On the other hand, very promising results were presented recently by Yu et al. with a new technique to manufacture highly aligned short fiber composites [29] and well dispersed intermingled short fiber hybrids [30], which showed pseudo-ductile failure. In this study we focused on interlayer hybrids, which are simpler to manufacture by stacking layers of different prepregs together. The potential for achieving pseudo-ductility was demonstrated earlier by the authors [31] and [32] using standard thickness E-glass and emerging thin-ply carbon prepregs suitable for suppressing unstable delamination in UD interlayer hybrid composites during and after the fragmentation of the low strain material of the hybrid. Although the initial results were promising, the extent of pseudo-ductile strains was moderate because of the limited glass/epoxy strength.

The recently introduced thin ply composites have generated high interest and have been studied extensively [33], [34], [35], [36], [37], [38], [39], [40], [41] and [42] because of their unique potential to allow for highly dispersed lay-up designs which results in favorable damage suppression properties. The general conclusion of the cited studies is that the onset of damage is delayed and the strength is usually increased in unnotched thin-ply composites because premature matrix cracking and delamination can be suppressed due to the lower energy release rates with thin plies, but the final failure in return becomes more brittle. Thin carbon layers in interlayer hybrids on the contrary show favorable progressive damage (i.e. fragmentation) by suppressing unstable delamination after the first carbon layer fracture in a glass–carbon hybrid laminate due to low energy released [31]. The unique, so-called stable pull-out damage type refers to the stable delamination of the carbon layer fragments from the undamaged glass layers. Stable pull-out takes place after the onset of carbon layer fragmentation, but well before final failure and therefore it is a key enabling damage mechanism of pseudo-ductile thin-ply UD hybrid composites together with fragmentation.

The aim of this research is to develop a set of immediately applicable pseudo-ductile composite materials exploiting the demonstrated thin-ply hybrid concept [31], the recently developed analytical modelling and design tools [43] and [44] and a comprehensive new set of constituent materials with a wide range of properties. The developed new UD pseudo-ductile composites are expected to be suitable for some specific applications where tensile loading is highly dominant and also provide a strong foundation for further development of more versatile pseudo-ductile materials.

MATERIAL AND CONFIGURATION DESIGN

This section gives details of the overall concept, the applied materials and the design considerations to assure a stable pseudo-ductile failure of the hybrid laminates.

Concept

Thin-ply UD interlayer hybrid composites have recently demonstrated the potential for pseudo-ductility [31] through fragmentation and stable pull-out of a low strain (carbon)layer in the center of the laminate from the outer high strain (glass) layers. In addition, the hybrid composites have about 5% lower density and typically minimum 20% higher initial modulus than the baseline glass/epoxy. However in the demonstration phase of the research, standard E-glass was applied for the high strain layers, which did not have a high enough strain margin especially when hybridized with high strength carbon fibers, therefore no increase in stress was achieved after pseudo-yielding (after the knee point marked with (1) on Fig. 1). Advanced design tools, higher performance S-glass and a variety of the available thin carbon prepregs were utilized in the present study resulting in a range of pseudo-ductile responses, with wide stress and strain margins between the initiation of damage and final failure.

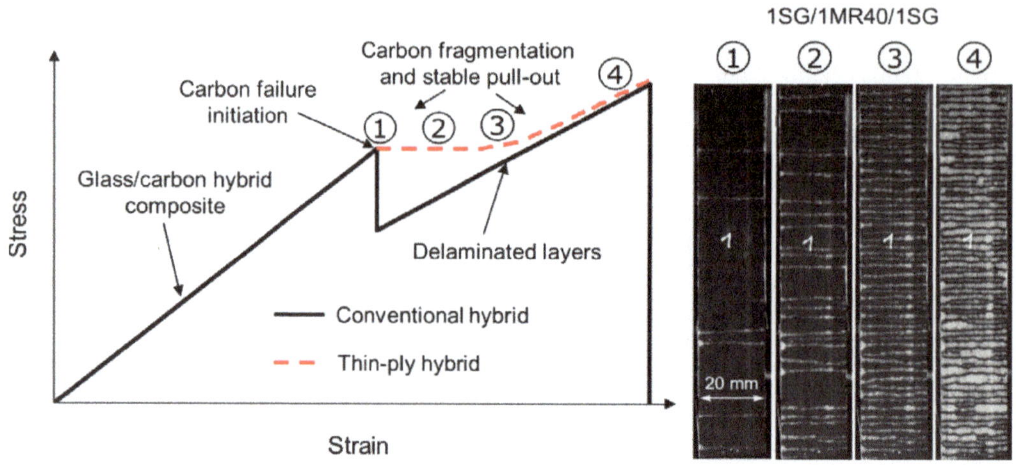

Figure 1. Schematic of the stress–strain response of conventional and thin-ply interlayer hybrid composites and typical appearance of thin-ply hybrid specimens at successive damage phases (dark areas show bonded, light areas show damaged glass/carbon interface through the translucent outer glass layer).

Fig. 1 shows the expected stress–strain response and the typical appearance of a UD thin-ply interlayer glass/carbon hybrid composite specimen at successive damage phases (carbon fragmentation and stable pull-out). The key feature of the expected response compared with our previous results which offered a relatively short, flat stress plateau, is the second rising part after a much longer plateau. This beneficial extra feature is achievable with advanced high strength S-glass prepregs, which can provide a stress margin and extended strain margin before final failure.

Materials

The set of available thin-ply UD prepreg materials was found to be very limited. Therefore the design of the configurations was significantly limited by material availability especially by the lack of thin S-glass in a tough epoxy matrix. The materials considered for design, and used for the experiments were standard thickness S-glass/epoxy prepregs supplied by Hexcel, thin S3-glass/epoxy from North Thin Ply Technology and various thin carbon/epoxy prepregs from SK Chemicals and North TPT. The epoxy resin systems in the prepregs were the aerospace grade 913 (Hexcel), ThinPreg 120 EPHTg-402 (North TPT) and K50 (SK chemicals). All resins in the hybrid laminates were 120 °C cure epoxies, which were found to be compatible, although no details were provided by the suppliers on the chemical formulation of the resins. Good integrity of the hybrid laminates was confirmed during test procedures and no phase separation was observed on cross sectional micrographs. Basic properties of the applied fibers and prepreg systems can be found in Table 1 and Table 2.

Table 1. Fiber properties of the applied UD prepregs based on manufacturer's data (carbon fiber types: SM – standard modulus, IM – intermediate modulus, HM – high modulus and UHM – ultra-high modulus).

Fiber type	Manufacturer	Tensile modulus	Strain to failure	Tensile strength	Density
		[GPa]	[%]	[GPa]	[g/cm³]
Pyrofil TR30 carbon	Mitsubishi Rayon	234 (SM)	1.9	4.41	1790
Pyrofil MR40 carbon	Mitsubishi Rayon	295 (IM)	1.5	4.41	1760
Pyrofil HS40 carbon	Mitsubishi Rayon	455 (HM)	1.0	4.61	1850
Torayca T1000 carbon	Toray	294 (IM)	2.2	6.37	1800
Torayaca M55JB carbon	Toray	540 (HM)	0.8	4.02	1910
Granoc XN80 carbon	Nippon GFC	780 (UHM)	0.5	3.43	2170
S3 UHM glass	AGY advanced materials	99	–	3.30	2830
FliteStrand S ZT S-glass	Owens corning	88	5.5	4.8–5.1	2450

Table 2. Cured ply properties of the applied UD prepregs.

Prepreg material	Manufacturer	Fiber areal mass[a]	Cured ply thickness[b]	Fiber volume fraction[a]	Initial modulus[b]
		[g/m²]	[μm]	[%]	[GPa]
TR30/epoxy	SK chemicals	21 [31]	28.9 [31]	41 [31]	101.7 [8]
MR40/epoxy	SK chemicals	50	61.4	45	134.6
HS40/epoxy	SK chemicals	65	70.3	50	229.2
T1000/epoxy	North TPT	28	32.3	48	143.3
M55/epoxy	North TPT	30	30.5	52	280.0
XN80/epoxy	North TPT	50	50.5	46	357.5
S3-glass/epoxy	North TPT	85	49.2	61	61.7
S-glass/epoxy	Hexcel	190	155.1	51	45.6 [45]

Values with references were determined experimentally by our group previously.

[a]Based on manufacturer's datasheet.

[b]Calculated using manufacturer's data.

Design of Hybrid Laminates

The following design criteria were identified and published earlier [31] to assure stable pseudo-ductile failure for UD glass/carbon interlayer hybrids. These criteria were adopted here for the preliminary design of similar interlayer hybrid configurations with new sets of constituent prepregs:

- The outer, high strain layers need to be thick and strong enough to take the full load after low strain material fracture and pull-out with a sufficient margin. This is required to account for stress concentration in the glass layer due to the carbon layer fracture which is not considered in this approach, but shown to be moderate for similar interlayer hybrid configurations [43]. Formula (1) gives the minimum strength of the high strain layer for given layer thicknesses and initial moduli, which is usually determined by the available prepreg types.

$$\sigma_{1b} > \frac{\sigma_{2b}(2E_1t_1 + E_2t_2)}{2E_2t_1} \quad (1)$$

where E_1 is the initial modulus of the high strain (glass) layers, E_2 is the initial modulus of the low strain (carbon) layer, t_1 is the thickness of one high strain layer, t_2 is the thickness of the low strain layer as shown on Fig. 3, σ_{1b} is the breaking stress of the high strain layers, σ_{2b} is the breaking stress of the low strain layer.

- The energy release rate (G_{II}) at the expected failure strain of the low strain (carbon) layer must be lower than the mode II fracture toughness (G_{IIC}) of the interface to avoid delamination of the central low strain layer after its first fracture. This criterion assures the condition for the multiple fractures (i.e. fragmentation) and stable pull-out of the low strain layer.

$$G_{IIC} > G_{II} = \frac{\varepsilon_{2b}^2 E_2 t_2 (2E_1t_1 + E_2t_2)}{8E_1t_1} \quad (2)$$

where ε_{2b} is the failure strain of the low strain layer.

The preliminary design resulted in the material configurations summarized in Table 3. The fracture toughness (G_{IIC}) of the glass/carbon composite interface was measured earlier in [46] for UD E-glass/TR30 carbon hybrid specimens, where the central carbon plies were cut perpendicular to the fibers before lay-up and curing. The hybrid specimens were made with the same 913 resin in the E-glass prepreg as that in the S-glass prepreg applied in this study. Based on our previous results obtained from specimens comprising the same resin in the glass prepregs, 1.0 N/mm was used as an estimated G_{IIC} for all S-glass configurations. A lower G_{IIC} value was assumed for the S3G/M55/S3G configuration where the S3-glass prepreg was made with North TPT's resin which exhibited lower toughness during the tests of similar interlayer hybrids of other materials comprising the same resin system. A change from stable to unstable failure was observed in configurations having $G_{II} > 0.5$ N/mm therefore this value was adopted as an approximation for G_{IIC}. Most of the predicted G_{II} energy release rates in the table are lower than the critical value (G_{IIC}), therefore stable, pseudo-ductile failure is expected for these specimen types. The two configurations with three and four TR30 carbon layers (the one with four TR30 plies is a scaled version of the one with two TR30 plies) have energy release rates higher than the estimated fracture toughness of the carbon/glass layer interfaces and so were not expected to show pseudo-ductility. These were tested to confirm the assumed energy release rate values and presented to give a more comprehensive set of failure behaviors of UD interlayer hybrids.

Table 3. Specimen types tested within the present study (specimen type designation: SG-S-glass, S3G-S3-glass. Subscripts indicating the number of plies in a block are only added where multiple plies are stacked together. Relative carbon layer thickness was normalized with the full specimen thickness).

Spec. Type	Fiber areal mass	Nominal thickness	Relative carbon layer thickness	Calculated G_{II} at carbon fiber failure strain	Approx. G_{IIC}
	[g/m²]	[mm]	[–]	[N/mm]	[N/mm]
SG/TR30/SG	190/21/190	0.339	0.085	0.306 at 1.9%	1
SG/TR30$_2$/SG	190/42/190	0.368	0.157	0.715 at 1.9%	1
SG/TR30$_3$/SG	190/63/190	0.397	0.218	1.225 at 1.9%	1
SG$_2$/TR30$_4$/SG$_2$	380/84/380	0.736	0.157	1.430 at 1.9%	1
SG/MR40/SG	190/50/190	0.372	0.165	0.736 at 1.5%	1
SG/HS40/SG	190/65/190	0.381	0.185	0.861 at 1%	1
SG$_2$/HS40/SG$_2$	380/69/380	0.691	0.102	0.632 at 1%	1
SG/T1000/SG	190/28/190	0.343	0.094	0.744 at 2.2%	1
S3G/M55/S3G	85/30/85	0.129	0.236	0.328 at 0.8%	0.5
SG/XN80$_2$/SG	190/100/190	0.411	0.246	0.802 at 0.5%	1

A novel representation of the damage modes of interlayer hybrid composites developed recently by the authors [43] and [44] was applied to the tested configurations of this study and presented in Fig. 2 to show their expected damage modes. Each map divides all possible configurations of a material pair into four groups associated with four possible damage sequences of UD interlayer hybrid composites i.e. (1) premature glass failure: the whole hybrid specimen fails at first carbon fracture, (2) unstable delamination: the layers delaminate at first carbon fracture, (3) carbon layer fragmentation: the energy released at first carbon layer fracture is not enough to drive unstable delamination, so other fractures take place in the carbon layer until they saturate, (4) carbon fragmentation and stable delamination: the energy release rate exceeds the critical value before final failure, therefore the fragmented carbon segments are pulled-out stably from the glass layers. The maps can easily be used as a design tool to achieve optimal hybrids with desired damage modes. Full details of the possible damage scenarios and the construction of the maps can be found in [43] and [44].

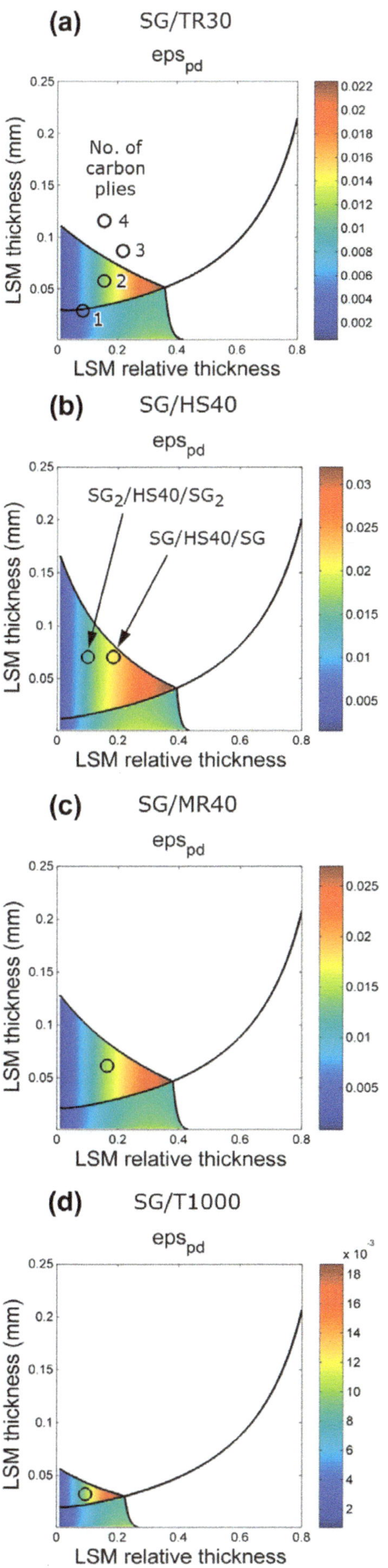

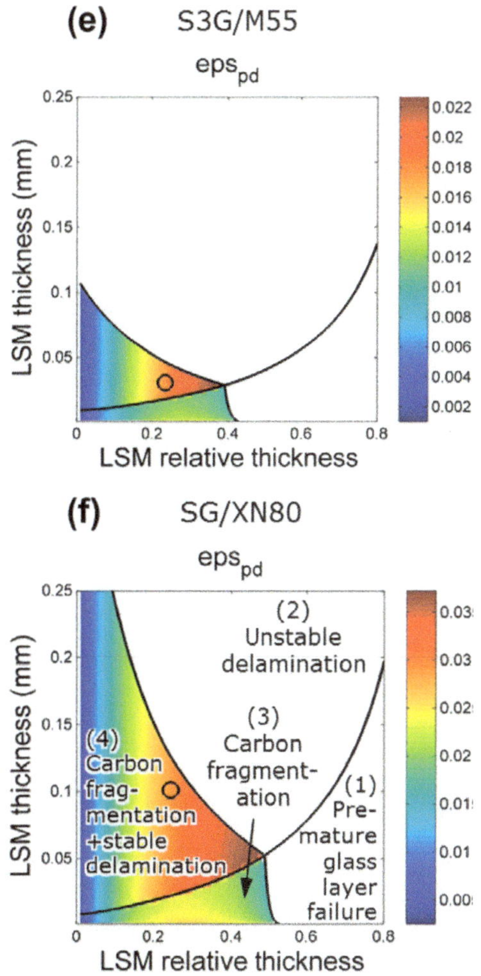

Figure 2. Damage mode maps of the designed material combinations with S-glass and various carbon plies (eps$_{pd}$ – pseudo-ductile strain, LSM – low strain material (i.e. carbon), part f) indicates four regions of the map with their associated failure processes).

Each damage mode map in Fig. 2 has four regions with the expected damage modes explained briefly in Fig. 2f. The horizontal axis of each map shows the carbon layer thickness relative to the full thickness and the vertical one shows the absolute thickness of the carbon layer so every point on the map represents a specific hybrid configuration. The maps show the expected damage modes and the achievable pseudo-ductile strains for the selected configurations marked with circles. The colored regions of the map indicate favorable pseudo-ductile damage process and the white regions show either premature glass layer failure or catastrophic delamination. The pseudo-ductile strain is defined between a point at the initial slope line at failure stress and the final failure strain as shown in Table 4.

Table 4. Results summary of the specimen types tested (specimen type designation: SG-S-glass, S3G-S3-glass. Relative carbon layer thickness was normalized with the full specimen thickness. Numbers in brackets indicate the coefficients of variation in [%]).

Spec. Type	Relative carbon layer thickness	Initial modulus	Modulus increase (to pure glass/ epoxy[a])	Pseudo-yield/ drop strain[b]	Pseudo-yield/ plateau stress[b]	Approx. final failure strain	Pseudo-ductile strain[c]
	[-]	[GPa]	[%]	[%]	[MPa]	[%]	[%]
SG/TR30/SG	0.0852	50.18 (2.2)	9.98	2.38	1142	3.7	0.91
$SG/TR30_2/SG$	0.1570	53.36 (3.6)	16.94	2.21	1129	3.4	1.21
$SG/TR30_3/SG$	0.2184	56.76 (1.3)	24.40	1.86	-	3.5	-
$SG_2/TR30_4/SG_2$	0.1570	53.40 (1.2)	17.03	1.90	-	3.7	-
SG/MR40/SG	0.1652	59.89 (2.0)	31.26	1.72	972	3.5	1.44
SG/HS40/SG	0.1847	78.59 (2.9)	72.24	1.11[d]	893[e]	3.7	2.12[e]
$SG_2/HS40/SG_2$	0.1017	65.12 (1.0)	42.71	1.28	768	3.7	1.64
SG/T1000/SG	0.0943	54.21 (1.1)	18.81	2.61	1358	3.4	0.80
S3G/M55/S3G	0.2364	112.89 (4.3)	82.97	0.85	852	3.0	2.06
$SG/XN80_2/SG$	0.2457	124.14 (6.4)	172.07	0.52	520	3.6	2.64

[a]Baseline S-glass composite modulus: 45.6 GPa (see Table 2)
[b]*Pseudo-yield points* were defined as the intersection of the test curve with a straight line parallel to the initial slope of the stress-strain graph with an offset of 0.1% strain (similar to the offset yield point or proof stress in metals terminology).
[c]*Pseudo-ductile strain* was defined between the strain of a point on the initial slope line at the *failure stress* (defined at the point where a 5% reduction in stress after the maximum has occurred) and the *strain at the failure stress*.
[d]Determined at first load drop
[e]Load drops observed on the plateau (see Fig. 5)

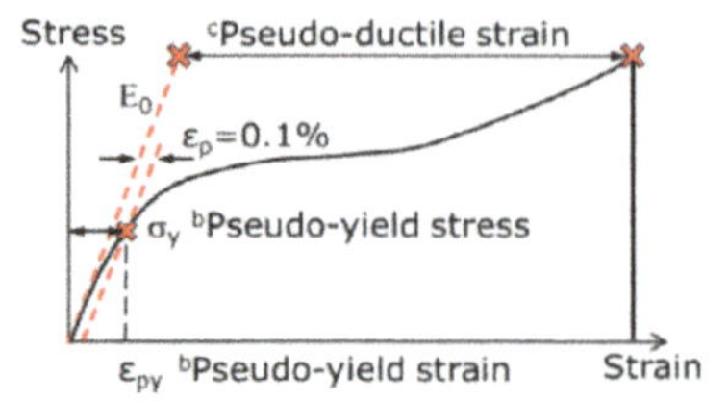

Each damage mode map in Fig. 2 was generated assuming a G_{IIC} = 1 N/mm fracture toughness (except for the S3-glass/M55 configuration with G_{IIC} = 0.5 N/mm) and the first carbon layer fracture at the failure strain of the corresponding constituent fibers (see Table 1).

The glass layer failure was predicted using a statistical strength distribution based on the fiber failure strain and typical Weibull parameters for glass fibers. Stress concentrations around the fractured low strain layer were taken into account as explained in [43], but those at the end-tab region were not considered, therefore the predicted pseudo-ductile strains represent upper bounds.

Eight of the ten different configurations marked with circles on the damage mode maps are expected to fail in a stable pseudo-ductile way while the remaining two S-glass/TR30 configurations with thick carbon layers are included in Fig. 2a for completeness only. The majority of the pseudo-ductile configurations fall into the most advantageous region (4) (carbon layer fragmentation + stable pull-out), where the highest pseudo-ductile strain can be achieved. Only the SG/TR30/SG configuration is at the border of regions (3) and (4) in Fig. 2a, therefore there is a risk of limited pseudo-ductile strain here, because only fragmentation, but no stable delamination is predicted. If there is no stable delamination around a crack in the carbon layer, the stiffness reduction will be minor, and a high stress-concentration may arise and break the surrounding glass layers resulting in the final failure of the whole specimen earlier than in a stably delaminating configuration.

EXPERIMENTAL

Specimen Geometry

The specimens tested within the experimental part of the study were UD, parallel edge tensile specimens. Nominal specimen dimensions were 240/160/20/h mm overall length/L_f-free length/W-width/h-variable thickness respectively. Fig. 3 shows the geometric parameters on the side and top view schematics of a tensile specimen.

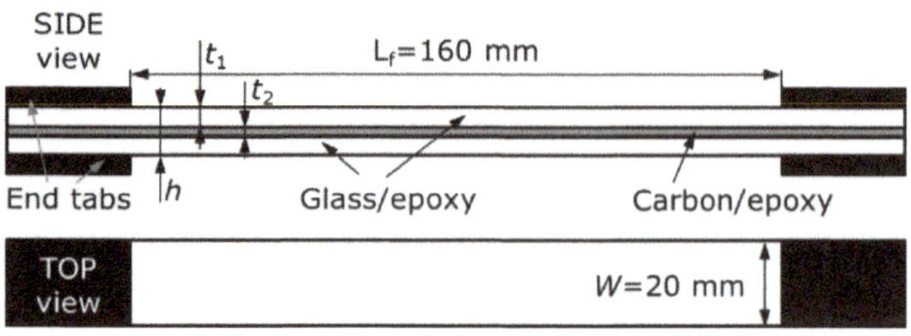

Figure 3. Schematic of the specimen geometry.

Specimen Manufacturing

The interlayer hybrid specimens were made by stacking the specified prepreg layers on top of each other, vacuum bagging the composite plate and curing it in an autoclave according to the longest of the recommended cure cycles of the constituent prepregs at their common 120 °C cure temperature and 0.7 MPa pressure. The individual specimens were fabricated with a diamond cutting wheel. Finally 40 mm long cross-ply glass/epoxy tabs were bonded to the ends of the specimens.

Test Method

Testing of the parallel edge specimens was executed under uniaxial tensile loading and displacement control at a crosshead speed of 2 mm/min on a computer controlled Instron 8801 type 100 kN rated universal hydraulic test machine with a regularly calibrated 25 kN load cell and wedge type hydraulic grips. Strains were measured using an Imetrum videogauge system, with a nominal gauge length of 130 mm. A minimum of five specimens were tested from each configuration.

Results and Discussion

Fig. 4 shows the stress–strain response of all four configurations comprising high strength (standard modulus) TR30 carbon plies. The thinnest specimens made with only one thin carbon ply had a low relative carbon layer thickness, therefore relatively low initial modulus and moderate decrease in the slope of the final rising part of the stress–strain curves was observed. Short plateaus in between the successive quasi-linear parts of the curve and smooth transitions between them were observed, which suggest a very dispersed, fragmentation dominated failure, where several short cracks develop in parallel, instead of larger ones running across the whole specimen width. This fragmentation mechanism was confirmed by the observed narrow, non-straight cracks, which appeared in the carbon ply just before the first smooth knee (between 2% and 2.5% strain) in the curve. These cracks were visible through the translucent glass ply, while growing simultaneously and gradually covering the whole specimen with an approx. 1 mm spacing. This slow damage accumulation and the corresponding limited pseudo-ductility is in agreement with the predicted damage mode (Fig. 2a). The reason for the observed benign damage accumulation is the low G_{II} of this configuration due to the very low carbon layer thickness.

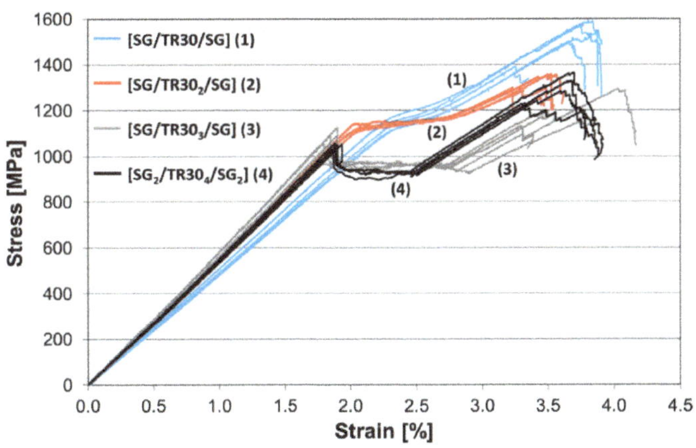

Figure 4. Tensile stress–strain graphs of S-glass/TR30 carbon configurations.

The SG/TR30$_2$/SG configuration showed a favorable pseudo-ductile failure producing a pronounced knee around the failure strain of the carbon, which can be interpreted as pseudo-yielding. A wide, slightly rising plateau and finally a second rising part can be observed on the stress–strain graph, providing sufficient stress margin after the first failure event before final failure. The specimens showed a striped pattern with an average spacing of around 2–3 mm due to the well predicted fragmentation resulting in long, straight cracks typically across the whole specimen width and stable pull-out of the carbon layer fragments (see Fig. 2a). The initial modulus of this configuration was significantly higher than that of the single ply one due to the higher relative carbon layer thickness.

The SG/TR30$_3$/SG and the SG$_2$/TR30$_4$/SG$_2$ configurations showed load drops, which rendered them unsuitable for pseudo-ductility, but the test results were consistent with our expectations based on the calculated energy release rates included in Table 3. Both specimen types delaminated to a certain extent (~30–50 mm) immediately after the first fracture in the carbon layer, as predicted by the damage mode map (Fig. 2a), and then the pull-out of the carbon layer continued stably indicated by the flat parts of the graphs after the load drops. Fig. 4 demonstrates the significant difference in the behavior of the same hybrid material combination due to an increase from two to three thin central carbon plies. The reason is that three carbon plies released enough energy to delaminate the layers at the first carbon fracture, while the pull-out of two plies remained stable. The completely different graphs of the two scaled thickness specimen types SG/TR30$_2$/SG and SG$_2$/TR30$_4$/SG$_2$ also highlight the effect of ply thickness in interlayer hybrid composites. It is also interesting to note, that the first knee of the curves corresponding to the carbon layer failure strain was shifted significantly towards higher strains if thin central layers were incorporated. This is attributed to the hybrid effect which is discussed in [47]. Fig. 4 reveals that the "width" of the pseudo-ductile stress plateau as well as the initial modulus of the hybrid material is controlled by the relative carbon layer thickness, while the failure type is strongly affected by the absolute thickness of the carbon layer (through G_{II}) for a given glass thickness (see the configurations with 1–3 carbon plies).

Fig. 5 shows the stress–strain graphs of two configurations comprising high modulus HS40 carbon prepreg. Favorable pseudo-ductile behavior was predicted for both configurations through the damage mode maps in Fig. 2b although the thinner laminate was close to the upper boundary of region (4) due to having the highest sub-critical energy release rate of all tested specimen types at the carbon layer failure strain (Table 3). Furthermore, the test results indicated that the average strain at the first fracture of the carbon layer (accompanied by a minor load drop) was significantly higher (around 1.1%) than that quoted by the fiber manufacturer (1%). An updated energy release rate calculation using the experimental carbon layer failure strain yielded an energy release rate G_{II} = 1.04 N/mm right at the limit for delamination. The small load-drops on the stress–strain graphs were observed during the tests to correspond to successive limited delaminations immediately after each carbon layer fragmentation. The overall behavior of the SG/HS40/SG configuration is very attractive with high initial modulus (up to 75% higher than that of UD S-glass/epoxy) and a 1% strain wide plateau, therefore it was developed further in [45] with a special architecture where the carbon ply was periodically cut perpendicular to the fiber direction to initiate stable pull out and avoid delamination. The thicker SG$_2$/HS40/SG$_2$ configuration had a lower energy release rate (G_{II} = 0.835 N/mm) even though it was updated with the experimentally observed 1.15% strain for carbon fracture. This configuration therefore showed a stable, pseudo-ductile failure, but a shorter plateau and decreased initial modulus, due to the relatively low carbon proportion. A slightly thinner carbon ply could have been beneficial for this material combination but our material choice was limited.

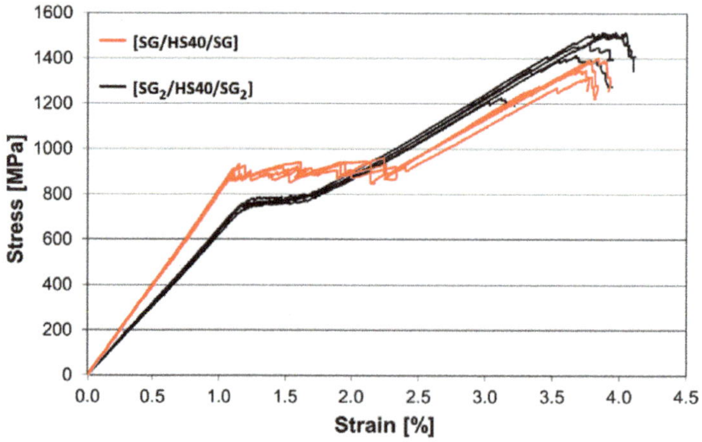

Figure 5. Tensile stress–strain graphs of S-glass/HS40 carbon configurations.

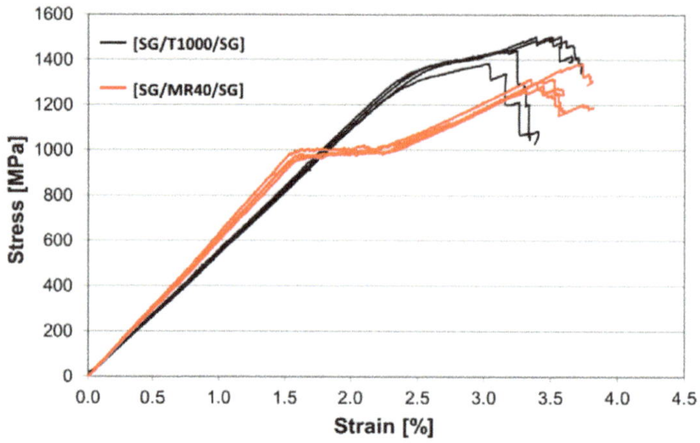

Figure 6. Tensile stress–strain graphs of S-glass/IM carbon configurations.

Fig. 6 shows the results of configurations made with intermediate modulus carbon fibers, both showing favorable pseudo-ductile stress strain responses in agreement with their damage-mode maps (Fig. 2c and d). The key difference between the similar modulus carbon fiber types applied, was their failure strain. The very high failure strain of T1000 fibers resulted in a high pseudo-yield stress, but limited pseudo-ductile strain, while the earlier failure of the MR40 fibers provided a more balanced overall shape with a wide plateau and a lower pseudo-yield stress. This figure clearly shows that there is a trade-off between pseudo-ductile strain and pseudo-yield stress of the hybrid configurations and confirm the findings of our previous paper [44] presenting parametric studies of UD hybrid configurations based on an analytical model [43]. The G_{II} at the lower failure strain of the MR40 carbon fibers remained sub-critical even with a high relative carbon layer thickness resulting in a high initial modulus. On the other hand the initial modulus of the SG/T1000/SG type was lower because the outstanding (approx. 2.3%) failure strain of the T1000 fibers resulting in high G_{II} (according to Eq. (2)) limited the critical carbon ply thickness allowable for progressive failure by fragmentation and stable pull-out. This is also highlighted by the very limited absolute and relative thicknesses of the carbon layer allowable for pseudo-ductile failure modes on the corresponding damage-mode map (see the small colored area in Fig. 2d).

The favorable pseudo-ductile stress–strain responses of the configurations comprising the highest modulus carbon fibers are presented in Fig. 7 and are in agreement with the predictions of Fig. 2e and f. The initial moduli of the S3G/M55/S3G and SG/XN80$_2$/SG type specimens were both outstanding and comparable to that of HS carbon/epoxy non-hybrid composites (113 and 124 GPa respectively). The relative carbon layer thickness and the carbon fiber modulus were higher for the XN80, but the difference in stiffness was reduced by the high volume fractions of the M55 and S3-glass fibers and the higher initial modulus of the S3-glass compared to the S-glass fibers. The pseudo-ductile failure mode area on Fig. 2e was limited by the relatively low failure strain of the S3-glass and the low G_{IIC} of this configuration. The first fracture of the XN80 carbon layer was early therefore the pseudo-yield stress was relatively low. On the other hand the pseudo-ductile strain was the highest (up to 2.6%) for this configuration. The S3G/M55/S3G type generated a more balanced stress–strain response with high initial modulus, yield stress and pseudo-ductile strain. However, the S3-glass fiber properties resulted in slightly lower final failure strain than that of the S-glass configurations, which limited the achievable pseudo-ductile strain. The set of graphs corresponding to the S3G/M55/S3G configuration on Fig. 6show minor variation of the stress along the plateau, which corresponds to small load drops due to limited localized interfacial damage which appeared instantaneously after the carbon layer cracks. This is in agreement with the corresponding G_{II} = 0.328 N/mm relatively close to the critical value (0.5 N/mm for this material combination, see Table 3), which indicates that there is a small risk of delamination for this configuration. The less stable nature of the fragmentation and delamination process observed for the hybrids containing high modulus carbon might also be affected by the various dynamics of the fracture mechanisms depending on the stiffness ratio of the adjacent layers.

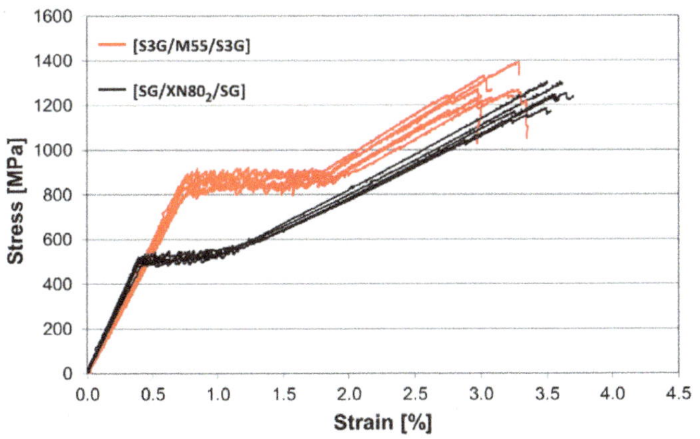

Figure 7. Tensile stress–strain graphs of S-glass/HM and UHM carbon configurations.

Table 4 summarizes the results of the tested configurations and gives the definitions of the pseudo-ductility parameters. The initial modulus values were evaluated with the nominal thicknesses (see Table 3) and the pseudo-ductility parameters were determined graphically on the aggregated test graphs of each series. The table reveals, that the final failure strains of all hybrid configurations is significantly lower than the quoted failure strain of the S-glass fibers, but we have no information on how the fiber strain was measured. This effect is commonly observed in any type of UD composites and usually attributed to stress concentrations at the grips, size effects and possible manufacturing and machining defects.

The evaluated pseudo-ductile strains were in good agreement with those predicted by the damage mode maps. This is remarkable given that the prediction of the damage mode maps were largely based on constituent fiber properties from the manufacturers' datasheets and the approximate fracture toughness of the layer interfaces. The graphs ofFig. 4, Fig. 5, Fig. 6 and Fig. 7 as well as Table 4 highlight that the stress–strain response of the tested pseudo-ductile hybrid specimen configurations were strongly affected by the failure strain of the carbon layer. The observed damage modes ranged from well distributed multiple crack formation (fragmentation) to single catastrophic delamination at the first crack (for non-ductile, thicker TR30 carbon configurations). The most important trends identified for the pseudo-ductile configurations of the same S-glass/epoxy high strain material hybridized with various carbon/epoxy low strain materials are presented in Fig. 8. Two groups of parameters were distinguished: (i) initial modulus and pseudo-ductile strain and (ii) pseudo-yield stress and strain, according to the similar trends they followed across the tested hybrid configurations. The correlations within the two groups are highlighted in the left axis plot of Fig. 8a and the right axis plot ofFig. 8b respectively. The right axis plot of Fig. 8a highlights the trade-off identified between the pseudo-ductile strains and the pseudo-yield stresses of the tested hybrids. The inverse correlation, found between the initial modulus and the pseudo-yield strain of the tested configurations, is shown in the left axis plot of Fig. 8b. This trend corresponds to the intrinsic trade-off between the tensile modulus and failure strain of different carbon fiber grades.

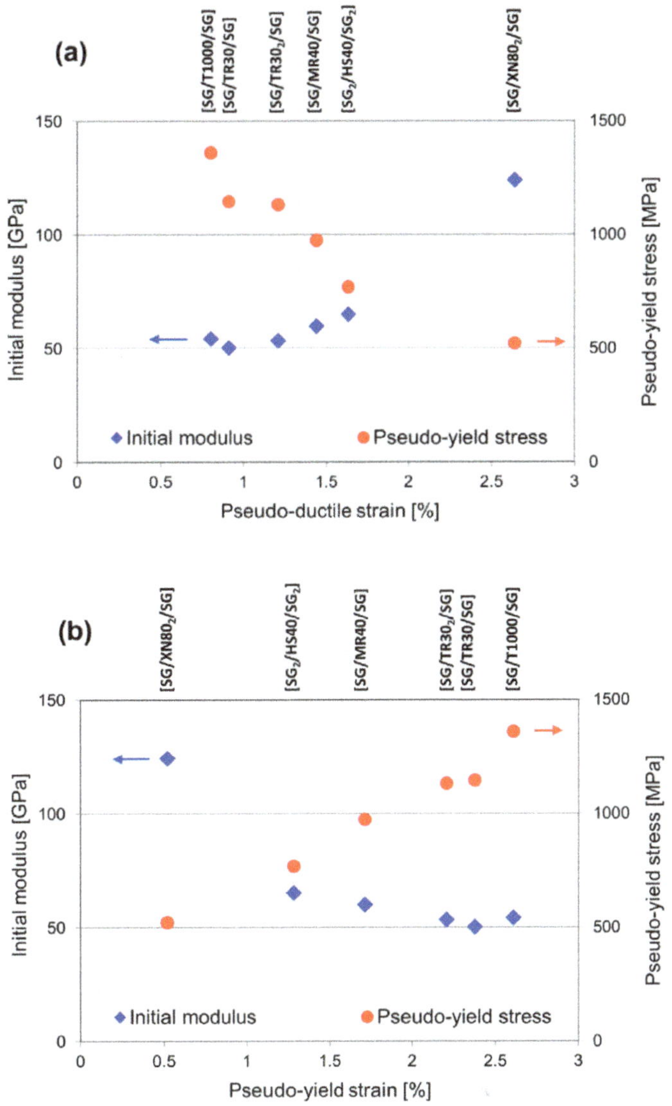

Figure 8. Representation of the trends identified in the test data.

Moreover, the highest modulus XN80 fibers failed so early, that it was possible to suppress unstable delamination (because of the low G_{II} at low strains) even with thicker carbon layers (high relative carbon thickness), which improved the initial modulus of the UHM carbon hybrid further. These conditions resulted in outstanding initial modulus and high pseudo-ductile strain, but relatively low pseudo-yield strain and stress for the UHM carbon hybrid configuration. This configuration demonstrated that it is possible to combine the stiffness of high strength carbon and the final failure strain of S-glass composites.

The other extreme was the configuration comprising the highest failure strain T1000 carbon fibers, where late damage initiation resulted in high pseudo-yield strain and stress, but limited pseudo-ductile strain. Only a low carbon thickness was allowable for this specimen type because of the high G_{II} at the onset of damage in the carbon, therefore the initial modulus increase compared to the glass/epoxy baseline was moderate. The thinnest TR30 carbon plies allowed for a series of configurations illustrating the design flexibility in tuning the failure sequence and stress–strain response of UD thin-ply hybrids by changing the relative and the absolute carbon thickness in the specimen types.

CONCLUSIONS

The following conclusions were drawn from the extensive study of thin-ply unidirectional carbon/glass fiber reinforced epoxy interlayer hybrid composites:

- A range of new material combinations providing favorable pseudo-ductile stress–strain responses were developed using various thin carbon/epoxy and standard thickness S-glass/epoxy prepregs, suitable for tensile load dominated applications.
- The SG/XN80$_2$/SG configuration provided the combination of high initial modulus up to 124 GPa, similar to that of high strength carbon/epoxy and a final failure strain similar to that of S-glass/epoxy with a pseudo ductile strain of 2.64%.
- The SG/MR40/SG configuration demonstrated balanced properties of 60 GPa initial modulus, 970 MPa pseudo-yield stress and 1.44% pseudo-ductile strain.
- The strong correlation between the initial modulus and the pseudo-ductile strain was confirmed experimentally for the tested hybrid configurations and it was attributed to the trade-off between the failure strain and tensile modulus of the carbon fiber types.
- The trade-off between the achievable pseudo-ductile strain and the pseudo-yield stress was also confirmed experimentally as a key feature which governs the design of pseudo-ductile hybrid laminates.
- A two-step procedure was presented involving simple analytical criteria and the novel damage mode maps which were applied here for the first time for the design of unidirectional interlayer hybrid composites, assuring favorable pseudo-ductile failure through carbon layer fragmentation and stable pull-out.
- Good agreement was demonstrated between the analytically predicted and experimentally observed failure modes and pseudo-ductile strains. The damage mode maps based on basic mechanical properties of the constituent plies provided insight into the important factors for optimal hybrid configuration design.

ACKNOWLEDGEMENTS

This work was funded under the UK Engineering and Physical Sciences Research Council Program Grant EP/I02946X/1 on High Performance Ductile Composite Technology in collaboration with Imperial College London. Gergely Czél acknowledges the Hungarian Academy of Sciences for funding through the Post-Doctoral Researcher Program fellowship scheme, the János Bolyai scholarship and the Hungarian National Research, Development and Innovation Office (NKFIH) for funding through Grant Ref. K 116070. The authors acknowledge Hexcel Corporation and North TPT for supplying materials for this research. Supporting data can be requested from the corresponding author.

REFERENCES

1. Chawla KK. Composite materials: science and engineering. New York: Springer-Verlag; 2012.
2. Mészáros L, Turcsán T. Development and mechanical properties of carbon fiber reinforced EP/VE hybrid composite systems. Periodica Polytech Mech Eng 2014;58:127–33.
3. Boncel S, Rajyashree M, Sundaram RM, Windle AH, Koziol KKK. Enhancement of the mechanical properties of directly spun CNT fibers by chemical treatment. ACS Nano 2011;5:9339–44.
4. Allaer K, De Baere I, Lava P, Van Paepegem W, Degrieck J. On the in-plane mechanical properties of stainless steel fiber reinforced ductile composites. Compos Sci Technol 2014;100:34–43.
5. Callens MG, Gorbatikh L, Verpoest I. Ductile steel fiber composites with brittle and ductile matrices. Compos A Appl Sci Manuf 2014;61:235–44.
6. Callens MG, Gorbatikh L, Bertels E, Goderis B, Smet M, Verpoest I. Tensile behaviour of stainless steel fiber/epoxy composites with modified adhesion. Compos A Appl Sci Manuf 2015;69:208–18.
7. Callens MG, De Cuyper P, Gorbatikh L, Verpoest I. Effect of fiber architecture on the tensile and impact behaviour of ductile stainless steel fiber polypropylene composites. Compos Struct 2015;119:528–33.
8. Fuller JD, Wisnom MR. Pseudo-ductility and damage suppression in thin ply CFRP angle-ply laminates. Compos A Appl Sci Manuf 2015;69:64–71.
9. Fuller JD, Wisnom MR. Exploration of the potential for pseudo-ductility in thin ply CFRP angle-ply laminates via an analytical method. Compos Sci Technol 2015;112:8–15.
10. Fuller JD, Jalalvand M, Wisnom MR. Combining fiber rotation and fragmentation to achieve pseudo-ductile CFRP laminates. Compos Struct 2016;142:155–66.
11. Pimenta S, Robinson P. Wavy-ply sandwich with composite skins and crushable core for ductility and energy absorption. Compos Struct 2014;104:111–24.
12. Qiana H, Bismarck A, Greenhalgh ES, Shaffer MSP. Carbon nanotube grafted carbon fibers: a study of wetting and fiber fragmentation. Compos A Appl Sci Manuf 2010;41:1107–14.
13. Grail G, Pimenta S, Pinho ST, Robinson P. Exploring the potential of interleaving to delay catastrophic failure in unidirectional composites under tensile loading. Compos Sci Technol 2015;106:100–9.
14. Pimenta S, Robinson P. An analytical shear-lag model for composites with 'brick-and-mortar' architecture considering non-linear matrix response and failure. Compos Sci Technol 2014;104:111–24.
15. Czel G, Pimenta S, Wisnom MR, Robinson P. Demonstration of pseudo-ductility in unidirectional discontinuous carbon fiber/epoxy prepreg composites. Compos Sci Technol 2015;106:110–9.
16. Hancox NL. Fiber composite hybrid materials. London: Applied Science Publishers Ltd.; 1981.
17. Summerscales J, Short D. Carbon fiber and glass fiber hybrid reinforced plastics. Composites 1978;9:157–66.
18. Short D, Summerscales J. Hybrids – a review Part 1. Techniques design and construction. Composites 1979;10:215–21.

19. Short D, Summerscales J. Hybrids – a review Part 2. Physical properties. Composites 1980;11:33–8.
20. Kretsis G. A review of the tensile, compressive, flexural and shear properties of hybrid fiber-reinforced plastics. Composites 1987;18:13–23.
21. Hayashi T. Development of new material properties by hybrid composition. 1st report. Fukugo Zairyo (Compos Mater) 1972;1:18–20.
22. Hayashi T, Koyama K, Yamazaki A, Kihira M. Development of new material properties by hybrid composition. 2nd report. Fukugo Zairyo (Compos Mater) 1972;1:21–5.
23. Bunsell AR, Harris B. Hybrid carbon and glass fiber composites. Composites 1974;5:157–64.
24. Manders PW, Bader MG. The strength of hybrid glass/carbon fiber composites. J Mater Sci 1981;16:2233–45.
25. Marom G, Fischer S, Tuler FR, Wagner HD. Hybrid effects in composites: conditions for positive or negative effects versus rule-of-mixtures behaviour. J Mater Sci 1978;13(7):1419–26.
26. Svensson N. Manufacturing of thermoplastic composites from commingled yarns – a review. J Thermoplast Compos Mater 1998;11(1):22–56.
27. Swolfs Y, Gorbatikh L, Verpoest I. Fiber hybridization in polymer composites: a review. Compos A Appl Sci Manuf 2014;67:181–200.
28. Diao H, Bismarck A, Robinson P, Wisnom MR. Pseudo-ductile behaviour of unidirectional fiber reinforced polyamide-12 composite by intra-tow hybridization. In: Proceedings of ECCM 15 conference. Venice; June 2012.
29. Yu H, Wisnom MR, Potter KD. A novel manufacturing method for aligned discontinuous fiber composites (high performance-discontinuous fiber method). Compos A Appl Sci Manuf 2014;65:175–85.
30. Yu H, Longana ML, Jalalvand M, Wisnom MR, Potter KD. Pseudo-ductility in intermingled carbon/glass hybrid composites with highly aligned discontinuous fibers. Compos A Appl Sci Manuf 2015;73:35–44.
31. Czél G, Wisnom MR. Demonstration of pseudo-ductility in high performance glass–epoxy composites by hybridization with thin-ply carbon prepreg. Compos A Appl Sci Manuf 2013;52:23–30.
32. Jalalvand M, Czél G, Wisnom MR. Numerical modelling of the damage modes in UD thin carbon/glass hybrid laminates. Compos Sci Technol 2014;94:39–47.
33. Sihn S, Kim RY, Kawabe K, Tsai SW. Experimental studies of thin-ply laminated composites. Compos Sci Technol 2007;67:996–1008.
34. Yokozeki T, Aoki Y, Ogasawara T. Experimental characterization of strength and damage resistance properties of thin-ply carbon fiber/toughened epoxy laminates. Compos Struct 2008;82:382–9.
35. Yokozeki T, Kuroda A, Yoshimura A, Ogasawara T, Aoki T. Damage characterization in thin-ply composite laminates under out-of-plane transverse loadings. Compos Sci Technol 2010;93:49–57.
36. Saito H, Morita M, Kawabe K, Kanesaki M, Takeuchi H, Tanaka M, et al. Effect of ply-thickness on impact damage morphology in CFRP laminates. J Reinf Plast Compos 2011;30:1097–106.
37. Arteiro A, Catalanotti G, Xavier J, Camanho PP. Notched response of non-crimp fabric thin-ply laminates. Compos Sci Technol 2013;79:97–114.
38. Arteiro A, Catalanotti G, Xavier J, Camanho PP. Notched response of non-crimp fabric thin-ply laminates: analysis methods. Compos Sci Technol 2013;88:165–71.
39. Amacher R, Cugnoni J, Botsis J, Sorensen L, Smith W, Dransfeld C. Thin ply composites: experimental characterization and modeling of size-effects. Compos Sci Technol 2014;101:121–32.
40. Guillamet G, Turon A, Costa J, Renart J, Linde P, Mayugo JA. Damage occurrence at edges of non-crimp-fabric thin-ply laminates under off-axis uniaxial loading. Compos Sci Technol 2014;98:44–50.
41. Borg C. An introduction to spread tow reinforcements: Part 1 – manufacture and properties. Reinf Plast 2015;59:194–8.
42. Ohlsson F. An introduction to spread tow reinforcements. Part 2: design and applications. Reinf Plast 2015;59:228–32.
43. Jalalvand M, Czél G, Wisnom MR. Damage analysis of pseudo-ductile thin-ply UD hybrid composites – a new analytical method. Compos A Appl Sci Manuf 2015;69:83–93.
44. Jalalvand M, Czél G, Wisnom MR. Parametric study of failure mechanisms and optimal configurations of pseudo-ductile thin-ply UD hybrid composites. Compos A Appl Sci Manuf 2015;74:123–31.
45. Czél G, Jalalvand M, Wisnom MR. Demonstration of pseudo-ductility in unidirectional hybrid composites made of discontinuous carbon/epoxy and continuous glass/epoxy plies. Compos A Appl Sci Manuf 2015;72:75–84.
46. Czél G, Jalalvand M, Wisnom MR. Development of pseudo-ductile hybrid composites with discontinuous carbon- and continuous glass prepregs. In: Proceedings of ECCM-16 conference. Seville; June 2014.
47. Wisnom MR, Czel G, Swolfs Y, Jalalvand M, Gorbatikh L, Verpoest I. Hybrid effects in thin ply carbon/glass unidirectional laminates: accurate experimental determination and prediction. Compos A Appl Sci Manuf; in press.

CHAPTER 4

FULLY REVERSIBLE REINFORCEMENT OF SOFTWOOD BEAMS WITH UNBONDED COMPOSITE PLATES

Marco Corradi, Antonio Borri, Giulio Castori, and Emanuel Speranzini

Keywords: GFRP plates; Softwood; Screwed connection; Composite materials; Bending tests; FEM analysis

INTRODUCTION

Softwood is from gymnosperm plants and it is the basis of approx. 85% of the world's production of wood elements. Softwood as a traditional building material has been extensively used from antiquity to the present and, among softwoods, fir wood is characterized by low weight density and good performance in terms of tensile strength, is likely to distort during seasoning. Knots or grain deviation are the main causes of tension failures. Timber construction constitutes a significant part of the infrastructure in many countries: its extensive use is essentially due to its excellent workability, good mechanical properties and low weight density. Splits caused during seasoning and natural defects may highly affect its mechanical properties and particularly cause high decreases of capacity. This reduction of the tensile strength may be as high as 90% [1].

Softwood beams are usually replaced or reinforced with traditional methods involving the use of standard building materials such as steel or aluminum plates, or composite materials. Timber reinforcement is often necessary for civil infrastructures: approx. 47% of US timber bridges is structurally deficient according to the National Bridge Inventory[2].

The application of composites for strengthening of softwood beams is not new. FRPs (Fiber Reinforced Polymers) have high tensile strength and stiffness. The structural use of Glass and Carbon FRP composites (GFRP and CFRP, respectively) is becoming common not only for new timber members, but also for reinforcement of structural elements belonging to the architectural heritage. Composites are usually used where at least two of their beneficial properties, e.g. high tensile strength and ease of application, may be exploited. In these situations, the total cost of using composite materials is similar to metallic alternatives such as steel and aluminum plates or replacement.

There are three "traditional" methods for reinforcing timber beams with FRPs: (1) Bonding of consolidated (pultruded) laminates [3], [4] and [5]; (2) Resin infusion of fabric reinforcement into grooves cut in the wood [6], [7], [8], [9], [10] and [11]; and (3) Wet lay up of FRP sheet reinforcement using epoxy adhesives [12], [13], [14] and [15]. According to the above procedures in the last two decades, FRP composites have been diffusively used in bridge decks, trusses, timber floors, etc. [16], [17], [18] and [19].

However the wide choice of composite products and their scattered mechanical properties can lead to serious problems for the designer. The selection of the reinforcement layout and the most appropriate material should be based on an accurate examination of the timber beams to be strengthened in order to avoid ineffective interventions [20] and [21]. The long-term durability of some FRPs also needs to be demonstrated [22] and [23].

Important issues remain to be solved. For example the use of FRP composites to reinforce timber beams without organic oil-based adhesives (e.g. epoxy resins) is less established. Recently, the use of natural fibers with non-organic matrixes or mechanical metal connectors has been investigated [24], and it aims at developing an interesting competitor to the use of organic oil-based fibers (e.g. CFRP) or resins, which present

Citation: Marco Corradi, Antonio Borri, Giulio Castori, Emanuela Speranzini, Fully reversible reinforcement of softwood beams with unbonded composite plates, http://dx.doi.org/10.1016/j.compstruct.2016.04.014.

problems of limited durability, low reversibility and poor performance at high temperatures [25]. Governmental and local conservation bodies do not often authorize an extensive use of organic adhesives on listed timber structures and this highly limits the use of composite materials on historic constructions. Ethical guidelines for conservation works on historic constructions often list the minimal intervention and the use of appropriate materials and fully-reversible methods [26].

In the field of green building, there are several positive aspects in this research. For example, any disposal process requires sorting materials based on composition and nature and, because timber is doubtless the ultimate green building material, its preservation and use is also desirable [27].

The reinforcement method proposed in this research meets the above requirements. The results of flexure tests on firwood beams reinforced using GFRP plates, applied on the tension side without the use of an organic adhesive, are presented in this paper. Plates have been fixed to the beam's tension surface using metal screws or bolts [28] and [29]. Reinforcement can be easily removed, if needed, because no organic adhesives have been used for the application of the GFRP plates.

MATERIAL CHARACTERIZATION

Timber

A total of 41 sharp-edged softwood beams were tested of which 28 beams were reinforced with varying amounts of tensile reinforcement. The type of softwood used in the investigation is fir wood (Abies Alba). The wood stock was firstly assessed by visual and mechanical grading. Bending tests were conducted on beams with two different dimensions: 95 × 95 × 2000 mm (26 beams) and 200 × 200 × 4000 mm (15 beams) (Fig. 1). Small beams are composed of solid softwood while large beams are made of laminated timber, also called, glulam. For small beams the moisture content and weight density were 14.31% (SD = 0.89%) and 417 kg/m^3 (standard deviation SD = 24 kg/m^3), respectively. European standard EN 13183-1 [30] was used to measure the moisture content. For large beams moisture content and weight density were 11.31% (SD = 0.37%) and 430.8 kg/m^3 (SD = 18 kg/m^3), respectively.

The mechanical properties of timber in terms of Young's modulus and compressive strength are presented in Table 1. The compressive strength varied in the range of 34.54–40.43 MPa, and the Young's modulus from 9.54 to 11.91 GPa.

a)

b)

Figure 1. (a) Small softwood solid beams. (b) Large softwood glulam beams.

Table 1. Properties of the timber.

	Small beams	Large beams
Wood species	Abies Alba	
Wood type	Solid	Glulam
Weight density (kg/m^3)	417 (24.2)	430.8 (18.2)
Moisture content (%)	14.31 (0.89)	11.31 (0.37)
Specimen dimensions (mm)	20 × 20 × 60	20 × 20 × 60
Number of tested specimens	10	10
Compressive strength (MPa)	36.90 (2.06)	37.94 (2.56)
Young's modulus (GPa)	10.55 (1.81)	11.9 (1.55)

SD in ().

GFRP Plate

The GFRP plates consisted in high-volume fraction high-strength unidirectional glass fiber in a polyester resin. Plates are produced by Fibrenet SpA under the commercial name of Fbprofile. The producer data sheet reports a Young's modulus and compressive strength of 32.6 GPa and 395 MPa, respectively. Results of mechanical characterization basically confirmed these values: according to ASTM D 3039 [31] standard Young's modulus was 31.57 GPa (SD = 2.458 GPa) with a tensile strength of 368.8 MPa (SD = 30.1 MPa). The manufacture of the GFRP plate is by pultrusion process (Table 2).

Table 2. Properties of the GFRP plate.

Thickness (mm)	9.5
Sample size	10
Tensile strength (MPa)	368.8 (30.1)
Weigth density (kg/m^3)	1779 (51.8)
Young's modulus (MPa)	31575 (2458)
Strain at failure (%)	1.19

SD in ().

For small softwood beams, GFRP plates were reduced to a length of 1400 mm and were symmetrically applied on the beam's tension side using different types of steel screws. These beams were reinforced with a single GFRP pre-drilled plate (Fig. 2).

Figure 2. Application of a pre-drilled GFRP plate on small softwood beams.

For reinforcement of large beams, two overlapping pre-drilled rectangular plates of dimensions 3600 × 80 × 9.5 mm (length × width × height) were used. GFRP plates were epoxy glued together and connected to the timber surface using metal screws or bolts. The mechanical properties of these plates are the same of the ones used for reinforcement of small beams.

EXTERNAL STRENGTHENING

Strengthening was performed with the application of the GFRP plates before the bending loads were applied. The timber surface was cleaned by air jet to rid it of loose particles and dust. The adhering face of the GFRP plate was also cleaned by acetone. Pre-drilled GFRP plates were fixed to the timber surface using commercially available metal screws (Fig. 3a) applied according to different configurations.

For the geometrical arrangement of the screws on small beams, four configurations have been used. According to the first arrangement, (Fig. 4a), 28 woodscrews (50 mm length, thread diameter 4 mm) were transversally placed at a center-to-center distance of 100 mm. All screws were positioned 25 mm from plate's edges. According to the second arrangement (Fig. 4b), 28 screws of the same type were placed diagonally (45°) at a center-to-center distance of 100 mm. Two or four U-shaped steel brackets have been used to increase the efficiency of the connection (3rd and 4th arrangement, respectively,Fig. 4c and d). The GFRP plate was epoxy glued to the steel bracket (Fig. 5).

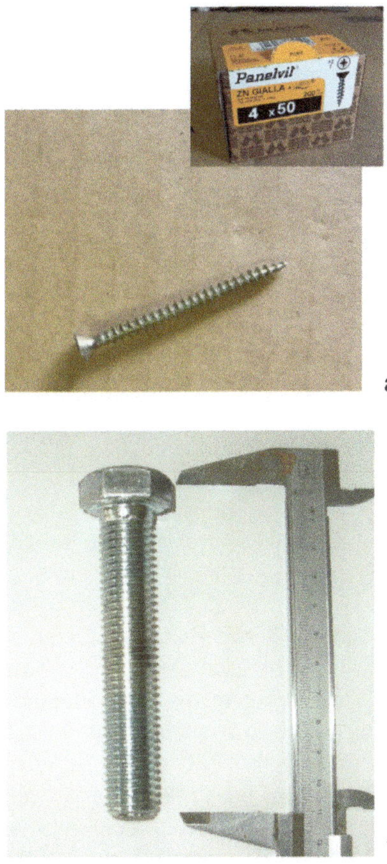

a)

b)

Figure 3. Detail of the used fully threaded woodscrews for reinforcement (a) length L = 50 mm. diameter ϕ = 4 mm; (b) length L = 100 mm. diameter ϕ = 18 mm.

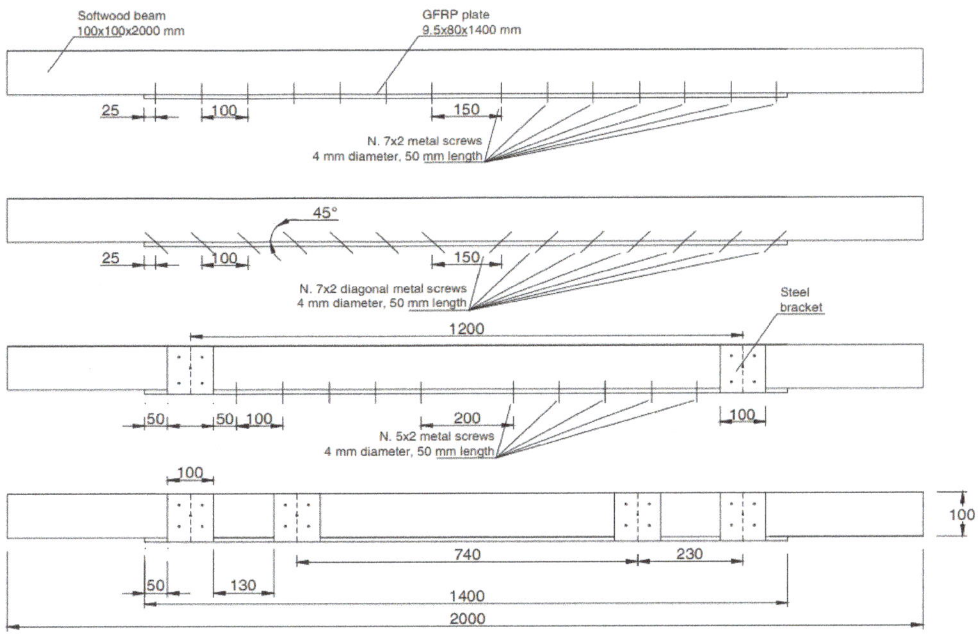

a)

b)

c)

d)

Figure 4. Detail of the method of connection of the GFRP plate to the wood tension side (small beams): (a) 28 trasversal screws (1st arrangement); (b) 28 diagonal screws (2nd arrang.); (c) 2 steel brackets and 20 trasversal screws (3rd arrang.); (d) 4 steel brackets (4th arrang.) [dimensions in (mm)].

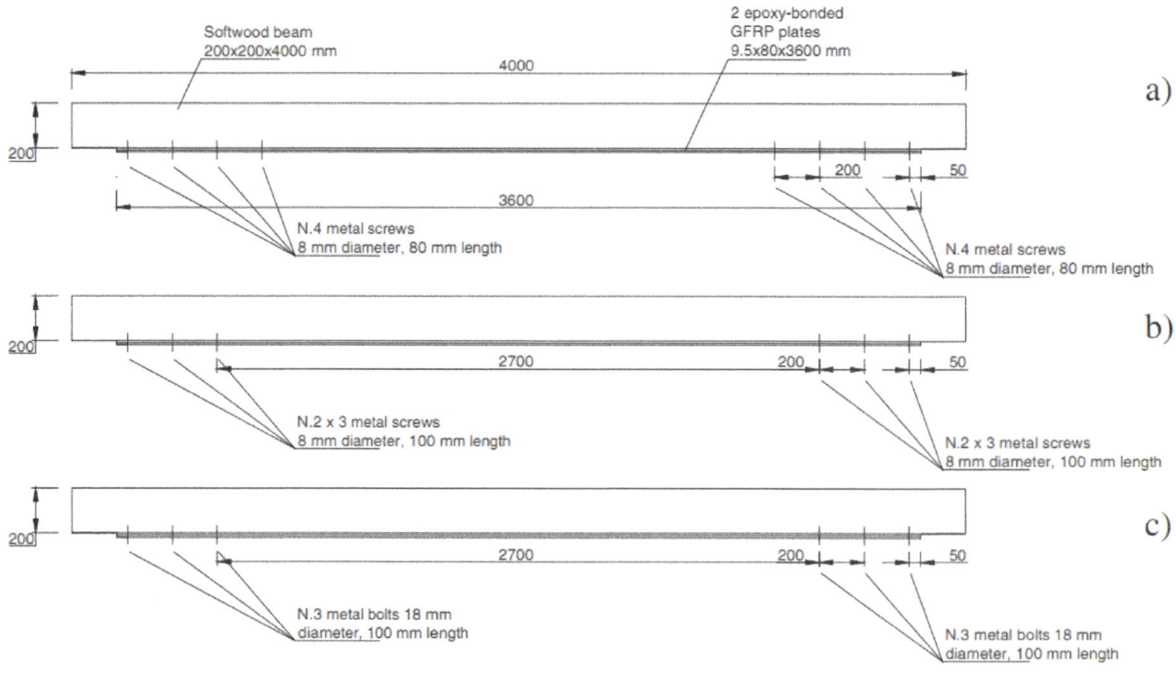

Figure 5. Detail of the method of connection of the GFRP plate to the wood tension side (large beams): (a) No. 8 8 mm-diameter screws (5th arrang.); (b) No. 12 8 mm-diameter screws (6th arrang.); (c) No. 6 18 mm-diameter screws (7th arrang.) [dimensions in (mm)].

For large beams, 8 mm-diameter metal screws or 18 mm-diameter bolts were used (80 and 100 mm in length, respectively). Three arrangements were adopted as shown in Fig. 5. According to the first configuration (5th arrangement) 8 8 mm-diameter (length = 80 mm) screws were applied (Fig. 5a). Screws have been placed at a center-to-center distance of 200 mm. For the second configuration (6th screw arrangement), 6 8 mm-diameter (length = 100 mm) screws were used for each plate end, for a total of 12 screws (Fig. 5b). The last configuration (7th screw arrangement) is similar to first one: 6 18 mm-diameter metal bolts were used (Fig. 5c). For all arrangements, screws or bolts were positioned 50 mm from GFRP plate's edge.

With the aim of comparing the effectiveness of the reinforcement technique, a limited number of both small and large softwood beams were strengthened by bonding the GFRP plate using an epoxy-resin. By ensuring perfect adherence between GFRP plate and timber, it was possible to define an upper limit to the capacity of reinforced beams.

Finally, for a small number of beams, steel brackets (Fig. 6 and Fig. 7) were used to increase the level of connection between and timber material and the GFRP plate. For each beam, 2 steel brackets were applied near the supports where the shearing force reaches the maximum value. GFRP plate was epoxy-glued to the internal surface of the steel brackets and by drilling holes on them it was possible to apply a larger number of metal screws. The aim was to cause a better distribution of the shear loads and a reduction of stress concentration around the holes and of slippage between the GFRP and timber (Fig. 8 and Fig. 9).

a)

b)

Figure 6. The steel brackets used for reinforcement on small (a) and large (b) softwood beams.

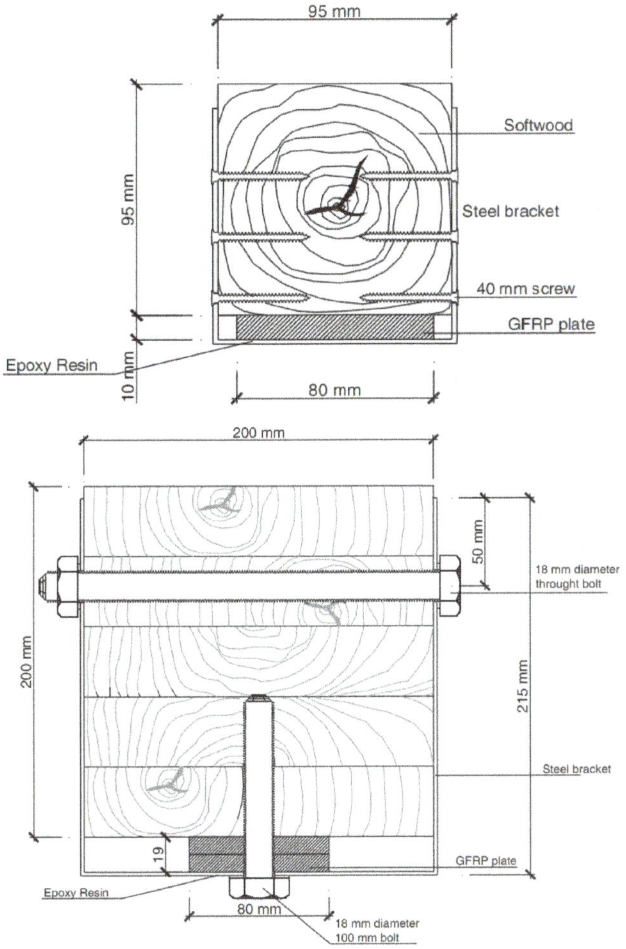

Figure 7. Detail of the connection at a steel bracket.

Figure 8. The four small beams after the application of the reinforcement and of two steel bracktes.

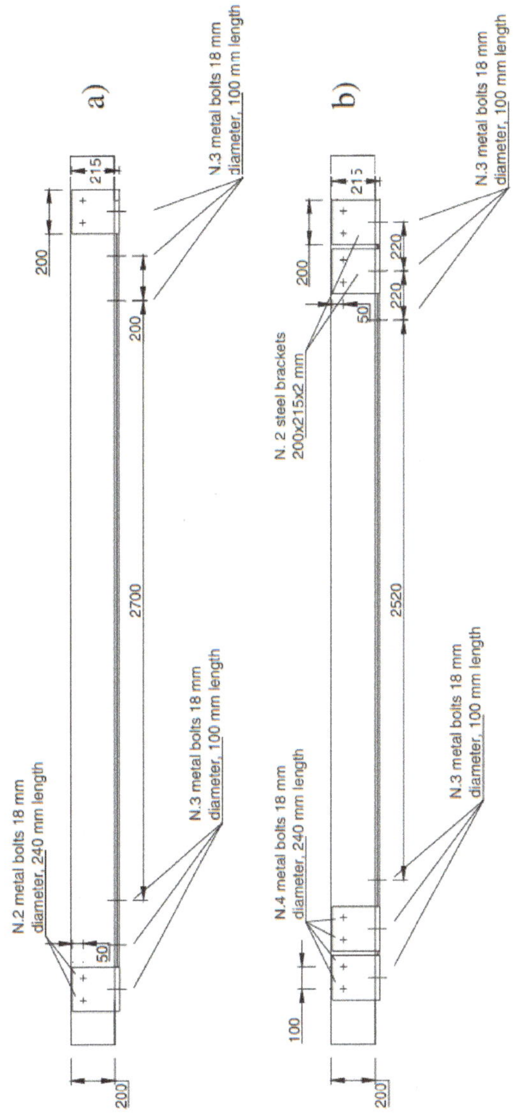

Figure 9. Detail of the method of connection of the GFRP plate to the wood tension side (large beams): (a) 2 steel brackets and 6 trasversal bolts (8th arrang.); (b) 4 steel brackets and 6 trasversal bolts (9th arrang.) (dimensions in mm).

SHEAR CONNECTORS

In this investigation the connection between the softwood beams and the composite plates was done using commercially available woodscrews or bolts. The natural consequence of this connection is the shear flow between the two structural components. If there is no connection or the connectors fail, the softwood beam and the GFRP plate would bend and slip reciprocally. The presence of a shear connection prevents slippage between the two materials and provides the means to achieve the reinforcement action, thus increasing both flexural strength and stiffness of softwood beams. If perfect bonding (zero slip) occurs the two components behave as one and the effect of the reinforcement is maximum.

Because the four point bending test was used for both small and large beams, the longitudinal shear forces V_{AB} and V_{CD} are constant between the end-supports and the loading points:

$$V_{AB} = -V_{CD} = \frac{P}{2} \tag{1}$$

where P is the total value of the bending force applied.

Using the elastic theory of the beam, because normal strains and stresses vary linearly from the neutral axis to the extreme tension or compression fiber, the shear flow P_{Sd} at any level of a generic cross-section is:

$$P_{Sd} = \frac{V_{AB}Q^*}{I_{id}} \times i \tag{2}$$

where Q^* is the first moment about the neutral axis, I_{id} is the second moment of area of the effective cross-section of the reinforced beam and i is the center-to-center distance between two shear connectors. For both I_{id} and Q^* the area of the GFRP plate was taken as its transformed area using a modular ratio of 4.55. By using and adapting the formulation given in Eurocode 4 for design of composite steel and concrete structures [32], the resistance of the shear connector is the smallest of the following:

$$P_{Rd} = 0.8 f_u \frac{\frac{\pi d^2}{4}}{\gamma_V} \tag{3}$$

$$P_{Rd} = 0.29 \times \alpha \times d^2 \frac{\sqrt{(f_{kw} \times E_w)}}{\gamma_V} \tag{4}$$

where f_u is steel connector's ultimate tensile strength, d is the diameter of the shear connector, γ_V is a partial factor for design shear resistance (usually 1.25), f_{kw} and E_w are the ultimate (parallel to grain) compression strength and Young's modulus of wood, respectively.

Eq. (3) derives from the design of shear connectors in concrete and it is used to study the effect of embedment of the connector. According to this theory:

$$\begin{cases} \alpha = 0.2\left[\left(\frac{h}{d}\right) + 1\right] & if\ 3 \leqslant \frac{h}{d} \leqslant 4 \\ \alpha = 1 & if\ \frac{h}{d} > 4 \end{cases} \tag{5}$$

where h is the height of the shear connector. The total number of shear connectors needed is:

$$n_p = \frac{V_{AB}}{P_{Rd}} \tag{6}$$

The above theory has been used to estimate the needed number of connectors and it is based on the maximum shearing force values calculated by increasing of 50% the maximum bending forces applied on un-reinforced beams. However clear limitations are present in this theory, making its application questionable and only possible for a preliminary design. The main limitations are:

- The above formulas are based on the elastic theory of bending. This is acceptable for the GFRP plate, but not for timber. Non-linearity can occur (especially for low-grade timber) in wood due to yielding on the compression side. This will cause a shift of the neutral axis toward the compression side, a significant increment in deflections and shear flow.
- The effectiveness of the reinforcement mainly depends on the bonding properties between the GFRP plate and the softwood beam and not on the resistance of the shear connector. Small values of slip may compromise the reinforcement's effectiveness.

METHODOLOGY AND TEST RESULTS

Four series of flexure tests were performed on unreinforced and reinforced beams (Table 3). It was decided to undertake four point bending tests, according to UNI EN 408 standard [33].

Table 3. Test matrix.

Index	Number of softwood beams	Reinforcement
UNS_series	10	Un-reinforced
RES_series	16	GFRP plate
UNL_series	3	Un-reinforced
REL_series	12	GFRP plate

For small beams, a distance between the loading heads of 576 mm and a span of 1728 mm were used. In order to minimize local wood crushing, the beams were subjected to a double-point loading using two 42 mm-diameter steel cylinders by means of a compression hydraulic jack.

The simply-supported large softwood beams were monotonically loaded over a span of 3900 mm in four-point bending until failure occurred. A spreader steel H-shaped beam was used to apply the load to the beams, 1040 mm apart. Lateral supports were also used to preclude the lateral buckling of the beams. A load cell inside the 500 kN hydraulic actuator recorded the applied load.

For both small and large softwood beams, three inductive transducers (LVDT) were used to measure deflection at ¼, midspan and ¾ of the span. LVDT transducers were installed near the neutral axis (approx. the center of the beam's height). Displacement-controlled loading with a crosshead speed of 2 mm/min was used.

The bending strength f_m was calculated according to:

$$f_m = a \frac{P_{max}}{2W} \tag{7}$$

where P_{max} is the maximum value of the load; a is the distance between a loading head and the nearest end support (576 and 1430 mm, respectively for small and large beams) W is the modulus of resistance of the section (142.9 and 1333 cm^3). The global modulus of elasticity was calculated with the following equation:

$$E_{m,g} = \frac{l^3(P_2 - P_1)}{bh^3(d_2 - d_1)}\left[\left(\frac{3a}{4l}\right) - \left(\frac{a}{l}\right)^3\right] \tag{8}$$

where b is width of softwood beam cross section, l is the beam's span; $P_2 - P_1$ is an increment of the bending load on a straight-line portion of the load vs deflections response curve; $d_2 - d_1$ is the deflection increment corresponding to $P_2 - P_1$.

The beams' stiffness k_1 and k_2, measured in the elastic range and at maximum load, respectively, was measured with the following:

$$\begin{cases} k_1 = \frac{(P_E - P_i)}{(d_{P_E} - d_{P_i})} \\ k_2 = \frac{(P_{max} - P_i)}{(d_{P_{max}} - d_{P_i})} \end{cases} \tag{9}$$

where P_E was 7 and 50 kN, for small and large softwood beams, respectively. P_i was a preload, applied to remove non-linear effects from the calculation of the beam's stiffness, of 2 and 5 kN, for small and large beams. d_{PE} and d_{Pi} are the corresponding mid-span deflections at P_E and P_i.

Effect of reinforcement and post-elastic behavior were analyzed using the following indices:

$$\begin{cases} \delta_1 = \dfrac{P_{max}}{P_{max\,un}} \\ \delta_2 = \dfrac{d_{P_{max}}}{d_{P_{max}\,un}} \\ \delta_3 = \dfrac{k_1}{k_{1,un}} \end{cases} \quad (10)$$

$$\lambda = \frac{k_2}{k_1} \quad (11)$$

where $P_{max,}un$,$d_{Pmax,}un$ and $k_{1,un}$ are the mean maximum load, the corresponding mid-span vertical deflection and stiffness measured for un-reinforced beams.

For small beams, the test setup is shown in detail in Fig. 10.

Figure 10. Small softwood beams: four-point bending.

Un-reinforced Beams

For small beams, ten un-reinforced elements were subjected to flexure in four-point-bending. Load was applied incrementally until failure. Test results on unreinforced beams were used to evaluate the reinforcement's effectiveness through a comparison with the results of identical tests performed on small reinforced beams. For all specimens without reinforcement the letter designation used is UNS and UNL, for small and large beams respectively. For unreinforced beams different modes of fracture on the tension side were recorded: knot influenced, simple tension, cross grain tension (Fig. 11). Grain deviation or the presence of knots greatly influenced the crack propagation. Results of flexure tests on un-reinforced beams are shown in Table 4. The mean bending strength was 23.91 MPa (SD = 7.86 MPa) and the Young's modulus $E_{m,g}$ 5932 MPa (dev. 907 MPa). In terms of capacity, the mean maximum load applied P_{max} was 11.86 kN and the corresponding mid-span deflection d_{Pmax} was 31.67 mm. The Coefficient of Variation (CoV) of the beam capacity is very high (CoV = 32.8%).

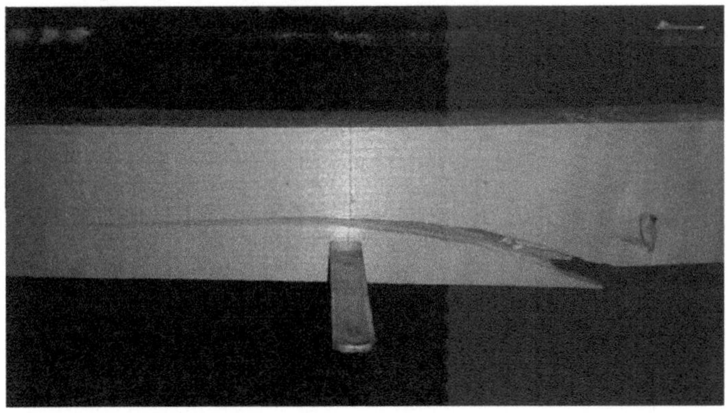

Figure 11. Typical failure along the grain.

Table 4. Test results (small beams).

Index	Connection Timber-Plate	Maximum load P_{max} (kN)		Bending strength f_m (MPa)	Deflection at maximum load d_{Pmax} (mm)	
UNS_1	–	12.92		26.04	38.5	
UNS_2	–	7.44		14.99	23.49	
UNS_3	–	14.70		29.63	36.65	
UNS_4	–	8.41		16.95	22.70	
UNS_5	–	14.25	11.86	28.72	42.90	31.67
UNS_6	–	12.91	(3.90)	26.02	36.30	(9.59)
UNS_7	–	15.76		31.76	41.60	
UNS_8	–	15.38		31.00	32.60	
UNS_9	–	12.78		25.76	29.20	
UNS_10	–	4.06		8.18	12.80	
RES_11	28 screws	18.41		37.10	43.88	
RES_12	L = 50 mm,	14.66	18.96	29.55	32.47	44.55
RES_15	ϕ = 4 mm	23.81	(4.60)	47.99	57.31	(12.43)

RES_13	28 screws*	15.38		31.00	26.90	
RES_14	L = 50 mm,	22.15	18.84	44.64	45.81	34.92
RES_16	ϕ = 4 mm	19.00	(3.39)	38.29	32.05	(9.78)
RES_17	4 steel	17.91		36.10	26.25	
RES_18	Brackets	26.13	20.77	52.66	79.72	49.51
RES_19		18.52	(3.74)	37.33	61.56	(25.66)
RES_20		20.50		41.32	30.12	
RES_21	2 steel	15.88		32.01	22.57	
RES_22	Brackets +	19.45	18.79	39.20	37.29	31.63
RES_23	20 screws	13.05	(5.93)	26.30	26.59	(8.38)
RES_24	L = 50 mm	26.76		53.93	40.06	
RES_25	Epoxy	25.13	26.76	50.65	20.06	22.89
RES_26		28.39	(2.31)	57.22	25.72	(4.00)

*applied diagonally, SD in ().

Load–displacement response (Fig. 12) is initially linear. As the bending load increases, softwood begins to yield in compression and cracking occurs on the tension side. Beams UNS_2, UNS_4 and UNS_10 failed at a low load level due to the presence of a large defect (knot) on the tension side. For unreinforced beams, standard deviation (SD) of results in terms of capacity load was very high for the influence caused by the presence of natural timber defects.

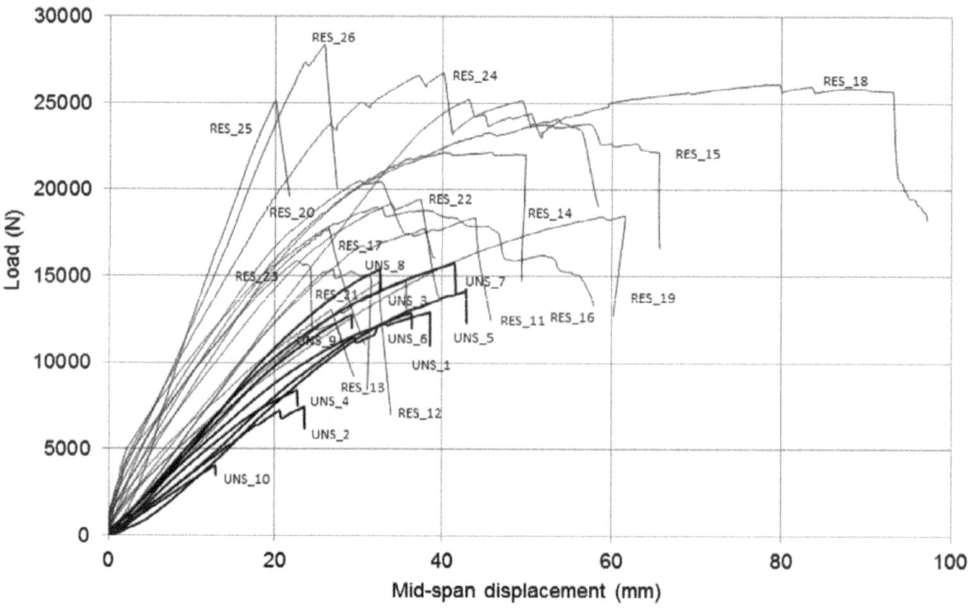

Figure 12. Load vs. mid-span displacement for unreinforced and reinforced small softwood beams.

Because of large softwood beams are made of glulam, the quantity and influence of defects is smaller compared to solid wood beams. This caused an increase of the mechanical properties (tensile strength and Young's modulus) and a reduction of the CoV. The mean capacity and bending strength was 71.87 kN (SD = 14.58 kN) and 38.54 MPa (SD = 7.82 MPa). The CoV was equal to 20.28%. The crack usually ran horizontally along the grain and vertically between the load points where the bending moment is maximum.

The un-reinforced small and large softwood beams exhibited almost linear elastic behavior with brittle tensile failure. A very limited deviation from the value of 1 was calculated for the coefficient λ, given by the ratio between the stiffness in an elastic range and at maximum load ($\lambda = 1$ denotes no variation in stiffness during loading and a perfect linear elastic behavior). For small and large beams, the value of λ was 0.812 and 0.88, respectively.

GFRP Reinforced Beams

Small Softwood Beams

Reinforced beams were tested with the same test procedure used for unreinforced ones. For each of the 16 small beam tests, graphs of mid-span deflection versus vertical load have been drawn. These are presented in Fig. 12. Numerical results are reported in Table 5 and Table 6. For screw arrangement No. 1 and 2, it can be seen that the measured ultimate (maximum) load increased significantly after the application of the GFRP reinforcement, with an increment in capacity of 59.9 and 58.9%, respectively. The different orientation of the screws (vertical in Arrangement No. 1, and diagonal in Arrangement No. 2) did not cause a significant variation of the beam capacity and in its increment. The failure began with the displacement of the fasteners (screws) and of timber material. This caused a progressive loss of effectiveness of the GFRP reinforcement due to the slippage phenomena between the two materials.

Table 5. Effect of reinforcement (small beams).

	Number of tested beams	Maximum loadP_{max} (kN)	$\delta_1 = \frac{P_{max}}{P_{max\,un}}$ (-)	Deflection at maximum load d_{Pmax}(mm)	$\delta_2 = \frac{d_{Pmax}}{d_{Pmax\,un}}$ (-)
Unreinforced	10	11.86		31.67	
1st Arrangement	3	18.96	1.599	44.55	1.407
2nd Arrangement	3	18.84	1.589	34.92	2.944
3rd Arrangement	4	18.79	1.584	31.63	2.667
4th Arrangement	4	20.77	1.751	49.41	4.166
Epoxy resin	2	26.76	2.256	22.89	1.930

Table 6. Stiffness properties of reinforced and unreinforced small beams.

	Number of tested beams	k_1 (N/mm)		$\delta_3 = \frac{k_1}{k_{1,un}}$ (-)	k_2 (N/mm)		$\lambda = \frac{k_2}{k_1}$ (-)
Unreinforced	10	451.6	(85.8)	–	366.6	(54.8)	0.812
1st Arrangement	3	633.1	(185.8)	1.402	200.8	(63.6)	0.317
2nd Arrangement	3	898.1	(284.1)	1.989	279.7	(67.0)	0.311
3rd Arrangement	4	861.5	(300.7)	1.908	418.5	(74.5)	0.486
4th Arrangement	4	745.0	(233.9)	1.650	252.4	(108.3)	0.339
Epoxy resin	2	1273	(212.1)	2.818	831.3	(12.9)	0.653

SD in ().

By comparing the reinforced beams with control beams, it can be seen that the load at which the beams failed was not much higher for the reinforced beams, suggesting that the internal forces were only little shared between the GFRP plate and the timber. The use of an epoxy resin to bond the GFRP reinforcement caused a higher increase in beam capacity (+126%) and demonstrated the importance to achieve a perfect bonding between the two materials. The load deflection plot is shown for beams reinforced with epoxy-glued and screwed GFRP plates in Fig. 13. The composite action of the epoxy-glued GFRPs can be clearly seen here in the form of elevated stiffness (slope in the linear-elastic phase) and maximum bending load (beam's capacity). It is evident that the use of an epoxy resin is more reliable that the mechanical fasteners because it causes an higher increase in both stiffness and capacity. However the long term behavior of an epoxy bond should be investigated.

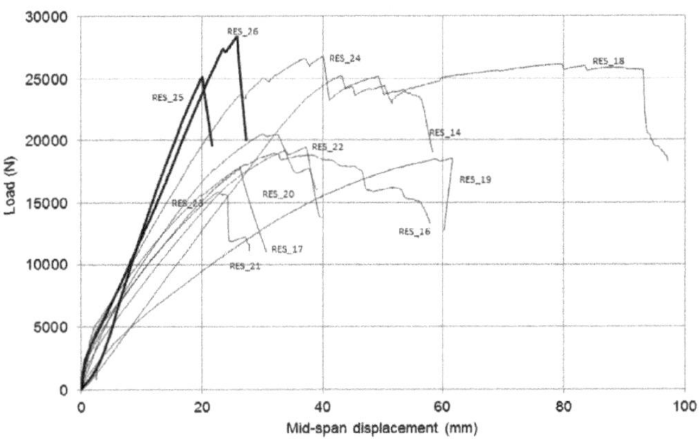

Figure 13. Load vs. mid-span displacement for reinforced small softwood beams: epoxy-bonded reinforcement (black curves). screwed reinforcement (grey curves).

In order to prevent the displacement of the fasteners and/or of timber, steel brackets have been applied near the supports to facilitate the transmission of the internal shearing force between the timber and the GFRP reinforcement. The most part of these timber beams failed because of timber cracking on the tension side; it was also observed that GFRP plates did not crack. For this reason a residual bending strength has been detected following timber failure. The beams reinforced according to the 3rd arrangement (2 steel brackets) evidenced an average bending capacity of 18.79 kN, with an increment of 58.4% compared to unreinforced beams. However by comparing this value with the result of small beams reinforced without the use of steel brackets a negligible increase of the bending strength has been recorded.

Finally the application of 4 steel brackets according to the 4th Arrangement produced a significant increase of both the bending strength (75.1%) and mid-span deflection at maximum load (316.6%) compared with control beams.

An important consequence of the application of the GFRP reinforcement is the increment in the beams's stiffness, measured using the coefficient δ_3 in Eq. (10). Values of δ_3, varying between 40.2 and 98.9%, have been measured for small beams reinforced according to the four configurations. It is worth to compare these values of δ_3 with the ones calculated for beams reinforced using an epoxy resin to bond the GFRP plates. The use of an epoxy resin allowed to achieve the condition of no-strain (zero-slippage) between the two materials (softwood and GFRP) and the maximum possible value of stiffness. This represents an upper bound of the stiffness value and its increment compared to control beams. Beams reinforced using an epoxy resin exhibited a stiffness of 1273 N/mm with an increment 181.8% compared with control beams. On the other hand, beams reinforced using screwed connections exhibited values of the stiffness approx. 30–35% smaller compared to this upper bond value denoting the above mentioned displacement phenomena of the fasteners.

The reinforced small softwood beams exhibited plastic elastic behavior with brittle tensile failure of timber material. A high deviation from the value of 1 was calculated for the coefficient λ. This coefficient varied between 0.317 (1st arrangement) and 0.653 (use of epoxy resin). The high deviation from the linear behavior in mainly produced by two causes: the displacement of the fasteners (screws) and of timber material and yielding phenomena of timber in compression. It is not easy to quantify the two causes: however the sole contribution of the wood yielding can be determined from the results of beams reinforced using an epoxy resin: for these beams λ = 0.653, denoting a significant contribution of the displacement phenomena of beams reinforced using screws (λ = 0.317–0.486).

Large Softwood Beams

As expected, the application of the GFRP plate also increased the bending strength of large beams, but results highly differed depending on the arrangement, diameter and number of screws used. A summary of the test results, obtained from the four-point bending tests, is presented in Table 7 and Fig. 14. Reference UNL denotes a test on an un-reinforced softwood beam while the reference REL is used for GFRP-reinforced beams.

Table 7. Test results (large beams).

Index	Connection Timber-Plate	Maximum load P_{max} (kN)		Bending strength f_m (MPa)	Deflection at maximum load- d_{Pmax} (mm)	
UNL_1	–	59.39		31.85	44.95	
UNL_2	–	68.32	71.87	36.64	55.88	62.12
UNL_3	–	87.89	(14.58)	47.13	85.54	(21.00)
REL_4	8 screwsL = 80 mm	74.64	71.59	40.03	68.69	64.70
REL_5	ϕ = 8 mm	68.53	(4.32)	36.75	60.71	(5.64)
REL_6	Epoxy	82.52	102.52	44.25	78.61	95.13
REL_7		122.52	(28.28)	65.70	111.65	(23.36)
REL_8	12 screwsL = 100 mm	80.92	89.99	43.39	66.37	79.27
REL_9	ϕ = 8 mm	99.06	(12.83)	53.12	92.17	(18.24)
REL_10	6 screwsL = 100 mm	93.26	89.74	50.01	73.15	69.69
REL_11	ϕ = 18 mm	86.22	(4.98)	46.24	66.23	(4.89)
REL_12	6 screwsL = 100 mm	116.87	92.73	62.67	105.84	75.94
REL_13	ϕ = 18 mm + 2 brackets	68.59	(34.14)	36.78	46.04	42.28
REL_14	6 screwsL = 100 mm	93.37	95.64	50.07	64.97	69.29
REL_15	ϕ = 18 mm + 4 brackets	97.90	(3.20)	52.50	73.6	(6.10)

SD in ().

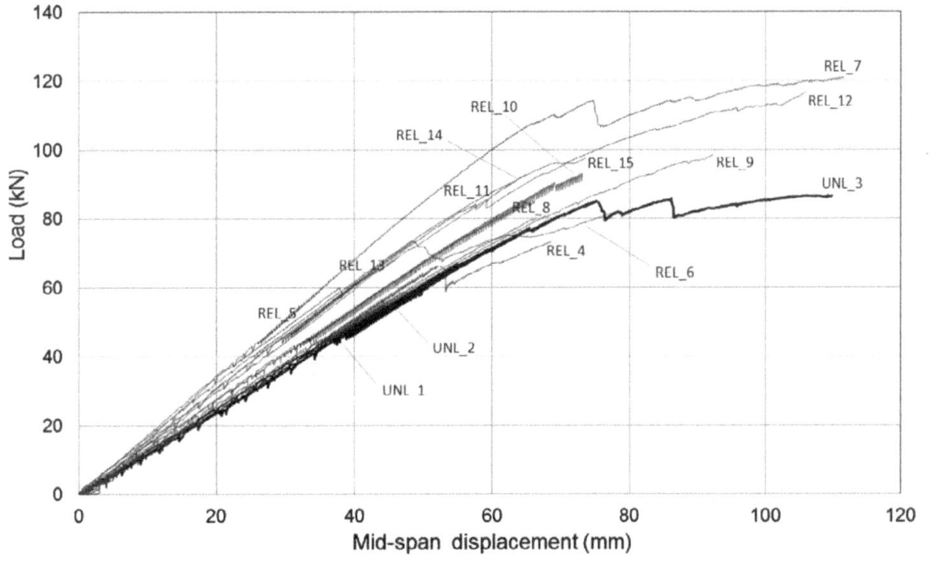

Figure 14. Load vs. mid-span displacement for unreinforced and reinforced large softwood beams.

The application of only 8 8 mm-diameter screws (5th arrangement) did not cause an appreciable increment both in capacity (−0.04%) and stiffness (5.7%). The displacement of both the screws and timber material caused the detachment of the GFRP reinforcement and compromised its reinforcing action (Fig. 15).

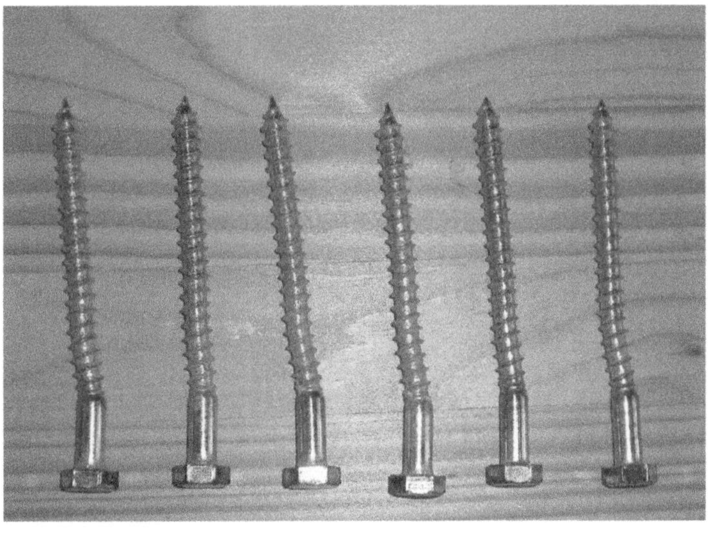

Figure 15. Detail of the displacement of the metal 8 mm-diameter screws (5th arrangement).

By increasing the number of screws, the effectiveness of the reinforcement also increased, but very little. The application of GFRP plates according to the 6th screw arrangement (by using 12 8 mm-diameter screws) caused an increase of the beam's capacity δ_1 and stiffness δ_3 of 25.2 and 5.6%, respectively. The failure again began with the displacement of the screws and of timber material, producing slippage phenomena that compromised the effectiveness of the reinforcement. All beams reinforced according to the 5th and 6th arrangements demonstrated tensile failures initiated by defects in the timber material. Results also indicate that the reinforcement produced an increment of the mid-span deflection δ_2 at maximum load between 4.1% and 27.6%, compared to control un-reinforced beams (Table 8 and Table 9).

Table 8. Effect of reinforcement (large beams).

	Number of tested beams	Maximum loadP_{max} (kN)	$\delta_1 = \frac{P_{max}}{P_{max\,un}}$ (−)	Deflection at maximum load d_{Pmax}(mm)	$\delta_2 = \frac{d_{P_{max}}}{d_{P_{max\,un}}}$ (−)
Unreinforced	3	71.87		62.12	
5th Arrangement	2	71.59	0.996	64.70	1.041
6th Arrangement	2	89.99	1.252	79.27	1.276

7th Arrange-ment	2	89.74	1.249	69.69	1.122
8th Arrange-ment	2	92.73	1.290	75.94	1.222
9th Arrange-ment	2	95.64	1.331	69.29	1.115
Epoxy resin	2	102.5	1.427	95.13	1.531

Table 9. Stiffness properties of reinforced and unreinforced large beams.

	Number of tested beams	k_1 (N/mm)		$\delta_3 = \frac{k_1}{k_{1,un}}$ (−)	k_2(N/mm)		$\lambda = \frac{k_2}{k_1}$ (−)
Unreinforced	3	1202	(25.84)	–	1058	(239.7)	0.880
5th Arrangement	2	1270	(62.30)	1.057	1144	(82.52)	0.901
6th Arrangement	2	1269	(26.66)	1.056	1146	(111.9)	0.904
7th Arrangement	2	1367	(32.53)	1.137	1286	(27.79)	0.941
8th Arrangement	2	1510	(43.49)	1.256	1261	(253.3)	0.835
9th Arrangement	2	1506	(70.29)	1.253	1356	(59.82)	0.901
Epoxy resin	2	1695	(145.6)	1.410	1070	(29.27)	0.631

SD in ().

By replacing screws with larger diameter bolts (diameter 18 mm, length 100 mm) it was possible to reduce the stresses both in the metal and in the bottom timber lamination (7th arrangement). Beams reinforced according to this arrangement exhibited a more linear elastic behavior caused by the reduction of the displacement phenomena in the area around the holes. Yielding of timber material in compression and the low bearing resistance at the joint bolt-timber produced a plastic behavior for high values of the bending load. Failure was first produced by shear rupture of the metal bolts (Fig. 16) and subsequently timber cracking. All beams failed in the pure moment region (between the points of application of the bending loads). Photographs of typical fractures are shown in Fig. 17.

Figure 16. Shear failure of metal bolts (7th arrangement).

By applying steel brackets near the lateral supports (Fig. 18), it was possible to increase the level of connection between the GFRP reinforcement and softwood. Two or four steel brackets were positioned approximately 200 mm inside the lateral supports. Each 2 mm-thick steel bracket was connected to the softwood beam using two 18 mm-diameter through bolts (Fig. 19 and Fig. 20) and one 18 mm-diameter 100 mm length bolt. The application of the GFRP reinforcement according to the 8th arrangement (two steel brackets) produced an increase in the beam capacity δ_1 of 29%. An increment of 31.1% was recorded for the 9th arrangement. The increment in stiffness δ_3 was similar for both these two reinforcement arrangement (25.6% and 25.3%, respectively).

Figure 17. Detail of tensile failure in timber.

Figure 18. Detail of the steel brackets applied on large beams.

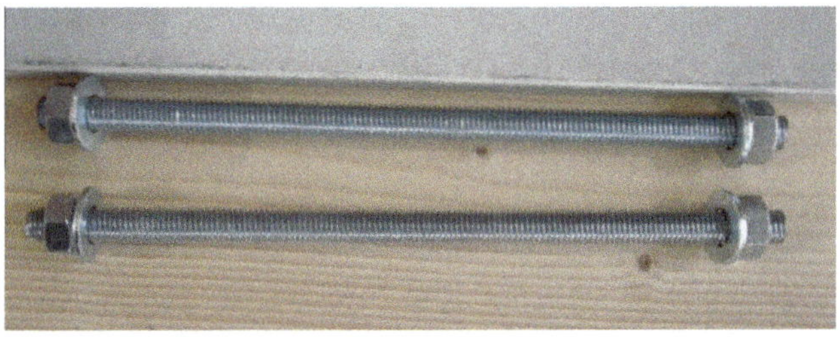

Figure 19. 18 mm-diameter steel though bolts used to connect the steel bracket with softwood beams.

Figure 20. 4-point bending test on a GFRP reinforced beam (with steel brackets).

Similarly to the previous tests on small beams, for two large beams the GFRP plate was applied using an epoxy adhesive. This produced an increment of the beam capacity $\delta_1$and stiffness δ_3 of 42.7% and 41%, respectively and it represents an upper bound for the reinforcement effectiveness (Fig. 21).

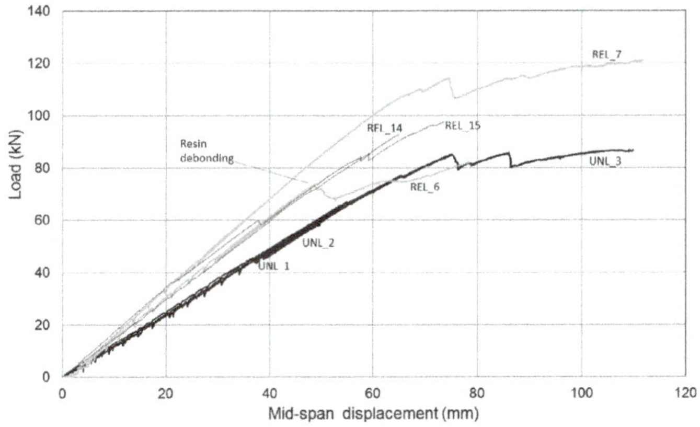

Figure 21. Load vs. mid-span displacement for reinforced large softwood beams: epoxy-bonded reinforcement (bold grey curves), screwed reinforcement (grey curves), unreinforced beams (black curves).

For tests REL_7 and REL_8 the strains of the GFRP plate were determined by one strain gauges (gage factor 2.05, resistance 120 Ohms) applied in plate's center at mid-span (1800 mm from the end of the GFRP plate). Gauge was placed parallel to the long plate axis. For test REL_7 (epoxy-bonded plate) a maximum strain of 0.554% was measured. This generates in the plate a tensile stress of 174.9 MPa equal to 47.5% of the plate tensile strength. Conversely, strain data from REL_8 (screwed connection made with 12 8 mm-diameter screws) show only a tensile stress 36.7 MPa. This low value is again the consequence of displacement phenomena producing slippage between softwood and GFRP plate.

NUMERICAL ANALYSIS

Model Development

Non-linear static analysis has been carried out on a complete three-dimensional FE model of both unreinforced and FRP-reinforced softwood beams. Simulation was made using commercial software Ansys, ver. 15.0 [34]. The numerical model was built to accurately reproduce the geometry of the specimens. To this end, minor regularizations were made and the presence of imperfections was disregarded. The geometry of the timber beams were firstly reconstructed by means of CAD tools, next the volumes were imported and modeled using Solid45 elements (three dimensional eight-node hexahedron isoparametric elements), which are defined by eight nodes with three degrees of freedom at each node and orthotropic material properties. The average size of the hexahedron elements was chosen so as to have 10 × 10 elements across the specimen cross section: this allows the more critical details to be captured avoiding shear lock effects.

For the reinforcement, the Shell181 element was chosen to model the GFRP plate. This element material was also assumed to be orthotropic and the stress–strain relationship was assumed to be linear up to failure. GFRP plates were connected to joint points on the beam. The same mesh size was used to model the GFRP plate and softwood beam in order to provide full overlap at the joints. Fig. 22 shows the finite element model consisting of 25,974 nodes and 25,360 elements, with 76,992 degrees of freedom.

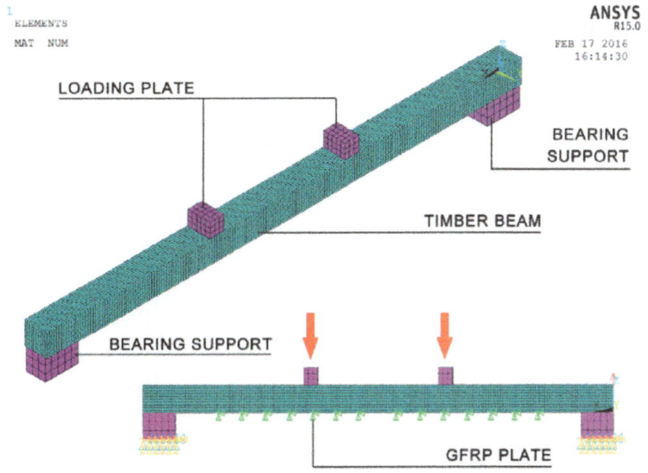

Figure 22. FEM model: mesh discretization.

Values concerning the physical properties of timber were established on statistical analysis of test data found in the literature [35] and from previous mechanical characterization. The final inputs had the following values: E_x = 10,500 MPa;E_y = 330 MPa; E_z = 330 MPa; υ_{xy} = 0.041; υ_{yz} = 0.350; υ_{xz} = 0.033; G_{xy} = 630 MPa;G_{yz} = 630 MPa; G_{xz} = 630 MPa; ρ = 420 kg m^{-3}. A linear elastic response was assumed for the GFRP material, with the Young's modulus and the Poisson's ratio equal to 80% of the experimentally measured values (Table 2), considering the variance of the results. To increase the reliability of the proposed FEM, unilateral contact interfaces were used for the simulation of the contacts between the specimen and the loading plates and bearing supports, respectively. The modeling of the contacts requires the use of specific flexible/flexible elements. Specifically a unilateral contact law was applied in the normal direction of each interface, indicating that no tension forces can be transmitted in this direction and a gap may appear if the stresses become zero. For the behavior in the tangential direction, it was taken into account that sliding may or may not occur, by using the Coulomb friction model with a friction coefficient equal to 0.4.

Analysis Results

In order to find the actual stress field at maximum load, a finite element analysis was conducted, in which the timber beams were subjected to both self-weight and a uniform load pressure under different load stages.

The results of the finite element analysis are summarized in Fig. 23, showing the contour plots of the maximum principal stress (maximum tensile stress) on the tests specimens. To show the efficiency and accuracy of the proposed FEM model, the predictions of the ultimate load capacity are compared with experimental results in Table 10.

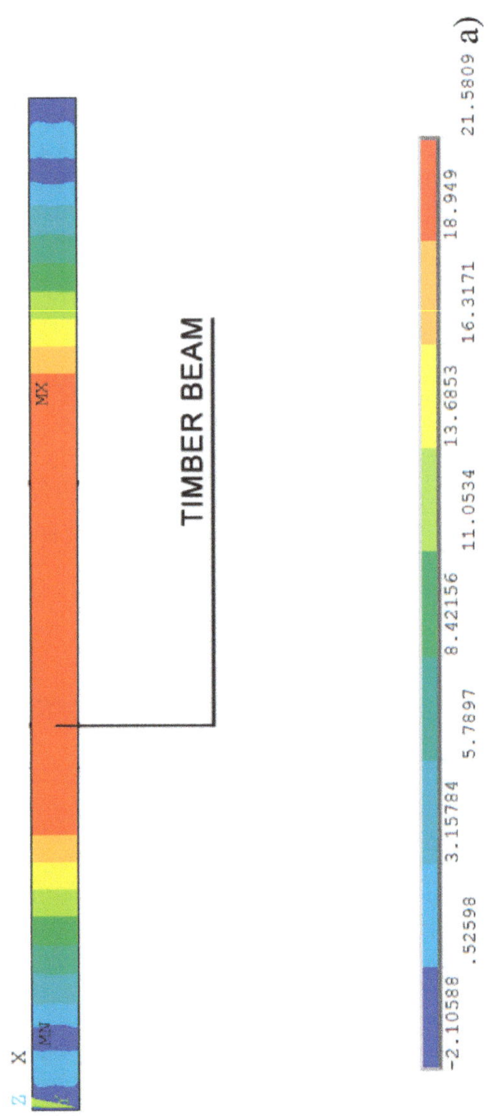

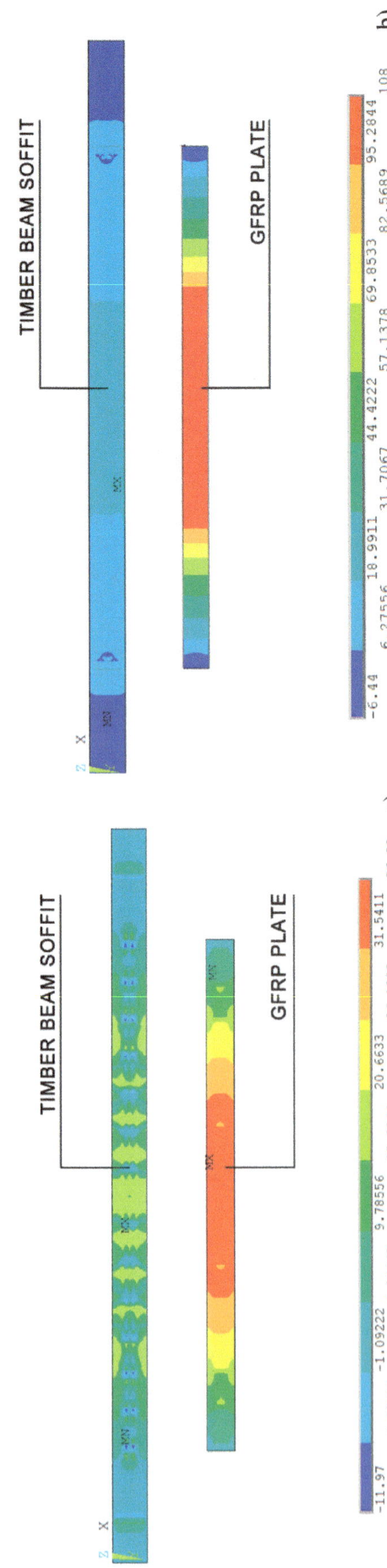

Figure 23. Contour plots of the maximum principal stress (maximum tensile stress): (a) unreinforced beam; (b) epoxy-bonded GFRP-reinforced beam; (c) unbonded GFRP-reinforced beam (1st Arrangement) (dimensions in [MPa]).

Table 10. Experimental versus predicted ultimate load capacities.

	Experimental load capacity ($P_{max,ex}$) (kN)	Predicted load capacity ($P_{max,th}$) (kN)	$\frac{P_{max,ex}}{P_{max,th}}$
Unreinforced	11.86	13.80	0.86
1st Arrangement	18.96	17.00	1.12
Epoxy resin	26.76	24.00	1.12

This comparison reveals good agreement between the theoretical predictions and experimental data for the collapse mode and for the corresponding load-carrying capacity. The unreinforced beam was predicted to fail in tension in a brittle manner, as was also determined experimentally, and the deviations between the calculated and measured values was found to be no more than 14%. A good agreement between theoretical and experimental results was also detected for both epoxy-bonded and unbonded GFRP-reinforced beams. The error of the model was in fact 12% in both cases. As for the failure mode (Fig. 24), the FEM model properly simulated the experimental behavior of the beams, which failed because of timber cracking on the tension side (suggesting that the internal forces were not completely shared between the GFRP plate and the timber).

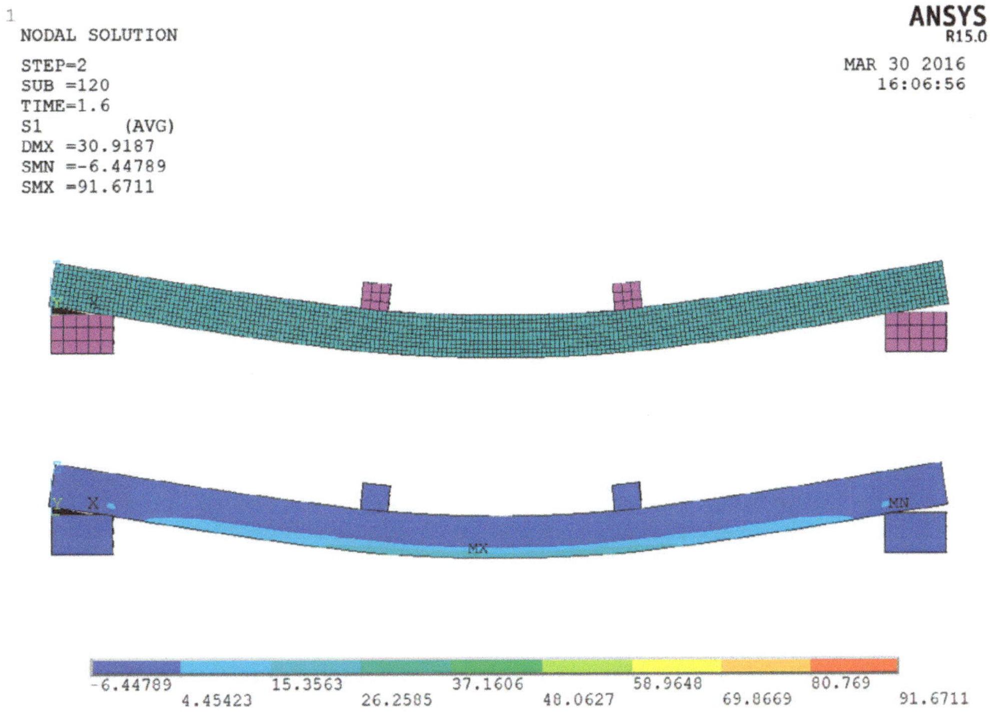

Figure 24. Deflection and failure mode of epoxy-bonded GFRP-reinforced beam (dimensions in [MPa]).

CONCLUSIONS

The mechanical behavior of low-grade (softwood) solid and laminated (glulam) beams reinforced using unbonded pultruded GFRP plates has been investigated. The pultrusion process is ideally suited to the economic production of prismatic composite profiles and the GFRP application without polymeric adhesives may be of interest to avoid irreversible interventions and to guarantee a more durable mechanical connection between timber substrate and reinforcement. The composite plates were anchored at the bottom of softwood beams using metal screws or bolts.

The fastener's and timber displacement, induced by both the low quality of commercially available fasteners and the limited parallel-to-grain timber compression strength, partially compromised the effectiveness of the reinforcement. Different fasteners' configurations were investigated. In order to reduce the stress concentration near the connection, the fasteners' number and/or diameter was increased: in this way it was possible to achieve an increase of bending strength up to 58.9% and of the flexural stiffness up to 98.9%. A further increment in both capacity and stiffness was measured when steel brackets were applied near the beam's end-supports to increase the level of connection between GFRP plate and timber.

The typical failure mode of GFRP-reinforced beams initiated from the displacement of the metal fasteners and timber material and subsequently resulted in timber cracking on the tension side without any significant damage to the GFRP plate. Strain measurements on the GFRP pultruded plates were always well below the plate's tensile strength.

The effectiveness of the reinforcement was studied by deriving an upper bound for capacity and stiffness (by preventing slippage between GFRP plate and softwood beams). This was achieved by using an epoxy adhesive to bond the two materials.

Initial prediction from the numerical model showed good agreement with test results for capacity, mid-span deflection and stiffness. However the modeling method of the connections between the metal fasteners and softwood needs to be further investigated in order to take into account the local effects in terms of shear stresses and displacements.

ACKNOWLEDGEMENTS

The authors acknowledge the support of the Lastru laboratory at the University of Perugia (Italy) for the use of the test equipment. This project was sponsored by the Italian Ministry of Education and Research [ReLUIS (2014) Linea di ricerca WP1 e WP2]. The investigation was carried out with the help of the undergraduate students G. Capruzzi, M. S. Bernardi and S. Rumori, and the graduate research assistant A. Molinari. The authors also express their gratitude to Fibrenet for providing the composite materials.

REFERENCES

1. Thelandersson S, Larsen HJ. Timber engineering. John Wiley & Sons; 2003.
2. FHWA. US Department of Transportation, Seventh annual report to the congress on highway bridge replacement and rehabilitation program; 1986.
3. Corradi M, Borri A. Fir and chestnut timber beams reinforced with GFRP pultruded elements. Compos Part B-Eng 2007;38:172–81.
4. Raftery GM, Harte AM. Low-grade glued laminated timber reinforced with FRP plate. Compos Part B-Eng 2011;42(4):724–35.
5. Nowak TP, Jasien´ ko J, Czepizak D. Experimental tests and numerical analysis of _ historic bent timber elements reinforced with CFRP strips. Constr Build Mater 2013;40:197–206.
6. Chajes MJ, Kaliakin VN, Meyer AJ. Behaviour of engineered wood-CFRP beams. In: Saadatmanesh H, Ehsani MR, editors. Proc., 5th Int. Conf. on Camp. in Infrastr. Tucson, Arizona: Univ. of Arizona; 1996. p. 870–7.
7. Gentile C, Svecova D, Rizkalla SR. Timber beams strengthened with GFRP bars: development and applications. J Compos Constr 2002;6:11–20.
8. Borri A, Corradi M, Grazini A. A method for flexural reinforcement of old wood beam with CFRP materials. Compos Part B-Eng 2005;36:143–53.
9. Micelli F, Scialpi V, La Tegola A. Flexural reinforcement of glulam timber beams and joints with carbon fiber-reinforced polymer rods. J Compos Constr 2005;9:337–47.
10. Raftery GM, Whelan C. Low-grade glued laminated timber beams reinforced using improved arrangements of bonded-in GFRP rods. Constr Build Mater 2014;52:209–20.
11. Righetti L, Corradi M, Borri A. Bond strength of composite CFRP reinforcing bars in timber. Materials 2015;8:4034–49.
12. Plevris N, Triantafillou TC. FRP-reinforced wood as structural material. J Mater Civ Eng 1992;4:300–17.
13. Radford DW, Van Goethem D, Gutkowski RM, Peterson ML. Composite repair of timber structures. Constr Build Mater 2002;16:417–25.
14. Hay S, Thiessen K, Svecova D, Bakht B. Effectiveness of GFRP sheets for shear strengthening of timber. J Compos Constr 2006;10:483–91.
15. Schober KU, Harte AM, Kliger R, Jockwer R, Xu Q, Chen JF. FRP reinforcement of timber structures. Constr Build Mater 2015;97:106–18.
16. Davanos JF, Zipfel MG, Quiao P. Feasibility study of prototype GFRP-reinforced wood railroad crosstie. J Compos Constr 1999;3(2):92–9.
17. Lopez-Anido R, Hu H. Structural characterization of hybrid FRP-Glulam panels for bridge decks. J Compos Constr 2002;6(3):194–203.
18. Jankowski LJ, Jasien´ ko J, Nowak TP. Experimental assessment of CFRP reinforced wooden beams by 4-point bending tests and photoelastic coating technique. Mater Struct 2010;43(1–2):141–50.
19. Jasien´ ko J, Nowak TP, Bednarz Ł. Baroque structural ceiling over the Leopoldinum Auditorium in Wrocław University: tests, conservation, and a strengthening concept. Int J Archit Herit 2014;8(2):269–89.
20. Tampone G. Il restauro delle strutture di legno. Hoepli Editore, 1996 [in Italian].
21. Sandanus J, Sogel. Refurbishment of significant timber structures in Slovakia. J Civil Eng Arch 2014;8(8).
22. Tascioglu C, Goodell B, Lopez-Anido R, Peterson M, Halteman W, Jellison J. Monitoring fungal degradation of E-glass/phenolic fiber reinforced polymer (FRP) composites used in wood reinforcement. Int Biodeterior Biodegrad 2003;51:157–65.
23. Borri A, Castori G, Corradi M, Speranzini E. Durability analysis for FRP and SRG composites in civil applications. Key Eng Mater 2015;624:421–8.
24. Borri A, Corradi M, Speranzini E. Reinforcement of wood with natural fibres. Compos Part B-Eng 2013;53:1–8.
25. Papanicolaou CG, Triantafillou TC, Papathanasiou M, Karlos K. Textile reinforced mortar (TRM) versus FRP as strengthening material of URM walls: out-of-plane cyclic loading. Mater Struct 2008;41:143–57.
26. Feilden B. Conservation of historic buildings. Routledge; 2007.
27. Wang L, Toppinen A, Juslin H. Use of wood in green building: a study of expert perspectives from the UK. J Clean Prod 2014;65:350–61.
28. Blass HJ, Bejtka I. Reinforcements perpendicular to the grain using self-tapping screws. In: The 8th world conference on timber engineering, Lahti, Finland, vol. 1; 2004.
29. Munafò P, Stazi F, Tassi C, Davì F. Experimentation on historic timber trusses to identify repair techniques compliant with the original structural–constructive conception. Constr Build Mater 2015;87:54–66.
30. EN 13183-1:2002. Moisture content of a piece of sawn timber. Determination by oven dry method.
31. ASTM D3039:2009. Standard test method for tensile

properties of fiber-resin composites.

32. Eurocode 4, EN, C., 1994. 1-1, 2004. Design of composite steel and concrete structures. Part 1-1: general rules and rules for buildings.

33. EN 408:2010. Timber structures. Structural timber and glued laminated timber: determination of some physical and mechanical properties.

34. ANSYS Inc. AUTODYN user manual version 15; 2013.

35. EN 338:2009. Structural timber – strength classes.

CHAPTER 5

HYBRID EFFECTS IN THIN PLY CARBON/GLASS UNIDIRECTIONAL LAMINATES: ACCURATE EXPERIMENTAL DETERMINATION AND PREDICTION

M.R. Wisnom, G. Czél, Y. Swolfs, M. Jalalvand, L. Gorbatikh, and I. Verpoest

Keywords: A. Carbon fibres; A. Laminates; B. Fracture; B. Fragmentation; Hybrid effect

INTRODUCTION

Ever since Hayashi reported in 1972 that the failure strain of the carbon fiber layers in a carbon/glass hybrid composite was 40% higher than in the reference carbon fiber composite [1] there has been much interest and controversy over the so called hybrid effect. In a recent review Swolfs et al. concluded that the effect for tensile failure strain is well established, with a typical range of 10–50% for traditional hybrid composites such as carbon/glass [2]. The basic mechanisms responsible for the hybrid effect were reviewed, with the most significant considered to be thermal residual stresses, altered failure development due to statistical effects on formation of clusters of fiber breaks and dynamic stress concentrations. Phillips [3] documented the scientific discussions shortly after the discovery of the hybrid effect. A number of other researchers have made notable contributions [4], [5], [6], [7] and [8], with a gradual improvement in the understanding and predictive modelling of the hybrid effect in the seventies and eighties.

However, there are many difficulties in measuring the baseline tensile failure strain of a high strength unidirectional (UD) composite against which the strain at failure of the hybrid is compared to determine the hybrid effect. The strain to failure of the baseline low elongation material is referred to here as the reference strain to failure, and it may be significantly underestimated due to issues with the test methods such as stress concentrations at the load introduction regions. These effects may be responsible for some of the variability and high values claimed in certain cases for the hybrid effect. In this paper a new method is used to deduce the reference strain to failure of the low strain component of the hybrid and it is demonstrated that this method gives substantially higher and more realistic failure strains than conventional approaches.

Tests reported here with thin ply carbon/glass laminates enable the hybrid effect to be accurately determined. It is demonstrated that there is an enhancement to the failure initiation strain, the point when the carbon starts to fracture, but only when the carbon plies are very thin. In addition there is a second strength enhancement mechanism that arises if multiple fractures of the carbon can occur stably, and in this case the strength is effectively the point where a sufficient number of fractures have occurred to significantly reduce the stiffness of the hybrid rather than the first carbon fracture. This illustrates that the hybrid effect is intimately linked to variability of strength, confirming the view of Manders that "the hybrid effect arises from a failure to realize the full potential strength of the fibers in all-carbon fiber composites, rather than from an enhancement of their strength in the hybrids" [9].

This paper describes the accurate determination of the baseline carbon fiber strain to failure and then quantifies the magnitude of the hybrid effect for initiation of failure in carbon/glass hybrids with different carbon ply thicknesses. The additional hybrid effect due to multiple fractures is then demonstrated and quantified. Models for predicting both effects are described, and it is shown that they are able to capture and explain the observed phenomena.

Citation: M.R. Wisnom, G. Czél, Y. Swolfs, M. Jalalvand, L. Gorbatikh, I. Verpoest, Hybrid effects in thin ply carbon/glass unidirectional laminates: Accurate experimental determination and prediction, Composites Part A: Applied Science and Manufacturing, Volume 88, September 2016, Pages 131-139, ISSN 1359-835X, http://dx.doi.org/10.1016/j.compositesa.2016.04.014.

ACCURATE DETERMINATION OF BASELINE STRAIN TO FAILURE

Standard straight-sided specimens tend to underestimate the true ultimate strength of unidirectional composite materials. Stress concentrations arise where the load is introduced [10], but with very careful tapering of the specimens by chamfering plies through the thickness, gauge section failures can be achieved with significantly higher strains to failure [11] and [12]. In most reported hybrid composite studies, there were no such efforts to avoid tab failure, and so premature failure and decreased strains may have been recorded as the baseline. In this paper a different approach is adopted based on [13] to determine the baseline carbon fiber strength from carbon plies sandwiched between glass plies. The surface glass protects the interior carbon, and finite element analysis has revealed that there are no stress concentrations in the carbon layer, but only in the glass layer. This promotes carbon layer failure initiation in the gauge section, as has been verified experimentally [13]. Crucially, it will be shown later that when the carbon plies are not too thin, there is no significant strain enhancement in the presence of glass, and the tests can provide accurate values for the baseline strength of the carbon fiber composite.

The different failure mechanisms in carbon/glass laminates have been studied, and found to depend on the ratio of carbon to glass thickness and also the absolute thickness of the carbon [14]. With the right thicknesses it has been demonstrated that following first fracture of the carbon, the glass does not fail, but delamination occurs between the carbon and glass. This requires that the proportion of carbon is lower than the limit for premature glass failure and the absolute thickness is high enough to generate delamination rather than ply fragmentation. Delamination can be observed visually due to the translucence of the glass. The specimens are initially black due to the carbon, but after delamination, light is reflected from the interface, and the specimens appear yellow. It can also be seen clearly where failure has occurred, which is typically in the gauge section. This delaminating hybrid specimen is the method used here to determine the baseline strength of the carbon.

The basic properties of the fibers and prepregs applied are summarized in Table 1 and Table 2. Specimens were vacuum bagged and cured in an autoclave at 125 °C for one hour. Although the resins of the glass and carbon fiber prepregs were different, the recommended cure cycles were the same for both manufacturers, and there were no issues with resin incompatibility. Specimens were well consolidated, although the volume fractions were relatively low. Nominal thicknesses of each specimen type (based on the ply thickness data of Table 2) were used for data evaluation to compensate for the small variation (less than 3% CV) in thickness within the specimen series which is attributed to the variation in resin content.

Table 1. Fiber properties of the applied UD prepregs (based on manufacturer's data).

Fiber type	Manufacturer	Elastic modulus (GPa)	Strain to failure (%)	Tensile strength (GPa)	Density (g/cm^3)
Pyrofil TR30 carbon	Mitsubishi rayon	234	1.9	4.4	1.79
FliteStrand S ZT S-glass	Owens corning	88	5.5	4.8–5.1	2.45

Table 2. Cured ply properties of the applied UD prepregs.

Prepreg material	Fiber mass per unit area	Cured ply thickness (mm)	Fiber volume fraction (%)	Initial elastic modulus	Strain to failure[c]
	(g/m^2) (CV (%))[a]			(GPa) (CV (%))	(%) (CV (%))
TR30 carbon/ epoxy	21.2 (4.0) [16]	0.029 [16]	41 [16]	101.7 (2.75)[15]	1.50 (7.5) [15]
S-glass/913 epoxy	190[b]	0.155[b]	50[b]	45.7 (3.0) [17]	3.98 (1.1) [17]

Values with references were determined experimentally.

[a]Coefficient of variation.

[b]Based on manufacturer's data.

[c]From conventional tensile tests.

The tensile testing of the 16 ply all carbon (16C) specimens is described in [15]. The specimen size was 180/100/10 mm overall length/free length/width. The interlayer hybrid specimens had a three layer sandwich structure as shown in Fig. 1, with thin (0.029 mm) carbon plies between standard thickness (0.155 mm) glass plies. The nominal sizes were 240/160/20 mm overall length/free length/width respectively. Ideally the same dimensions would have been used for both specimen types. The volume of the carbon in the hybrid specimens is 60% and 80% of that in the all carbon ones. The effect of such differences in stressed volume is considered in Section 3, and shown to be relatively small. Un-chamfered 40 mm long cross-ply S-glass/epoxy end-tabs were used on all specimens.

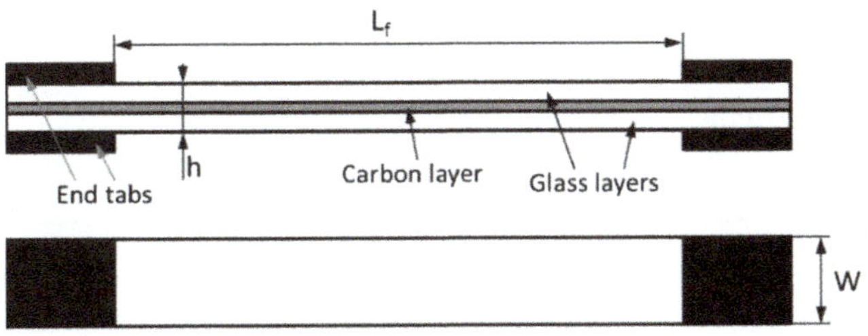

Figure 1. Schematic of an interlayer hybrid composite specimen.

The experimental results of the thin-ply UD hybrid composites analyzed here were generated as part of a wider experimental program, reported in [18] with emphasis on their pseudo-ductility. Testing of the parallel edge specimens was executed under uniaxial tensile loading and displacement control using crosshead speeds of 1 mm/min for the all carbon specimens and 2 mm/min for the longer, lower modulus hybrid specimens, giving similar times to failure. A computer controlled Instron 8801 universal hydraulic test machine with a regularly calibrated 100 kN Instron Dynacell load cell and wedge type hydraulic grips was used for the tensile tests. Strains were measured using an Imetrum videogauge system, with a nominal gauge length of 130–140 mm. The test results are summarized in Table 3. They are very consistent, with low coefficients of variation of around only 2%.

Table 3. Tested baseline configurations and results (specimen type designation: C- thin carbon and SG- standard thickness S-glass ply with the numbers corresponding to the number of plies).

Specimen type	No. of spec. tested	Nominal thickness (mm)	Carbon/epoxy failure strain at load-drop (%) (CV (%))	Initial compressive thermal strain in carbon (%)	Corrected carbon failure strain (%)
16C	10 [15]	0.464 [15]	1.500 (7.5) [15]	–	–
2SG/4C/2SG	4	0.736	1.900 (1.5)	−0.023	1.877
1SG/3C/1SG	5	0.397	1.859 (2.1)	−0.020	1.839

The average strain at failure as measured at the point of the load drop when unstable delamination occurs is 1.900% and 1.859% for the cases with 4 and 3 carbon plies. These specimens behaved similarly and the difference between them is not statistically significant. There is however a small effect of thermal residual stresses, which can be corrected based on the measured fiber direction moduli of 45.7 GPa and 101.7 GPa and typical expansion coefficients of 3.85×10^{-6} K^{-1} and 0.67×10^{-6} K^{-1} for the glass and carbon plies respectively, with 100 °C difference between the cure and room temperatures. This gives small residual strains of −0.023% and −0.020% in the carbon layers. The effect of resin shrinkage would be negligible, as it normally contributes very little – less than 5% of the residual stress in one study on cross-ply laminates [19]. An even lower effect would be expected in unidirectional hybrids where stresses are driven by the difference in fiber expansion coefficients rather than by matrix contraction. The corrected elastic tensile strains at failure for the two cases are therefore 1.877% and 1.839%. The mean value of 1.858% is believed to be representative of the true ultimate strain of the carbon fiber plies since it corresponds to visually observed catastrophic failure away from any stress concentrations. This value is not significantly influenced by a hybrid effect, as shown experimentally in Section 3 and by modelling in Section 5. It also corresponds closely to the 1.9% fiber strain quoted by the manufacturer on the datasheet.

On the other hand the experimental value of 1.50% from the all carbon specimens is clearly too low, as will be shown later, and is believed to have been affected by premature failure due to the specimen design and gripping conditions. However, if this value had been used as a baseline, following the approach used in many other previous studies, then it might have been concluded that there was a hybrid effect of about 24% in these cases with three or four carbon plies.

ACCURATE MEASUREMENT OF HYBRID EFFECT AND INFLUENCE OF PLY THICKNESS

Two further series of tests on hybrid laminates similar to those described in Section 2were carried out but with thinner carbon layers. The reduced energy release rate due to the thinner plies meant that when the carbon fractured, complete delamination of the whole specimen did not occur. The materials were the same standard thickness S-glass/epoxy and thin carbon/epoxy, with layups (1SG/1C/1SG) and (1SG/2C/1SG).

These thin ply hybrids developed progressive multiple fragmentations of the carbon, leading to a pseudo-ductile stress–strain response with a plateau, analogous to yielding in metals, as shown for a typical specimen in Fig. 2. The failure initiation strains of the carbon were taken at the knee points, which can be easily established from the intersection of straight lines fitted to the elastic and plateau regions, as shown in the inset to Fig. 2. In fact there is a small amount of non-linearity before this point, which will be considered in Section 4.

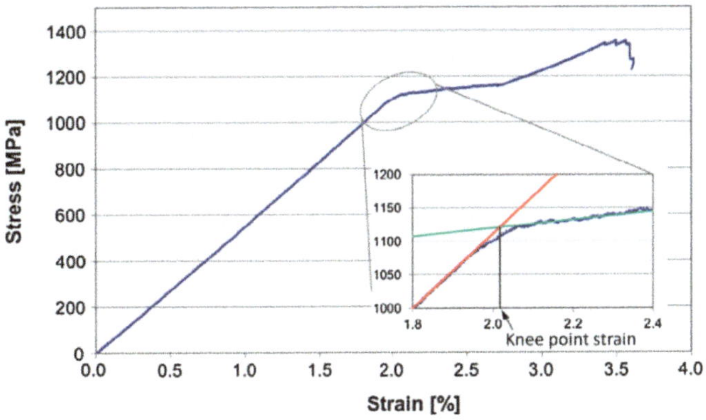

Figure 2. Typical non-linear response of (1SG/2C/1SG) laminate.

Table 4. Tested thin carbon layer configurations and results.

Specimen type	No. of spec. tested	Nominal thickness (mm)	Knee point strain (%) (CV (%))	Initial compressive thermal strain in carbon layer (%)	Knee point strain corrected with thermal strain (%)
1SG/1C/1SG	5	0.339	2.253 (1.4)	−0.026	2.227
1SG/2C/1SG	5	0.368	2.027 (1.6)	−0.023	2.004

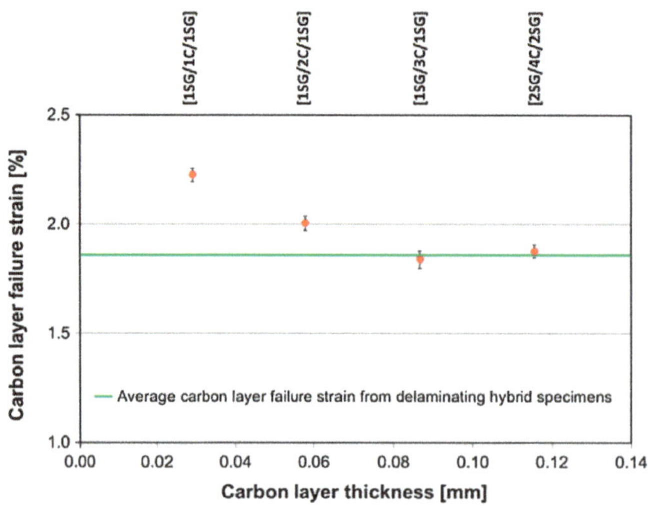

Figure 3. Dependency of carbon layer failure strain on the thickness showing a hybrid effect.

The failure strains of the single and double carbon plies are given in Table 4. Taking account of the calculated initial strains due to thermal residual stresses gives corrected elastic tensile strains at failure of 2.227% and 2.004% respectively. Fig. 3 reveals the comparison of results against the strain at failure established from the previous tests where delamination occurred after first carbon layer failure. The strain at failure of the single carbon ply is 11.1% higher than that of the double ply, which in turn is a further 7.9% higher than the average of the cases with 3 and 4 plies that delaminated. This effect cannot be explained by residual stresses since the absolute increase in residual strain in the single compared with the double carbon ply case is only 0.003%. Another notable point is the low coefficients of variation of less than 2% in the knee point strains. It is well known that there is a size effect in tensile strength of carbon fibers due to the higher probability of finding a cluster of weaker fibers in a larger volume of material [20]. The single ply has half the volume of the double ply case, and might therefore be expected to have a higher strain at failure due to the statistical size effect. However the magnitude of this effect is relatively small. The Weibull modulus of the carbon fiber/epoxy material is not known, but in a previous study of tensile tests with IM7 carbon fibers, a modulus of 41 was determined for the composite from the reduction in strength of scaled specimens with increasing size [21]. For a volume ratio of 2, this would lead to a relative enhancement of strain of 1.7%, small compared with the 11.1% observed experimentally. Even if the Weibull modulus of the carbon ply is lower than 41, it cannot explain the much higher strain at failure of the single ply case.

The substantial increase in strain for the single carbon ply case is believed to be caused by the restraint from the adjacent glass plies which inhibits the formation of clusters of carbon fiber breaks, as discussed in Section 5. This constitutes a substantial hybrid effect of 19.8%, due to the delay in failure of the carbon fibers. The smaller increase in strain at failure for the two ply case represents a hybrid effect of 7.9%, but beyond this thickness there does not appear to be any further effect, given the similar results with 3 and 4 plies. This justifies the use of the thicker specimens to establish the baseline reference strength.

HYBRID EFFECTS RELATED TO INITIATION AND MULTIPLE FRACTURES

In the previous section, the failure strains of the carbon layers were based on the knee in the stress–strain response. However, careful examination of the tests revealed that first failure initiation in the carbon layer actually occurred earlier. The point of first fracture of the carbon layer was established visually from studying the videos of each specimen taken with the strain measurement system. The carbon layer cracks could be detected because of the associated delamination and translucency of the glass layers, and the strains were determined from the video extensometer readings using the test time to match the videos to the strains. These results are summarized in Table 5, with the corrections for thermal residual stresses given above included.

Table 5. Summary of measured carbon layer failure strains, corrected for residual stresses.

Layup	Strain at first carbon layer fracture [Absolute %] (CV (%))	Strain at knee point [Absolute %] (CV (%))
1SG/1C/1SG	2.166 (2.5)	2.227 (1.4)
1SG/2C/1SG	1.963 (2.3)	2.004 (1.6)
1SG/3C/1SG	1.839 (2.1)	–
2SG/4C/2SG	1.877 (1.5)	–

The strain at the knee points is higher than at failure initiation by an additional 0.060% for the single and 0.041% for the double carbon ply case. Although these differences are relatively small, they are consistent from specimen to specimen, and the coefficients of variation are low.

This indicates that the hybrid effect can be split into two components, the delay in initiation due to the constraint on cluster formation, and an additional element corresponding to multiple fragmentation. This arises from the fact that the knee point marking the effective strength of the carbon ply requires sufficient fractures in the carbon ply to reduce the modulus rather than corresponding to the formation of the very first critical cluster. These two components are displayed as A and B in Fig. 4 and the hybrid effects are summarized in Table 6. The hybrid effects were calculated relative to the average failure strain of the three and four carbon ply hybrids, which was 1.858%. For the single ply case the hybrid effect is 16.6% based on the initiation of failure, with a further 3.2% based on multiple fragmentation. For the two ply case the corresponding values of the hybrid effect are 5.7% for initiation and a further 2.2% at fragmentation.

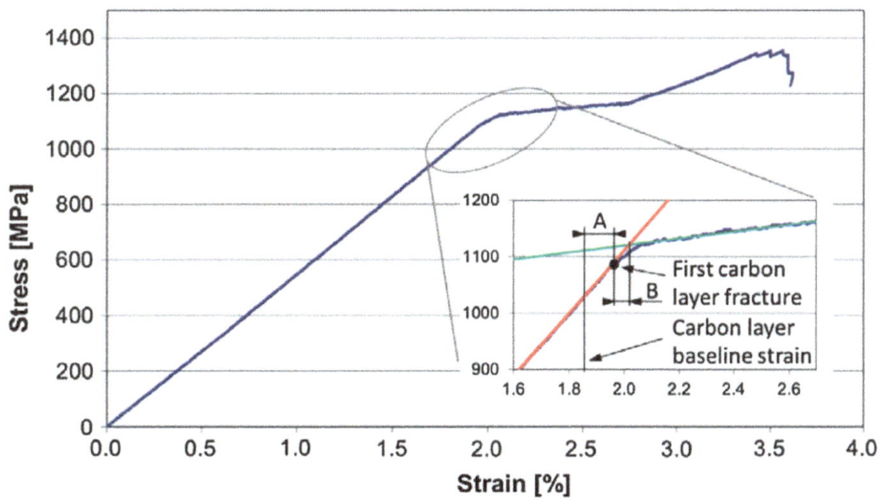

Figure 4. Two components of the hybrid effect: delay in the initial cluster formation (A) and additional element due to established multiple fragmentation (B).

Table 6. Hybrid effects for initiation and fragmentation (strains corrected for residual stresses).

Layup	Strain at first carbon layer fracture [Absolute %] (CV (%))	Strain at knee point (start of fragmentation) [Absolute %] (CV (%))	Initiation hybrid effect [Relative %] (A)	Fragmentation hybrid effect [Relative %] (B)	Total hybrid effect [Relative %] (A + B)
1SG/1C/1SG	2.166 (2.5)	2.227 (1.4)	16.6	3.2	19.8
1SG/2C/1SG	1.963 (2.3)	2.004 (1.6)	5.7	2.2	7.9
Baseline – average of 3 and 4 carbon ply delaminating hybrids	1.858	–			

MODELLING THE HYBRID EFFECT

The mechanism of the hybrid effect on initiation due to the constraint on cluster formation can be illustrated using the statistical strength model of Swolfs et al. [22], [23] and [24]. This captures the distribution of individual fiber strengths together with the stress transfer and associated stress concentrations at the fiber breaks. The additional effect due to multiple fragmentation within the plies can be illustrated using the ply level finite element model of Jalalvand et al. [14] which considers the distribution of strength for the whole carbon ply, and stress transfer at the ply rather than individual fiber level.

Increase in Failure Initiation Strain

The four tested lay-ups of carbon/S-glass hybrids are analyzed in this section. These cases are labelled based on the number of carbon fiber plies in the middle, ranging from "1 carbon ply hybrid" to "4 carbon ply hybrid".

Very local load sharing and hexagonal packing were assumed, meaning that the load from a broken fiber is shed only to its 6 nearest neighbors. While the experimental V_f is different in the carbon (41%) and glass (51%) layers, such a difference in V_f is difficult to work with in a hexagonal packing. These different fractions would cause changes in the fiber spacing at the interface between dissimilar layers, and would be expected to have a negligible influence on the outcome of the model. Therefore, an overall volume fraction of 50% was chosen.

The carbon fiber was assumed to be transversely isotropic with a longitudinal fiber stiffness of 234 GPa. The other elastic constants of the fiber were the same as in Swolfs et al. [24] and [25], whereas the coefficients of thermal expansions were already mentioned in Section 2. The S-glass fiber was assumed to be isotropic with a longitudinal stiffness of 88 GPa and a Poisson's ratio of 0.22. The matrix was assumed to be isotropic, well-bonded and linear elastic with a stiffness of 3.4 GPa and a Poisson's ratio of 0.4. Table 7 summarizes the values used and where they came from. A FE model with hexagonal packing and 50% fiber volume fraction was used to calculate the ineffective length [25] and [27]. This is defined here as the length over which the stress in a broken fiber is less than 90% of the nominal stress [26]. The value was 101 μm and 53 μm for the carbon and glass fiber, respectively. Combining these values with the assumptions of a linear stress recovery profile and very local load sharing yields the entire stress redistribution around a fiber break.

Table 7. Parameters used in modelling failure initiation.

Property	Value	Source
Carbon fiber		
Longitudinal elastic modulus	234 GPa	Data sheet
Weibull modulus	5	Typical value [28]
Reference gauge length	10 mm	Chosen value
Weibull scale parameter	3029 MPa	Fitted
Ineffective length	101 μm	Predicted by FE model
S-glass		
Elastic modulus	88 GPa	Data sheet
Poisson's ratio	0.22	Typical value [28]
Weibull modulus	4	Chosen value
Reference gauge length	50 mm	Chosen value
Weibull scale parameter	1520 MPa	Chosen value
Ineffective length	53 μm	Predicted by FE model
Matrix		
Elastic modulus	3.4 GPa	Data sheet
Poisson's ratio	0.40	Typical value [28]
Model		
Overall fiber volume fraction	50%	Based on actual V_f
Layer thickness	24 μm, 57 μm, 90 μm and 122 μm (1, 2, 3 and 4 carbon ply hybrid)	Chosen to be close to experiments
Model width	2 mm	Chosen value
Model length	10 mm	Chosen value
Total number of fibers	From 2337 to 4463	Results from width, layer thickness and V_f
Number of simulations/ configuration	200	Chosen value

Since the layer thickness is an important parameter in the hybrid effect, simply using the manufacturer's data sheet was insufficiently accurate. Optical microscopy images were hence used to determine the most appropriate layer thickness to model. Accurately measuring the layer thickness in the 1 carbon ply hybrid is difficult as the boundary between the layers is not clearly defined. Therefore, measurements were performed on hybrid composites with the number of carbon fiber plies varying from 1 to 4, giving an average layer thickness of 29μm. In a hexagonal packing with a V_f of 50%, this corresponds to a layer of about three fibers thick. Similarly, the average layer thickness of the glass fiber plies was determined to be 155μm.

The variation in layer thickness leads to a change in the fiber dispersion, i.e. the fineness or degree of mixing of the constituents. This variation needs to be taken into account, as it is known to have a significant influence on the hybrid effect for initial failure strain[22] and [2]. Therefore, the layer thickness of the 1 carbon ply hybrid was varied between 1 and 5 fibers thick (see Fig. 5a). The number of fibers over the thickness was always an odd number. The thickness of the 2, 3 and 4 carbon ply hybrid was randomly varied between 5–9, 9–13 and 13–17 fibers respectively. This gave averages of 3, 7, 11 and 15 fibers per layer, with thicknesses of 24μm, 57μm, 90μm and 122μm for the 1, 2, 3 and 4 carbon ply hybrid, respectively.

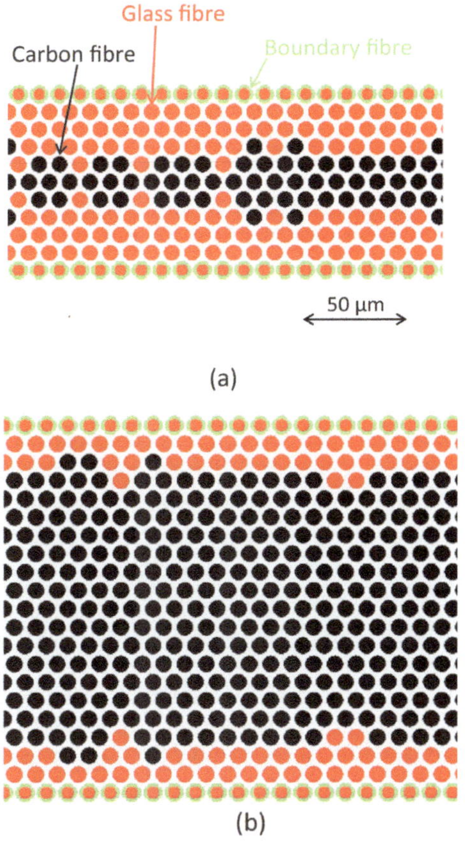

Figure 5. Modelled geometry of the carbon/glass hybrid composites: (a) 1 carbon ply hybrid, and (b) 4 carbon ply hybrid. The width corresponds to 200μm out of the 2 mm total width modelled.

The glass fiber layers above and below the carbon layer were modelled to be at least three fibers thick, all along the width of the model. The glass layers are thicker in reality, but their thickness does not affect the modelling predictions. This was confirmed by modelling thicker glass fiber layers, which had no influence on the predicted failure strains. This is mainly due to the very local load sharing assumption, which causes only stress concentrations on the 6 nearest neighbors. The thickness of the glass fiber layer would however have a small influence on the thermal residual stresses, but these were not taken into account in the model, and were subtracted from the experimental results.

Boundary fibers were added to make it more representative of larger sample sizes [24]. These fibers have the same properties as the other fibers, but are not allowed to break. Their absence could make a small difference at the left and right edges of the model (see Fig. 5). The boundary fibers on the top and bottom, however, do not make a difference as the critical cluster will not develop in the glass fiber layers. The modelled width was 2 mm, while the length was 10 mm. The number of fibers ranged from 2337 for the 1 carbon ply hybrid to 4463 for the 4 carbon ply hybrid. Each model contained close to 1700 glass fibers, as the glass fiber layers always had the same thickness (see Fig. 5).

The Weibull distributions for both fiber types are unknown. Therefore, the following strategy was used to set up the parameters of a unimodal Weibull distribution. The carbon fiber Weibull modulus m was set to 5, which is a reasonable value for this carbon fiber type [29] and [30]. The gauge length L_0 was set to 10 mm, which corresponds to the actual length of the model. The Weibull scale parameter σ_0 was chosen in such a way that the predicted failure strain of the 4 carbon ply hybrid corresponded to the experimental value of 1.858%. By setting σ_0 to 3029 MPa, a failure strain of the carbon fiber plies in the 4 carbon ply hybrid of 1.856 ± 0.043% was achieved. After correcting for thermal residual stresses, this becomes 1.877 ± 0.043%. The Weibull modulus of the S-glass fibers was set to 4, as that is a reasonable value for glass fibers [31]. σ_0 and L_0were set to 1520 MPa and 50 mm respectively to yield a reasonable failure strain of 3.9%.

A total of 200 simulations was performed for every configuration. Failure of the carbon fiber layer was detected as an exponential and unstable increase in the number of fiber breaks on successive iterations within one strain increment. This corresponds to the propagation of a critical cluster in the carbon fiber layer. The failure strain of the reference all-carbon and all-glass composites was determined using a model with the same dimensions as the 4 carbon ply hybrid (see Fig. 5b). By using boundary fibers, the exact size of these reference composites will only have minor influence on their predicted failure strain [24]. Thermal residual stresses were not added to the predicted values, although this is in principle possible. This ensures consistency with the experimental results, from which the thermal residual strains were subtracted (see Table 6).

The hybrid effect was calculated as the relative failure strain increase compared to the all-carbon fiber reference composite. The modelled failure strain of 1.836 ± 0.04% for the all-carbon composite was used as the reference value. This is slightly lower than the failure strain of the four carbon ply hybrid in the model, which was 1.856 ± 0.043%, suggesting that the 3 and 4 ply carbon hybrids may still include a very small hybrid effect. The experimental values for comparison were taken from the results presented earlier for the visual determination of the initial failure, after correction for the thermal residual stresses (see Table 6).

Fig. 6 reveals an increase in the predicted failure strain with decreasing number of carbon fiber plies. As described in detail in Swolfs et al. [22], this can be attributed to the increasing restrictions on forming clusters of fiber breaks.

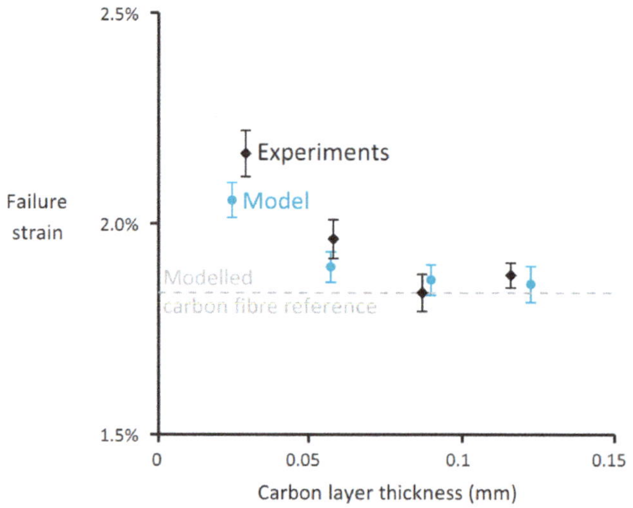

Figure 6. Comparison between model failure strain predictions and experimental results as a function of the carbon fiber layer thickness. The error bars indicate the standard deviations. Residual stresses were not taken into account in the model, and were subtracted from the experimental results.

There is a reasonable agreement between the model predictions and experiments. For the 1 ply carbon hybrid, the model underestimated the hybrid effect: 11.9% compared to 16.6% in the experiments based on initiation, as presented in Section 4. For the 2 carbon ply case the predicted hybrid effect is 3.3% compared to 5.7% in the experiments. The differences may be attributed to:

- *Dynamic Effects:* this may further increase the hybrid effect, but its relative importance is unclear from the literature [2] and [32].
- *Uncertainties in the Input Data:* the Weibull modulus is known to be crucial for the hybrid effect [2], but is also very difficult to measure [24]. A smaller Weibull modulus would lead to larger hybrid effects in the model, and bring the modelling results closer to the experimental values.
- *Possible Size Scaling Effects:* the model is significantly smaller than the actual tensile samples. The effect of size scaling on the hybrid effect has not yet been investigated in the literature.

Another key conclusion is that the 4 carbon ply hybrid has nearly the same failure strain as the carbon fiber reference composite. This means that carbon fiber plies with a thickness above 100μm will have a negligible hybrid effect. Sandwiching carbon fiber plies in between glass fiber plies is therefore a reliable method for measuring the carbon fiber composite failure strain.

Higher Carbon Layer Failure Strain at the Start of the Fragmentation Process

The effect of multiple fragmentation in the carbon ply is analyzed with the approach proposed by Jalavand et al. [14]. This method has been used to study the effect of laminate configuration on UD hybrid tensile response and was able to accurately predict the damage processes of different UD hybrid specimens. A finite element model at the ply level is used with cohesive elements to represent interfacial failure between plies, and multiple cohesive elements embedded within the carbon plies normal to the tensile load to represent their statistical strength distribution. An identical strength distribution for the carbon layer has been used in all different hybrid laminates and the different damage developments were predicted accurately. The details of the model can be found in[14] and [33].

The same modelling approach was applied to predict fragmentation growth and its effect on the stress–strain curve of the 1SG/1C/1SG and 1SG/2C/1SG laminates. The modulus and ply-thickness of the S-glass/epoxy and carbon/epoxy layers were chosen according to Table 2. The Weibull modulus is not known and so the same random distribution with$m = 41$, as in Fig. 4 of [14], was applied to simulate the strength variability of the TR30 carbon/epoxy layer. The weakest point from the finite number of values in the strength distribution has a 1.92% failure strain, which is based on the carbon layer fracture of the 2EG/3C/2EG laminate (EG stands for E-glass) presented in Table 6 of [16]. This failure strain is only 0.02% higher than that of 2SG/4C/2SG as presented in Table 3 and therefore, using the same strength distribution as in [14] is plausible. Thermal residual strains were not taken into account, which is acceptable because the difference in strains for the 1SG/1C/1SG and 1SG/2C/1SG laminates is only 0.002% according to Table 3. The first failure in the finite element analysis occurs at the same overall extension in both 1SG/1C/1SG and 1SG/2C/1SG laminates since this model does not account for the hybrid effect on initiation. But the subsequently predicted fragmentation rate which determines the stress–strain curve of each laminate will be different. This initial failure strain is slightly different from the 1.86% failure strain of the laminates with 3 and 4 layers of carbon presented earlier. But there was no data available on the strength variability of this prepreg, and therefore the same strength and Weibull modulus already used for the analysis of E-glass/TR30 carbon hybrid laminates [14] as given above were employed here. The model simulates the full non-linear response of the hybrid, giving similar curves to the ones obtained experimentally. The knee point on the stress–strain response can be established by fitting lines to the data in exactly the same way as in Fig. 2. For the case with a single ply, the first carbon fracture at 1.92% does not produce a significant change in stiffness, and the knee in the stress–

strain curve is delayed until a strain of 2.04%, as shown in Fig. 7a. The reason for such delay is that the change in slope corresponds to the point at which multiple fiber fractures build up throughout the whole laminate rather than just the very first failure. However, for the case with 2 carbon plies, the knee is at 1.97%, fairly close to the first initiation of failure at a strain of 1.92%. Carbon layer fractures have a more pronounced effect on the average stress–strain curve of this hybrid laminate due to the higher contribution of the thicker carbon layer to the overall stiffness of the laminate, and the lower number of fractures required to establish the fragmentation process since the critical length of the ply increases with ply thickness. The model shows that the process zone around the tip of the carbon layer fragments in the double carbon ply is twice as long as for the single ply case and therefore, the number and rate of crack formation are lower in the thicker carbon case.

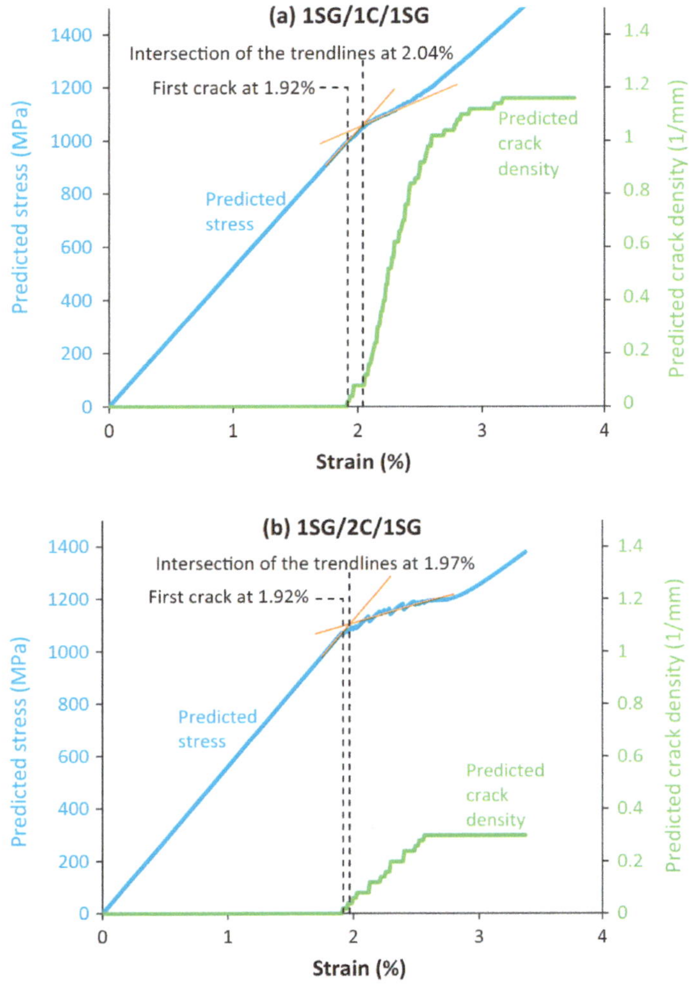

Figure 7. Predicted stress–strain and crack density curves for different carbon ply thicknesses showing the delay in establishing ply fragmentation in the thinner ply case.

The development of carbon layer fragmentation and associated translaminar crack density growth is also shown in Fig. 7(a) and (b) for both specimens. The initial fractures do not have much effect on the stress–strain curve, and there is a significant change in slope only when a regular process of steady fragmentation is established. The maximum predicted crack density of the 1SG/1C/1SG specimen is 1.16 mm^{-1} while it is only 0.33 mm^{-1} (3.5 times less) for the 1SG/2C/1SG laminate because of the greater ply-level critical length and local delaminations in the thicker laminate. This means that the number of cracks required to achieve saturation in the 1SG/1C/1SG sample is significantly larger than in the 2-ply case. The crack density growth has a small rise at the very beginning which is distinct from the main fragmentation process initiating at 2.04%. Therefore, the apparent slope change is not obvious until this strain value. Since the ratio of carbon in the 1SG/2C/1SG specimen type is much higher than that for the 1SG/1C/1SG type, the effect of each carbon layer fragmentation event is accompanied by more noticeable stress fluctuations on the overall stress–strain curve. As a result, the intersection of the initial slope and secondary average lines is at 1.97%, much closer to the first fiber failure strain. These results are aligned with the experimental results in Table 6. They can be compared by taking the difference between the knee point and the first carbon layer fracture, which should not depend on the initiation hybrid effect. The difference in the model is 6.2% and 2.6% in relative terms for the single and double carbon layers respectively, compared with equivalent experimental values of 3.2% and 2.2%. The values are higher than measured, but the model correctly predicts the larger effect in the specimen with one carbon layer and explains the mechanism in terms of the larger number of cracks required to have an effect on the apparent stress–strain curve. Note that the fracture initiation strain is the same in both cases as this is an input parameter to the model and the effect of ply thickness on the initiation hybrid effect cannot be predicted with this approach. On the other hand the model presented in Section 5.1 is able to predict the difference in initiation strains, but cannot represent the ply fragmentation process. The models are therefore complementary to each other in explaining the hybrid effects observed experimentally.

Further Increase in Carbon Layer Failure Strain during Fragmentation

An analytical model of the fragmentation process was also established [34] which revealed that for a fixed carbon ply failure strain, fragmentation occurs at constant stress, giving a flat plateau. However the stress level required for multiple cracking increases during the fragmentation process due to the statistical distribution of carbon fiber strengths. As a result, the stress slightly rises along the plateau until the ply fragmentation becomes saturated. At this point, there is a second knee as the stress in the glass starts to rise with the load contribution of the carbon layer remaining constant due to the inability to transfer any further load across the interface. At the second knee point, the strain varies along the specimen both in the carbon and glass layers and the average strain in the carbon is no longer equal to the external extension. The analytical model [34] allows to relate the strength of the carbon at the fragmentation saturation point, S_C, to the overall laminate-level stress, σ, according to Eq. (1):

$$\sigma = \frac{S_C}{E_C}\frac{E_G t_G + E_C t_C}{t_C + t_G}, \qquad (1)$$

where E and t are the modulus and thickness of the glass/epoxy (*G*) and carbon/epoxy (*C*) layers.

To find the strain at the point of the last fracture in the carbon layer, ϵ_C, which corresponds to fragmentation saturation, Eq. (1) is rewritten as :

$$\epsilon_C = \sigma\frac{t_C + t_G}{E_G t_G + E_C t_C} \qquad (2)$$

This is the maximum value in the middle of the fragments of the carbon layer when saturation is complete, i.e. at a crack density of 1.16 mm^{-1} for the 1SG/1C/1SG specimen. The strain at fragmentation, ϵ_C, calculated by this equation was validated using the FE model discussed in Section 5.2 and the relative difference found to be less than 0.2%. Therefore, this simple equation can be applied to find the strength of the carbon/epoxy layer at the second knee point. At the start of the plateau the difference in the overall measured extension and strain in the carbon ply is negligible. However, as the carbon layer fragmentation gradually introduces gaps between the ends of the fragments, the average strain in the carbon layer increasingly deviates from the overall extension.

The average overall stress at the end of the plateau in the experimental results determined from the intersection of lines fitted to sections of the test graphs before and after the knee point is equal to 1202 MPa for the 1SG/1C/1SG sample. Putting this value into Eq. (1), with the moduli and thicknesses from Table 2, the value of strain in the carbon layer is found to be 2.381% for the single ply case. The overall stress of 1162 MPa for the 1SG/2C/1SG specimens gives a strain of 2.131% at fragmentation saturation at the end of the plateau. The effect of thermal residual strains after fragmentation saturation is less than in the laminate without any fracture of the carbon layer since fragmentation allows the stresses to relax via interfacial deformation. So the compressive thermal strains quoted in Table 4 are the upper bound, but to be conservative, they are subtracted from the values found from Eq. (2). As a result, the failure strain of the 1SG/1C/1SG laminate at fragmentation saturation is at least 2.352% and the failure strain of the 1SG/2C/1SG is at least 2.108%.

It is worth mentioning that the thermal residual strains at fragmentation saturation are location dependent and may vary between zero and the values in Table 4 along each fragment. The maximum strains are compared in Table 8 with the results presented earlier in Table 6. The strains are more than 5% higher than the values at the start of the plateau. This can be considered as a further type of hybrid effect over and above what is presented in Table 6, as it is possible to go beyond the early failures and reach the strains of the stronger parts of the material, exploiting a greater proportion of the carbon strength distribution in these thin-ply hybrid configurations.

Table 8. Summary of strains at different stages of failure of the carbon ply (corrected for residual stresses).

Layup	Measured strain at first carbon layer fracture	Measured strain at first knee (start of fragmentation)	Calculated strain in carbon layer at second knee (end of fragmentation)
	[Absolute %] (CV (%))	[Absolute %] (CV (%))	[Absolute %] (CV (%))
1SG/1C/1SG	2.166 (2.5)	2.227 (1.4)	2.352 (2.2)
1SG/2C/1SG	1.963 (2.3)	2.004 (1.6)	2.108 (0.4)
Baseline – average of 3 and 4 carbon ply delaminating hybrids	1.858	–	–

Fig. 8 summarizes the overall results for the 1SG/1C/1SG specimens, showing the increasing strains in the carbon layer as the failure process and fragmentation progresses. A typical stress–strain response is plotted, and the strains in the carbon layer at the different stages are labelled on the plot. These values are based on the average measured strains for different specimens including corrections for thermal stresses and so do not correspond exactly to the points on the graph. The final value at the end of fragmentation is calculated from Eq. (2), representing the maximum value of the varying strain along the length of the carbon fragments at the point of saturation. It should be noted that this is lower than the measured overall strain on the outer surface of the specimen.

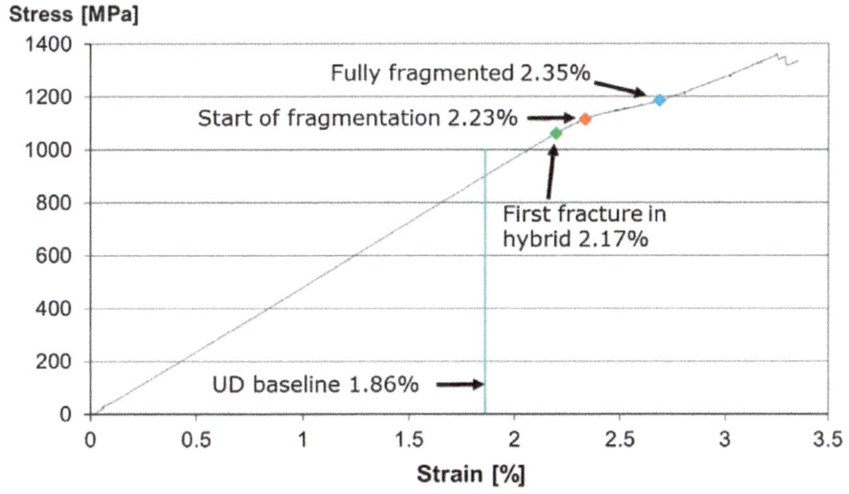

Figure 8. A typical stress–strain curve of the 1SG/1C/1SG hybrid specimen with carbon failure strains highlighted at different stages. Labelled values are averages calculated from all tested specimens rather than measured directly on the specimen shown.

CONCLUSIONS

Baseline failure strains of carbon/epoxy composites can be significantly underestimated in conventional unidirectional tests, leading to overestimation of hybrid effects. A specimen with carbon plies sandwiched between glass plies has been found to give consistent gauge length failures away from stress concentrations at the grips, and can be used to determine reference failure strains.

The magnitude of the hybrid effect depends on the ply thickness. For the carbon/S-glass–epoxy there is a hybrid effect of up to 20% at the first knee on the stress–strain curve when the carbon is only 29μm thick, but there is no significant effect for plies over 80μm thick. The hybrid effect can be separated into different parts: an effect on initiation of failure, due to constraint at the fiber level, and another effect due to delay in establishing stable fragmentation as a result of constraint at the ply level. There is a further increase in strain during fragmentation as increasing stress is required to form successive critical clusters in the carbon layer. Models have been presented which are able to illustrate all of these mechanisms, and give reasonable quantitative results for the hybrid effects.

These results illustrate how the strain in the carbon layers increases as the failure progresses from first ply fracture to establishment of progressive ply fragmentation to saturation of the fragmentation process. They demonstrate that different values for the hybrid effect are possible depending on how the point of failure is defined. At initiation, delay in failure is possible because of the constraining effect at the fiber level on the development of critical clusters of fiber breaks due to the limited number of fibers through the thickness of the ply. During fragmentation there is a constraint at the ply level that limits the development of further clusters of fiber breaks, allowing greater advantage to be taken of the variable strength of the ply. In both cases the hybrid effect occurs because of the possibility of making use of a greater proportion of the distribution of the strength of the carbon fibers.

ACKNOWLEDGEMENTS

This work was partly funded under the UK Engineering and Physical Sciences Research Council (EPSRC) Program Grant EP/I02946X/1 on High Performance Ductile Composite Technology in collaboration with Imperial College, London. Gergely Czél acknowledges the Hungarian Academy of Sciences for funding through the Post-Doctoral Researcher Programme fellowship scheme, the János Bolyai scholarship and the Hungarian National Research, Development and Innovation Office - NKFIH, for funding through grant OTKA K 116070. Ignaas Verpoest is holder of the Toray Chair in Composite Materials at KU Leuven. The authors acknowledge Hexcel Corporation for supplying materials for this research. Supporting data can be requested from the corresponding author.

REFERENCES

1. Hayashi T. On the improvement of mechanical properties of composites by hybrid composition. In: Proc 8th int reinforced plastics conference. p. 49–52.
2. Swolfs Y, Gorbatikh L, Verpoest I. Fiber hybridisation in polymer composites: a review. Composites Part A 2014;67:181–200.
3. Phillips LN. The hybrid effect – does it exist? Composites 1976;7(1):7–8.
4. Bunsell AR, Harris B. Hybrid carbon and glass fiber composites. Composites 1974;4:157–64.
5. Aveston J, Sillwood JM. Synergistic fiber strengthening in hybrid composites. J Mater Sci 1976;11(10):1877–83.
6. Zweben C. Tensile strength of hybrid composites. J Mater Sci 1977;12 (7):1325–37.
7. Xing J, Hsiao GC, Chou TW. A dynamic explanation of the hybrid effect. J Compos Mater 1981;15(Sep):443–61.
8. Fukuda H, Chou TW. Stress concentrations in a

hybrid composite sheet. J Appl Mech-Trans ASME 1983;50(4A):845–8.

9. Manders PW. The strength of mixed fiber composites PhD thesis. Surrey, United Kingdom: University of Surrey; 1979.

10. De Baere I, Van Paepegem W, Degrieck J. On the design of end tabs for quasistatic and fatigue testing of fiber-reinforced composites. Polym Compos 2009;30:381–90.

11. Wisnom MR, Maheri MR. Tensile strength of unidirectional carbon fiber-epoxy from tapered specimens. In: 2nd European Conf. on composites testing and standardisation. Hamburg; September 1994. p. 239–47.

12. Khan B, Potter K, Wisnom MR. Suppression of delamination at ply drops in tapered composites by ply chamfering. J Compos Mater 2006;40:157–74.

13. Czél G, Jalalvand M, Wisnom MR. Hybrid specimens eliminating stress concentrations in tensile and compressive testing of unidirectional composites. Composites Part A [submitted for publication].

14. Jalalvand M, Czel G, Wisnom MR. Numerical modelling of the damage modes in UD thin carbon/ glass hybrid laminates. Compos Sci Technol 2014;94:39–47.

15. Fuller JD. Pseudo-ductility of thin ply angle-ply laminates PhD thesis. University of Bristol; 2015.

16. Czél G, Wisnom MR. Demonstration of pseudo-ductility in high performance glass–epoxy composites by hybridisation with thin-ply carbon prepreg. Composites Part A 2013;52:23–30.

17. Czél G, Jalalvand M, Wisnom MR. Demonstration of pseudo-ductility in unidirectional hybrid composites made of discontinuous carbon/epoxy and continuous glass/epoxy plies. Composites Part A 2015;72:75–84. http://dx.doi. org/10.1016/j.compositesa.2015.01.019.

18. Czél G, Jalalvand M, Wisnom MR. Design and characterisation of advanced pseudo-ductile unidirectional thin-ply carbon/epoxy–glass/epoxy hybrid composites. Compos Struct 2016;143:362–70.

19. Gigliotti M, Wisnom MR, Potter KD. Development of curvature during the cure of AS4/8552 [0/90] unsymmetric composite plates. Compos Sci Technol 2003;63:187–97.

20. Wisnom MR. Size effects in the testing of fiber-composite materials. Compos Sci Technol 1999;59:1937–57.

21. Wisnom MR, Khan B, Hallett SR. Size effects in unnotched tensile strength of unidirectional and quasi-isotropic carbon/epoxy composites. Compos Struct 2008;84:21–8.

22. Swolfs Y, McMeeking RM, Gorbatikh L, Verpoest I. The effect of fiber dispersion on initial failure strain and cluster development in unidirectional carbon/glass hybrid composites. Composites Part A 2015;69:279–87.

23. Swolfs Y, McMeeking RM, Verpoest I, Gorbatikh L. Matrix cracks around fiber breaks and their effect on stress redistribution and failure development in unidirectional composites. Compos Sci Technol 2015;108:16–22.

24. Swolfs Y, Verpoest I, Gorbatikh L. Issues in strength models for unidirectional fiber-reinforced composites. Compos Sci Technol 2015;114:42–9.

25. Swolfs Y, Gorbatikh L, Romanov V, Orlova S, Lomov SV, Verpoest I. Stress concentrations in an impregnated fiber bundle with random fiber packing. Compos Sci Technol 2013;74:113–20.

26. Rosen BW. Tensile failure of fibrous composites. AIAA J 1964;2:1985–91.

27. Swolfs Y, Gorbatikh L, Verpoest I. Stress concentrations in hybrid unidirectional fiber-reinforced composites with random fiber packings. Compos Sci Technol 2013;85:10–6.

28. Kelly A, Zweben C. Comprehensive composite materials. Elsevier; 2000.

29. Naito K, Yang J-M, Kagawa Y. Tensile properties of high strength polyacrylonitrile (PAN)-based and high modulus pitch-based hybrid carbon fibers-reinforced epoxy matrix composite. J Mater Sci 2012;47(6):2743–51.

30. Naito K, Yang JM, Tanaka Y, Kagawa Y. The effect of gauge length on tensile strength and Weibull modulus of polyacrylonitrile (PAN)- and pitch-based carbon fibers. J Mater Sci 2012;47(2):632–42.

31. Thomason JL. On the application of Weibull analysis to experimentally determined single fiber strength distributions. Compos Sci Technol 2013;77:74–80.

32. Xing J, Hsiao GC, Chou TW. A dynamic explanation of the hybrid effect. J Compos Mater 1981;15(SEP):443–61.

33. Jalalvand M, Hosseini-Toudeshky H, Mohammadi B. Numerical modeling of diffuse transverse cracks and induced delamination using cohesive elements. Proc Inst Mech Eng, Part C: J Mech Eng Sci 2013;227(7):1392–405.

34. Jalalvand M, Czél G, Wisnom MR. Damage analysis of pseudo-ductile thin-ply UD hybrid composites – a new analytical method. Composites Part A 2015;69:83–93.

CHAPTER 6 QUANTIFICATION OF MICRO-SCALE VARIABILITY IN FIBER BUNDLES

F. Gommer, A. Endruweit, and A.C. Long

Keywords: A. Carbon fibres; B. Microstructures; C. Statistical properties/methods; D. Optical microscopy

INTRODUCTION

Textile reinforcements utilized in the manufacture of composite components are based on filaments which are usually bundled in yarns and then processed into a fabric for easier handling. Due to this structure, fiber reinforced components are usually analyzed on different levels, each with its own variability. Composite parts are made from layers of textile reinforcement (macro-scale) which consist of individual fiber bundles (meso-scale). The bundles can consist of several thousand filaments (micro-scale). It was shown that the variability at each scale affects the properties of a fiber reinforced composite component [1] and [2]. Depending on the scale of analysis, properties at the smaller scales of the material are usually homogenized, and any variability at these scales is ignored [3]. Attempts have been made to incorporate statistical variability at the smaller scales in multi-scale modelling approaches [4]. The variabilities are, however, usually obtained from randomly generated distributions only and are uncorrelated to the actual variability in the material.

For numerical analysis, fiber reinforcements are often represented by an idealized geometry of the smallest repeat in form of a unit cell [5]. To add variability of fiber bundle paths at the meso-scale, representative volume elements can be utilized [6]. In this case, the bundle paths exhibit a defined degree of variability; however, the geometry at the boundaries is usually periodic. As composite components are made from several layers of fiber reinforcement, this can introduce additional sources of variability. The stacking process introduces variability in between layers such as nesting or fiber bundle shape changes. Attempts were made to incorporate these bundle deformations into a geometrical model [7]. For the analysis of properties at the meso-scale, the filament arrangements at the micro-scale within the fiber bundles are usually treated as uniform[8]. The properties at the micro-scale are then easily estimated from bulk parameters such as the overall fiber volume fraction, V_f [9] in the yarn. Treating the fiber bundles as homogenous material ignores, however, any information about the local arrangement[10] or waviness [11] and [12] of the filaments. It was demonstrated numerically that the macro- or meso-scale failure of a part may be initiated by local stress concentrations at the micro-scale [13]. The local arrangements of filaments in fiber bundles should therefore not be ignored, but a systematic description is currently not available.

This paper analyzes low crimp carbon fiber bundles and local effects of the reinforcement geometry on the micro-scale filament arrangements. Non-destructive examination methods such as micro-computed tomography are only feasible on very small samples and have a limited resolution [14] and [15]. Conventional optical microscopy is therefore used as an inexpensive alternative technique which also allows achieving a higher resolution. Automatic image analysis of optical micrographs of cross-sections of the material allows a detailed analysis in a time efficient manner [16]. The obtained data can be employed in multi-scale modelling approaches, for example, to improve the accuracy of local properties assigned to voxels in a finite element analysis [17].

Citation: F. Gommer, A. Endruweit, A.C. Long, Quantification of micro-scale variability in fiber bundles, Composites Part A: Applied Science and Manufacturing, Volume 87, August 2016, Pages 131-137, ISSN 1359-835X, http://dx.doi.org/10.1016/j.compositesa.2016.04.019.

MATERIALS AND METHODS

Fiber Reinforcement and Processing

Composite panels with dimensions 125 mm × 60 mm were made from a single layer of low-crimp plain weave reinforcement and a low viscosity epoxy resin system. The fabric consisted of aligned carbon fiber bundles with a filament count of 12 K (warp direction), stabilized by thin glass fiber weft yarns coated with a thermoplastic polymer (Fig. 1). The dry fabric was placed in a stiff metallic tool where the global fiber volume fraction, V_f, is determined by the cavity height and was estimated to be 0.45. The fabric was saturated with a liquid resin matrix, MVR 444 supplied by the Advanced Composites Group, by circumferential injection at 2 bar pressure. The resin was preheated to 70° before the injection pressure was applied which reduced the initial viscosity to approximately 0.025 Pa s.

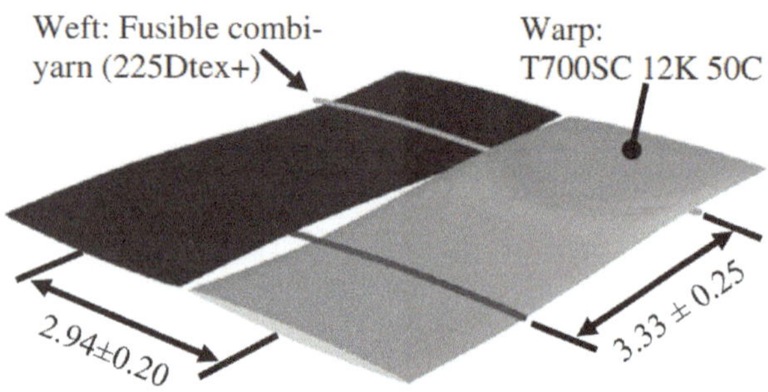

Figure 1. Unit cell model of low-crimp plain weave, generated with the open source software TexGen available from http://texgen.sourceforge.net/. The measurements given are in mm and correspond to Ends/10 cm = 34 ± 2.5 and Picks/10 cm = 30 ± 2.5.

Sample Preparation and Image Processing

Micrographic analysis of cross-sections of these specimens, cut perpendicular to the nominal fiber bundle path, allowed to identify the filament distribution within the carbon fiber bundles (Fig. 2). By successively removing thin layers of material at the surface of these specimens and acquiring new images, the filament arrangement can be determined at different positions along the fiber bundle path [18].

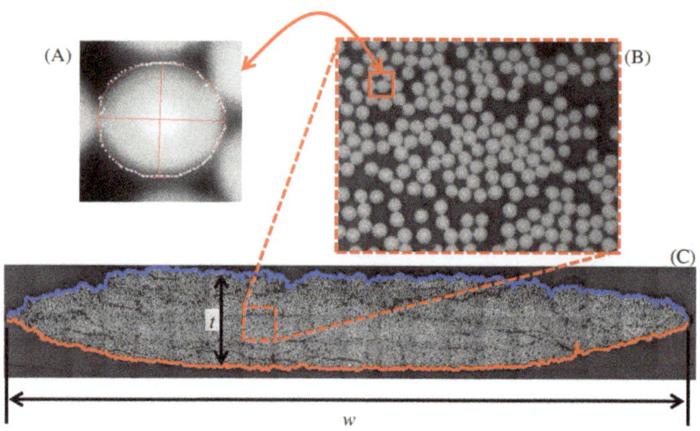

Figure 2. (A) Filament cross-section with identified perimeter determined by image processing [16]. (B) Micrograph of carbon fiber bundle cross-section. (C) Tiling a large number of overlapping micrographs enables to analyze an entire fiber bundle. The determined fiber bundle outline is shown with solid lines and the width, w, and thickness, t, are indicated. (For interpretation of the references to color in this figure legend, the reader is referred to the web version of this article.)

Stitching of overlapping images (Fig. 2C) allows large area images of fiber bundle cross-sections to be studied at high resolution [19] and [20]. A stitching code was implemented based on fast- Fourier correlations programmed in Matlab [21]. Micrographs were acquired at a high magnification which resulted in a pixel size of 0.09μm (Fig. 2B). All individual filament cross-sections were determined by means of edge detection based on color gradients in the images and fitting ellipses to the identified boundaries [16]. It was estimated that the fitted ellipse overestimates a filament cross-sectional area by about 3% on average.

Due to differences in material hardness between fibers and polymer matrix, the mechanical polishing process during sample preparation can cause filaments, located at the edge of the fiber bundle, to be abraded more than filaments in the bulk of the material. These filaments are therefore not within the limited depth of field of a conventional optical microscope imaging the bulk of the material. This results in blurred images of these filaments located at the bundle edge which can therefore not be detected with the employed image analysis method (Fig. 3). An additional false detection rate during image processing, e.g. over-/underestimation or non-detection of filament cross-sections within the bulk of the material, of approximately 2% was estimated. For these reasons, approximately only 11,500 filaments were detected during the image processing of a 12 K fiber bundle. For further interpretation of the results, the acquired micrographs and measured data of the detected filaments have been made publicly available [22].

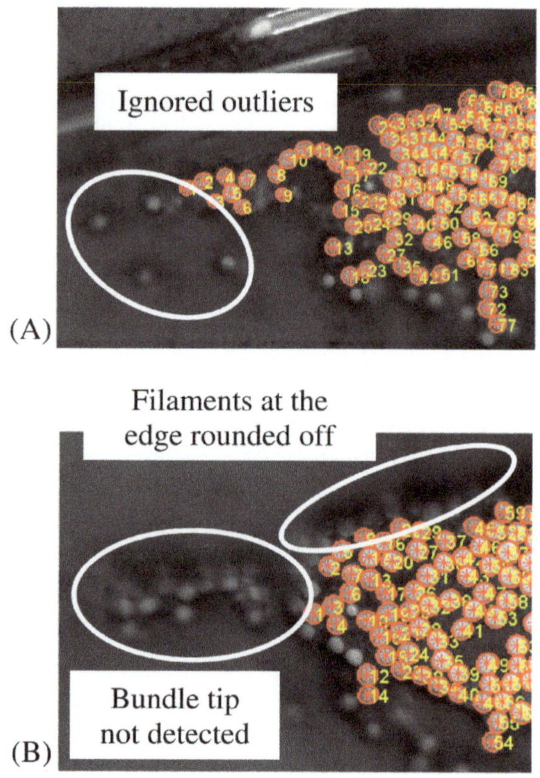

Figure 3. Tip of a fiber bundle at two different positions in a specimen; the detected fiber bundle boundaries are marked with red ellipses. (A) Filaments separated from the bundle are ignored. (B) Problem of bundle determination due to poor image contrast at the fiber bundle

tips and edges due to rounding effects. (For interpretation of the references to color in this figure legend, the reader is referred to the web version of this article.)

DEFINITION OF FIBER BUNDLE DESCRIPTORS

Fiber Bundle Boundary

The fiber volume fraction in a fiber bundle is defined by the combined cross-sectional area of all filaments divided by the entire area of the bundle. The cross-sectional areas of all filaments within a fiber bundle are determined during image analysis. The total bundle area is defined by a virtual boundary enclosing all filaments in the fiber bundle cross-section (Fig. 2C). To determine the envelop enclosing all filament cross-sections, the location of the top and bottom filament edge is determined over the width of the bundle. To avoid sudden changes in the estimated boundary, the filament locations are determined for a discrete step size. The width of this step size was selected to be approximately equivalent to the average diameter of the filaments. This process allows a more precise estimation of the actual fiber bundle area compared to, for example, fitting an ellipse enclosing all filaments [23].

The identified filament positions can be used to estimate the bundle width and thickness. The bundle dimensions are defined as the maximum distance along a reference direction between two filament centers on opposite sides of the fiber bundle plus one nominal filament diameter. Unlike the height, the exact determination of the bundle width usually depends strongly on the accurate detection of filaments at the narrow tips of the fiber bundles (Fig. 3). However, this accuracy may be limited due to the aforementioned problems related to the mechanical polishing process.

Fiber Bundle Volume Fraction

The filament arrangement within fiber bundles is not uniform. A qualitative measure of the variations in local filament densities can be achieved by measuring greyscale levels in vertical and horizontal slices [24]. However, for exact quantification of the local V_f, the exact description of all filament positions is necessary [16].

After determining all filament cross-sectional areas and the bundle envelope (Fig. 4A), the fiber bundle is segmented into a finite number of vertical slices. The width of these segments was selected to be equal to 1 pixel. The absolute number of filaments detected in each slice decreases at the tips of the fiber bundle as expected (Fig. 4B). If, however, the relative amount of filaments within the fiber bundle envelop is calculated, the V_f is more uniform (Fig. 4C). A decrease in V_f towards the tips of the fiber bundle [25] could not be found. The peaks in V_f at the bundle tips occur since only a single or a very small amount of vertically arranged filaments are present, and the step size to determine the bundle boundary (Section 3.1) is smaller than the respective filament diameters. Hence, a relatively large ratio of filaments can be present in an analyzed segment locally. Other artefacts, such as separations within the fiber bundle which may be caused by handling of the dry textile (Fig. 4A), decrease the absolute number of filaments locally and therefore also reduce the internal V_f in this area. Even though the V_f is relatively constant, gradual changes over the width of a fiber bundle may be present.

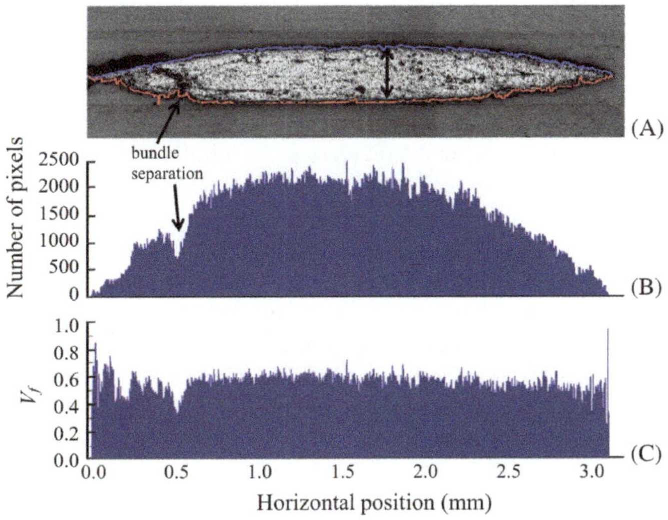

Figure 4. (A) Fiber bundle with estimated boundaries; (B) Total filament area in pixels within the fiber bundle cross-section, estimated for vertical slices of one pixel width (δc = 0.093μm); (C) Fiber volume fraction, V_f, within the boundaries of the fiber bundle. (For interpretation of the references to color in this figure legend, the reader is referred to the web version of this article.)

Local Fiber Volume Fractions

The cross-sectional area of a fiber bundle can be divided into an array of polygons by use of Voronoi cells [26]. Each point within the fiber bundle cross-section is associated to the center point of a filament closer than any other filament center. This results in a tessellation of the entire domain into multiple polygons (Fig. 5). The ratio of the determined filament cross-sectional area to the area of the enclosing polygon reflects the local V_f. Alternatively, the bundle could be divided into a regular grid and the V_f within each cell be calculated [27]. Unlike the Voronoi cell analysis, the V_f results would, however, depend on the selected grid size [28].

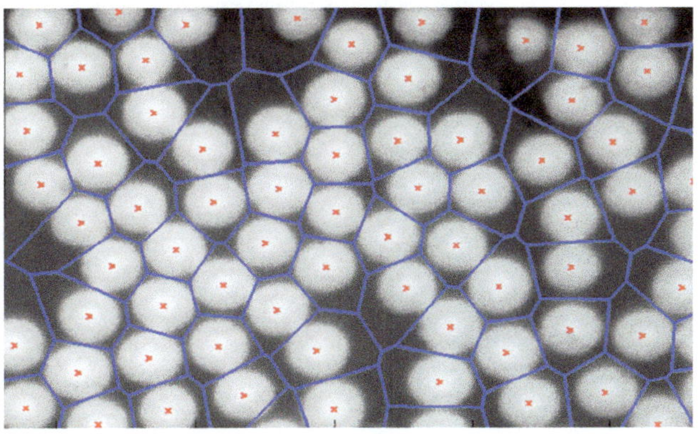

Figure 5. Section of a micrograph of a carbon fiber bundle overlaid with polygons as a result of a Voronoi cell tessellation. (For interpretation of the references to color in this figure legend, the reader is referred to the web version of this article.)

A limitation to the concept of Voronoi cell tessellation of a fiber bundle is that the analysis is based only on filament center positions. Varying fiber diameters or apparent ellipticity of the cross-sectional areas, e.g. as a result of filament waviness inside the fiber bundle which causes filaments not to be sectioned perpendicular to their path, are ignored. The determination of the local V_f could be refined if the actual cross-sectional area of the filaments are considered [29], i.e. the Voronoi cell in this case is not dependent on filament cross-sectional centers but based on their elliptical perimeters. In this case, the Voronoi cell boundary is not any longer located equidistant between two filament centers but located half way between the closest points of two filament boundaries. In addition, the number of strongly misaligned filaments which will result in more pronounced ellipticity of individual filament cross-sections will be limited for the analyzed uni-directional fiber bundles [11] and [12]. For analysis of the local V_f in this work, the error of ignoring the filament cross-sectional shapes for the Voronoi cell tessellation of a fiber bundle is therefore considered small as the average misalignment of filaments is small. Hence, the variation in ellipticity of the filament cross-sections is limited and the simpler Voronoi cell tessellation based on filament centers can be used.

Attempts were made to describe the packing configuration of filaments by the Voronoi cell tessellation [30]. Voronoi polygons sharing a side with each other represent filaments which are nearest neighbors. The average number of polygon edges is therefore equivalent to the average nearest neighbor count. This method ignores, however, varying distances between filaments. The average nearest neighbor count for any fiber bundle was estimated to be 5.7. This suggests that the filaments have a tendency to be arranged in a hexagonal array (number of edges = 6) which corresponds to values reported in the literature [31] and [32]. However, the filaments are not necessarily arranged hexagonally (Fig. 5). Analyzing the angular distribution of filament center points between nearest neighbors, it was demonstrated that the filaments at this low V_f have the tendency to be arranged randomly [16].

CHANGES IN FIBER BUNDLE COMPOSITION

Fiber Volume Fraction Variability

For two fiber bundles, a series of 12 slices (four of which are shown in Fig. 6) was imaged at successive positions over a length of about 5.3 mm. This corresponds approximately to the length of one unit cell (Fig. 1). During each re-polishing step, between 0.3 mm and 0.6 mm of material was removed. This step size is too large to reconstruct individual filament paths [11], however, it has been demonstrated that the majority of filaments in unidirectional fiber bundles only change positions within a limited cross-sectional area[12]. Hence, the effects on V_f distributions due to filament path variability between different bundle cross-sections will be limited as well.

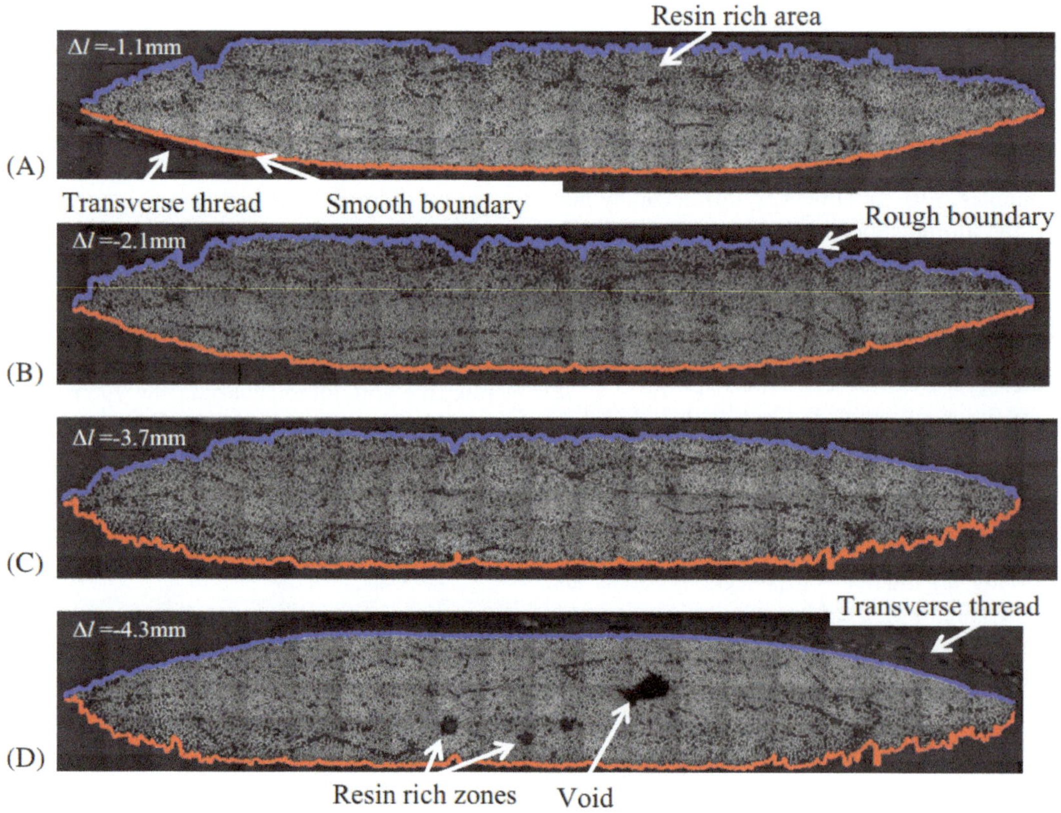

Figure 6. Micrographs of a carbon fiber bundle cross-sectional area at different position, Δl, along the bundle path. (For interpretation of the references to color in this figure legend, the reader is referred to the web version of this article.)

The filament cross-sections and fiber bundle boundaries were determined for each section, and the resulting V_f (Section 3.2) are shown in Fig. 7A. It can be shown that the average V_f is not constant and varies significantly over the analyzed bundle segments. In areas where a transverse binder yarn is present, V_f increases. As single layers of material were analyzed at a constant mold cavity height, the compaction force acting on the fiber bundles in this area increases. This results in a compaction of the filaments locally, and hence in a local increase in V_f. The value of V_f and the fiber bundle height have therefore opposite trends (Fig. 7A and B). In areas where the fiber bundles are only compacted by the mold tool, the V_f is lower and the bundle thickness higher. The maximum peak height of V_f depends also on the bundle width (Fig. 7C). At the same height, a more narrow fiber bundle will have a higher V_f. The changes in fiber bundle width are, however, gradual and hardly affected by local bundle compaction. This suggests that changes in width of a fiber bundle are more affected by the initial state during processing. A similar effect was found when analyzing the aspect ratios of bundle dimensions in a twill weave composite, which suggest the bundle width to be fairly constant over the length of two unit cells [33]. It should mentioned that the estimated standard deviation in V_f for the samples analyzed in this work seems to be independent of the slice position and ranges between 0.06 and 0.09.

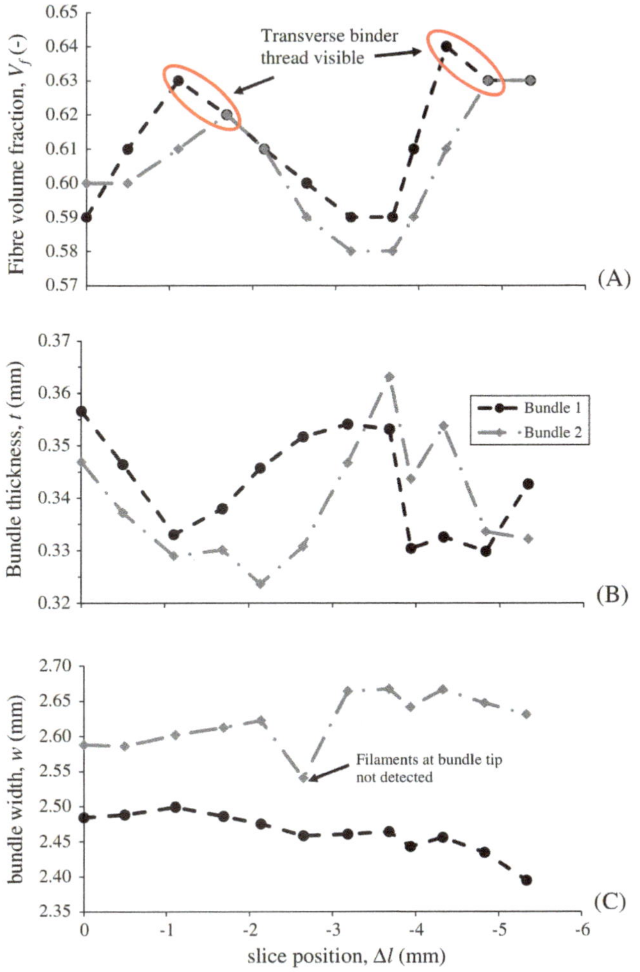

Figure 7. (A) Average fiber volume fraction, (B) thickness and (C) width of two fiber bundles at different positions, Δl, along the bundle path. (For interpretation of the references to color in this figure legend, the reader is referred to the web version of this article.)

The variation in fiber bundle dimension over approximately one unit cell length in this work was determined to be about 3.2% for the thickness and 1.3% for the bundle width. These variations are lower than presented for the randomly selected bundles found in twill weaves (14% and 6% for bundle thickness and width) [23] or non-crimp fabrics [34]. However, these changes in bundle dimensions will be affected by the fabric geometry and the amount of compaction [16]. It should be kept in mind that, depending on the bundle shape (e.g. lenticular), the bundle tips may not involve many filaments [35] and the width may therefore not significantly affect the average yarn V_f. For other properties such as permeability, the bundle tips may however significantly affect the resin flow.

Change in Local Fiber Arrangements

The surface of the determined fiber bundle envelope is irregular in areas of lower local V_f, e.g. where no transverse binder thread is present (Fig. 6). It does, however, gradually change and becomes smooth in areas where the transverse thread is crossing. The interface between the fiber bundles and the thin glass threads is subjected to a higher normal force during mold closure which results in a local compaction of filaments. The filaments at the perimeter of the fiber bundle are able to move and rearrange along the path of the continuous glass thread. As the glass threads are coated with a thermoplastic material and have a low fiber count, movement of the filaments within these binder yarns are thought to be insignificant.

Analyzing changes in local V_f on bundle cross-sections by means of Voronoi cell analysis, a higher V_f can be observed at the boundary where the transverse thread is present (Figs. 6D and 8D). This suggests that the filaments rearrange locally in response to the compaction force. It can also be seen, that this higher V_f is confined to a small area only and does not affect the filament arrangement in the entire fiber bundle. The affected zone has been estimated to have a width between 5 and 13 filament diameters for the analyzed samples. This zone may be larger if samples with a higher global V_f, i.e. larger amount of compaction, are analyzed.

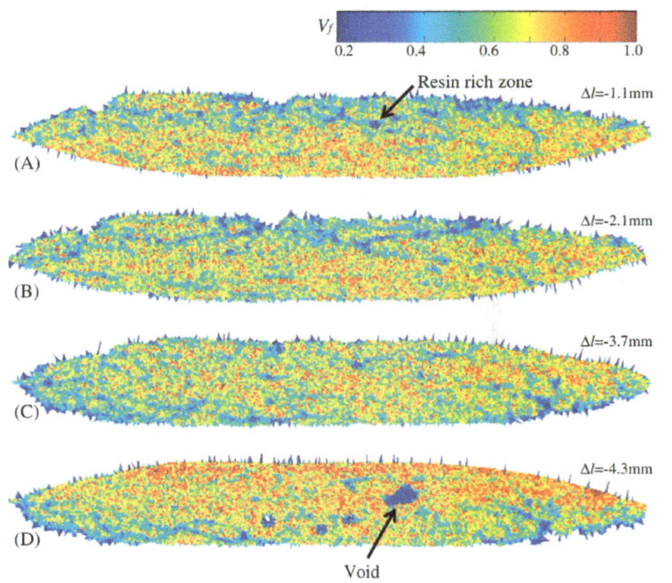

Figure 8. Local fiber bundle volume fractions based on Voronoi cell tessellation at different position, Δl, along a fiber bundle path. (For interpretation of the references to color in this figure legend, the reader is referred to the web version of this article.)

The filament arrangements in the center of the fiber bundle seem not to be affected by the local increase of V_f at the bundle perimeter. Areas of low or high V_f in the center of the fiber bundle change gradually independent of the relative position of the weft thread. Within the fiber bundle, filament clusters of high V_f are present and separated by areas of low V_f. For this case, the filament arrangements were quantified in terms of distance and angle distributions between nearest neighbors and have been presented in an earlier study [16]. In the case of liquid resin flow during composite processing, it is likely that these areas of lower local V_f will act as main flow channels [36].

Effects of Voids

In several bundle sections, voids were detected after abrading and re-polishing the sample surfaces (Fig. 6D). These voids are a result of air entrapment in the liquid resin matrix during the impregnation process. In maps of local V_f, these voids appear as areas of reduced V_f within the fiber bundle (Fig. 8D). As the voided area does not contain a matrix to fixate the filaments locally, filaments in this area may fracture during the polishing process. The voids therefore appear to contain no filaments when using optical microscopy, which may not be the case.

However, the trends in determined V_f in this work (Fig. 7A) seem not to be affected by the appearance of voids. It was observed that all detected voids originate in resin rich zones in between filaments (Fig. 9). The voids may change shape further along the fiber bundle path but they are surrounded by an area of increased V_f. These densely packed filaments are restricted in their movement [18]. This suggests that voids in the samples analyzed here are formed only in areas of the fiber bundle which contain none or just a very limited amount of filaments.

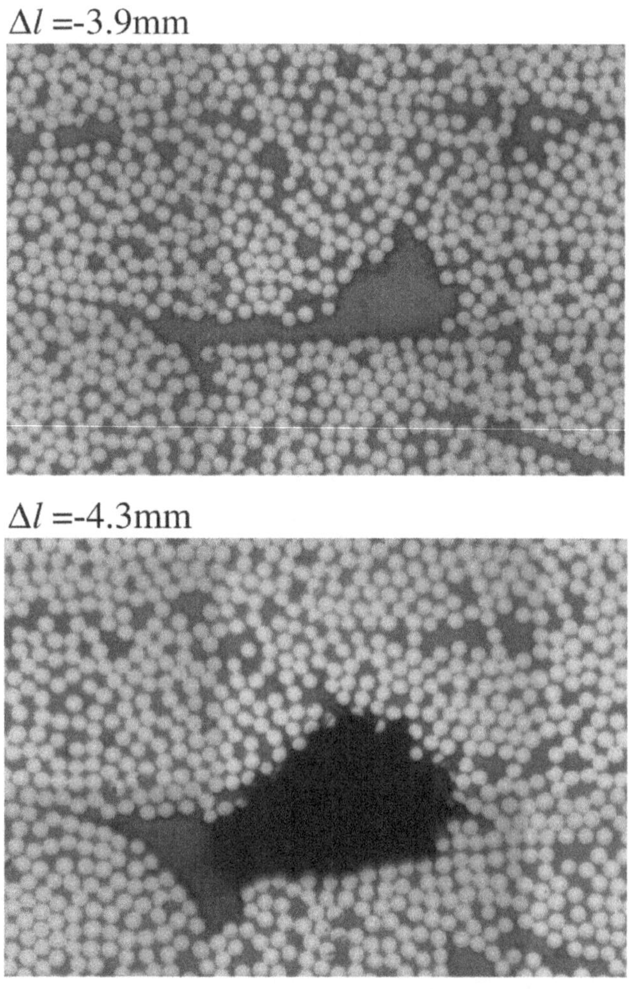

Figure 9. Micrographs of a carbon fiber bundle cross-section before and after a void was detected.

The additional edge of the void may lead to poor image contrast at its circumference due to rounding effects of the filaments during polishing. The effect seems, however, to be small when analyzing the changes in V_f over the width of a sample in adjacent bundle sections (Section 3.2). Therefore, no adjustments have been made to the analysis of the filament arrangements and these voids are treated as resin rich zones in the fiber bundle.

V_f Effects on Properties

The increase in V_f of the low crimp bundles at the crossing points with the binder threads will result in differences in local permeabilities [27]. Hence, the permeability in these zones will decrease which will increase the probability of gas entrapment during impregnating flow in these zones. On the other hand, the influence on the elastic mechanical properties in bundle direction is expected to be small. However, the observed increase in V_f at the crossing of two fiber bundles is probably present in the case of a similar transverse weft bundle as well. Numerical simulations have shown that the stresses and strains in the bundles transverse to the loading direction are increased. Using the example of plain weave unit cells, damage initiation was always observed at the interface of transverse fiber bundles [37] and [38]. The increased V_f at the interface will lead to a more brittle material which may exceed the matrix strength between individual filaments locally [13]. Incorporating the observation of increased V_f at the fiber bundle interface rather than using random distributions in numerical meso-scale models[37] will probably further highlight this effect. It is therefore likely, that this causes cracks in [24] and [39] or around [40] the transverse fiber bundles to form during mechanical loading.

CONCLUDING REMARKS

Changes in the random arrangement of filaments in fiber bundles were quantified along the bundle path. A local increase of the fiber volume fraction, V_f, towards the interface at cross-over points with transverse threads has been observed. In addition, local clusters of high V_f within a fiber bundle have been found which seem not to be affected by adjacent or interlaced fiber bundles. These clusters have different local filament arrangements and affect the fiber bundle properties differently. For example, the permeability in these high filament density clusters is decreased and may prevent full saturation with a liquid resin matrix during impregnation locally. If an area of low filament density is located within such a cluster of increased V_f, this may cause air entrapment in this area. In addition, larger areas of low V_f may act as main fluid transport channels which dominate the fiber bundle permeability. This could lead to local race tracking inside the fiber bundle leading to gas entrapment within the adjacent clusters of higher V_f.

In addition, the increased V_f at the interface between two intersecting fiber bundles will lead to additional stress concentrations during mechanical loading as a result of an increased difference in stiffness. When numerically analyzing the properties of a fiber bundle, the determined systematic and random changes in local V_f inside a fiber bundle can be taken into account for more accurate predictions.

The obtained results are specific for the analyzed fabric at a single global V_f and should be confirmed on other fabric styles, different V_f and layered materials. It is thought, however, that the effect of local compaction of a fiber bundle at the interface between interlaced fiber bundles will also be found under different conditions. This effect is expected to be more pronounced when layers of transversely arranged bundles are compacted within a mold tool. On the other hand, the local compaction effects will be less distinct when bundles are aligned in parallel and nesting is allowed. But even in this case, at an increased amount of compaction, similar local effects at the bundle interface are probable to occur as the filaments within individual fiber bundles are limited to move individually due to long range entanglements. Unlike in the analyzed case of a single bundle being compacted between two rigid plates, the compaction induced effects at the interface for layered materials will be shared between adjacent fiber bundles.

ACKNOWLEDGEMENTS

This work was supported by the Engineering and Physical Sciences Research Council[Grant Number: EP/IO33513/1], through the EPSRC Center for Innovative Manufacturing in Composites (CIMComp) and by an EPSRC Doctoral Prize award hosted by the University of Nottingham.

REFERENCES

1. Mrse A, Piggott M. Compressive properties of unidirectional carbon fiber laminates: II. The effects of unintentional and intentional fiber misalignments. Compos Sci Technol 1993;46(3):219–27.
2. Endruweit A, Long AC, Robitaille F, Rudd CD. Influence of stochastic fiber angle variations on the permeability of bi-directional textile fabrics. Compos Part A 2006;37(1):122–32.
3. Tan H, Pillai KM. Fast liquid composite molding simulation of unsaturated flow in dual-scale fiber mats using the imbition characteristics of a fabric-based unit cell 10.
4. Sriramula S, Chryssanthopoulos MK. Quantification of uncertainty modelling in stochastic analysis of FRP composites. Compos Part A 2009;40(11):1673–84.
5. Long AC, Brown LP. Modelling the geometry of textile reinforcements for composites: TexGen. In: Boisse P, editor. Composite reinforcements for optimum performance. Cambridge: Woodhead Publishing Ltd.; 2011. p. 239–64.
6. Zeng XS, Long AC, Gommer F, Endruweit A, Clifford M. Modelling compaction effect on permeability of 3D carbon reinforcements. In: 18th International conference on composite materials (ICCM18), Jeju Island, South Korea.
7. El Said B, Green S, Hallett SR. Kinematic modelling of 3D woven fabric deformation for structural scale features. Compos Part A 2014;57:95–107.
8. Park CH, Lebel A, Saouab A, Bréard J, Lee WI. Modeling and simulation of voids and saturation in liquid composite molding processes. Compos Part A 2011;42 (6):658–68.
9. Gebart BR. Permeability of unidirectional reinforcements for RTM. J Compos Mater 1992;26(8):1100–33.
10. Guild FJ, Summerscales J. Microstructural image analysis applied to fiber composite materials: a review. Composites 1993;24(5):383–93.
11. Gommer F, Wedgwood KCA, Brown LP. Stochastic reconstruction of filament paths in fiber bundles based on two-dimensional input data. Compos Part A 2015;76:262–71.
12. Czabaj MW, Riccio ML, Whitacre WW. Numerical reconstruction of graphite/ epoxy composite microstructure based on sub-micron resolution X-ray computed tomography. Compos Sci Technol 2014;105:174–82.
13. Trias D, Costa J, Mayugo JA, Hurtado JE. Random models versus periodic models for fiber reinforced composites. Comput Mater Sci 2006;38(2):316–24.
14. Requena G, Fiedler G, Seiser B, Degischer P, Di Michiel M, Buslaps T. 3DQuantification of the distribution of continuous fibers in unidirectionally reinforced composites. Compos Part A 2009;40(2):152–63.
15. Scott AE, Mavrogordato M, Wright P, Sinclair I, Spearing SM. In situ fiber fracture measurement in carbon–epoxy laminates using high resolution computed tomography. Compos Sci Technol 2011;71(12):1471–7.
16. Gommer F, Endruweit A, Long AC. Analysis of filament arrangements and generation of statistically equivalent composite micro-structures. Compos Sci Technol 2014;99:45–51.
17. Matveev MY, Long AC, Jones IA. Modelling of textile composites with fiber strength variability. Compos Sci Technol 2014;105:44–50.
18. Mansfield EH, Purslow FRS, Purslow D. The influence of fiber waviness on the moduli of unidirectional fiber reinforced composites. RAE Technical Report; 1974.
19. Eberhardt C, Clarke A. Fiber-orientation measurements in short-glass-fiber composites. Part I: automated, high-angular-resolution measurement by confocal microscopy. Compos Sci Technol 2001;61(10):1389–400.
20. Davidson NC, Clarke AR, Archenhold G. Large-area, high-resolution image analysis of composite materials. J Microsc 1997;185(2):233–42.
21. Luong B. FFT-based convolution. <www.mathworks.co.uk> [15.April 2013].
22. Gommer F. Measurement data: Quantification of micro-scale variability in fiber bundles. <https://drive.google.com/folderview?id=0B-w_YqRvIONTQnUxbXhSd0NJa0k&usp=sharing> [02.April.2016].
23. Olave M, Vanaerschot A, Lomov SV, Vandepitte D. Internal geometry variability of two woven composites and related variability of the stiffness. Polym Compos 2012;33(8):1335–50.
24. Summerscales J, Russel PM. Observations on the fiber distribution and fiber strain in a woven fabric reinforcement. Adv Compos Lett 2004;13(3):135–40.
25. Nakanishi Y, Matsumoto K, Kurashiki T, Zako M. Multiscale analysis of material damping properties for textile composites. In: 13th European Conference on Composite Materials (ECCM13), Stockholm.
26. Okabe A, Boots B, Sugihara K. Spatial tessellations. Concepts and applications of voronoi diagrams. J. Wiley and Sons; 1992.
27. Endruweit A, Gommer F, Long AC. Stochastic analysis of fiber volume fraction and permeability in fiber

bundles with random filament arrangement. Compos Part A 2013;49:109–18.

28. Silberschmidt VV. Effect of micro-randomness on macroscopic properties and fracture of laminates. J Mater Sci 2006;41(20):6768–76.
29. Hellström JGI, Frishfelds V, Lundström TS. Mechanisms of flow-induced deformation of porous media. J Fluid Mech 2010;664:220–37.
30. Pietsch T, Gindy N, Fahmi A. Nano- and micro-sized honeycomb patterns through hierarchical self-assembly of metal-loaded diblock copolymer vesicles. Soft Matter 2008;5:2188–97.
31. Paluch B. Analysis of geometric imperfections affecting the fibers in unidirectional composites. J Compos Mater 1996;30(4):454–85.
32. Zangenberg J, Larsen JB, Østergaard RC, Brøndsted P. Methodology for characterisation of glass fiber composite architecture. Plast Rubber Compos 2012;41(4–5):187–93.
33. Bale H, Blacklock M, Begley MR, Marshall DB, Cox BN, Ritchie RO. Characterizing three-dimensional textile ceramic composites using synchrotron X-Ray micro-computed-tomography. J Am Ceram Soc 2011;95(1):392–402.
34. Mattsson D, Joffe R, Varna J. Methodology for characterization of internal structure parameters governing performance in NCF composites. Compos B Eng 2007;38(1):44–57.
35. Yurgartis SW, Morey K. Measurement of yarn shape and nesting in plainweave composites. Compos Sci Technol 1993;46(1):39–50.
36. Yazdchi K, Srivastava S, Luding S. Micro-macro relations for flow through random arrays of cylinders. Compos Part A 2012;43(11):2007–20.
37. Matveev MY. Efects of variabilities on mechanical properties of textile composites PhD thesis. University of Nottingham; 2014.
38. Zhou Y, Lu Z, Yang Z. Progressive damage analysis and strength prediction of 2D plain weave composites. Compos B Eng 2013;47:220–9.
39. Ito M, Chou T-W. An analytical and experimental study of strength and failure behavior of plain weave composites. J Compos Mater 1998;32(1):2–30.
40. Li L, Lomov SV, Yan X, Carvelli V. Cluster analysis of acoustic emission signals for 2D and 3D woven glass/epoxy composites. Compos Struct 2014;116:286–99.

CHAPTER 7

NOVEL METHOD FOR FUNCTIONALIZING AND PATTERNING TEXTILE COMPOSITES

Dmitry S. Ivanov, Yann M. Le Cahain, Surush Arafati, Alexandre Dattin, Sergey G. Ivanov, and Andrey Aniskevich

Keywords: Carbon nanotubes and nanofibers, Multifunctional composites, Electrical properties, 3-D printing

INTRODUCTION

The study investigates potential of a novel manufacturing method that can create local patches or pins of matrix spanning through the thickness of a continuous textile preform. The method of liquid resin print was originally developed for stabilizing textile preform prior to the infusion [1]. It allows creating a stiff skeleton within a compliant preform which secures the material in a predefined shape through the infusion and consolidation stages. The process is implemented by injecting liquid reactive resin using a syringe with a needle inserted to a predefined depth into textile preform. Each injection creates a thin layer of liquid resin in the plane of preform. The resin flow around the needle tip occurs predominantly due to the capillary forces and lower in-plane permeability compared to the through-thickness permeability. A series of injections through-thickness connect these layers and, upon heat and pressure assisted consolidation, a solid patch is formed. The dimensions of the patch can be controlled through the volume of each injection. In addition, the bespoke printer allows creating patches arranged in any programed pattern and density.

Thus, the method is implemented in the spirit of additive manufacture. However, unlike the conventional 3D print process, the deposited resins are liquid and reactive. Printing liquid resin is possible due to the intrinsic architecture of textile medium which allows distributing the resin around the injection point and holding the fluid in place until the material is fully consolidated. The liquid state of the deposited resin gives other advantages. For example, as will be shown further in this paper, it easily allows introducing additives to the injected resin systems and tuning local material properties.

The ability of incorporating additives locally makes the process attractive for functionalization. The composites community has been consistently targeting at increasing efficiency of composites by enhancing structural composites with thermal, electrical, damping and other properties. Even though many different methods were recently developed to incorporate function-carrier additives into material there is still a challenge of creating functional reinforcements locally, precisely where needed and without introducing a major impact on the manufacturing process, internal architecture and structural composite properties.

The resin print process allows addressing some of these challenges integrally. After the local patches are created where required, any conventional liquid molding process can be applied to create a composite component. Away from the patches the reinforcement geometry remains unaffected and so overall the process complies with requirement of being minimally invasive. However, patching does create a number of unconventional structural features in the modified area whose impact on composite performance needs to be studied:

- *Fiber bridged interfaces between printed and infused zones:* since the patches are cured prior to infusion a weak interface between matrices of the cured and infused regions can be expected. Unlike other known and well-studied interfaces (e.g. fiber-to-matrix, ply-to-ply, grain-to-grain), patch interfaces are bridged by fibers and conduct the load normal to its boundaries. These interfaces present a particular interest in the

Citation: Dmitry S. Ivanov, Yann M. Le Cahain, Surush Arafati, Alexandre Dattin, Sergey G. Ivanov, Andrey Aniskevich, Novel method for functionalizing and patterning textile composites, Composites Part A: Applied Science and Manufacturing, Volume 84, May 2016, Pages 175-185, ISSN 1359-835X, http://dx.doi.org/10.1016/j.compositesa.2016.01.018.

context of damage accumulation process and their role is not yet known.

- *Thickness variation:* the patches are consolidated at a higher pressure level than used for resin infusion process under flexible bagging. This is needed to suppress voids resulting from the unsaturated capillary flow. As a result the composite thickness may (or may not depending on a reinforcement and preform type) be lower in the patched region. This creates a wavy surface and an additional fiber crimp that may negatively affect composite strength.
- *Varying matrix properties:* the interaction of various matrices with distinctly different properties affects stress distribution and load flow. It can be expected that this factor may play a role in damage accumulation process.

If the patch pattern is regular, these features introduce a new material scale intermediate between the component scale and characteristic scale of textile unit cell. At this scale the architecture is fully defined by the settings of the injection printer and can be adjusted to reach an optimum composite performance. The first step towards the optimization is to understand how these features affect strain distribution and failure behavior of the material. The paper reports an experimental characterization of the patched patterned composites using tensile tests with in-situ measurements of local strain evolution and post-mortem characterization of damage and failure.

The second challenge addressed here is to understand the potential for functionalization. The focus of this work is on through-thickness electrical conductivity. This property is critically important for lightning strike protection and release of static charge. Electrically conductive composites have also a potential for damage and strain sensing [2] and [3]. The commonly adopted approach for enhancing this property is to incorporate third phase into a composite: conductive particles. Carbon nano-tubes (CNT) tend to be particularly attractive as their high aspect ratio helps forming a percolation network at very low weight fractions. Among recent developments targeted specifically at improving conductivity there are methods such as growing CNT grown on fiber surfaces [4] and [5], integrating CNT in fiber coating or sizing [6] and [7], depositing CNT in powder form on fiber surface [8] and [9], and delivering CNT within the resin in a liquid molding process[10], [11], [12], [13] and [14]. These studies show that material can acquire electrical conductivity if percolation of CNT is achieved. Providing electrical percolation through the thickness of glass composites is most challenging since the formation of CNT networks is impeded by non-aligned non-conductive fibers [14].

Most of the known methods require significant changes to manufacturing process to enable functionalization. For instance it is known that growing CNT on fiber surfaces may impact on fiber strength and compressibility of the preform [15], resin infusion may be impacted by the high viscosity of CNT solution and filtration mechanisms [19]. The formation of agglomerates and re-agglomeration during mixing, cure and consolidation processes [11] and [16] significantly limits the weight fraction of CNT realistically achievable for RTM process. da Costa et al. [11] found that the suspensions of CNT with loading above 0.5 wt.% were not appropriate for liquid molding.

This paper aims to demonstrate that CNT can be incorporated in a radically new way. Liquid resin print may present a remedy for some of these problems by delivering required amount of CNT fraction precisely where needed without affecting the subsequent infusion process. This studied process has a potential to eliminate the issues associated with large flow lengths. The resin is deposited locally, point wise and hence, much higher volume fraction can be allowed. Mechanical testing was then used to assess how structural properties were affected by the new process.

MATERIALS AND COMPOSITES MANUFACTURING

Patterned samples were manufactured using a two-step manufacturing technique. In the first step, the reactive liquid resin containing dispersed CNT was printed into a dry multi-ply textile laminate. The print procedure was followed by curing and consolidation in a hot press to create cured patches. In the second step the material was infused using a variant of vacuum assisted Resin Infusion with Flexible Tooling (RIFT) with a conventional epoxy system. The composites with integrated CNT patches were compared with a reference composites obtained by RIFT alone. Carbon Five Harness Satin (5HS) and Glass Mock Leno (ML) woven composites were chosen as representative woven fabrics. For 5HS composites an extra configuration was considered: patterned with pure epoxy. This allowed distinguishing evolutions of conductivity of carbon fabric associated with consolidation and fiber fraction evolution from the added effect of CNT presence. This made five material configurations: (1) 5HS-I: reference composite infused using RIFT, (2) 5HS-P: with a pattern of patches printed using the same resin as for the subsequent RIFT, (3) 5HS-CNT: carbon composite with a pattern of CNT-enhanced resin, (4) ML-I: the reference glass composite infused using RIFT, (5) ML-CNT: glass composite with CNT-enhanced resin patches.

Materials

Two types of reinforcement were used in this study:

- Carbon woven 5-harness satin (5HS) fabric (Moldlife A0277/000, with an areal density of 285 g/m^2 per ply, 3 K yarns and a yarn spacing of 1.43 mm, Toray T300B fibers), the preforms were constructed of 18 plies laid-up in the warp direction.
- E-glass woven Mock Leno (ML) fabric (Hexcel HexForce 00695) with yarn spacing ranging from 1.2 mm to 2.4 mm, areal density 223 g/m^2. The fabric is sparse and has large inter-yarn openings of approximately 1.2 by 1.2 mm size. The glass preforms were constructed of 14 plies to have a comparable thickness with carbon samples.

The patterns of these fabrics, produced by WiseTex software [17], are shown in Fig. 2.

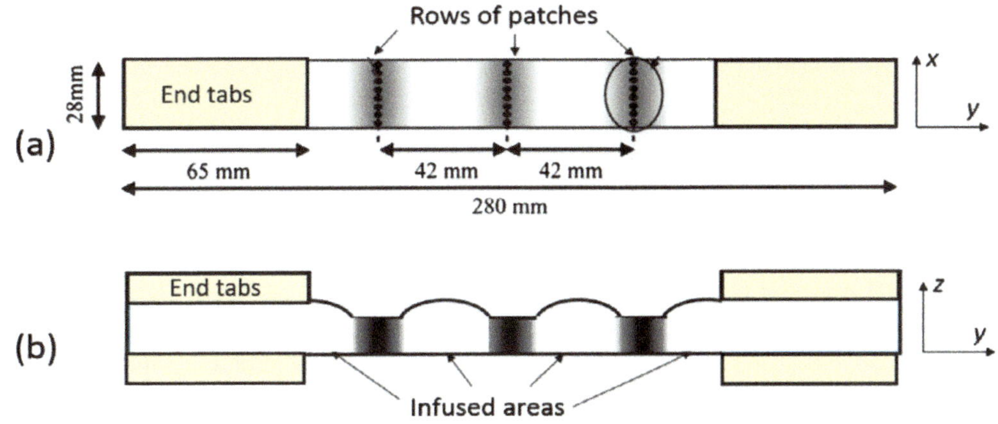

Figure 1. Dimensions and print patterns of specimen for tensile tests: (a) in-plane, (b) out-of plane (thickness variation in *y–z* plane is due to the difference in consolidation pressure in infusion and injection processes). (For interpretation of the references to color in this figure legend, the reader is referred to the web version of this article.)

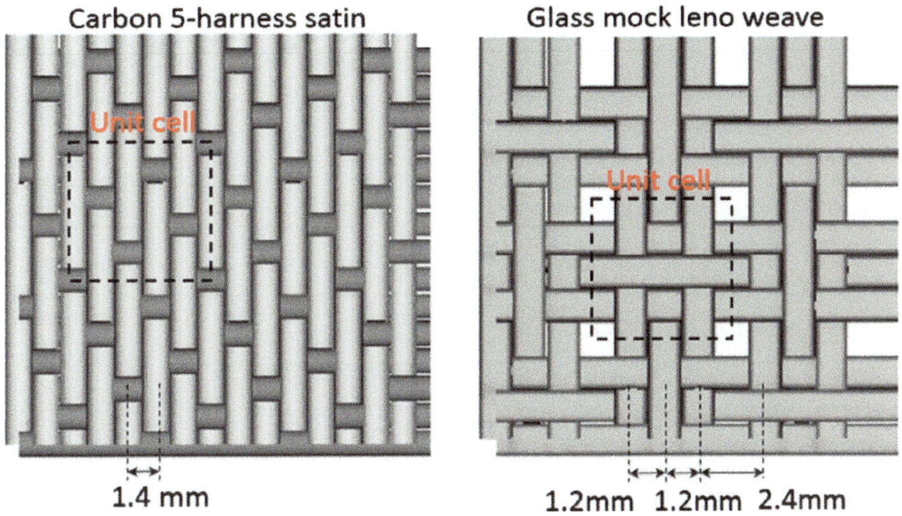

Figure 2. Architecture of two fabrics used in the study. The models are produced by WiseTex software [17]. (For interpretation of the references to color in this figure legend, the reader is referred to the web version of this article.)

Baseline matrix system used for the infusion and printing was Prime 20 (by Gurit) – low viscosity system specifically designed for liquid molding process. The resin was degassed for 4 h prior to mixing with hardener and around 20 min after it was mixed. The CNT enhanced resin was prepared using Nanocyl master batch Epocyl NC R128-02 (3% CNT weight fraction as suggested by Korayem et al. [18]), which was based on Bisphenol-A resin. The resin was diluted in the ratio 1:2, which exceeded the ratio recommended by manufacturer by a factor of three. As a carrier, Araldite LY 1564 resin with Hardener XB 3486 produced by Huntsman were used. The CNT dispersion procedure was arranged as follows: (a) mixing the resin with master batch in a high shear mixer Silverson L5M for 13 min at 3500 rpm, (b) pulse sonication (using high intensity VCX-750 Sonicator) of the obtained solution with 5 s on/3 s off program for 15 min and inputting energy of 852 J/g, (c) the resin was then degassed for 1 h, (d) the degassed resin was then mixed with hardener and, finally, (d) degassed for another 15 min before being injected. In general, this mixing procedure is common in the context of liquid molding procedures and allows achieving an acceptable level of CNT dispersion.

Injection Procedure

The print process was realized by means of a novel liquid resin print rig. It was built by modifying the RepRap Mendel 3D printer [1] where a thermoplastic extruder was substituted by a syringe holder (standard medical 10 ml syringes could be fitted, 21G needles with an outer diameter of 0.8 mm and inner diameter of 0.5 mm were used in this study). The printer has three translational degrees of freedom to move the syringe's needle and an independent controller enabling the injection of the specified amount of resin at a required speed and quantity. The rig is capable of implementing programed injections in any position both in-plane and through-thickness of a preform. The injection speed was set to be 2 ml/min. The pause time, set after every injection to let the resin flow, was 3.5 s.

The number of injections per thickness was chosen to have at least one injection per ply, i.e. 19 injections for 18-ply carbon preform and the same number of injections in the glass preform with fewer plies. The step between injections was chosen based on a bespoke measurement of preform thickness using the injection rig. To do that the needle was inserted all the way through the preform down to the tool side and then incrementally taken out with a fixed step until the needle tip could be seen. The needles were significantly tapered – 3 mm long which helped to minimize the damage to the fibers when inserting the needles but created a difficulty in impregnating the bottom ply. To avoid a non-uniform impregnation, the preform was placed on a soft silicon pad and the needle at its lowest position was inserted half-way through the taper into the silicon. The preforms were wrapped in a loose film and constrained by the tape to avoid unintentional movement when the preform was moved from the printer to the press where the patches were consolidated. This film did not aim at mitigating against preform deformation during consolidation and infusion. After the patches were cured the preform did not require any further constraining and presented integral interconnected material which could not be disintegrated or change dimensions in the following manufacturing processes. In a prospective print process film wrapping will not be necessary if the preform is consolidated in-situ. Rectangular 125 × 300 mm preforms were used for most of the injection and infusion procedures which in general allowed producing three tensile samples out of each plate.

The need to study the role of the interface dictated the simplest print pattern – two/three rows of patches were printed across the sample width – Fig. 1. Patches within the row were placed sufficiently close to be merged and form a continuously printed rib so that the loading in the tensile test is normal to the interface of the rib. The parameters of the injections, patches and pattern are given in Table 1. All the CNT-patterned samples were produced using the program A. For the only non-CNT patterned configuration (5HS-P) a slightly different program (B) was applied, which resulted in a larger patch size. This program was implemented for technical reasons but not to study the effect of injected volume on the mechanical performance.

Table 1. Parameters of injection program.

Parameters of each injection Program A/Program B		Parameters of patch Program A/ Program B		Parameters of pattern Program A/ Program B	
Injection step in through-thickness direction, mm	0.7	Distance between patches in a row, mm	7	Distance between the rows, mm	42
Number of injections per thickness	19	Number of injections in a row	15	Number of rows	2/3
Resin volume per injection, μl	8/6	Total injected volume per patch, μl	152/114	Row width/patch diameter, mm	24/18
Pause time after the injection, s	3.5				

The printing process was followed by a consolidation procedure, targeted at curing the patches and suppressing porosity. Since the injection flow is not vacuum assisted and driven by a relatively small pressure gradient, the flow was highly unsaturated and entrapped a noticeable fraction of voids behind the flow front. Upon applying pressure these voids tended to form micro intra-yarn porosity in dense fabrics and large inter-tow pores in open fabrics with large intra-yarn spacing. A consolidation procedure was set to improve the quality of the patches. The preform with injected patches was heated in oven at 60°C for 30 min prior to application of the consolidation pressure. This transformed the matrix to a state where it gelled but was not fully cured. In a follow-up compaction the high viscosity of resin reduced the flow lengths and prevented squeezing the resin out of the patch boundaries. Instead, the resin carried the major fraction of applied pressure which was sufficient to suppress the voids in the patch center. In the current version of the process all the patches were cured and consolidated at once in a hot press with pressure exceeding 6 bar applied for 5 h at 60°C. For an industrial environment, it would be possible to modify the process by adopting fast curing resins and applying local tooling for consolidation to make consolidation almost instantaneous but this discussion goes beyond the scope of the paper.

Infusion Procedure

The patterned rectangular preforms were impregnated in a resin infusion process with flexible bagging. The rectangular preforms were infused across their long sides. The injection port was placed outside the preform delivering resin directly to a distribution spring oriented along the preform. The distribution mesh was placed on the top and bottom of the composite preform conducting the resin in the width direction. Another spring was placed parallel to the first one on the other side of the preform to collect the propagated resin. A uniform flow front was achieved and the infusion was completed in roughly 20 min. Infused part was then cured in the oven at 50°C in 16 h in line with the specification of resin manufacturer. The same resin infusion procedure was applied to the reference and patterned samples.

Features of Patterned Composite

The micrography and dimensional assessment of patterned panels revealed the following features:

Thickness Variation

Thickness variation in the carbon preform between the patches (consolidated at high pressure) and the infused area (consolidated under vacuum bag) varies by roughly 15–20% – Table 2. The bottom surface, which was in contact with the tool, was flat while the top surface has acquired three (or two in the case of CNT patching) waves of a regular undulation – Fig. 1b. In the glass ML preform the thickness variation between the infused and printed patches was considerably lower due to the higher nesting of the plies and a weaker spring-back reaction of preform after consolidation.

Table 2. Thickness of patterned and infused samples.

Samples	5HS-I carbon infused	5HS-P carbon patterned without CNT	5HS-CNT carbon patterned with CNT	ML-I glass infused	ML-CNT[a] glass patterned with CNT
Patch region, mm	n/a	4.6 ± 0.2	4.6 ± 0.1	n/a	4.2 ± 0.2
Infused area, mm	5.3 ± 0.2	5.5 ± 0.3	5.3 ± 0.1	5.1 ± 0.1	4.8 ± 0.1

[a]The provided data for ML-CNT are measured after polishing the patch area.

Impregnation Quality and Patch Interfaces

On sample surface, the rows of patches formed well-defined regions of rectangular shape spanning across the samples width with darker sub-domains showing regions with higher fraction of CNT resin –Fig. 3b and c.

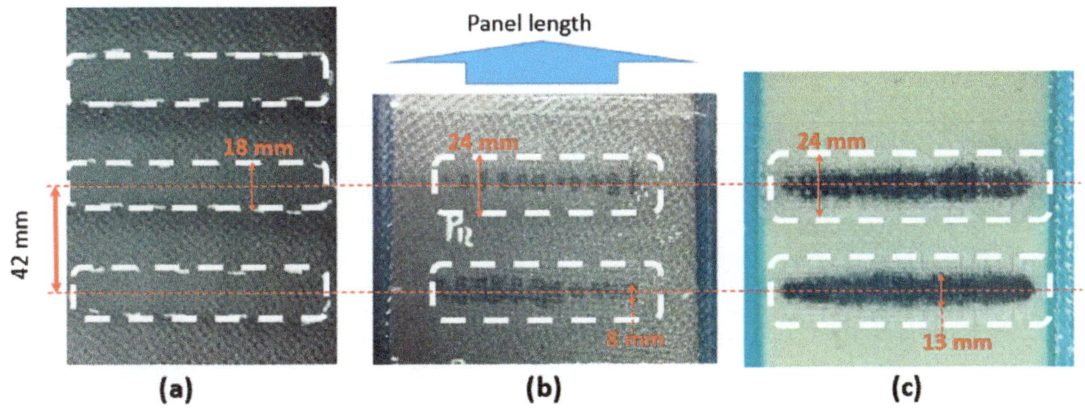

Figure 3. Rows of patches: (a) Prime 20 resin injected into carbon 5HS preform (Program A), (b) CNT enhanced resin injected into carbon 5HS preform (Program B), (c) CNT enhanced resin injected into glass Mock Leno preform (Program B). White dash contours show the visible external boundaries of the patch row. Darker areas in figures (b) and (c) reveal the CNT rich regions. (For interpretation of the references to color in this figure legend, the reader is referred to the web version of this article.)

In the patch regions a distributed micro-porosity was seen. Few larger isolated pores with a characteristic size under 0.2 mm could be found in the central infused area. The maximum porosity was found on the boundary between infused and printed region with a relatively good impregnation close to injection sites. The primary reason for the micro pore occurrence was the air entrapment behind non-uniform flow front occurring due to the dual scale flow. The consolidation procedure described above in Section 2.2 was aimed at suppressing the voids using compaction of nearly gelled resin. This did help to improve the impregnation quality in the center of the patch but could not fully suppress defects on the patch edges: pressure in resin inevitably decays at the patch boundaries.

Compared to pure epoxy patches the CNT patches revealed higher level of micro-porosity and occasionally the presence of bigger inter-yarn impregnation defects – Fig. 4. This feature was likely caused by sonication of CNT resin and subsequent degassing which struggled to eliminate all the voids. Apart from the non-uniformity of micro-pores no visible interface between the printed and infused regions could be seen – Fig. 5b. It can be suggested that the interface presented a finite volume where injected and infused resins shared a finite space rather than a 2D boundary between the distinctly different regions.

Filtering of CNT

Relatively high fraction of the CNT-enhanced resin led to a clearly observed filtering mechanism. The darker colored region (containing CNT) appeared to be approximately twice as small compared to the overall patch size both in glass and carbon preforms – Fig. 3. This implies that the content of the CNT in the close vicinity of injection is higher compared to the CNT content of injected resin. On the one hand, the length of CNT propagation (~8 mm for 5HS and ~13 mm for ML) could be attributed fabric architecture. The CNT flow length as comparable to the dimensions of unit repeats of textiles, which is 6.9 mm for carbon 5HS and 4.8 mm for glass ML samples. On the other hand, the flow length characterizes the dispersion quality. Fan et al. [19], suggested an experiment where a glass yarn was immersed in a CNT solution. They found that depending on dispersion quality the length for CNT propagation varied between 2 mm (large agglomerates) and 20 mm (fine dispersion).

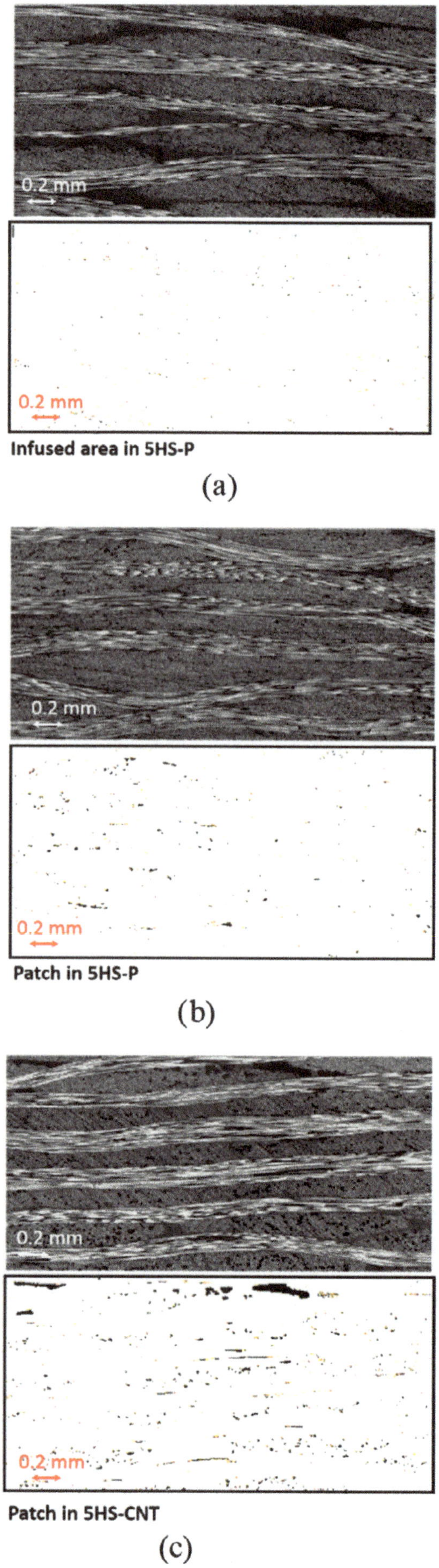

Figure 4. Three representative micrographs: (a) Infused area in patterned 5HS-CNT samples (similar micrographs are seen in 5HS-I and in the infused area of 5HS-P), (b) patch areas in composite patterned with Prime 20 resin (5HS-P), (c): patch areas in composite patterned with CNT-enhanced resin (5HS-CNT). Figures below show the same images with a different contrasts setting to highlight the porosity distribution. (For interpretation of the references to color in this figure legend, the reader is referred to the web version of this article.)

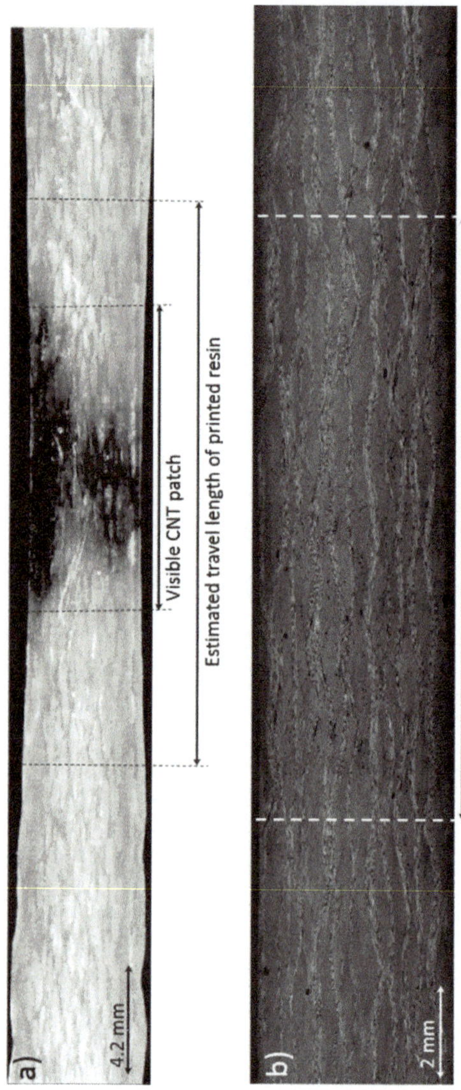

Figure 5. CNT patch in a ML glass preform: (a) Scan of 1.3 mm slice cut in the local proximity of the needle insertion, (b) micrograph of the correspondent CNT patch.

In the settings of a conventional liquid molding process high degree of filtering would be unsatisfactory for the process and would be evidence for insufficient dispersion of the CNT in the solution. In the resin print process, the higher loading may be beneficial for the local conductivity and creating a higher contrast in local material properties. A scan of a thin glass composite slice made in the vicinity of the injection, Fig. 5, revealed that CNT tended to fill inter-yarn space sometimes enveloping the yarns. It also showed a non-uniform distribution of injected resin through thickness which implied that the local inter-connectivity between the layer of liquid resin could be further improved by increasing the number of injections through thickness. The filtering was much more intensive in the carbon samples which may be explained by the fact that it had significantly smaller inter-yarn space and inter-bundle channels.

Samples for Tensile Testing

Glass woven EF1612 prepreg was used for end-tabs. The 8-ply stack was cured in an oven at 130 °C for 15 min in line with the specification of the manufacturer. The resultant thickness was 1.9 mm. The end tabs were glued to each panel and then cured at 80 °C for 2 h and then cut to 3 tensile test samples of 28 mm in width. The sample edges were polished to remove any defects possibly imposed by the diamond saw. At least three samples were prepared for all five materials configurations considered in the study.

TESTING OF PATTERNED COMPOSITES

Conductivity Measurements

In order to measure the conductivity the surface in the location of measurements (in the patch region) was slightly polished. This was required to eliminate a thin insulating layer of resin created during post-printing infusion and covering the patch surface; the conductivity was beyond the sensitivity of measuring device when this operation was not performed. Thickness measurements conducted on selected samples showed that polishing removed about 50 μm on each side. The authors believe that this was mostly a surface layer of resin left by the distribution mesh. Some fibers at surface could be affected by polishing but their fraction would be low to have any pronounced impact on the mechanical properties due to the considerable thickness of the laminates.

Once polished the surface was cleaned and a thin layer of conductive silver-based glue (L100 from Kemo Electronic) was applied on top and bottom sides of patched region to improve the contact between the electrodes and the composite surface. The diameter of the conductive spot was chosen to be 6 mm as it was comparable then to a characteristic CNT rich zone within the patch. The resistance was measured using a multimeter tester UNI-T UT33D and the conductivity was calculated by reversing the resistivity value and taking into account the spot area and the local composite thickness in the studied location. The conductivity was also measured in the syringes where resin remained after the injections. The cylindrical samples of CNT resin cured after injections at room temperature were cut from syringes to measure their conductivity. The top and bottom surfaces were polished and the conductive silver glue spots of the same dimensions as for the composite samples were applied on flat surfaces. The results are shown in Table 3.

Table 3. Conductivity of patterned samples and injected resins.

Fibers	T300B carbon		T300B carbon		E glass		No fibers	
Material	5HS-CNT (Batch 1)		5HS-P		ML-CNT (Batch 2)		CNT-epoxy resin (from syringes)	
Location	Row 1	Row 2	Row 1	Row 2	Row 1	Row 2	Batch 1	Batch 2
Thickness, mm	4.41	4.50	4.45	4.45	4.21	4.07	13.6	12.7
Resistance, Ω	2.1	2.5	2.6	2.1	$14.7 \cdot 10^3$	$15.1 \cdot 10^3$	$175 \cdot 10^3$	$8300 \cdot 10^3$
Conductivity, S/m	74.31	63.69	61.79	74.98	0.0101	0.0095	0.0027	0.0001

The through-thickness conductivity of the carbon samples with no CNT was high and the presence of CNT patches did not affect the resistance. The obtained values were comparable but somewhat higher than measured for flat laminates [20], which can be explained by higher alignment of crimped textile yarns with the measurement direction. The electrical conductivity of the reference non-functionalized glass samples was beyond the sensitivity of the tester and can be considered practically non-conductive. The conductivity was significantly improved through adding CNT.

Their absolute values were still significantly lower than that of the carbon composite. However, they were significantly (1–6 orders of magnitude) higher than the values previously reported elsewhere for CNT enhanced glass composites obtained with liquid molding technologies [10], [11] and [12] and comparable to the conductivity of the glass composite with CNT grown on fiber surface [5]. Comparing the through-thickness properties of glass composites to the correspondent value for the injected resin gave an insight on the reason for this unusually high performance. Against obvious expectations, the conductivity of glass composite appeared to be considerably higher – Table 3. The increase in conductivity was most likely related to the intensive filtering of carbon nano-tubes around the injection spot as evidenced by Figs. 3c and 5a. Filtering left much higher fraction of the CNT in the vicinity of the location where conductivity was measured. Apart from this direct optical observations it is suggested that the other factors could play a role as well:

- The orientation of the CNT by the resin flow in the needle helped to align the CNT in the thickness direction, and created a better percolation network compared to the reference resin sample where CNT orientation was random. A positive role of CNT orientation is known from the studies where alignment of nano reinforcement was implemented using electric or magnetic fields [21], [22] and [23].
- The resin in glass composite was compacted after gelling which could well lead to a squeezing flow and, as a result, to reconfiguration and merging of the CNT clusters whereas the resin in syringes was cured in oven without any pressure imposed.
- The needle insertion left traces in the preforms: thin channels comparable in dimensions of the needle – 0.5 mm [1]. This could further increase the probability of forming a continuous network in the through-thickness direction.

Mechanical Testing of Patterned Composites

The tensile tests were performed on a 250 kN Instron hydraulic machine. La Vision 3D Digital Image Correlation (DIC) system was used to assess the difference in strain distribution due to the variation in thickness and fiber volume fraction and measure the applied strain field calculated as an average strain over the sample surface. The flat sample sides were speckled by black spray on uniformly painted white background. The field view covered the area of ~512 pixels (28.0 mm) along the sample width and ~2010 pixels (110 mm) along the length. The characteristic speckle size was around 0.5 mm. The displacement was determined on the subset size of 17 pixels (equivalent to 0.9 mm) and the step between the subset was chosen to be 8 pixels (equivalent to 0.4 mm). The strains were measured both in the smooth patch and rough infused areas. The deformation of rough surface could contribute to the noise however, the use of stereo strain measurements helped to eliminate the impact of the out of plane roughness on the measured signal.

In further discussion the results of the tensile tests are shown in load–strain diagrams as opposed to traditional stress–strain presentation of test results. This was more appropriate when comparing samples with thickness variation and where the load normalized by cross-section area differed along the sample length. It is important to note that the amount of fibers in any cross-section along the length of the sample was the same in all the compared patterned vs. reference samples.

Carbon Composites

The carbon samples exhibited a linear load–strain response till failure. Failure of the reference infused 5HS samples occurred in the grips at applied strain in the range of 1.1–1.2%. The patterned samples ruptured predominantly in the patterned area (6 out of 8) with few samples failing in the interface region (2 out of 8 samples). Incorporation of the CNT did not seem to have a massive impact on damage mode or load/strain at failure –Fig. 6. However, there was a slight increase in average strain at failure – Table 4. The failure mode in all the cases was brittle and the fiber rupture zone was localized in a 5–10 mm zone showing traces of highly localized delamination's around the fracture (possibly occurring after the initial fiber rupture).

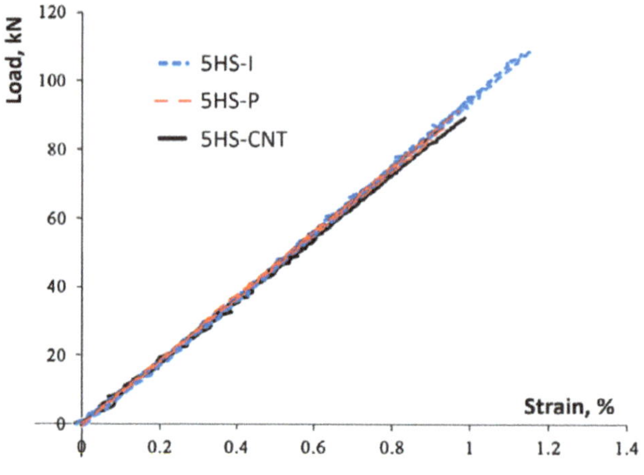

Figure 6. Mechanical response of the patterned 5HS-P and 5HS-CNT (failed in patch region) and reference/infused 5HS-I samples (failed in the grips). (For interpretation of the references to color in this figure legend, the reader is referred to the web version of this article.)

Table 4. Mechanical properties of tested samples.

Material	5HS-CNT	5HS-P	5HS-I	ML-CNT	ML-I
Load at failure, kN	88.2 ± 4.0	84.4 ± 3.8	104.3 ± 4.7	28.3 ± 0.8	32.0 ± 0.9
Strain at failure, %	0.96 ± 0.03	0.91 ± 0.04	1.10 ± 0.04	1.17 ± 0.03	1.23 ± 0.03
Data normalized with respect to the patch cross-section area					
Young's modulus, GPa	72.0 ± 0.8	71.8 ± 1.1	n/a	20.1 ± 0.4	n/a
Data normalized with respect to the cross-section of infused area					
Young's modulus, GPa	62.1 ± 1.6	61.0 ± 1.6	64.1 ± 0.8	17.8 ± 0.2	17.6 ± 0.3

The surface strain measurements revealed three distinctly different regions in the patterned samples: the infused region, the patched region and an interface between them – Fig. 7a. The highest strain, averaged over the examined region, was observed in the infused area and the lowest at the patch interfaces. The strain variation was pronounced – at failure the average strain in the infused region reached 1.2% whereas at the interfaces the strain hardly exceeded 0.8% reaching 40% difference. The average strain at failure in the patched region was 0.9–1.0%. Despite the high strain contrast, the failure location did not correlate with the location of maximum strain often occurring in the region of lower strain. Hence, in this case the strain in the damaged region could not be used as an indicator of failure risk.

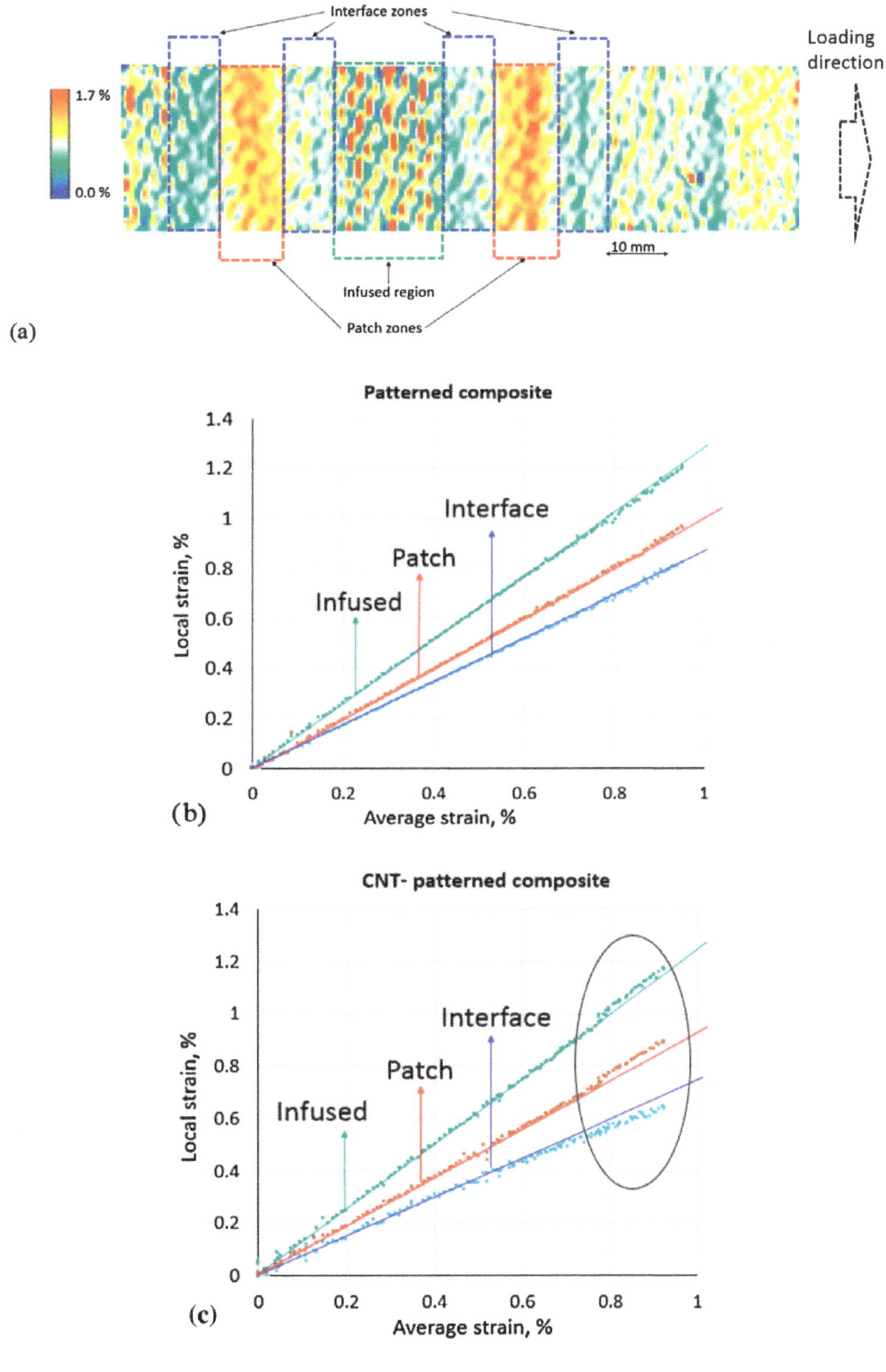

Figure 7. (a) Characteristic strain distribution in a patterned sample before failure (in loading–horizontal direction), (b) local strain history in the patch, infused and interface areas in composites patterned by Prime 20 resin (b) and CNT enhanced resin, (c) straight solid lines show the trends. Kinks (patched and infused zones) and non-linearity (interface zone) can be seen at ~0.8% in the CNT composite. Straight lines show linear trends and deviations from them. (For interpretation of the references to color in this figure legend, the reader is referred to the web version of this article.)

The infused region exhibited a much higher gradient of local strain clearly revealing the textile pattern, whereas strain in the patched region was much more uniform. This can be explained by the fact that the infusion process where the preform, being constrained by a flexible bagging, left a textile imprint on the composite surface, whereas the patterned region, consolidated under the press, tended to be much smoother. In addition to that the patch areas were slightly polished to remove the thin surface layer of resin for measuring conductivity.

The high variation of strain over the infused-printed areas is likely to be caused by the thickness variation which created additional local bending effects superimposed with the applied tension. Local strain in 90° weft yarn at failure load approached 1.7%. However, the average strain response in all the three regions appeared to be linear till failure for the case when patched and infused resin was the same. In CNT-enhanced patches at ~0.8% a non-linearity in the interface region and small but visible kinks were observed in the infused and patched zones, Fig. 7b, pointing at a damage event. To investigate the residual deformations indicating irreversible deformations or damage, several samples were loaded till 77 kN (90% of the apparent strength of the samples) and then unloaded at 1.5 mm/min while tracking the surface strain evolution. The residual strains, most likely localized in the 90 yarns, reached the value of 0.3% compared to the reference state prior to testing.

Glass Composites

The glass samples revealed a non-linear response related to intensive damage accumulation. Deviation from linear response started at approximately 0.25% of applied strain – Fig. 8a. The initial stiffness and the strain at the kink are the same for reference infused ML-I and patterned ML-CNT samples. The secant stiffness after the kink was slightly lower for the patterned samples pointing at higher intensity of damage events in the patterned sample. Both the sample types experienced the onset of fiber rupture at approximately the same applied strain level (1.23% for ML-I and 1.17% ML-CNT) but load at this drop is slightly higher for the reference samples (32.0 kN for ML-I and 28.3 kN for ML-CNT) – Table 4.

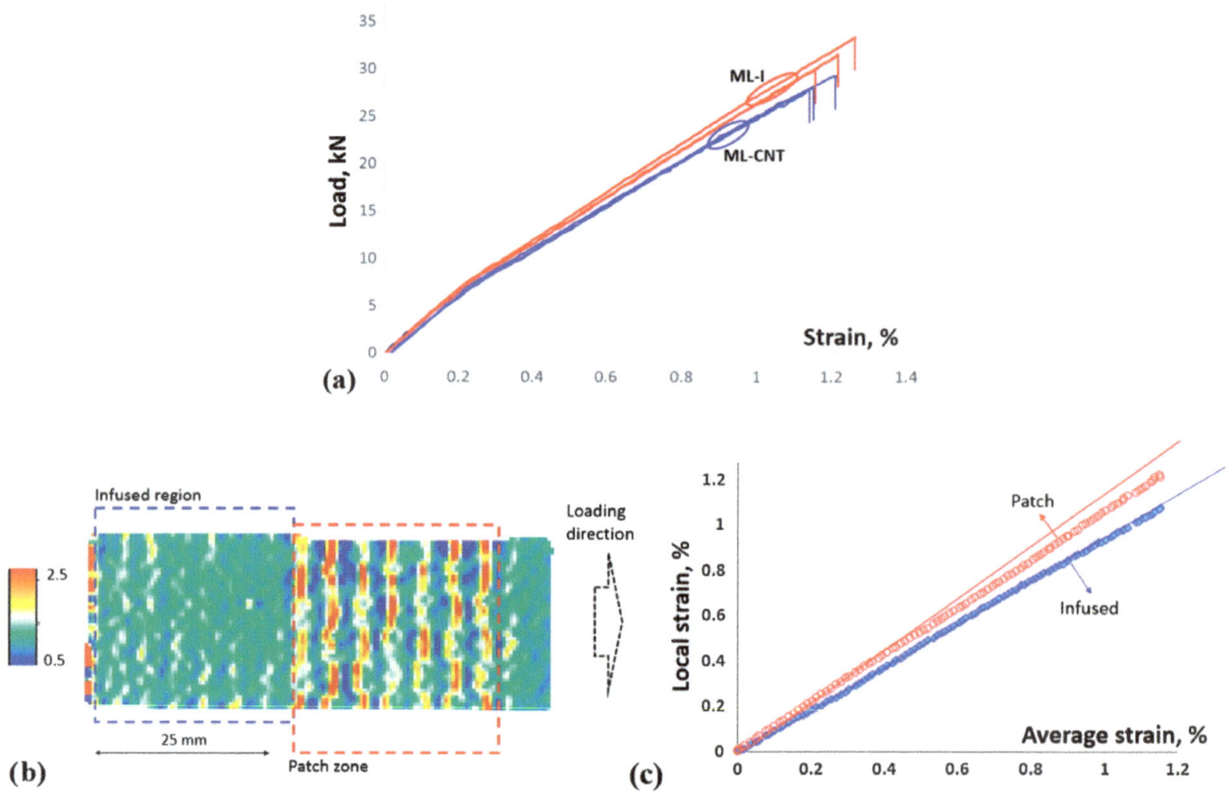

Figure 8. (a) Mechanical response of the patterned/patched ML-CNT (failed in patch region) and reference ML-I (failed in the grips) samples, (b) characteristic strain distribution in a patterned ML-CNT before failure (in loading–horizontal direction), (c) local strain history in the patch and infused areas in ML composites (straight lines are showing linear trends and a deviation from them). (For interpretation of the references to color in this figure legend, the reader is referred to the web version of this article.)

All but one ML-I failed in the grips and only one patterned sample failed outside the central region. The patterned samples revealed two different failure modes: in one case the fiber rupture started in CNT patch and propagated instantly through its thickness. In another case, fiber rupture in CNT patch spread over half of its thickness at which point a delamination developed and ran towards the second patch. There it stopped and fiber rupture continued in the second patch though the other half of the sample – Fig. 9. This failure mode resulted in two kinks on the load–displacement curve delaying the ultimate disintegration of the sample. Interestingly the apparent local strength, calculated as the failure load normalized by the thickness at the fiber rupture location, appeared to be noticeably higher for the CNT-enhanced samples: 244 MPa against 224 MPa.

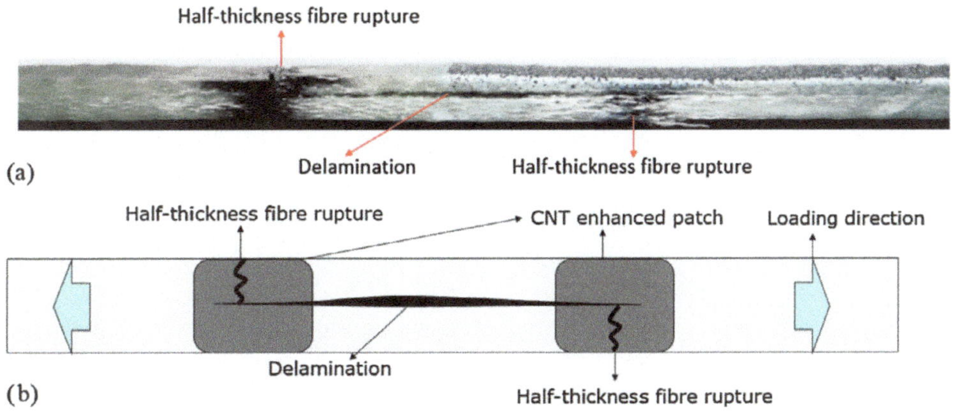

Figure 9. Side view of glass ML-CNT samples with delamination connecting two patches with half ruptures: (a) sample picture, (b) sketch highlighting the mechanisms of crack development. (For interpretation of the references to color in this figure legend, the reader is referred to the web version of this article.)

Analysis of surface strain distribution revealed two distinctly different region – in the patch and infused zones – Fig. 8b. Unlike what was seen in the carbon samples, the strain in the patch area was higher than in the infused region – Fig. 8c. There were two counteracting factors that could affect the surface strain distribution and lead to this difference. These were: (a) the contrast in stiffness between the patched and infused areas and (b) the bending effects associated with the thickness variation.

These effects had different impact on the strain in the glass and carbon samples. The bending effects in the carbon panels were more prominent due to greater thickness variation. On the flat side of the sample, where the measurements were taken, compressive strains associated with bending occur in the patch area. These strains are superimposed with the tensile strain resulting in the lower deformation on the patch surface. In contrast, the factors related to the stiffness of matrix, such as the presence of intra-yarn cracks, micro porosity, or nano additives, were more explicit in the glass samples. The impact of the latter became evident from the stiffness degradation trends. In the carbon–epoxy systems the material response was dominated by fibers and hence, linear till failure. In the glass–epoxy composites the contribution of non-fiber related factors was more distinct: for instance, intra-yarn damage accumulation led to a prominent reduction in stiffness.

The CNT's on the one hand and pores/cracks on the other hand had competing positive and negative impacts on patch stiffness in the glass samples. The fact that strain was higher in the patch area in the beginning of the test suggested that the porosity could have a larger influence on patch stiffness than the presence of CNT. The evolution of strain provided the insight on damage accumulation. The strain in the infused region remained strictly proportional to the applied strain whereas the rate of strain in the patch region was slowing after ~0.4% of applied strain. A deviation of the strain in the patch from linear trend indicated that damage accumulation was more intense in the infused area.

The optical observation of light transmitted through the sample performed in-situ during the test showed progressive of accumulation of damage seen as white areas on the images – Table 5. The most intensive occurrence of these zones was observed in the transition from patch to infused zones. The damage accumulation was localised as compared to the reference samples which exhibited a uniform evolution of its color throughout the test.

Table 5. Failure pattern at various levels of strain as seen in the transmitted through thickness light (tensile load is the vertical direction, full width of sample in the horizontal direction is shown, the scale bars shown in left column are applicable to all the images).

Strain, %	0.15	0.35	0.8	1.1
ML-CNT				

ML-I				

Micrographs of fractured samples away from the fiber rupture zone revealed different damage patterns in the patches and infused areas – Fig. 10. In the infused areas an intensive intra-yarn cracking was observed reaching up to 2–3 cracks per mm in some yarns. Few local isolated delamination's could also be observed. In the porous transition region, the intensity of micro-cracking was significantly lower, however, frequent intra-yarn in-plane cracks connecting chains of micro pores spanning through the entire yarn were found. These cracks were clearly seen as white areas occurring in the transmitted light images at the advanced deformation stages – Table 5. In the patch centre, i.e. CNT rich region, the intensity of intra-yarn cracking was lower (less than 1 crack per mm of a ply). It is interesting to note that this is in agreement with the observations of de Greef et al. [24] and [25] who found that the presence of CNT hinder the intra-yarn crack density accumulation.

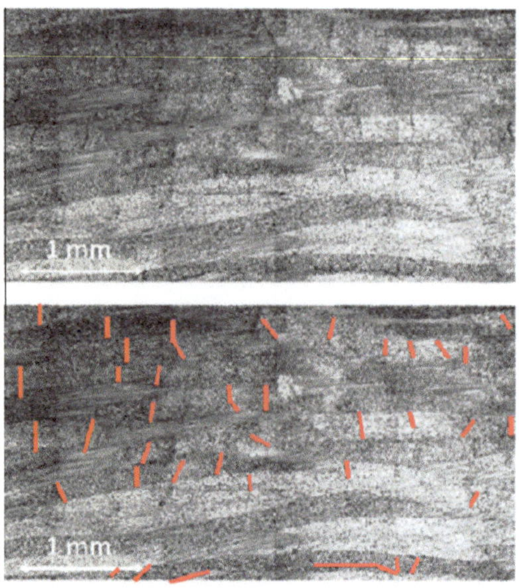

(a)

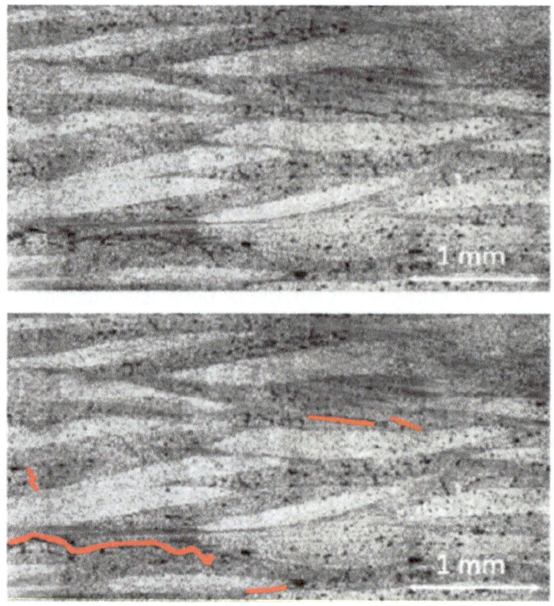

(b)

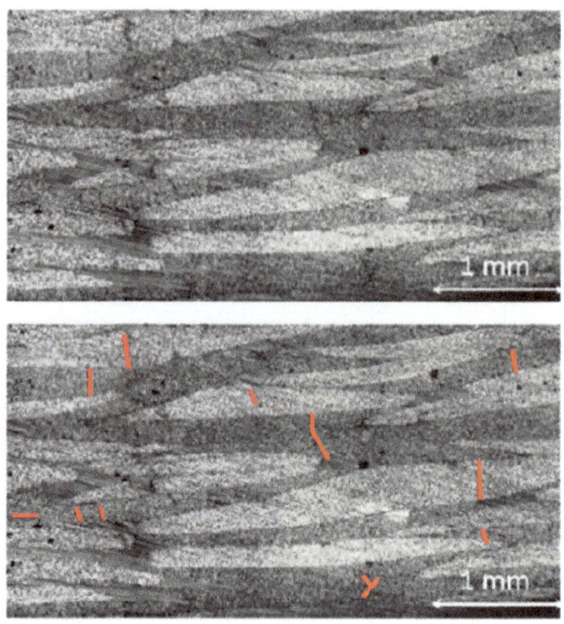

(c)

Figure 10. Representative micrographs of the ML sample brought to failure (cross-sections are at 45° to warp yarns): (a) infused area, (b) transition area, (c) patch center. Images below highlight crack distribution (in red). (For interpretation of the references to color in this figure legend, the reader is referred to the web version of this article.)

DISCUSSION

The tensile test revealed several important features in the behavior of patterned composites:

- The load in tensile test was applied perpendicular to the inter-patch interface, thus revealing the greatest impact that an interface may have on composite properties. Knowing the interface issues in other materials, it could be suggested that breaking integrity of a hypothetical crack-like interface should make a composite more compliant. However, for the considered systems no noticeable effect on stiffness of the patched composite was seen. Only a minor knock-down on secant modulus after failure initiation was detected in glass composites. As evidenced by the micrographs this compliancy was related to higher density of intra-yarn flat cracks in the plane of the composite panel.
- An early failure was observed in all the patterned samples which should most likely be attributed to the thickness variation and as a result to additional bending stress occurring in fibers. The patterned material failed systematically in the patch or in the interface regions with one sample showing the potential to confine macro delamination in between two rows of patches. Apparently the delamination-like flat cracks, initiated at the interface, triggered the macro-delamination's which propagated instantly between the two patches but was suppressed when entering another patch. This peculiar failure mechanism suggests that patterning may offer a new instrument for manipulating and controlling damage events. However, this failure mode was observed only once and more needs to be done to understand how to control this process consistently.
- Surface strain measurement revealed distinctly different regions pointing at various deformation and failure mechanisms in the patterned material. Partially this can be attributed to the waviness of the sample and bending deformation superimposed with the tensile stress. Apart from purely elastic effects, the analysis of the damage pattern in the glass samples indicated different failure mechanisms in the infused (intense intra-yarn matrix cracking), interface (intra-yarn crack in the plane of the sample) and patch areas (low crack density). The patch had both positive and negative effect on crack dynamics. On the one hand, the overall crack density appeared to be lower in the CNT enhanced region. On the other hand, the patch interface reveals a higher level of porosity which promotes a new type of cracking in that region that can trigger the delamination.

CONCLUSIONS

A novel method for functionalizing and patterning textile composites was demonstrated. It was shown that insulating glass composite can be made electrically conductive. The level of conductivity is drastically higher than what could be achieved by any previous liquid molding technologies. The major challenges of the functionalization in conventional processes, such as filtering of conductive particles, play to the benefit of the liquid resin print process since the volume fraction of CNT in the injected location increases and hence, a better conductive network can be formed. As opposed to other techniques, the process offers a possibility to add functional properties locally and in any pattern or configuration, which may be defined by requirements to structural or functional performance. This minimizes the interference to the subsequent infusion process and reduces the impact of CNT presence on the component performance.

This study showed examples of structures obtained with a novel manufacturing method which exhibited peculiar features: micro-porosity, thickness variation, internal boundaries, non-periodic pattern of the reinforcement, interaction of distinctly different matrices, etc. The impact of these features on the deformation and failure mechanisms was explored. It was found that the material modification had minor effect on the early stages of deformation but may affect (both positively and negatively) the damage accumulation process and the failure of the material. The main negative factor is earlier fiber failure caused by thickness variation. This feature however, as discussed in [1], has a potential of being controlled by organizing the structure (e.g. fiber bridging between the patches in quasi-isotropic structures) or modifying manufacturing process (equalizing pressure in post-injection consolidation and subsequent liquid molding process for instance by using rigid mold RTM). Summarizing the output of the mechanical testing we can claim that the print process creates a very promising method for creating structural components with added capabilities.

ACKNOWLEDGEMENTS

The work of Yann LeCahain and Alexandre Dattin was supported by FP7 Marie Curie CIG InterCom project 304062: "New inter-scale techniques for novel composite architectures". The work of Dmitry Ivanov and Surush Arafati was supported by EPSRCGrant EP/M009149/1: "New generation of manufacturing technologies: liquid print of composite matrices". The work of Andrey Aniskevich was supported by the Latvian State Research Program "IMATEH".

REFERENCES

1. Ivanov DS, White JAP, Hendry W, Mahadik Y, Minett V, Patel H, et al. Stabilizing textile preforms by means of liquid resin print: a feasibility study. Adv Manuf: Polym Compos Sci 2015;1:26–35.
2. Thostenson ET, Chou TW. Carbon nanotube networks: sensing of distributed strain and damage for life prediction and self-healing. Adv Mater 2006;18:2837–41.
3. Starkova O, Mannov E, Schulte K, Aniskevich A. Strain-dependent electrical resistance of epoxy/MWCNT composite after hydrothermal aging. Compos Sci Technol 2015;117:107–13.
4. Lin Y, Gigliotti M, Lafarie-Frenot MC, Bai J, Marchand D, Mellier D. Experimental study to assess the effect of carbon nanotube addition on the through-thickness electrical conductivity of CFRP laminates for aircraft applications. Compos B 2015;76:31–7.

5. Gurkan I, Cebeci H. Electrical and mechanical property investigation of fuzzy fiber-reinforced composites. In: Proceedings of 20th International Conference on Composite Materials, Copenhagen 19–24th July 2015.
6. Zhang J, Zhuang R, Liu JMader, Mader E, Heinrich G, Gao Sh. Functional interphases with multi-walled carbon nanotubes in glass fiber/epoxy composites. Carbon 2010;48:2273–81.
7. Gao L, Chou T-W, Thostenson ET, Godara A, Zhang Z, Mezzo L. Highly conductive polymer composites based on controlled agglomeration of carbon nanotubes. Carbon 2010;48(9):2649–51.
8. Herceg T, Greenhalgh E, Bismarck A, Shaffer M. High performance nano and hierarchical composites. In: Proceedings of 20th International Conference on Composite Materials, Copenhagen 19–24th July 2015.
9. Zhang H, Liu Y, Kuwata M, Bilotti E, Peijs T. Improved fracture toughness and integrated damage sensing capability by spray coated CNTs on carbon fiber prepreg. Composites Part A 2015;70:102–10.
10. Wichmann MHG, Sumfleth J, Gojny FH, Quaresimin M, Fiedler B, Schulte K. Glass-fiber-reinforced composites with enhanced mechanical and electrical properties – benefits and limitations of a nanoparticle modified matrix. Eng Fract Mech 2006;73:2346–59.
11. da Costa EFR, Skordos AA, Partridge IK, Rezai A. RTM processing and electrical performance of carbon nanotube modified epoxy/fiber composites. Composites Part A 2012;43:593–602.
12. Domingues D, Logakis E, Skordos AA. The use of an electric field in the preparation of glass fiber/epoxy composites containing carbon nanotubes. Carbon 2012;50:2493–503.
13. Böger L, Wichmann M, Meyer L, Schulte K. Load and health monitoring in glass fiber reinforced composites with an electrically conductive nanocomposite epoxy matrix. Compos Sci Technol 2008;68(7–8):1886–94.
14. Thostenson ET, Gangloff JJ, Li CY, Byun JH. Electrical anisotropy in multiscale nanotube/fiber hybrid composites. Appl Phys Lett 2009;95(7). 073111-1-3.
15. Lomov SV, Gorbatokh L, Houlle M, Kotanjac Z, Koissin V, Vallons K, et al. Compression resistance and hysteresis of carbon fiber tows with grown carbon nanotubes/nanofibers. Compos Sci Technol 2011;71:1746–53.
16. Aravand M, Lomov SV, Verpoest I, Gorbatikh L. Evolution of carbon nanotube dispersion in preparation of epoxy-based composites: from a masterbatch to a nanocomposite. Express Polym. Lett. 2014;8(8):596–608.
17. Verpoest I, Lomov SV. Virtual textile composites software Wisetex: integration with micro-mechanical, permeability and structural analysis. Compos Sci Technol 2005;65(15–16):2563–74.
18. Korayem AH, Barati MR, Simon GP, Zhao XL, Duan WH. Reinforcing brittle and ductile epoxy matrices using carbon nanotubes masterbatch. Compos A 2014;61:126–33.
19. Fan Z, Hsiao K-T, Advani SG. Experimental investigation of dispersion during flow of multi-walled carbon nanotube/polymer suspension in fibrous porous media. Carbon 2004;42:871–6.
20. Todoroki A, Tanaka M, Shimamura Y. Measurement of orthotropic electric conductance of CFRP laminates and analysis of the effect on delamination monitoring with an electric resistance change method. J Compos Sci Technol 2002;62:619–28.
21. Ma C, Zhang W, Zhu Y, Ji L, Zhang R, Koratkar N. Alignment and dispersion of functionalized carbon nanotubes in polymer composites induced by an electric field. Carbon 2008;46(4):706–20.
22. Domingues D, Logakis E, Skordos AA. The use of an electric field in the preparation of glass fiber/epoxy composites containing carbon nano-tubes. Carbon 2012;50:2493–503.
23. Choi ES, Brooks JS, Eaton DL, Al-Haik MS, Hussaini MY, Garmestani H. Enhancement of thermal and electrical properties of carbon nanotube polymer composites by magnetic field processing. J Appl Phys 2003;94(9):6034–9.
24. De Greef N, Gorbatikh L, Godara A, Mezzo L, Lomov SV, Verpoest I. The effect of carbon nanotubes on the damage development in carbon fiber/epoxy composites. Carbon 2011;49(14):4650–64.
25. De Greef N, Gorbatikh L, Lomov SV, Verpoest I. Damage development in woven carbon fiber/epoxy composites modified with carbon nanotubes under tension in the bias direction. Compos A 2011;42(11):1635–44.

CHAPTER 8

MICROMECHANICAL MODELLING OF DEFORMATION AND FRACTURE OF HYDRATING CEMENT PASTE USING X-RAY COMPUTED TOMOGRAPHY CHARACTERIZATION

Mingzhong Zhang and Andrey P. Jivkov

Keywords: B. Fracture; B. Microstructures; B. Mechanical properties; C. Micro-mechanics; X-ray computed tomography

INTRODUCTION

Cement composites are the most popular and widely used construction material in the world. The understanding and prediction of elastic properties and fracture behavior of such material are of significant technological importance and scientific interest, as they play a crucial role in the durability assessments and life extension decisions of reinforced concrete structures. The time evolution of elastic properties and fracture of such material are complicated since they strongly depend on its heterogeneous and multiphase microstructure, which generally ranges over several length scales from nanometers to meters and develops over time as a result of continued hydration of the cement [1]. At meso-scale, concrete can be considered as a three-phase composite consisting of aggregate, bulk cement matrix and interfacial transition zone (ITZ) between them. ITZ is actually a "special" cement paste with higher porosity and larger pores relative to the bulk cement matrix [1] and [2]. Therefore, as the most complicated and basic component in cement-based materials, the mechanical properties and fracture of cement paste accounting for microstructure should be studied in detail firstly.

Because of its significance, a lot of efforts, both experimentally and numerically, have been made to study the evolution of elastic properties and cracks of hydrating cement paste in the past decades. Among them, micromechanical modelling approaches show a great advantage because some main microstructure information can be taken into account. These approaches can be classified into three general categories: analytical models, continuum-based numerical models and discrete lattice models.

With respect to analytical models, Bernard et al. [3] proposed a multistep micromechanics homogenization approach based on effective medium approximation scheme in combination with a kinetics model of the four main hydration reactions of ordinary Portland cement to predict the time evolution of elastic properties of cement paste. Cement paste is regarded as a composite of calcium-silicate-hydrate (C–S–H) matrix with large calcium hydroxide (CH) crystals, cement clinker inclusions and pores, which have different mechanical properties. Based on the same procedure, Constantinides and Ulm [4] estimated the effect of two types of C–S–H, i.e., high-density C–S–H and low-density C–S–H, on the macroscopic elastic properties of hydrating cement paste with water-to-cement (w/c) ratio of 0.5 in a quantitative manner. Sanahuja et al. [5] developed a similar homogenization scheme to estimate the evolution of effective elastic moduli of hydrating cement paste that is modelled as a four-phase material made up of anhydrous cement grains, inner hydration products surrounding the anhydrous particles, outer hydration products and pores. The simulated Young's moduli of cement paste during hydration were in excellent agreement with the experimental

Citation: Mingzhong Zhang, Andrey P. Jivkov, Micromechanical modelling of deformation and fracture of hydrating cement paste using X-ray computed tomography characterisation, Composites Part B: Engineering, Volume 88, 1 March 2016, Pages 64-72, ISSN 1359-8368, http://dx.doi.org/10.1016/j.compositesb.2015.11.007.

results available in the literature. Zheng et al. [6] and Zhao and Chen [7] predicted the elastic moduli of hardened cement paste at various curing ages with various w/c ratios by using a simple three-step analytical homogenization scheme. In this scheme, hardened cement paste is modelled as n-layer composite including spherical unhydrated cement particles, C–S–H, CH and capillary pores. Pichler and Hellmich [8], and Pichler et al. [9]modelled cement paste as a two-phase composite material composed of hydrate foam and spherical cement inclusions, and used the Mori-Tanaka (M-T) [10] scheme to determine the effective elastic moduli of cement paste, which are considered as input to further investigate the mechanical properties of mortar. Although analytical models are able to provide simple expressions for estimation of elastic properties of cement paste, the proposed simplified multiphase or n-layer cement hydration models need further validation prior to applications of them.

Regarding continuum-based numerical models, finite element method (FEM) is the most widely used one. Haecker et al. [11] utilized a computer-based cement hydration model, CEMHYD3D [12], to generate the 3D digital microstructure of cement paste consisting of unhydrated cement, hydration products and pore space, and then quantitatively predicted the elastic moduli of cement paste as a function of degree of hydration by using three-dimensional (3D) FEM. The required individual phase moduli were obtained from literature. Smilauer and Bittnar [13] predicted the elastic properties of hydrating cement paste by combining the CEMHYD3D model, percolation theory and finite element analysis. The size of the representative volume element (RVE) of cement paste was found to be 50 × 50 × 50μm. Similarly, a combination of CEMHYD3D model and FEM was applied to estimate the evolution of effective elastic modulus and diffusion coefficient of cement paste with w/c ratio of 0.4 during hydration [14], [15] and [16]. Hain and Wriggers [17] obtained the 3D microstructure of hardened cement paste from X-ray computed microtomography (micro-CT) with a spatial resolution of 1.0μm/voxel and carried out microstructure-based modelling of effective elastic properties of cement paste via homogenization. A series of numerical-statistical analyses were performed to determine the RVE size for measuring elasticity of cement paste. Recently, Huang et al.[18] proposed a statistical numerical homogenization procedure through FEM to study the effective elastic properties of cement paste based on X-ray micro-CT images with a resolution of 0.95μm/voxel. The mean values of the simulated elastic properties were in good agreement with the experimental data. Although the continuum approach is capable of predicting the mechanical properties of cement paste based on microstructure, it is inherently incapable of dealing with fracture process in cement paste, e.g., crack initiation, propagation and coalescence. In addition, the spatial resolution of X-ray CT was limited to 1.0μm that is not high enough to obtain a realistic microstructure of cement paste. Therefore, X-ray CT with a higher resolution is required.

In regard to discrete lattice models, they offer a promising modelling strategy to investigate the mechanical properties and explain fracture processes in cement paste because of its unique feature of representing a material by a lattice system of beam or spring elements [19]. Qian et al. [20] further developed and applied the lattice beam model proposed by Schlangen and van Mier [21] based on the simplest regular lattice with cubic cells to predict the mechanical properties and fracture of hydrating cement paste. The used 3D microstructures of cement pastes at different curing ages were simulated by using a cement hydration model, HYMOSTRUC3D [22], [23], [24] and [25]. However, the simulated stress–strain response of cement paste under uniaxial tension seems too brittle and remains not realistic [26]. Additionally, this lattice model is not physically realistic in terms of shape of the represented phases in cement paste and unable to provide a linear elastic response with an appropriate elastic modulus and Poisson›s ratio [27]. To overcome such limitations, Jivkov and Yates [27] proposed a novel bi-regular lattice beam model based on truncated octahedral cells, namely, site-bond model, to study the elasticity of solids and further successfully utilized to predict the mechanical behavior, and micro-crack population and damage evolution in cement-based materials considering microstructural characteristics [19] and [28]. Nevertheless, this model has a shortcoming that the relationship between the global elasticity and properties of local beam elements cannot be analytically derived [29]. In order to improve this aspect, Zhang and Jivkov [29] reformulated the site-bond model by replacing beam elements with spring elements and predicted the damage evolution in hardened cement paste.

The main purpose of this work is to further develop the proposed site-bond model and apply it to simulate the elastic properties and fracture behavior of hydrating cement paste taking into account its microstructural information. The microstructural characteristics, such as pore size distribution, porosity, and volume fraction and size distribution of anhydrous cement grains are derived from high-resolution X-ray micro-CT images along with a series of image processing and analyses. The 3D microstructure of cement paste comprised of anhydrous cement grains, hydration products and pores is represented by an assembly of truncated octahedral cells with sites at cell centers and bonds linking the neighboring cells. The simulated site-bond assembly is then subjected to uniaxial tensile loading and stress–strain response of cement paste is obtained. The fracture process is simulated by the removal of failed bonds. The simulation results are compared with the available experimental data.

MICROSTRUCTURAL CHARACTERIZATION OF CEMENT PASTE USING X-RAY MICRO-CT

X-ray Micro-CT Imaging

The 3D microstructure of hardened cement paste used in this work was obtained from X-ray micro-CT scans. ASTM type I Portland cement was used. The w/c ratio of the cement paste specimen was 0.5 (mass ratio). After drill mixing in a plastic beaker, small parts of the paste were poured into the syringe and then injected into a micro plastic tube with an internal diameter of 250μm. The specimen was stored in the standard curing room with a relative humidity of 95% and temperature of 20 °C and scanned at 1, 7 and 28 days. A high resolution Xradia MicroXCT-200 CT scanner at the Beckman Institute for Advanced Science and Technology at the University of Illinois at Urbana–Champaign was used for acquiring projection images with a source power of 10 W at 50 kV. The sample was rotated on the stage from −91° to 91° at an increase of 0.25° generating 729 shadow projections with a pixel size of

0.5μm. To the best of the authors› knowledge, this resolution is by far the highest one reported for 3D non-destructive imaging of cement-based materials. Each projection image was averaged from frames within an exposure time of 12 s. The image reconstruction was performed using the algebraic method implemented in the Xradia reconstruction software that results in a 3D stack of virtual 8-bit cross-section images with grayscale values from 0 to 255. The reader is referred to[30] for further information on the X-ray micro-CT testing.

A cylindrical region of interest (ROI) with 200μm in diameter and 100μm (200 slices) in thickness was extracted from the center of the reconstructed 3D images where the cement paste is considered to be the most homogeneous. Fig. 1 shows the micro-CT grey images of ROI of cement paste with curing ages of 1, 7 and 28 days, respectively. The dark phase stands for capillary pores. The white phase corresponds to anhydrous cement grains (unhydrated cement particles). The grey phases most probably represent hydration products. The hydration of cement proceeds with increasing curing age, as a result of which cement (i.e. white phase) and water (i.e. dark phase) are continuously consumed and an increasing number of hydration products (i.e. grey phase) appear in cement paste. In order to identify these three phases in cement phase and quantify their content, a series of image processing and analysis are required, which are described in detail in the following section.

Figure 1. Micro-CT images of hydrating cement pastes with 200μm in diameter and 100μm with curing ages of (a) 1 day; (b) 7 days; (c) 28 days. The dark, white and grey phases represent capillary pore, anhydrous cement grain and hydration product, respectively.

Microstructural Characterization

As discussed in Ref. [30], many approaches have been proposed and developed for image segmentation. Here the global thresholding method is applied and implemented by using ImageJ, a public domain Java-based image processing program (National Institutes of Health, USA). Based on the grey-level histogram two threshold grey values are defined to segment pores from anhydrous cement grains and hydration products, and subsequently segment anhydrous cement grains from hydration products. Fig. 2illustrates the identified capillary pores and anhydrous cement grains in cement paste at curing age of 7 days, which is shown in Fig. 1b. As a result, the original grey scale image is converted into a ternary image consisting of pore, anhydrous cement grain and hydration product. With this ternary image, the microstructural characteristics, such as porosity, grain density of anhydrous cement, size distributions of pore and anhydrous cement grain of ROI can be gained in a quantitative manner after a series of image analysis. More details about the image processing and analysis can be found in a previous work [29] and [31].

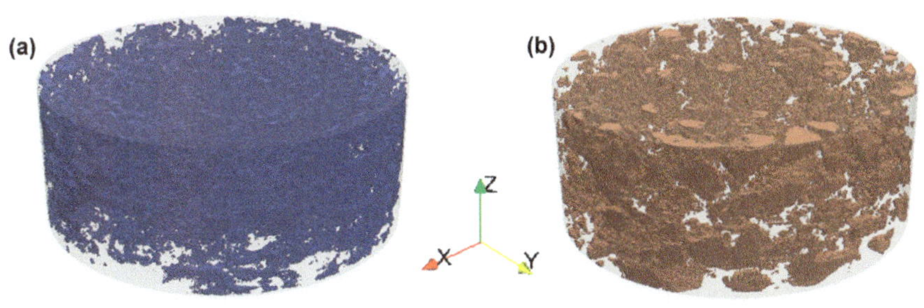

Figure 2. Different phases in cement paste at 7 days of curing: (a) capillary pores; (b) anhydrous cement grains.

The porosity and volume fraction of anhydrous cement grains of cement paste at 1, 7 and 28 days are found to be 25.69% and 22.34%, 18.71% and 15.67%, and 12.58% and 8.65%, respectively. Besides pore and anhydrous cement, the other volume is occupied by hydration products. It should be mentioned that the individual phase with size less than 0.5μm cannot be identified separately due to the limited spatial resolution. As curing age increases, both porosity and volume fraction of anhydrous cement grain of cement paste decreases, while the volume fraction of hydration product increases. Fig. 3 shows the obtained porosity for each 2D image of cement paste along Z-direction from top to bottom (200 slices, 100μm). It indicates that 28-day old cement paste has less fluctuation in porosity along the thickness of sample compared to cement paste at 1 and 7 days.

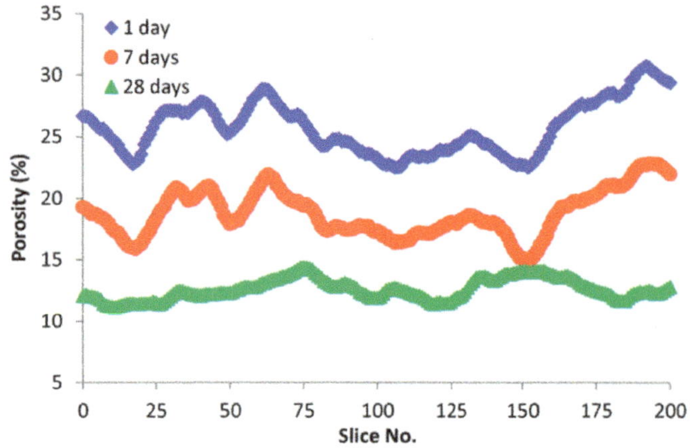

Figure 3. Porosity obtained for each 2D image of cement paste along Z-direction.

Assuming spherical shapes for pore and particle, the experimental size (i.e. equivalent radius) distributions of pore and unhydrated cement particle are determined by summarizing the volume of each separate cluster in ternary image of cement paste. Fig. 4 shows the obtained pore and unhydrated cement particle size distributions of cement pastes with curing ages of 1, 7 and 28 days. It is obvious that both pore size and particle size of unhydrated cement are reduced with increasing curing age due to the progress of cement hydration. In addition, cement paste at 1 day contains more big pores and unhydrated cement particles compared to 7- and 28-day old cement pastes. In order to further illustrate the pore size distribution of cement paste, the histogram of pore size distribution of cement paste at 28 days of curing with cumulative probability distribution for pores with radii less than 5μm is depicted in Fig. 5. It can be seen clearly that most of pores in the sample are small pores. Pores with radii less than 5μm occupy 95% probability and the remaining 5% probability is left for pores with radii larger than 5μm.

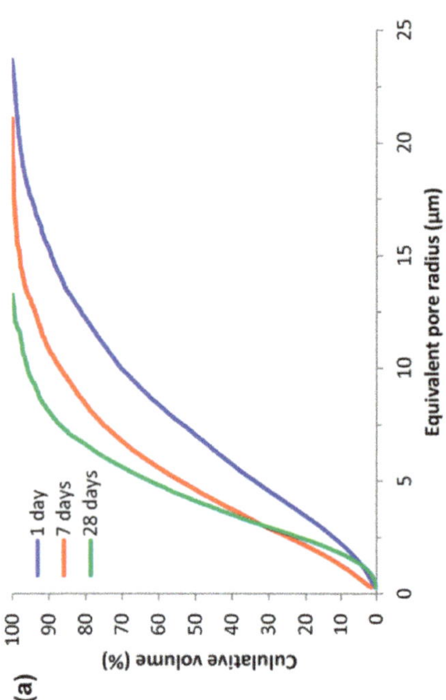

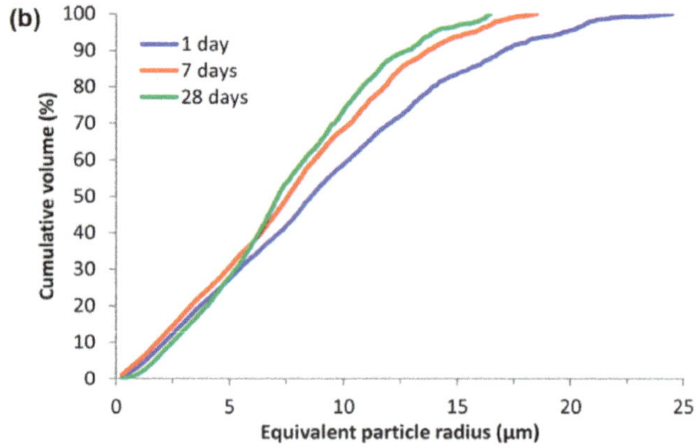

Figure 4. Size distributions of pore and unhydrated cement particle of cement paste with various curing ages: (a) pore size distribution; (b) particle size distribution of unhydrated cement.

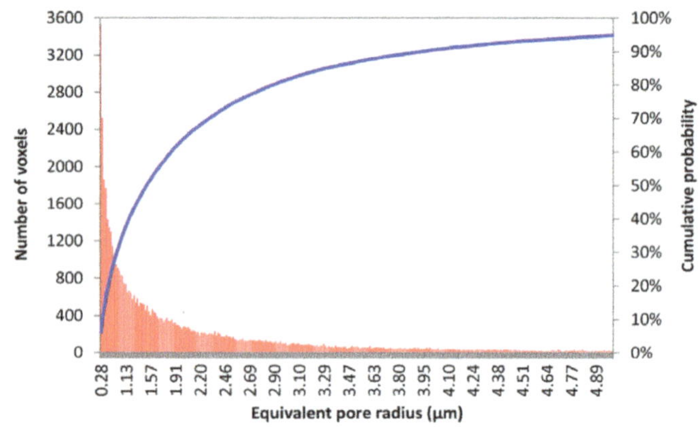

Figure 5. Pore size distribution of cement paste at 28 days of curing.

MICROMECHANICAL MODELLING

Herein, a microstructure-informed site-bond model is proposed and developed for micromechanical modelling by incorporating the obtained experimental microstructural characteristics of cement paste into basic site-bond model for elasticity. The theoretical background of site-bond model, and modelling and simulation procedure are described in detail below.

Basic Site-bond Model

In the basic site-bond model, a material volume is represented by an assembly of truncated octahedral cells, each of which consists of six square and eight regular hexagonal faces, as shown Fig. 6a. The truncated octahedron is considered to be the better choice for a regular representation of the microstructure compared to other regular cell shape, such as cube, regular hexagonal prism and rhombic dodecahedron, the number of cell faces of which are 6, 8 and 12, respectively. The cell center is regarded as a site linking to its neighboring sites by fourteen bonds. These bonds are classified into two groups: B_1 includes six bonds in principal directions (normal to the square faces) andB_2 consists of eight bonds in octahedral direction (normal to the hexagonal faces) as illustrated in Fig. 6b. The bonds are modelled with a bundle of elastic-brittle normal and shear springs to represent the relative deformations between adjacent cells, as plotted in Fig. 6c. By assuming shear springs have identical stiffness coefficients, four spring constants can be derived as follows based on the equivalence of the strain energy density of the discrete cell to the continuum system [32]:

$$k_n^p = \frac{(3-4\nu)EL}{8(1+\nu)(1-2\nu)}; \quad k_n^o = \frac{(1+8\nu)EL}{8(1+\nu)(1-2\nu)}; \quad k_s^p = k_s^o = \frac{(1-4\nu)EL}{8(1+\nu)(1-2\nu)} \qquad (1)$$

where k_n^p and k_s^p denote the stiffness coefficients of normal and shear springs in principal direction, k_n^o and k_s^o represent the stiffness coefficients of normal and shear springs in octahedral direction, E and ν are Young›s modulus and Poisson›s ratio of the material respectively, L is the spacing between two adjacent sites in principal directions.

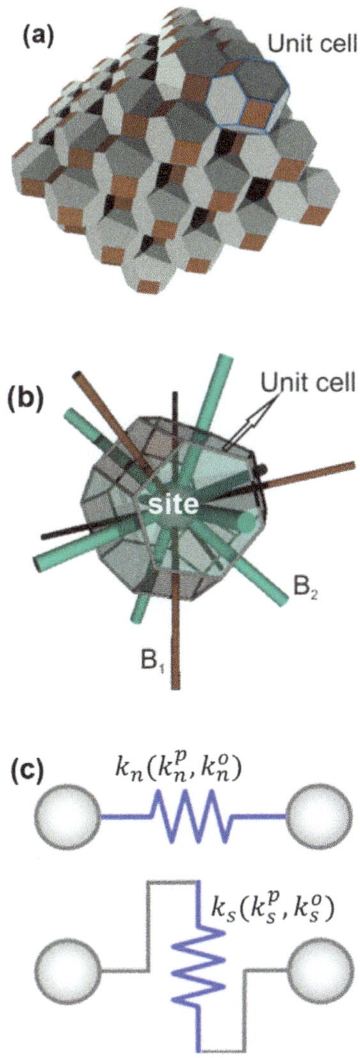

Figure 6. Cellular lattice structure: (a) site-bond assembly; (b) unit cell with bonds; (c) normal and shear springs.

The analytically derived spring stiffness coefficients were validated by benchmark simulations including uniaxial tension and plane-strain tension in previous work [32]. It was shown that the simulated overall mechanical constants, such as elastic modulus, Poisson's ratio and shear modulus are in good agreement with the theoretical values. In addition, the proposed lattice is able to represent any isotropic elastic materials with Poisson's ratio in the range from −1 to 1/2, which covers all engineering materials.

Microstructure-informed Site-bond Model

The microstructure-informed site-bond model is constructed by incorporating the microstructural characteristics of cement paste into the basic site-bond model based on a random allocation of features, pores and anhydrous cement grains, with sizes belonging to the experimentally measured distributions. Firstly, anhydrous cement grains of different sizes are randomly distributed to all sites of the cellular structure according to the experimental particle size distribution plotted in Fig. 4. A generator of uniformly distributed random numbers $0 \leq r < 1$ is used to assign grains with different sizes to individual sites. For each site, a random number r is generated and the assigned grain size is $R_i = F-1(r)$, where F(Ri) is the cumulative probability of grain sizes shown in Fig 5. As a result, each cell has one spherical anhydrous cement grain with size Ri. Secondly, the cell size, L , is calculated from the volume fraction of anhydrous cement grains, ϕ_a, and the volume of the cellular structure with N cells via $\frac{\phi_a N L^3}{2} = \sum_i \frac{4\pi R_i^3}{3}$. Lastly, pores are assigned randomly to the bonds using the same method described above. The assignment continues until the prescribed porosity in the modelled material volume is reached. Fig. 7 illustrates a comparison of the simulated and experimental pore and anhydrous cement size distributions for cement paste at 28 days of curing. This indicates that the simulated size distributions of pore and unhydrated cement particles fit the measured size distributions obtained from X-ray micro-CT very well. The remaining volume represents hydration product. Thus, the microstructure-informed site-bond model made up of pore, anhydrous cement grain and hydration product is established to incorporate the microstructure of cement paste into the basic site-bond model.

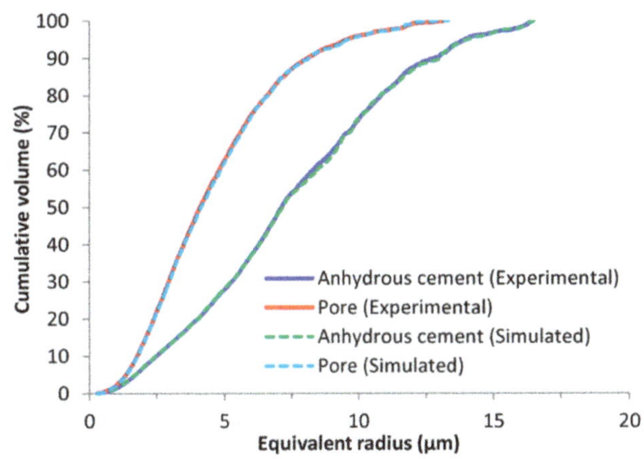

Figure 7. Comparison of the simulated and experimental size distributions of pore and unhydrated cement particle of cement paste at 28 days of curing.

As introduced in Section 3.1, the stiffness coefficients of normal and shear springs in each bond are functions of macroscopic mechanical constants (Young's modulus E and Poisson's ratio ν) of the represented homogeneous solid and cell size L. Since the microstructure-informed site-bond model has multiphase and porous features, it is necessary to homogenize the elastic properties of anhydrous cement grain and hydration product in the modelled cellular structure and to take into account pore effect. For this purpose, the effective Young's modulus Eeff and Poisson's ratio veff of a composite of anhydrous cement grain and hydration product, and effective cell size Leff are defined and used instead of E, v, and L to calculate stiffness coefficients of springs according to Eq. (1).

Taking 28-day old cement paste for example, its porosity and volume fraction of anhydrous cement grains are 12.58% and 8.65%, respectively, as presented in Section2. The remaining 78.77% is hydration product. For a composite of anhydrous cement grain and hydration product, anhydrous cement grain only occupies 9.9% of the volume. For this case, the Mori-Tanaka (M-T) method is appropriate for deriving effective elastic properties of the composite, since it has been shown to provide good estimates of the effective moduli of a two-phase composite, particularly when the volume fraction of the inclusion (herein, it is anhydrous cement grain) is low and the shape of the inclusion is spherical [10]. The M-T method is expressed by the following equations:

$$K_{eff} = K_h + \frac{\phi_a(K_a + K_h)(3K_h + 4G_h)}{3K_h + 4G_h + 3(1-\phi_a)(K_a - K_h)} \quad (2)$$

$$G_{eff} = G_h + \frac{5\phi_a G_h(G_a - G_h)(3K_h + 4G_h)}{5G_h(3K_h + 4G_h) + 6(1-\phi_a)(G_a - G_h)(K_h + 2G_h)} \quad (3)$$

$$E_{eff} = \frac{9K_{eff}G_{eff}}{3K_{eff} + G_{eff}}; \quad \nu_{eff} = \frac{3K_{eff} - 2G_{eff}}{2\left(3K_{eff} + G_{eff}\right)} \quad (4)$$

where Keff and Geff represent the effective bulk and shear moduli of the composite of anhydrous cement grain and hydration product, ϕ_a denotes the volume fraction of anhydrous cement of the composite, Ka, Kh, Ga and Gh are the bulk and shear moduli of anhydrous cement and hydration products, respectively.

The presence of a pore in a bond affects both the elastic constants and the failure propensity of the bond. The pore modifies the bond length to an effective value, Leff, which can be calculated from the effective volume of the polyhedron between two neighboring sites, Veff, i.e. the volume of the polyhedron V0 less the volume of the poreVp, as shown in Fig. 8. Due to the different shape of polyhedrons in principal and octahedral directions, their volumes are different: Vp,0 = $L^3/24$ for the polyhedron in principal direction, and Vo,0 = $3L^3/32$ for the polyhedron in octahedral direction. Thus, the corresponding effective cell sizes can be written as follows

$$L_{p,eff} = \sqrt[3]{24(V_{p,0} - V_p)}; \quad L_{o,eff} = \sqrt[3]{32(V_{o,0} - V_p)/3} \quad (5)$$

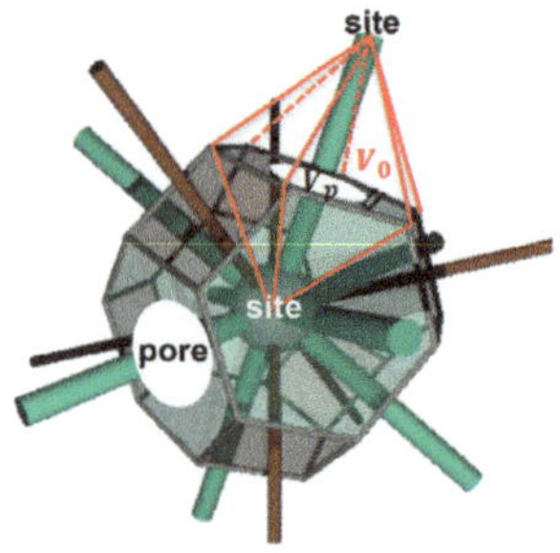

Figure 8. Schematic illustration of microstructure-informed site-bond model with anhydrous cement grains at sites and pores on surfaces.

The spring constants of each bond, accounting for the pore associated with it, are calculated by substituting Eqs. (4) and (5) into Eq. (1) to replace E, v and L with Eeff, vefffand Leff.

Failure Criterion

For microstructure-informed site-bond model, the spring elastic-brittle behavior is represented by linear elastic and linear softening branches separated by a damage initiation point. The elasticity is dictated by Eq. (1) for given macroscopic elastic properties and microstructure-related length scale. The softening branch approximates the local post-initiation resistance of the matrix to failure. The area under the softening branch represents the additional energy required to separate two cells along their common interface, from the initiation of damage to final failure. The constitutive models for normal and shear springs are illustrated in Fig. 9, where utmax and usmax are the deformation of the normal and shear springs at damage initiation with corresponding maximum normal Ftmax and shear forces Fsmax; wtmax and wsmax represent the deformation of the normal and shear springs at failure (set to 2utmax and 2usmax, respectively); $K_n=F_{t,max}/U_{t,max}$ is the stiffness coefficient of the normal spring and$K_s=F_{s,max}/U_{s,max}$ is the stiffness coefficient of the shear spring. For normal springs, failure in compression is not allowed, which is achieved by setting the critical compressive deformation as $u_{c,max}=-10u_{t,max}$ and the maximum compressive force as $F_{c,max}=-10F_{t,max}$.

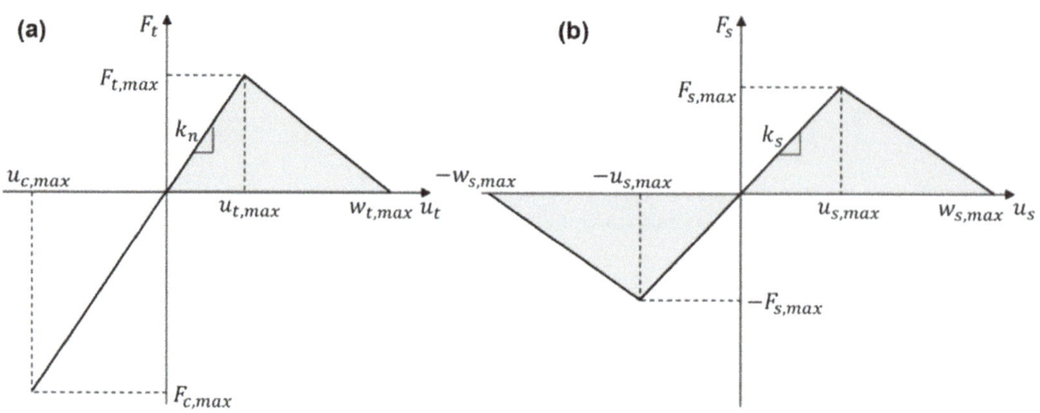

Figure 9. Constitutive law and failure criteria for microstructure-informed site-bond model: (a) normal spring; (b) shear spring.

In this study, the energy-based failure criterion is applied. The shaded area in Fig. 9represents the critical local failure energy, Gf, of a bond. This is calculated via $G_f=2\gamma A_e$, where γ a prescribed surface energy (material parameter), and Ae is the effective face, i.e. the original face area A_0 less the cross-section area of the pores assigned to the face,Ap. Due to the existence of pores with different sizes on bonds (faces), the critical failure energies of bonds vary from the maximum failure energy $2\gamma A_0$ for bonds with no pores to zero for pore area close to the intact area of the face, e.g. square face or hexagonal face. If the pore area in a bond is equal to or larger than the corresponding face area, this bond is removed before loading.

Modelling and Simulation

In this work, a material volume of size 20L × 20L × 20L is used. The corresponding cellular structure is made up of 17,261 cells; the site-bond model has the same number of sites and 113,260 bonds. The size of unit cell L can be determined based on the obtained microstructural information of cement paste from X-ray micro-CT following the procedure described in Section 3.2. As such, the experimental microstructure of cement paste is mapped to and represented by the microstructure-informed site-bond model. Table 1shows the used elastic constants and surface energies of different phases in cement paste. Herein, it should be pointed out that the hydration product in Table 1 is considered as a composite of C–S–H and CH. The volume fractions of these two phases in hydration product would change with time. However, C–S–H and CH are difficult to be detected and identified with each other by using X-ray micro-CT due to the similarity in density. Therefore, the mechanical properties of hydration product are assumed constant for simplicity in this work. Uniaxial tension is considered. The loads are applied via prescribed displacements to the boundary sites on top surface of the lattice system. The bottom surface is fixed and the other four surfaces are free. A schematic illustration of uniaxial tensile test on the microstructure-informed site-bond model for cement paste at 28 days of curing is given in Fig. 10. The bonds are presented in different colours (in web versaion) with dark red representing bonds with larger pores and blue denoting bonds with smaller pores. It can be seen that the spatial distribution of pore size is random. Most of larger pores are located around the center between the top and bottom surfaces.

Table 1. Elastic constants and surface energy of different phases (Zhang and Jivkov, 2014).

Phase	E (GPa)	v	K (GPa)	G (GPa)	γ(mJ/m^2)
Anhydrous cement grain	139.9	0.30	116.7	53.8	–
Hydration product	29.2	0.24	18.7	11.8	50
Pore	0	0	0	0	–

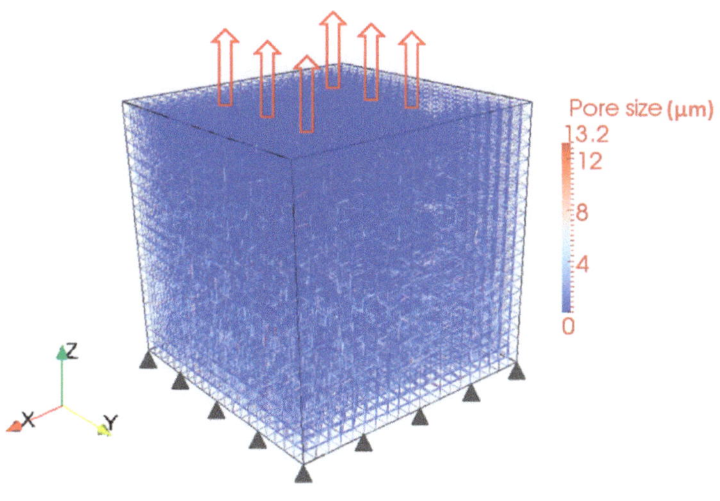

Figure 10. Uniaxial tensile test on the microstructure-informed site-bond model for cement paste at 28 days of curing.

At each load step, the forces and deformations are obtained and the bond energies are calculated. The bond fails when its energy reaches the assigned critical failure energy, mimicking the creation of a micro-crack along the interface between the two cells joined by the bond. Upon failure the bond is removed from the site-bond assembly. The simulation is continued until the assembly is disintegrated or a prescribed strain is reached. Thus, the fracture process can be obtained by the generation and evolution of micro-crack population under loading step by step.

RESULTS AND DISCUSSION

Elastic Properties

Fig. 11 shows the simulated stress–strain response for cement pastes at 1, 7 and 28 days under uniaxial tension. It can be seen that the initial elastic part is perfectly linear, as represented by black dashed lines. Beyond the linear response, "graceful" quasi-brittle non-linearity prior to the peak stress is observed. The last point of the stress–strain curve corresponds to final failure. Compared to 7- and 28-day old cement pastes, cement paste with curing age of 1 day appears to be much more brittle, which can be attributed to the higher porosity and larger pore size at 1 day than those at 7 and 28 days, as shown in Fig. 3 and Fig. 4. Additionally, for 28-day old cement paste, an obvious softening branch can be found before the final failure.

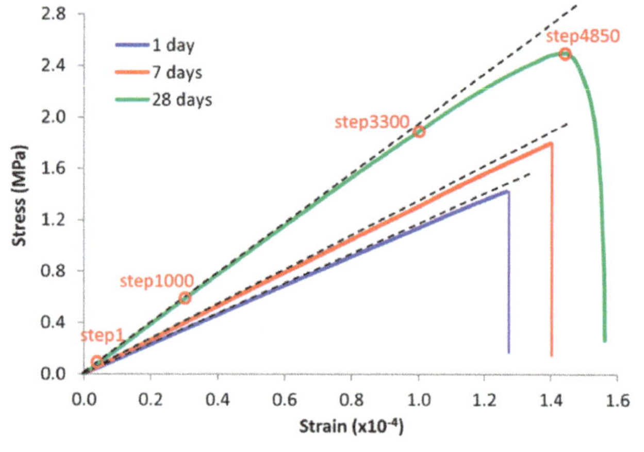

Figure 11. Stress–strain response of cement pastes with various curing ages under uniaxial tension.

From this curve, several important macroscopic parameters, such as elastic modulus, tensile strength, strain at failure point and fracture energy of cement paste, can be acquired. These parameters play crucial roles in the prediction of mechanical behaviors and fracture process of cement-based materials at meso-level, i.e. mortar and concrete (see e.g. Ref. [33]). The global elastic modulus corresponds to the slope of the stress–strain curve at the linear elastic stage. The tensile stress and strain at failure point are considered to be the tensile strength and peak strain, respectively. Fig. 12 shows the time evolution of Young's modulus and tensile strength of cement paste. As expected, there is an increase in both Young's modulus and tensile strength with curing, which is a result of the continued cement hydration reaction and microstructure development of cement paste during curing. For the purpose of validation, the available experimental data taken from Refs. [4], [5], [11], [16], [34] and [35] are plotted together in Fig. 12. It should be mentioned that the w/c ratio of cement paste specimens in their experiments was 0.5, which is the same as that used in this work. It can be seen that the simulated Young's moduli show a good agreement with the experimental data, in particular for the cement paste specimens with curing age of 1 and 28 days. In addition, there exists an obvious variation of the measured Young's moduli of cement paste at 28 days by different researchers, which can be mainly attributed to the different types of cement used in their experiments. In Refs. [4] and [34], the used cement was ordinary Type I Portland cement, which is the same as that studied in the simulation. However, in Ref. [11] the used cement was ASTM D cement, the chemical composition of which is somewhat different from that of Type I cement. As can be seen from Fig. 12, there is an obvious increase in tensile strength of cement paste with increasing curing age from 1 day to 7 days, followed by a gradual increase as curing age changes from 7 days to 28 days. Moreover, the strain at failure point increases with increasing curing age as a result of an obvious decrease in both porosity and pore size of hardened cement paste.

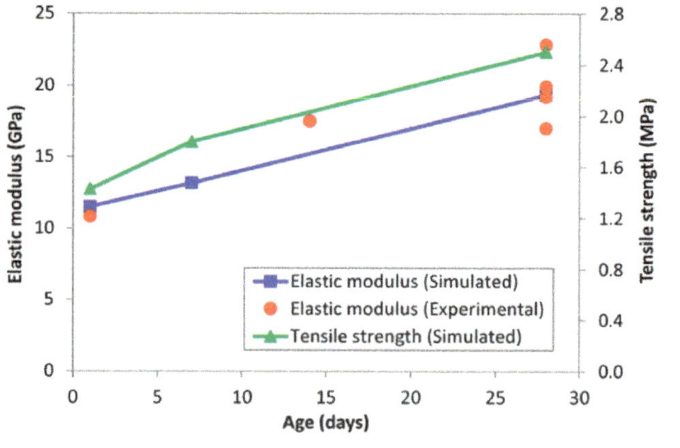

Figure 12. Evolution of elastic modulus and tensile strength of hardened cement paste: experimental data taken from Refs. [4], [5], [11], [16], [34] and [35].

The fracture energy that is an essential input to continuum-based fracture models, e.g. cohesive zone model, can be calculated as the area under the stress–displacement curve that is converted from the stress–strain response shown in Fig. 11. It denotes all energy lost in distributed micro-cracking up to eventual rupture divided by the cross-section area of cement paste specimen. Further insight into this process can be obtained from the development of fracture process as described below.

Fracture Processes

Fig. 13 shows the fracture processes in 28-day old cement paste under uniaxial tension (see Fig. 10) at loading steps of 1, 1000, 3300 and 4850 corresponding to stress–strain curve in Fig. 11. As can be seen from Fig. 13a, several micro-cracks initiate around the middle of the specimen after loading. These form by failure of bonds with large pores; compare with Fig. 10. Thus micro-crack initiation is triggered firstly by large pores in this region resulting in low critical failure energies of these bonds. Bond failures progress with the load increase, Fig. 13b, but the micro-crack population remains relatively discrete: isolated failures, no observable localization. Comparison with Fig. 11 suggests that isolated failures have a negligible contribution to the nonlinear stress–strain response. With further increase in loading, new micro-cracks are gradually generated, Fig. 13c and d. This is accompanied now by propagation and coalescence of existing micro-cracks forming clusters of free surfaces. The formation of larger free surfaces results in the clear non-linearity observable in Fig. 11. The peak in the stress–strain curve is reached via an avalanche-type failure, typical for tensile response, causing the separation of the whole structure into a collection of disjoint regions.

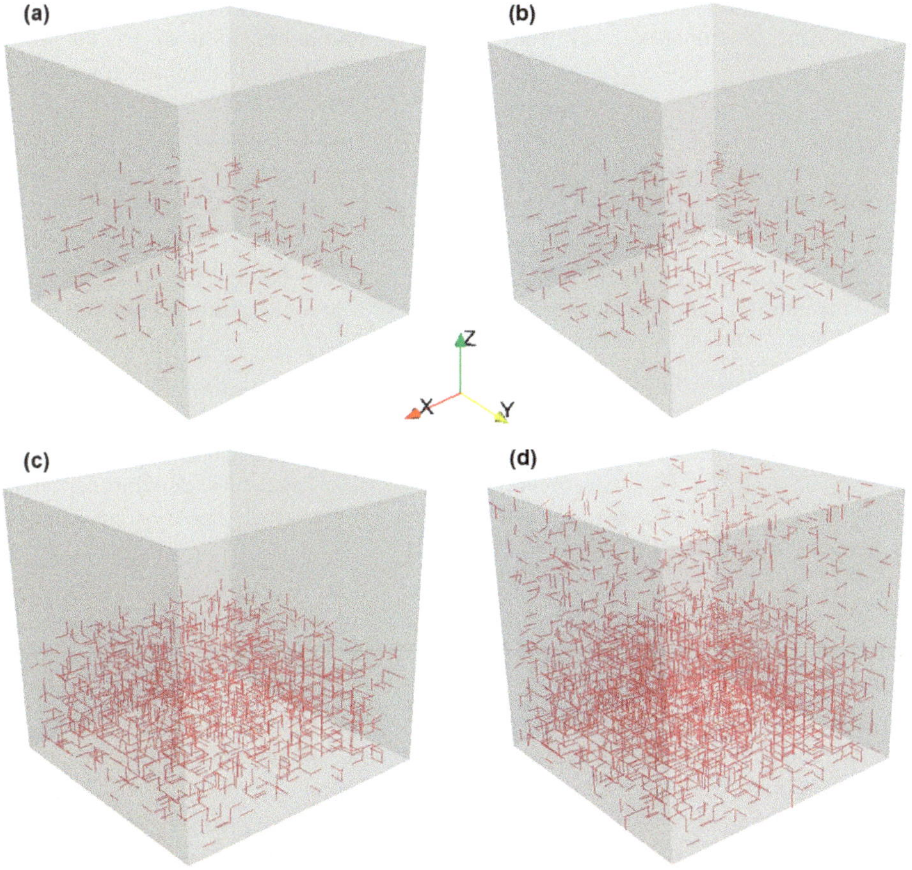

Figure 13. Fracture process in 28-day old cement paste under uniaxial tension corresponding to the stress–strain curve in Fig. 11: (a) step1; (b) step1000; (c) step3300; (d) step4850.

CONCLUSIONS

This work presents a micromechanical modelling of mechanical properties and fracture of hydrating cement paste based on proposed microstructure-informed site-bond method. The microstructure information of cement pastes at different curing ages obtained from high resolution X-ray micro-CT is incorporated into the site-bond assembly. The deformation and failure properties of the bonds are determined from microstructure-dictated length and energy parameters. The micro-crack evolution is simulated by removal of bonds. From the findings of the present study, the following conclusions can be drawn:

- Both porosity and pore size of cement paste decrease with increasing curing time. For cement paste with w/c ratio of 0.5, there is clear decrease in porosity from 25.69% to 12.58%, as age of cement paste increases from 1 day to 28 days. Meanwhile, particle size and volume fraction of anhydrous cement decrease as well.
- The microstructural evolution of cement paste due to cement hydration results in the development of mechanical properties (elastic modulus and tensile strength) of hardening cement paste.
- The simulated elastic properties of hydrating cement paste are in very good agreement with existing experimental data. Notably, the proposed site-bond models involve only data for sizes and interrelations of cement constituents without further empiricism.
- Micro-cracking initiates at pre-existing defects, e.g. large pores. As far as the micro-cracks remain isolated, i.e. in a discrete population, their effect on longer-scale behavior is negligible and might be undetected by macroscopic experiment.
- Propagation and coalescence of micro-cracks, forming larger internal free surfaces, causes the observable non-linear response prior to peak load. The final specimen disintegration appears as avalanche-type failure at peak load.

The damage evolution under more complex loadings, e.g. triaxial and bending, is of larger practical relevance and will be the focus of future works, following the methodology presented in Ref. [28]. Additionally, the obtained fracture energy of cement paste in this work will be used as input for modelling of mechanical properties and fracture behavior of cement mortar and concrete at meso-level based on a two-scale approach, which is a subject of ongoing work and will be presented in a future publication.

ACKNOWLEDGEMENTS

The authors acknowledge the support from EPSRC via grant EP/J019763/1, “QUBE: Quasi-Brittle fracture: a 3D experimentally-validated approach”, and from BNFL for the Research Center for Radwaste & Decommissioning.

REFERENCES

1. Zhang M, Ye G, van Breugel K. Multiscale lattice Boltzmann-finite element modelling of chloride diffusivity in cementitious materials. Part I: algorithms and implementation. Mech Res Commun 2014;58:53e63.

2. Ma HY, Li ZJ. Multi-aggregate approach for modeling interfacial transition zone in concrete. Aci Mater J 2014;111(2):189e99.
3. Bernard O, Ulm FJ, Lemarchand E. A multiscale micromechanics-hydration model for the early-age elastic properties of cement-based materials. Cem Concr Res 2003;33(9):1293e309.
4. Constantinides G, Ulm FJ. The effect of two types of C-S-H on the elasticity of cement-based materials: results from nanoindentation and micromechanical modeling. Cem Concr Res 2004;34(1):67e80.
5. Sanahuja J, Dormieux L, Chanvillard G. Modelling elasticity of a hydrating cement paste. Cem Concr Res 2007;37(10):1427e39.
6. Zheng JJ, Zhou XZ, Shao L, Jin XY. Simple three-step analytical scheme for prediction of elastic moduli of hardened cement paste. J Mater Civ Eng 2010;22(11):1191e4.
7. Zhao XH, Chen WF. The effective elastic moduli of concrete and composite materials. Compos Part B-Eng 1998;29(1):31e40.
8. Pichler B, Hellmich C. Upscaling quasi-brittle strength of cement paste and mortar: a multi-scale engineering mechanics model. Cem Concr Res 2011;41(5):467e76.
9. Pichler C, Metzler G, Niederegger C, Lackner R. Thermo-mechanical optimization of porous building materials based on micromechanical concepts: application to load-carrying insulation materials. Compos Part B Eng 2012;43(3):1015e23.
10. Mori T, Tanaka K. Average stress in matrix and average elastic energy of materials with misfitting inclusions. Acta Metall Mater 1973;21(5):571e4.
11. Haecker C, Garboczi EJ, Bullard JW, Bohn RB, Sun Z, Shah SP, et al. Modeling the linear elastic properties of Portland cement paste. Cem Concr Res 2005;35(10):1948e60.
12. Bentz DP. Three-dimensional computer simulation of portland cement hydration and microstructure development. J Am Ceram Soc 1997;80(1):3e21.
13. Smilauer V, Bittnar Z. Microstructure-based micromechanical prediction of elastic properties in hydrating cement paste. Cem Concr Res 2006;36(9): 1708e18.
14. Valentini L, Parisatto M, Russo V, Ferrari G, Bullard JW, Angel RJ, et al. Simulation of the hydration kinetics and elastic moduli of cement mortars by microstructural modelling. Cem Concr Comp 2014;52:54e63.
15. Bernard F, Kamali-Bernard S. Predicting the evolution of mechanical and diffusivity properties of cement pastes and mortars for various hydration degrees e a numerical simulation investigation. Comp Mater Sci 2012;61: 106e15.
16. Stefan L, Benboudjema F, Torrenti JM, Bissonnette B. Prediction of elastic properties of cement pastes at early ages. Comp Mater Sci 2010;47(3): 775e84.
17. Hain M, Wriggers P. Numerical homogenization of hardened cement paste. Comput Mech 2008;42(2):197e212.
18. Huang J, Krabbenhoft K, Lyarnin AV. Statistical homogenization of elastic properties of cement paste based on X-ray microtomography images. Int J Solids Struct 2013;50(5):699e709.
19. Jivkov AP, Engelberg DL, Stein R, Petkovski M. Pore space and brittle damage evolution in concrete. Eng Fract Mech 2013;110:378e95.
20. Qian Z, Schlangen E, Ye G, van Breugel K. Prediction of mechanical properties of cement paste at microscale. Mater Construc 2010;60(297):7e18.
21. Schlangen E, van Mier JGM. Experimental and numerical analysis of micromechanisms of fracture of cement-based composites. Cem Concr Comp 1992;14:105e18.
22. van Breugel K. Numerical-simulation of hydration and microstructural development in hardening cement-based materials .1. Theory. Cem Concr Res 1995;25(2):319e31.
23. van Breugel K. Numerical-simulation of hydration and microstructural development in hardening cement-based materials .2. Applications. Cem Concr Res 1995;25(3):522e30.
24. Ye G, van Breugel K, Fraaij ALA. Three-dimensional microstructure analysis of numerically simulated cementitious materials. Cem Concr Res 2003;33(2): 215e22.
25. Zhang M, Ye G, van Breugel K. Microstructure-based modeling of water diffusivity in cement paste. Constr Build Mater 2011;25(4):2046e52.
26. Bernard F, Kamali-Bemard S, Prince W. 3D multi-scale modelling of mechanical behavior of sound and leached mortar. Cem Concr Res 2008;38(4): 449e58.
27. Jivkov AP, Yates JR. Elastic behavior of a regular lattice for meso-scale modelling of solids. Int J Solids Struct 2012;49(22):3089e99.
28. Jivkov AP. Structure of micro-crack population and damage evolution in quasi-brittle media. Theor Appl Fract Mec 2014;70:1e9.
29. Zhang M, Jivkov AP. Microstructure-informed modelling of damage evolution in cement paste. Constr Build Mater 2014;66:731e42.
30. Zhang M, He Y, Ye G, Lange DA, van Breugel K. Computational investigation on mass diffusivity in Portland cement paste based on X-ray computed microtomography (mu CT) image. Constr Build Mater 2012;27(1):472e81.
31. Ma HY, Li ZJ. Realistic pore structure of Portland cement paste: experimental study and numerical simulation. Comput Concr 2013;11(4):317e36.
32. Zhang M, Morrison CN, Jivkov AP. Meso-scale site-bond model for elasticity: theory and calibration. Mater Res Innov 2014;18:982e6.
33. Chaudhuri P. Multi-scale modeling of fracture in concrete composites. Compos Part B Eng 2013;47:162e72.
34. Sun Z, Ye G, Shah SP. Microstructure and early-age properties of portland cement paste - effects of connectivity of solid phases. Aci Mater J 2005;102(2): 122e9.
35. Boumiz A, Vernet C, Tenoudji FC. Mechanical properties of cement pastes and mortars at early ages e evolution with time and degree of hydration. Adv Cem Based Mater 1996;3(3e4):94e106.

CHAPTER 9

AN EXPERIMENTAL INVESTIGATION INTO QUASI-STATIC AND FATIGUE DAMAGE DEVELOPMENT IN BOLTED-HOLE SPECIMENS

O.J. Nixon-Pearson and S.R. Hallett

Keywords: A. Carbon fibre; B. Delamination; B. Fatigue; D. X-ray computed tomography (CT)

INTRODUCTION

Fiber-reinforced composites laminates are increasingly being used to manufacture load bearing primary structures in the aerospace sector as composites offer a much better strength to weight ratio than metals. Joints which use fasteners are very commonplace in the assembly of aircraft structures. The complex failure modes of composites materials and the requirement for fasteners however demand a more rigorous approach to design than for traditional alloys. Bolt filled-hole, along with open-hole tension tests are widely used in industry to aid the understanding of the behavior of the tensile strength of bolted joints.

Factors affecting the characteristics of open and filled-hole laminate strengths include, ply orientation, washer pressures and washer sizes. There is still rather limited understanding of the mechanisms and causes of the differences in failure strength between the filled-hole and open-hole specimens. Previous work in this area includes an experimental study carried out to investigate the effects of clamp-up on the net-tension (quasi-static) failure of composite plates with bolt filled-holes by Yan et al. [1]. Bolted joints (100% bolt bearing load) and non-bolt bearing load (100% bypass tensile load) conditions were investigated. For bolted joints which failed in a net-tension mode, the bolt clamping force increased the strength of the joint irrespective of the layup. There have also been other quantitative studies into the tensile strength and fatigue behavior of bolted composite joints with single and multiple numbers of bolts [2], [3] and [4].

Persson and Eriksson [3], for example, investigated various different conditions of multiple row bolted joints to include geometry (width, spacing, and edge distances), laminate configuration, fastener type, and environmental conditions. They found that the most significant factors were the ratio of bolt diameter to specimen width, and the ratio of hole diameter to laminate thickness when using single-lap joints containing four fasteners and two rows. However very little work is available in the literature on the interactions between the sub-critical damage and the bolt clamp-up forces.

Previous work has investigated open-hole tension damage mechanisms in detail under static loading [5] and fatigue [6], [7] and [8]. These studies found that delamination plays a significant role in cases with thick ply blocks, where the hole diameter to ply block thickness ratio is relatively small. This results in increasing failure stresses for increasing in-plane dimensions using constant thicknesses. In contrast for cases with dispersed plies and repeated sublaminates through the thickness the reduction in the amount of damage propagating through the gauge section for configurations with larger hole-diameters means that there is a reduced amount of stress relief at the hole-edge, leading to earlier fiber-dominated failure.

Citation: O.J. Nixon-Pearson, S.R. Hallett, An experimental investigation into quasi-static and fatigue damage development in bolted-hole specimens, Composites Part B: Engineering, Volume 77, August 2015, Pages 462-473, ISSN 1359-8368, http://dx.doi.org/10.1016/j.compositesb.2015.03.051.

The effect of out of plane compressive stress on enhancing delamination strength is well known in composites [2]. It is thus expected that when the same configurations as tested in open-hole tension are subject to bolt clamping forces, this will have a direct effect on the damage formation and hence ultimate strength.

Since the understanding of formation and propagation of damage is of great importance to the overall failure of open-hole tension specimens, particularly in fatigue, this was investigated and characterized using X-ray Computed Tomography (CT) scanning in Refs. [6] and [8]. In other earlier fatigue related work, Spearing and Beaumont [9]concentrated on tension–tension fatigue of notched Carbon/Epoxy (T300/914) and Carbon/Polyetheretherketone (PEEK) laminates using an R ratio of 0.1. NDT (non-destructive testing) techniques included X-radiography to produce images of the damage. It was also shown how prolonged exposure to the zinc iodide dye penetrant can accelerate the growth of damage in the specimens. Other early work was carried out by Mohlin et al. [10] and [11] using tetrabromoethane (TBE) enhanced X-ray radiography to study delamination growth in notched/carbon epoxy laminates under compressive fatigue loading. The main limitations of this early work is the 2D nature of damage characterization, and the inability to separate out the global damage into individual delamination interfaces and ply cracking orientations. Fatigue damage in composites is distinctly three dimensional in character and therefore more recent techniques such as X-ray computed tomography (CT) are much more suitable for the evaluation of the micromechanical behavior of fatigue damage development. Work by Scott et al. [12]and Moffat et al. [13] has proven the capability of this technique to identify damage progression in notched carbon-epoxy specimens under quasi-static loading to failure.

In this paper the role of bolted clamp-up forces is examined in regard to its effect on the formation of damage in bolted-hole tension tests both in static and fatigue. The damage is characterized through X-ray CT scanning of tests interrupted at various stages of the damage formation and results are compared against previous open-hole tension fatigue work [6], [8] and [14].

EXPERIMENTAL PROCEDURE

The material used is Hexcel's carbon fiber-epoxy unidirectional (UD) prepreg system, IM7/8552, with a nominal ply thickness of 0.125 mm [15]. The specimen dimensions are shown in Fig. 1. Holes were drilled using a brad-point (also known as lip and spur), tungsten carbide drill bits. The lip and spur (also known as the brad point) drill bit, is optimized for drilling wood. Conventional drill bits may wander prior to initial contact with a flat surface, the sharp point of the spur pushes into the material keeping it in line. When drilling wood across the grain there is a tendency for the long fibers to pull-out around the circumference of the hole-edge rather than being cleanly cut. This also applies to composite materials which like wood contains many fibers. The lip and spur drill bit cuts the outer periphery of the hole prior to the inner cutting edges, planning the base of the hole. Thus the lip ensures that the fibers at the hole-edge are cut cleanly rather than pulling them out and splintering them.

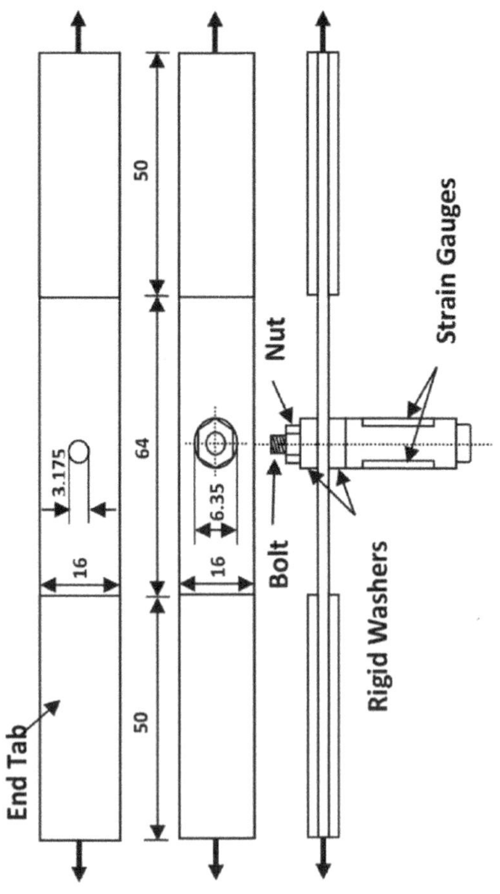

Figure 1. Open and bolted hole specimen dimensions.

Drilling parameters used in this paper were 1500 rpm (spindle speed), with a feed rate of 1 mm/min. If drill speed is too high then the propensity to generate heat is increased, if too low then the risk of back face delamination damage is enhanced. The specimen was clamped to a back plate in order to minimize back-face breakout. The drilling method used here is consistent with previous open-hole quasi-static and fatigue work carried out by the authors [6] and [8].

Quasi-static open-hole tests carried out by Green et al. [5] used reamed holes when preparing specimens of the same configuration. X-ray images in Ref. [16] show minimal damage in areas not affected by mechanical loading. X-ray images in Refs. [6] and [8]show similar low levels of drilling damage and more importantly, close agreement was obtained between the quasi-static tensile results from Ref. [5] and those obtained in Refs. [6] and [8], thus indicating that hole drilling quality was sufficient so as not to influence the results.

The ratio of width to hole diameter chosen represents the minimum distance needed to allow the elastic stress states to return sufficiently close to their original values at the edges, thereby giving dimensions offering the lowest failure forces, and using the least amount of material. The geometry used here had a W/d ratio of 5 to ensure consistency with previous extensive work on open-hole tensile specimens [5], [6], [7] and [8], which was based on an industrial test standard. This is slightly lower than the recommended W/d ratio of 6 in the ASTM D6742 standard.

In order to measure the clamping stress around the hole, a bush made from 6082-T6 aluminum alloy, with strain gauges bonded at opposite sides was used. These were calibrated so as to accurately calculate the clamping stresses and to ensure that there was no bending of the bush. Two "rigid" 4 mm thick aluminum washers of diameter 6.35 mm were then placed adjacent to the composite (Fig. 1) to apply the contact load. This gives a washer diameter to hole diameter ratio of 2:1. A4 class 70 stainless steel M3 bolts were used, with a measured thread diameter of 2.95 mm (using digital Vernier callipers). The hole diameter of the specimens was 3.175 mm (1/8 inch), thus the bolt thread diameter is a clearance fit with respect to the specimen holes so as to eliminate contact stress at the hole edge as an experimental variable.

The hole-diameter of the washer and bush is 3.1 mm with a maximum allowable shift of 0.15 mm with respect to the bolt. Care was taken when tightening to ensure that the washer remained central with the hole.

Quasi-isotropic carbon/epoxy laminates, with stacking sequence $[45_2/90_2/-45_2/0_2]_S$ and $[45/90/-45/0]_{2S}$ were used. The first, with plies blocked together is termed 'ply-level scaled', and the second, with repeated sub-laminate units is termed 'sub-laminate-level scaled'. The load bearing plies in the 0° direction are not at the surface and are thus more protected from impact damage, which is consistent with standard industry practice. These configurations were chosen as the open-hole ply-level specimens exhibited delamination dominated failure in quasi-static testing, with sub-laminate level specimens showing a fiber dominated pull-out type failure [5]. Bolt pressures of 23 MPa and 70 MPa were used for quasi-static testing, with 70 MPa used in fatigue, as the greater out-of-plane stresses are likely to produce the most notable effects. These bolt pressures were chosen from Ref. [1] as they represent pressures which gave the most notable effects for similar configurations. Tightening torque can be estimated from equation (1):-

$$T = K \cdot d \cdot F_{bolt} \quad (1)$$

where T is the torque applied, K is the "nut factor" for which a typical value of 0.2 is used, d is the bolt diameter and F_{bolt} is the axial bolt force. From the washer pressure and area the bolt force can be calculated as 554.8 N and 1688.4 N for the 23 and 70 MPa respectively, giving a tightening torque of 0.327 and 0.996 Nm. These values were low enough to be torque tightened using a hand-held hex key.

Using the bolt nominal stress area of 5.03 mm^2 the bolt stresses can be calculated to be 110.3 MPa and 335.7 MPa for the 23 and 70 MPa cases, well below the bolt yield stress of 450 MPa. Five quasi-static tests were carried out for each layup and washer pressure using a crosshead speed of 1 mm/min in all cases. The fatigue tests were run at various percentages of the average static failure load. Fatigue tests were run in load control at constant amplitude, with an R ratio of 0.1 and a frequency of 5 Hz. Each test was run to 10^6 cycles unless failure (defined as a 15% loss of stiffness) occurred prior to reaching this limit. This failure criterion was chosen by the authors and used in previous work[6] and [8] as it consistently represents the level of stiffness loss that corresponds to the change to an almost vertical slope in the stiffness versus number of cycles-curve.

10^6 cycles is commonly used in the aerospace industry as appropriate for the number of load cycles experienced during the useful life of an aircraft [17]. Fatigue tests have previously been carried out at 40%, 50%, 60%, 70% and 80% of the nominal static failure load P_{max} for ply level open-hole tests [6], and here 50%, 60%, 70% and 80% severities for bolt-filled ply-level specimens are used. For sub-laminate specimens, open-hole tests previously used 55%, 60%, 65%, 70% 80%, 85% and 90% of static failure load, and here 80%, 85% and 90% severities have been used for the bolt-filled sublaminate-level specimens. To characterize the effect of damage propagating under fatigue from the hole and free edge, an effective stiffness, E_{eff}, is calculated by using equation (2) below[6] and [8]:

$$E_{eff} = \frac{(P_{max} - P_{min})}{A(\varepsilon_{max} - \varepsilon_{min})} \approx \frac{P_{max}}{A(\varepsilon_{max} - \varepsilon_{min})} \quad (2)$$

where P_{max} is the load at peak amplitude, P_{min} is the load at the trough, A as the gross cross sectional area of the specimens and ε is the strain measured using a clip gauge extensometer recorded over a gauge length of 50 mm, symmetric about the hole for each specimen. The clip gauge and the relative stiffness change is used to identify the point of critical propagation of damage. The absolute value of stiffness is thus not of great significance, it is the normalized relative stiffness that is more important in the construction of the S–N curves.

The quasi-static and fatigue damage of selected bolted hole specimens were analyzed using X-ray CT and compared to previous work on the damage behavior of open-hole specimens [6].

NDT: 3D X-ray Computed Tomography

X-ray CT is rapidly becoming an increasingly important tool in the micromechanical damage analysis of materials [18] enabling routine non-destructive 3D imaging in the micron range.

To enhance X-ray contrast these samples were soaked in X-ray dye penetrant comprising 250 g zinc iodide, 80 ml distilled water, 80 ml isopropyl alcohol and 1 ml kodak photoflow, over a period of 2 days. Zinc iodide was used because it is relatively non-toxic and is reasonably X-ray opaque.

Micro X-ray CT scanning was performed at the UK National Composites Centre (NCC) in Bristol using a Nikon XT H 225 kV electron beam machine. Each specimen is mounted on the rotation stage and positioned between the X-ray source and the 2000 × 2000 16 bit pixel panel detector. A cone of X-rays is emitted from the target and passes through the specimen to the detector. During the scan, the specimen is incrementally rotated through 360° about the Z-axis with an X-ray image radiograph (projection) collected at each orientation. The individual set of radiograph projections are reconstructed into a 3D volume using a filtered back-projection reconstruction algorithm. For successful reconstructions, the width of the specimen (i.e. the dimension normal to the rotation axis) must always lie within the X-ray cone as it is rotated. The scanning voltage was 80 kV (320 μA) and the exposure time for each radiograph was 250 ms. The Metris CT Pro software was used to reconstruct each scan volume with post-processing and analysis performed using the VG studio max 2.1 and Avizo® 7 software packages. This enables visualization of the volumes of a given sample as a 3D map in which the grey scale value corresponds to the local X-ray absorption coefficient of the material. There are two peaks on the histogram of the grey scale intensity which indicates the solid material (fibers and matrix), and air (matrix cracking, delaminations, voids and other defects). Hence the matrix cracks and delaminations could be segmented by defining certain thresholds around the appropriated grey scale values [19].

For fatigue damage, a semi-automatic process was used for the analysis. A region of interest was placed around the approximate region of damage, if there was only localized damage in the specimen then a region close to the hole-edge was selected. For extensive delamination damage, a region of interest was selected to include the entire length of the specimen. A 3D seed growth tool was used in order to capture some excess damage which didn›t absorb the penetrant and to be able to capture the damage with greater precision. This tool enables segmentation of voxels by addition of all the adjacent points of a defined seed point which are within a tolerance with respect to the mean grey value of the selection [12].

QUASI-STATIC TESTING OF BOLTED-HOLE SPECIMENS

The initial bolted ply-level static tests were carried out using specimens manufactured from the same plate as the third batch of open-hole specimens used in the fatigue work described in Ref. [6]. Likewise the initial bolted sublaminate static tests used some specimens from the same plate as the second batch of open-hole sublaminate-level fatigue specimens from Ref. [6]. Therefore the static open-hole strengths were already characterized and documented, allowing for a straightforward comparison for the bolted results. A fourth batch of ply-level specimens were produced, which was used to carry out further bolted hole static testing and subsequent fatigue testing. A static interrupted test was loaded to 80% of the failure strength of a ply-level bolted specimen from the fourth batch to observe any subcritical damage. Another batch of specimens was produced for the sublaminate-level bolted fatigue tests. Table 1 shows all quasi-static bolt-filled hole results and their corresponding failure mode. D denotes delamination failure, P denotes pull-out.

Table 1. Summary of bolted results with their open-hole counterparts across all batches.

Layup	Batch number	Washer pressure (MPa)	Strength, MPa (CV%)	Number of specimens tested	Failure mode
Ply-level	3	Open-hole	395.7 (5.43%)	4	4xD
Ply-level	3	23	476.8 (4.4%)	5	2xP, 3xD
Ply-level	3	70	505.1 (2.9)	5	5xP
Ply-level	4	70	495.3 (4.5%)	3	3xP
Sub-laminate	2	Open-hole	531.8 (3.13%)	4	4xP
Sub-laminate	2	23	526.2 (2.0)	5	5xP
Sub-laminate	2	70	544.1 (3.1%)	5	5xP
Sub-laminate	3	70	561.3 (4.9%)	3	3xP

For the ply-level specimens (which have an open-hole failure mode by delamination) using a washer pressure of 23 MPa, an average failure strength of 476.8 MPa was recorded with a CV of 4.4%. This is remarkably higher than the open-hole strength of 395.7 MPa (with a CV of 5.4%) from the same batch of material, giving an increase in strength of 17.3%.Two of the five bolted specimens failed via pull-out, and three out of five specimens failed by delamination. For the ply-level specimens using a washer pressure of 70 MPa, an average failure strength of 505.1 MPa with a CV of 2.9% was recorded for 5 specimens. The next batch of specimens tested with the same 70 MPa washer pressure gave a failure strength of 495.3 MPa with a CV of 4.5% for 3 specimens. For the 70 MPa case pull-out failure occurred for all ply-level specimens tested. The batch 4 result was used to set the fatigue severities used in the following section.

Fig. 2 shows stress-crosshead displacement curves from three ply-level tensile tests; two for a washer stress of 23 MPa, one of which failed by pull-out and the other by delamination, and one for a washer stress of 70 MPa. The 23 MPa delamination case is an example of a test with load drops, due to the delamination occurring asymmetrically either side of the hole. After the first load drop, caused by delamination back to the grips, the load began to increase again until recovered to a similar level, when the second delamination propagated, causing the second load drop. After this the load increased again until fiber failure in the remaining 0° ligaments was observed. Although the specimens were loaded until failure of the 0° plies, the specimen had lost its structural integrity during the first major delamination, which is classed as the failure point. Applying a washer pressure around the hole-edge, increases the stresses required for delamination. There is thus a higher overall level of stress in the plies at the point at which delamination initiates compared to the open hole case. Once delamination starts to propagate it can happen that for some cases the load redistribution is sufficient to cause fiber failure, resulting the pull-out type failure. At this level of clamping pressure the specimens are thus in a transition regime between the two modes.

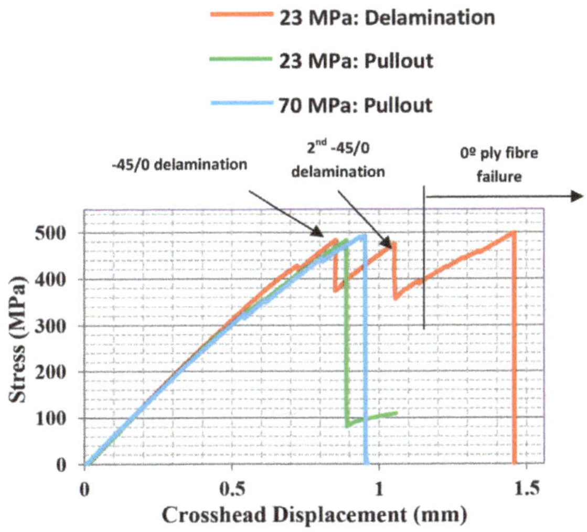

Figure 2. Ply-level tests with a bolt-filled hole.

When the washer pressure is increased to 70 MPa the enhancement of the delamination stress is such that fiber failure stress is consistently reached prior to full width delamination and pull-out failure occurs in all cases. The load curve for these specimens, as shown in Fig. 2, is very similar to the 23 MPa pull-out case.

There are some small drops in load prior to failure, which is due to subcritical damage in the form of matrix cracks and small localized delaminations. This cumulatively adds to the non-linearity of the curve but the load drops are not large enough to exceed the static failure criterion of a 5% load drop. Readers are referred to references [5] and [16] for a more extensive discussion on the differences between the delamination and pull-out failure mechanisms. Fig. 3 shows some of the clearest examples of (a) pullout failure, and (b) delamination failure from the open-hole tensile strength scaling study in Ref. [5].

(a)

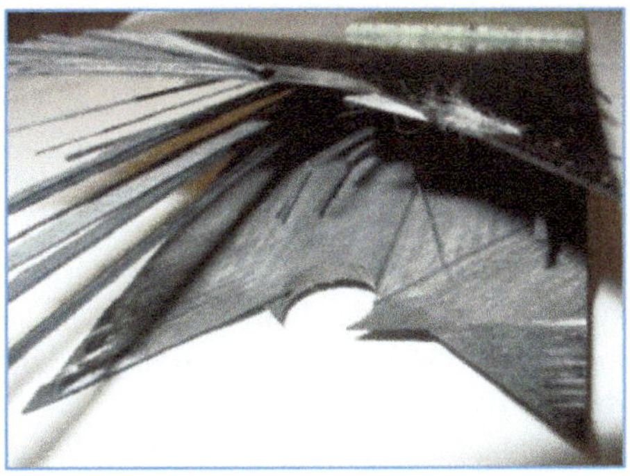

(b)

Figure 3. Examples of pullout failure: (a), and delamination failure: (b) from Ref. [5].

Fig. 4 shows X-ray CT scans of a ply-level 70 MPa washer pressure test loaded quasi-statically to 80% of the failure strength and then interrupted. Each image in the table shows the respective plies through the laminate thickness on one side of the center-line for a single specimen, along with an image of all of the damage at every interface. In some plies there is some "shadowing" from the adjacent interfaces due to the minimum resolution that could be obtained from the CT scans. The damage is mostly in the form of extensive matrix cracks in the surface 45° and 90° plies, which are dispersed throughout the entire length of the specimen. There is very little damage in the −45° and 0° plies with no observable delaminations. This differed substantially from the equivalent open-hole test which is also included in Fig. 4 where the matrix splits were only observed local to the hole at 80% of the failure load [6]. When comparing the magnitude of the peak loads of the bolted hole specimens at 80% severity (0.8 × 505 MPa = 404 MPa) to the open-hole strength of 395 MPa it is clear that the interrupted load of the bolted specimen is greater than the failure strength of the open-hole specimens. A greater amount of matrix cracking therefore occurs than was observed in the ply-level open-hole static interrupted test at 80% σ_{UTS}[6].

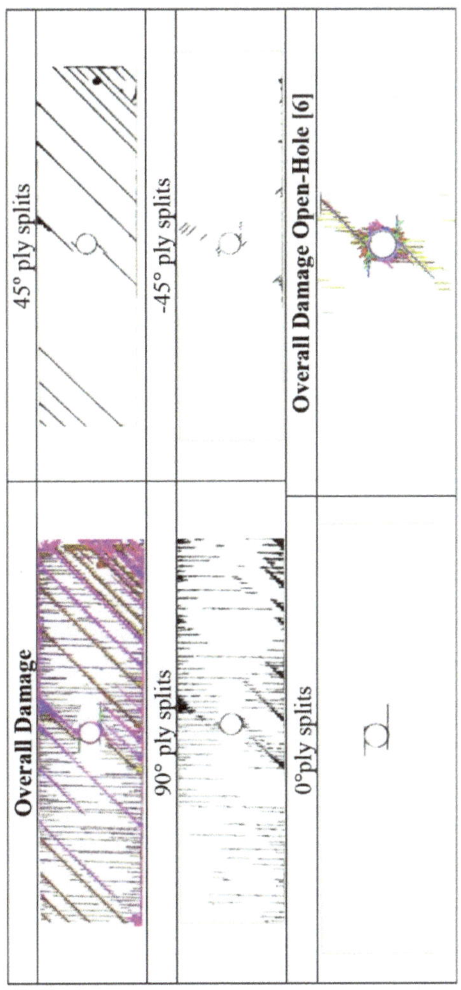

Figure 4. CT images of a static ply-level interrupted test at 80% strength for washer pressure of 70 MPa compared with the equivalent open-hole test.

The failure strengths of the bolt filled sublaminate specimens using a washer pressure of 23 MPa were found to be 526.2 MPa with a CV of 2.0%, and for the washer pressure of 70 MPa, 544.1 MPa with a CV of 3.1% for 5 specimens in each case. A new batch of sublaminate level specimens (batch number 3) was manufactured for fatigue testing using a washer pressure of 70 MPa only. The quasi-static failure strength for this batch was 561.3 MPa with a CV of 4.9% over three specimens using a washer pressure of 70 MPa. This result was used to set the severity for fatigue tests where only the 70 MPa washer pressure is used.

For bolt-filled hole sublaminate tests at both washer pressures, there was very little difference from the open-hole strength of 531.8 MPa with a CV of 3.13%. All specimens for both washer pressures failed by pullout as per the open-hole case. Fig. 5 shows an example of the typical stress displacement curve as well as variation of the washer pressures on the secondary axis for a bolted-hole static test at 70 MPa. Typically the strains on each side of the bush (washer pressures) are no more than 5% apart. As the load increases up to failure the washer pressures decreases slightly due to Poisson's contraction of the specimen. When the specimens fail, a small abrupt decrease in the washer pressure is observed as the damage causes the properties of the material below the washer to degrade.

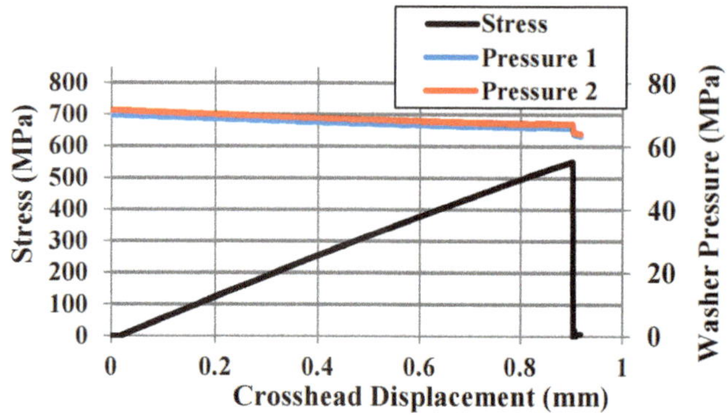

Figure 5. An example of a sublaminate-level test with a bolt-filled hole and washer pressure of 70 MPa.

A statically loaded test was interrupted at 90% of the failure strength for the sub-laminate level bolted-hole specimen using a 70 MPa washer pressure. There was very little sub-critical damage observed (Fig. 6). The damage was mostly small matrix splits in the 90° plies around the hole-edge and also the edges of the specimen. The open-hole equivalent showed a very similar level of damage.

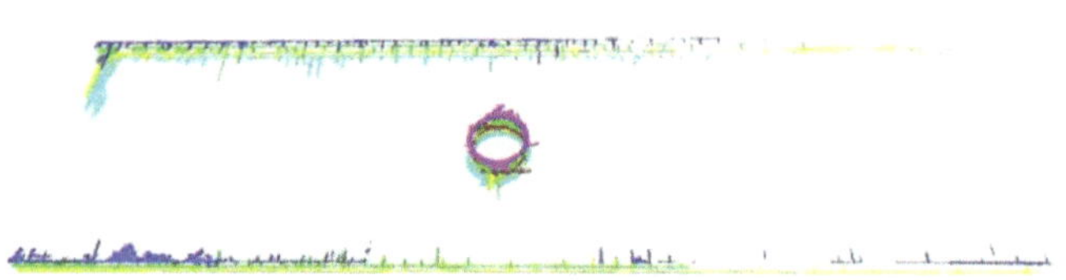

Figure 6. Sub-critical damage in a bolted quasi-static sublaminate test interrupted at 90% strength.

To summarize, only if the laminates are prone to failure by delamination, does the static failure strength increase. The bar chart in Fig. 7 shows the effect of the bolt-filled holes and clamping pressure compared to that of the open-hole configuration. The figure shows that the greatest increase in the failure strength occurs for the ply-level specimens when the washer pressure is first introduced at 23 MPa. The failure strength is now much closer to the strengths of the sublaminate configurations. Inhibiting the delaminations by using a clamping pressure increases the stresses required for delamination to the point at which pull-out failure is consistently observed. The sublaminate configuration was least affected by the bolt clamp-up in terms of the failure strength and mode of failure in static tests, however this configuration had previously shown a failure mode change from pull-out to delamination in open-hole fatigue testing. It was thus of interest to test both cases in fatigue for the bolted configuration. Although the average sublaminate level quasi-static strength for 23 MPa washer pressure is slightly lower than the open-hole quasi-static strength, the scatter for both configurations is greater than the difference in the strengths and so in terms of statistical significance the results can be considered to be the same. For the 70 MPa washer pressure sublaminate configuration the batch to batch variation in the average strengths is greater than the difference from the open-hole quasi-static and 23 MPa washer pressure strengths. Thus no overall trend in the sublaminate level configuration results could be observed, unlike for the ply-level configuration, and it therefore can be concluded that the washer pressure has no measurable effect on the sublaminate quasi-static strengths.

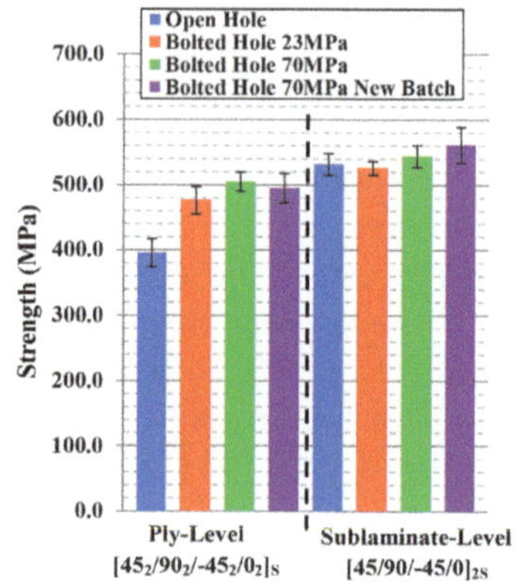

Figure 7. Bar chart for all configurations for ply and sublaminate tests with their failure strengths.

BOLTED HOLE FATIGUE TESTING

The bolted case was studied under fatigue loading to investigate possible differences due to the effects of clamp-up. For both the ply-level and sublaminate specimens the fatigue tests were carried out using a clamp-up pressure of 70 MPa only. From the quasi-static testing of the ply-level case this condition gave the most consistent change in failure mode from open hole to bolted testing. For the sublaminate case, although very little change was observed between the open hole and bolted condition in static tests it was postulated that if any differences were to be observed in fatigue, these would occur at the highest clamping pressure, minimizing delamination initiation around the hole. Open-hole tension fatigue tests using ply-level specimens gave delamination as the dominant failure mode [6]. Sublaminate-level tests gave a change in failure mode from pull-out, in the static case, to delamination dominated failure in all but the highest of severities of fatigue loads. Currently there has been very little work carried out on the effects of clamp-up pressure under fatigue loading conditions. A new batch of ply and sublaminate-level specimens were needed for the fatigue study on the effects of a bolt clamp-up stress. The results for three static tests on the new batches of specimens showed excellent consistency with previous results. The Weibull distribution is commonly used in survival analysis, reliability engineering and in failure analysis [20]. The probability function of a Weibull random variable is shown in equation (3):

$$f(x;\alpha,\beta) = \begin{cases} \frac{\beta}{\alpha}\left(\frac{x}{\alpha}\right)^{\beta-1} e^{-\left(\frac{x}{\alpha}\right)^{\beta}}, & x \geq 0 \\ 0, & x < 0 \end{cases} \quad (3)$$

where α is the scale parameter, β is the shape parameter (also known as the Weibull modulus). The Weibull random variable, x, in this instance is the number of cycles Nf. The fit of data is assessed using a Weibull plot. Equation (3) is integrated to give the cumulative density function in Equation (4) which is rearranged to give (5).

$$P_x = 1 - e^{-\left(\frac{x}{\alpha}\right)^{\beta}} \quad (4)$$

$$1 - P_x = e^{-\left(\frac{x}{\alpha}\right)^{\beta}} \quad (5)$$

$$R_x = 1 - P_x \quad (6)$$

The probability of failure is given by Px, and the probability of survival is given by Rx (6). Taking the natural logarithm of each side in equation (5) gives equation (7) in the linear regression form of y=mx+c respectively, thus giving $N_f = \ln x$, m=β, and $c = -\beta \ln \alpha$.

$$\ln\left(\ln\left(\frac{1}{1-P_x}\right)\right) = \beta \ln(x) - \beta \ln(\alpha) \quad (7)$$

The method of obtaining Probability of survival vs number of cycles plots is as follows:

- For each fatigue severity or stress level, the number of cycles to failure must be known.
- For a given severity a serial number must be assigned to each failure value (i = 1, 2, 3 ...).
- Each value of failure probability used the Bernard's Median Rank formula defined in equation (8):

$$MR = \frac{i - 0.3}{n + 0.4} \quad (8)$$

  where i is the serial number assigned to a failure within a particular fatigue severity, and n corresponds to the number of specimens in the given severity.

- Values for ln(ln(1−MR)), and ln(N_f) were calculated, and plotted on y and x axes respectively, in order to obtain parameters á and â via linear regression analysis (omitted here for simplification).
- Once parameters á and â are known, equation (9) is used for the calculation of life for a given probability of failure as shown in Fig. 9 and Fig. 16a, b.

$$N_{R_x} = \alpha \cdot (-\ln R_x)^{1/\beta} \quad (9)$$

For further details of other successful applications of Weibull statistics to fatigue data see Ref. [20]. The main advantages of using a Weibull analysis is its ease of application due to simple functions. It also provides some aid to physical rules regarding failure. The disadvantages of using this method are that it becomes more unreliable with smaller sample sizes and with specimens that fail in different modes, as in this paper. The ASTM E739-10 standard [21] gives reasonable guidelines for the minimum sample sizes required for a given test. The sample sizes in this paper are representative of preliminary and exploratory coupon level experiments (sample sizes ranging from 3 to 11 specimens depending on severity and configuration). Design allowables and reliability data typically require sample sizes between 12 and 24 specimens.

Bolt-filled Hole Ply-level Results

A minimum of 3 specimens were tested at 80%, 70% and 60% severity with a single run-out test at 50% severity. The results show a significant increase in the number of cycles to failure for the severities tested in comparison to the open-hole specimens. Fig. 8 shows a clear shift in the SN curve, giving failure at a higher number of cycles for a given severity for the bolted case. Fig. 9(a) and (b) show the Weibull distribution for open- and bolted-hole ply level specimens respectively. The plots show a good fit for the open-hole data which contains a wide enough sample size for accurate Weibull analyzes with reasonable fits for bolted specimens.

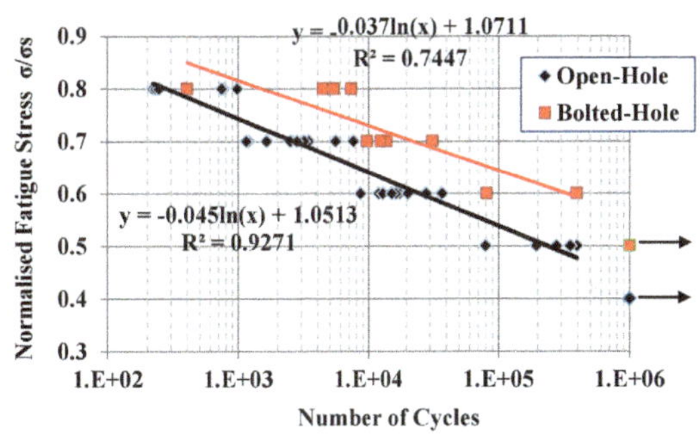

Figure 8. The relationship between open-hole and bolted-hole in fatigue loading for ply-level specimens.

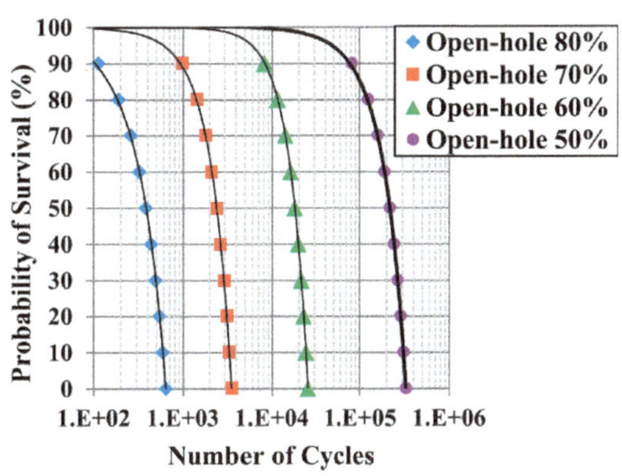

(a)

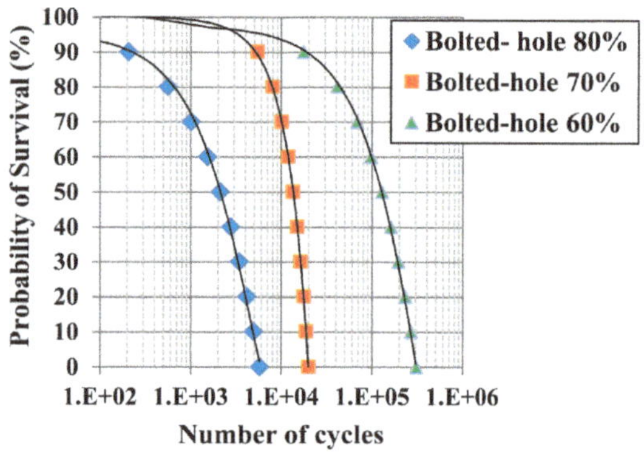

(b)

Figure 9. Probability of survival graphs for ply-level (a) open- and (b) bolted-hole test programmes.

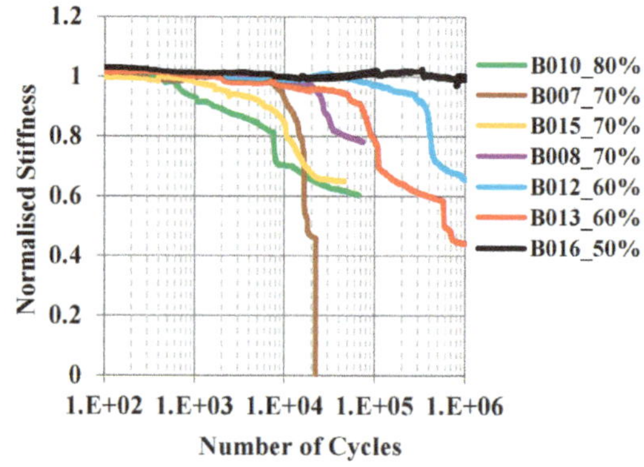

Figure 10. Initial results obtained for the bolted ply-level fatigue tests.

This is due to the clamp-up pressure which delays the onset of delamination from the hole-edge. For the majority of the bolted specimens with the ply-level configuration, once damage starts to propagate there is an initial decrease in the stiffness which satisfies the failure criterion, but the effective modulus curve then levels out at a plateau at approximately half way through the total stiffness decrease (Fig. 10). This can be attributed to the clamp-up pressure which inhibits the propagation of the 2nd asymmetric delamination of the -45/0 interface. There are some delaminations initiating from the specimen edges. Fig. 10 shows the fatigue results for selected ply-level bolted-hole cases which are representative stiffness loss curves, including some of the samples that were CT scanned.

In the specimen naming convention, B denotes that the specimen is bolted, this is followed by the test number and severity. All specimens tested failed by delamination at the 15% stiffness drop threshold.

In the case of specimen B007-70%, this failed in a similar manner to the open-hole ply-level specimens but after delamination has occurred fiber-failure ensues at the second plateau when only the 0° plies are carrying the load. This phenomenon was commonly observed in the open-hole sublaminate fatigue tests, but never in the ply-level open-hole cases, where the second plateau is continuous without any fiber failure.

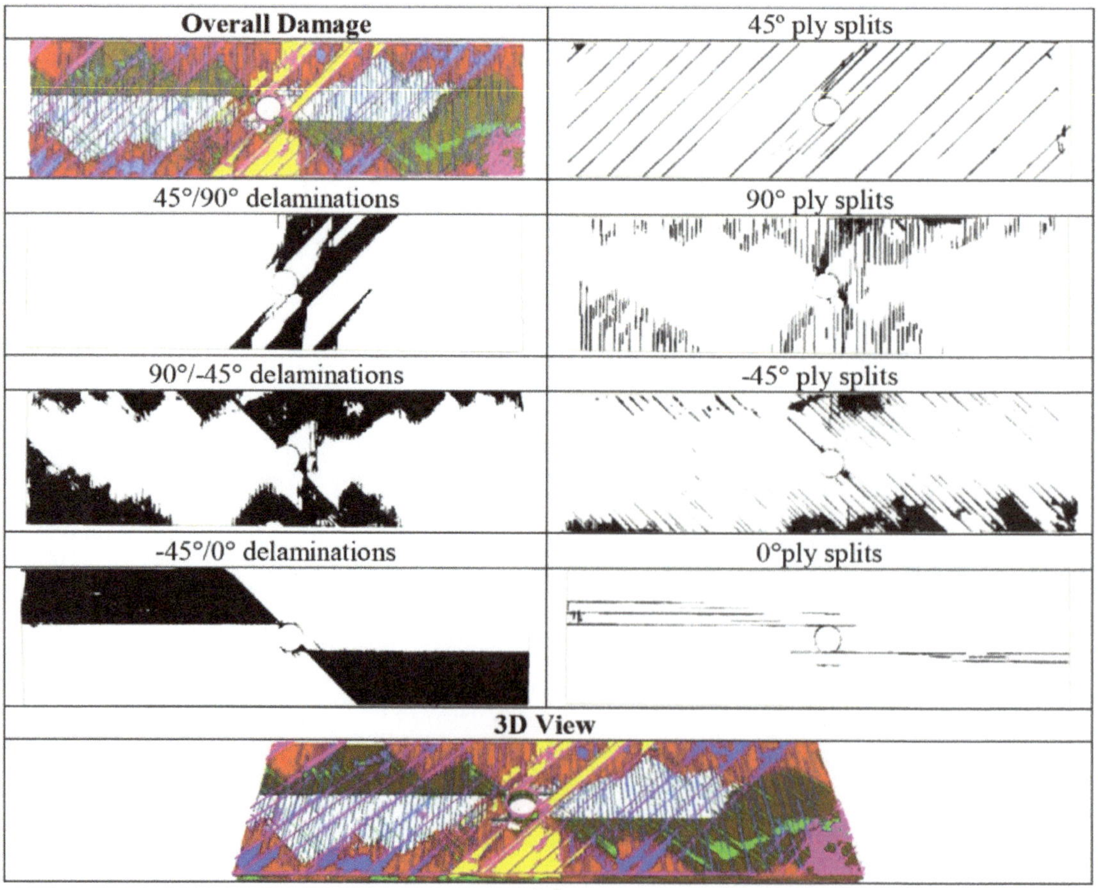

(a)

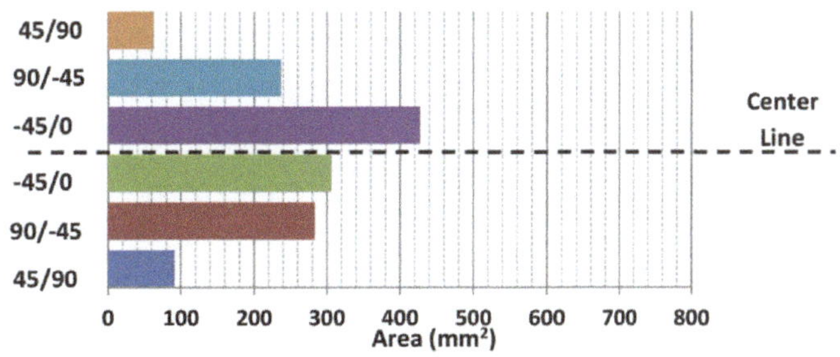

(b)

Figure 11. Specimen B012 – 60% severity: (a) CT images, and (b) Damage area quantification through the thickness.

In all but the test case of specimen B007-70%, the stiffness decreases from the initial plateau and levels out at a normalized stiffness of around $E/E_0 = 0.7$. The second drop in the stiffness occurs after a sizable delay, as seen in specimen B013 at 60% severity. To get a better understanding of the damage, specimen B012 and B013 (60% severity tests) were analyzed by X-ray CT. Specimen B012 reached 10^6 cycles whilst still on the first plateau (normalized stiffness of ~0.7) when the test was stopped at 10^6 cycles. Specimen B013, tested at the same severity by contrast had reached the second plateau (normalized stiffness of ~0.5) after the same number of cycles. Fig. 11 shows the extent of the damage in specimen B012. This epitomizes the damage seen in all specimens at $E/E_0 = 0.7$.

Fig. 12 shows the completely damaged specimen B013 at 60% severity. Firstly it is notable that the bolted –hole fatigue damage differs somewhat from open-hole fatigue tests. There is significant edge delamination in the 90/-45 interface due to the surface plies being more constrained around the hole by the bolt clamping force thus delamination can't open up so easily, causing them to initiate from the edges as well. Secondly the matrix cracks are much more dispersed throughout the laminate in particular cracks in the surface 45° and 90° plies are not local to the hole as in open-hole tests. Another notable difference is the complete absence of delamination which propagates out from the hole in between the parallel matrix cracks of the 0° plies as seen in open-hole tests. Comparing specimen B012 in Fig. 11 to B013 in Fig. 12, it can be seen that the delay in the stiffness loss giving rise to the intermediate plateau at around $E/E_0 = 0.7$, is due to hindrance of the second asymmetric -45/0 delamination from the compressive bolt forces around the hole. This is common for all severities. The damage seen in the other interfaces appear to be similar for both specimens. Using image processing software, bar charts were produced which accurately quantifies the extent of planar delamination areas at each interface through the thickness of the laminate. The 45/90 and 90/-45 interfaces have reached a saturation limit in which the delamination of the -45/0 interface is responsible for the majority of the stiffness degradation.

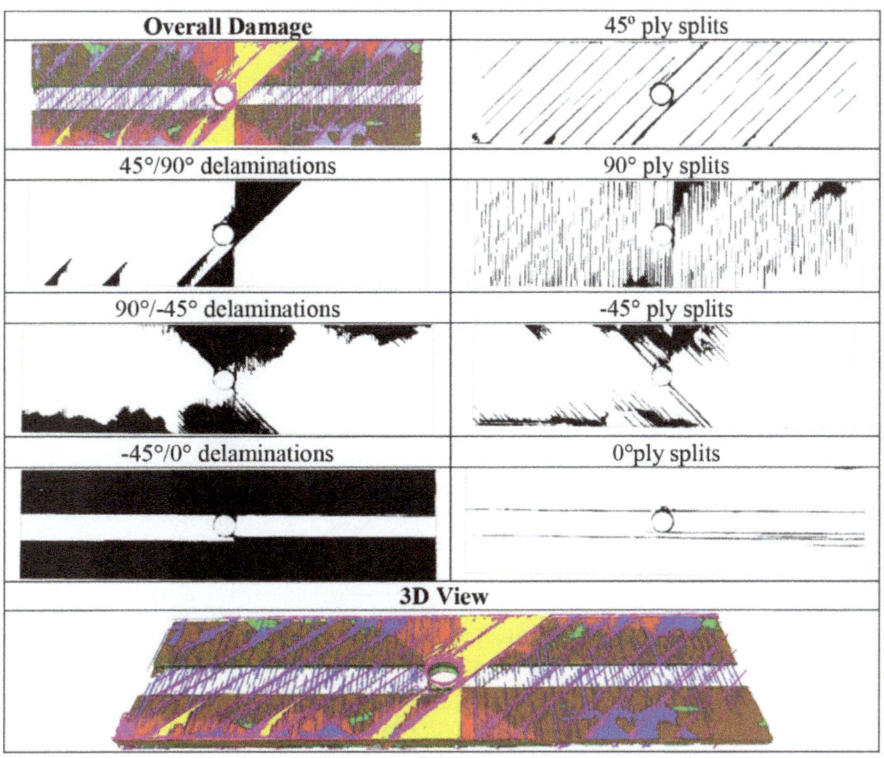

(a)

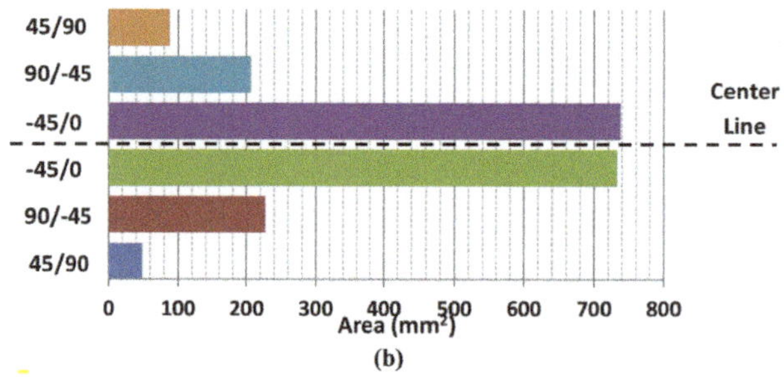

Figure 12. Specimen B013 – 60% severity: (a) CT images, and (b) Damage area quantification through the thickness.

A single specimen was tested at 50% severity which reached 10^6 cycles with no measurable stiffness loss (run-out), and was also evaluated using X-ray CT. Fig. 13illustrates the damage in this specimen. The dominant failure process responsible for the majority of the stiffness reduction in failed specimens is the delamination of the -45/0 interface, which is notably absent in this case. There is an extensive amount of matrix cracking throughout the specimen and a significant level of delamination in the 90/-45 interface emanating from the specimen edges. There is a no delamination in the 45/90 interface. Run-outs for the open-hole ply level configuration occurred at 40% severity with much less observable damage [6].

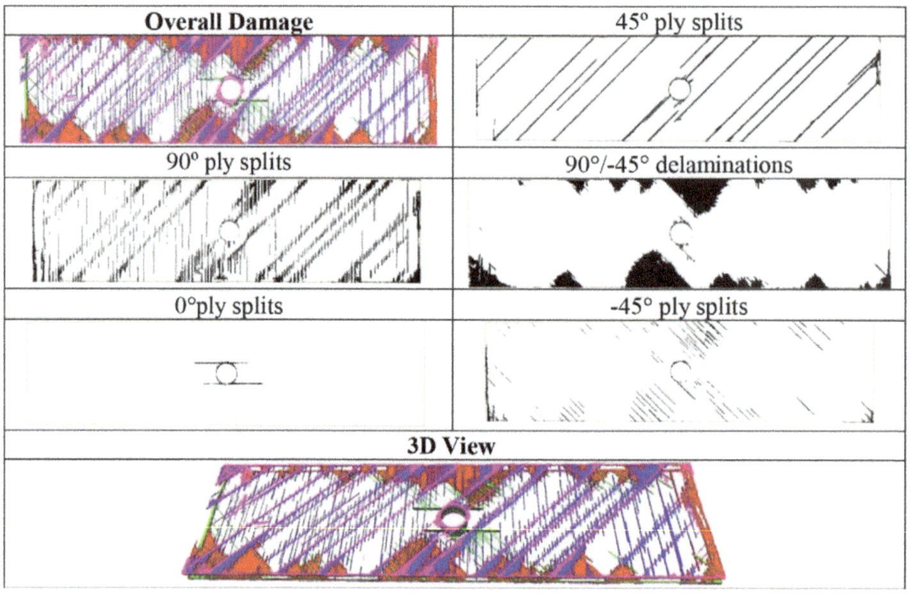

Figure 13. CT images of a runout specimen at 50% severity for a ply-level test at 70 MPa.

Fig. 14 shows a stiffness curve together with washer pressure for a typical bolt-filled ply-level fatigue test. It shows that there is a similar relationship between the bolt pressure and number of cycles as there is for stiffness loss, and thus the damage of the specimen. The washer pressure drops at around 10^5 cycles, which implies a loss of integrity of the material beneath it and that damage had initiated around the hole at this point, despite the clamping pressure. The fact that the damage around the hole is coincident with the loss in stiffness suggests that it is the principal failure mode and that the large amount of free-edge delamination seen in the CT scans is secondary.

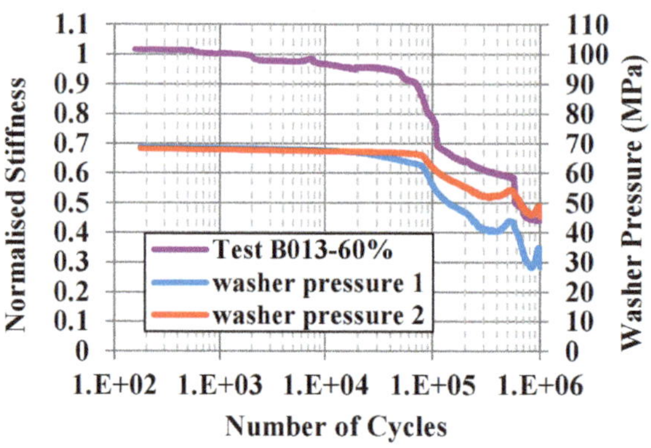

Figure 14. Washer Pressures (primary axis) and stiffness (secondary axis) for the ply-level bolted test B013-60%.

Bolted Hole Sublaminate-level Fatigue Results

The SN curve for the sublaminate bolted hole results (Fig. 15) also shows a substantial increase in the number of cycles to failure in comparison to the open hole specimens for the severities tested. All bolted sublaminate specimens tested failed by pull-out under fatigue loading, whereas for open-hole specimens they failed by delamination in all but the highest severities. The bolt clamp-up thus completely inhibits the delamination failure mode in this case, causing the substantial rise in fatigue life. There is notably less scatter than for the ply-level case as the failure mode is fiber-dominated. Fig. 16(a) and (b) shows the open- and bolted-hole Weibull probability curves respectively. The curve for 85% severity crosses over with the curve for 80% severity, due to an outlier at 85% severity which failed at around 2E4 cycles. The Weibull curves appear to be rather sensitive to scatter and outliers, furthermore they become less reliable for the bolted configurations due to relatively fewer specimens tested in comparison to open-hole configurations.

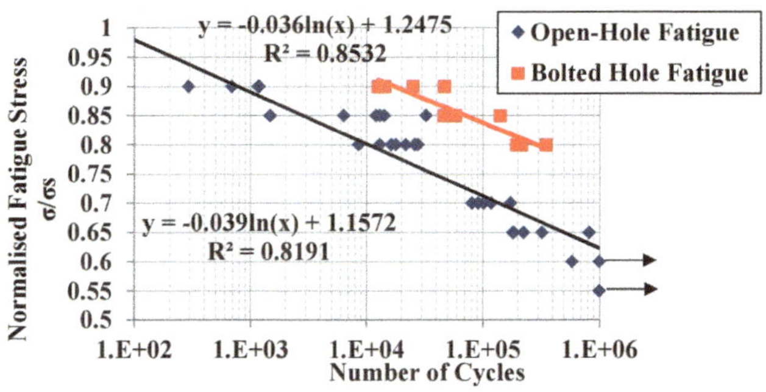

Figure 15. The relationship between open-hole and bolted-hole in fatigue loading for sublaminate-level specimens.

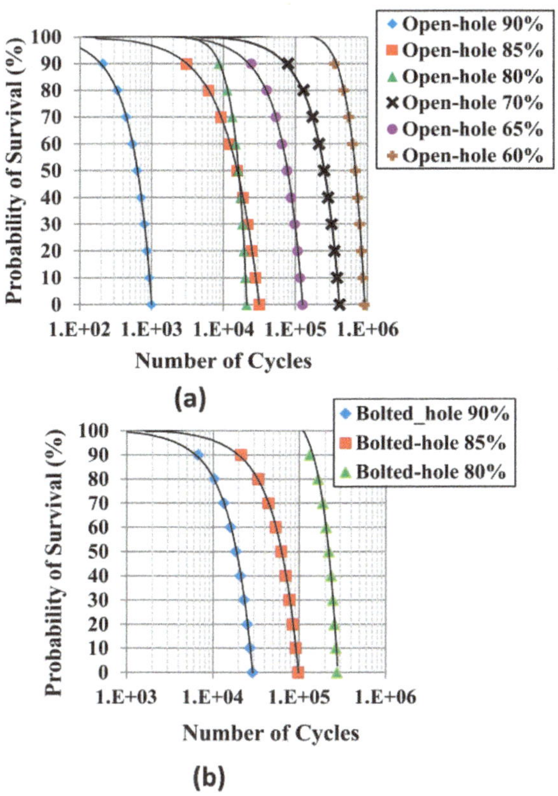

Figure 16. Probability of survival graphs for sublaminate level (a) open- and (b) bolted-hole test programs.

Although there is a subtle increase in the failure strength in the static case using the higher washer pressure, the effects seen in fatigue are much greater. A plot of normalized stiffness versus number of cycles for a specimen at each severity tested is shown in Fig. 17. In general each test shows a small degradation in the fatigue performance prior to failure by fiber-fracture and pull-out. In each case the failure criterion is met when fiber-failure ensues rather than through gradual stiffness loss prior to pull-out.

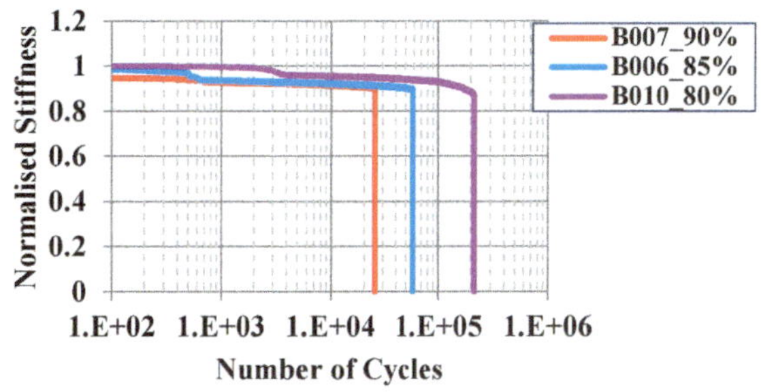

Figure 17. Stiffness loss curve for each of the three severities tested for the bolted sublaminate specimens.

Fig. 18 shows a typical relationship between stiffness loss and washer pressure for a bolt-filled fatigue test (sublaminate specimen B010 with an 80% severity). The stiffness gradually degrades as the number of cycles increases from the start of the test until pull-out failure occurs at 213,795 cycles. There is a minor decrease in the washer pressures measured from the compressive strains on both sides of the bush until failure of the specimen. As the specimen fails there is a sharp drop in the washer pressure down to 32 and 36 MPa for washer pressure 1 and 2 respectively. This suggests that there is some damage developing around the clamped hole, as the fatigue test is progressing.

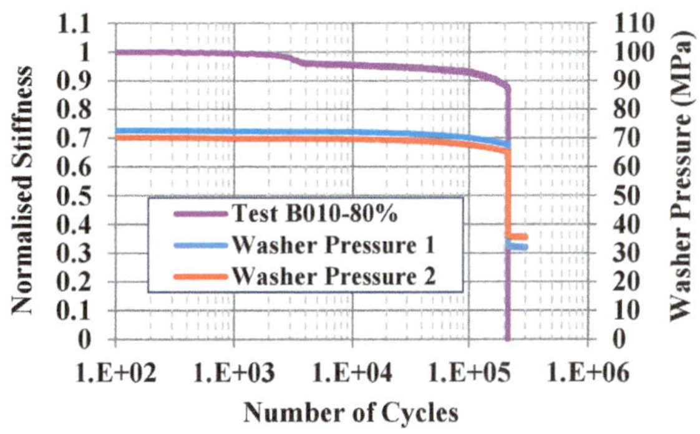

Figure 18. Washer Pressures (primary axis) and stiffness (secondary axis) for the sublaminate-level bolted test B010-80%.

Fig. 19 shows the extent of damage post failure in a test at 80% severity. There is extensive edge delamination at the -45/90 interface but no evidence of damage in the -45/0 interface. The CT slice showing fiber-failure of the central 0° plies is shown without segmentation for clarity.

Fig. 20 shows the extent of zero degree splitting in a typical fatigue failure compared to the quasi-static failure with the same washer pressure of 70 MPa. It shows that the extent of splitting under fatigue loading is much greater for fatigue tests than for quasi-static. The effect of this would be to increase the fatigue life due to the notch blunting further delaying the onset of fiber failure.

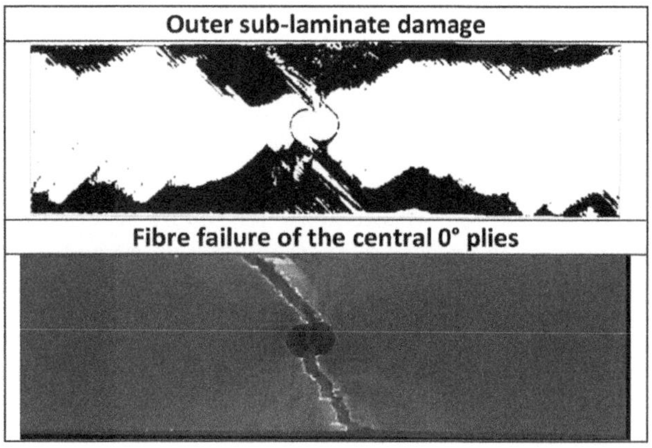

Figure 19. Damage in the surface sublaminate, with the un-segmented image of the central 0 plies showing fiber-failure.

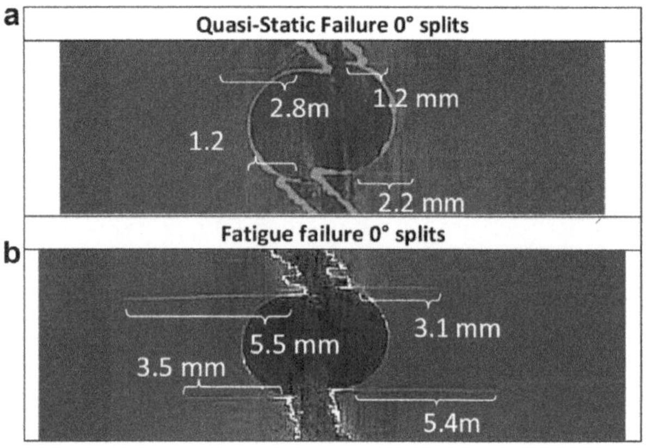

Figure 20. Comparison of zero degree split lengths obtained from CT scans for (a) the static failure, and (b) a typical fatigue failure at 80% severity.

CONCLUSIONS

An experimental program was undertaken to show how damage evolves in bolted composite specimens containing a centrally located hole with a bolted clamping force. This paper shows how the damage mechanisms can change under quasi-static and fatigue loading and how they compare to previous work on specimens containing an open-hole. Also it is shown how the bolted clamping stress inhibits delamination. The enhancement of through-thickness properties is a phenomenon previously reported in the literature, but not extensively for this configuration. The different specimen layups and loading were compared in terms of their overall failure mode and sequence of damage events leading to overall failure. Quasi-static testing of bolted specimens was carried out using two levels of bolt clamping pressures for two different layups. One bolt pressure was chosen for the fatigue study to observe any effects of the through thickness forces on the damage mechanisms and fatigue life. The following points contain some reiteration drawn from previously published material in Ref. [6] which is included for completeness and direct comparison with the bolted testing program. The key features of the specimen failures can be summarized as follows:-

- The ply level open-hole specimens showed the same overall failure mode of delamination under fatigue and quasi-static tensile loading with run-outs at 40% severity.
- The sublaminate open-hole specimens showed pull-out failure under quasi-static loading, which changed to overall failure by delamination under fatigue in all but the highest severities with run-outs at 55% severity.
- The ply-level specimens showed the greatest increase in the bolted quasi-static strength for the lower washer pressure of 23 MPa, 3 out of 5 specimens failed by pull-out, and 2 by delamination.
- At the higher washer pressure of 70 MPa, there was a smaller increase in the ply-level quasi-static strength with all specimens failing by pull-out.
- The sublaminate specimens showed only a minor increase in the quasi-static strength under bolted clamping pressures.
- Under fatigue, the ply-level case with a bolt-filled hole showed greater fatigue lives for all severities tested. The through thickness clamping force also delayed the onset of the second asymmetric -45/0 delamination with run-outs at 50% severity.
- The bolted sublaminate case gave greater fatigue lives at each severity tested.
- In all cases the increase in fatigue life was as a result of the clamping stresses delaying the onset or suppressing delamination.

It has been shown in this paper how X-ray CT scanning can be used to provide new information on the detail of damage formation during fatigue loading. This has been used to examine the role of bolt clamp up forces on the damage, in particular delamination. Previous studies have only highlighted the changes of overall strength due to bolting, but here this has been directly related to and explained by the details of the damage formed during both quasi-static and fatigue loading. In the future this information could be used to develop and validate numerical models for prediction of strength and joint performance. Furthermore this paper indicates some of the advantages and disadvantages of using Weibull statistics for sample sizes which are representative of preliminary and exploratory coupon level fatigue testing. The open-hole ply-level configuration gave a good fit of data as the sample size was adequate with consistent failures modes throughout all specimens and severities. The open-hole sublaminate configuration contained fiber dominated failure modes at the highest of severities with delamination dominating severities lower than 85%, which reduces the reliability of the Weibull analysis. For bolted hole configurations a smaller set of data is presented and compared with open-hole data, thus producing more variable Weibull fits.

ACKNOWLEDGEMENTS

This work was supported by the Engineering and Physical Sciences Research Councilthrough the EPSRC Centre for Doctoral Training in Advanced Composites for Innovation and Science [grant number EP/G036772/1].

REFERENCES

1. Yan Y, Wen WD, Chang FK, Shyprykevich P. Experimental study on clamping effects on the tensile strength of composite plates with a bolt-filled hole. Compos Part A: Appl Sci Manuf 1999;30(10):1215e29.

2. Schon J, Nyman T. Spectrum fatigue of composite bolted joints. Int J Fatigue € 2002;24(2e4):273e9.
3. Persson E, Eriksson I. Fatigue of multiple-row bolted joints in carbon/epoxy laminates: ranking of factors affecting strength and fatigue life. Int J Fatigue 1999;21(4):337e53.
4. Starikov R, Schon J. Local fatigue behavior of CFRP bolted joints. Compos Sci € Technol 2002;62(2):243e53.
5. Green BG, Wisnom MR, Hallett SR. An experimental investigation into the tensile strength scaling of notched composites. Compos Part A-Appl Sci Manuf 2007;38(3):867e78.
6. Nixon-Pearson OJ, Hallett SR, Withers PJ, Rouse J. Damage development in open-hole composite specimens in fatigue. Part 1: experimental investigation. Compos Struct 2013;106(December):882e9.
7. Nixon-Pearson OJ, Hallett SR, Harper PW, Kawashita LF. Damage development in open-hole composite specimens in fatigue. Part 2: numerical modelling. Compos Struct 2013;106(December):890e8.
8. Nixon-Pearson OJ, Hallett SR. An investigation into the damage development and residual strengths of open-hole specimens in fatigue. Compos Part A: Appl Sci Manuf 2015;69(0):266e78.
9. Spearing SM, Beaumont PWR, Kortschot MT. The fatigue damage mechanics of notched carbon-fiber PEEK laminates. Composites 1992;23(5):305e11.
10. Mohlin T, Carlsson L, Blom AF. An X-ray radiography study of delamination growth in notched carbon/epoxy laminates. In: Proceedings of testing, evaluation and quality control of composites. Guildford, England; 1983. p. 85e92.
11. Mohlin T, Blom AF, Carlsson L, Gustavsson AI. Delamination growth in notched graphite/epoxy laminates under compression fatigue loading. ASTM STP 876. In: Johnson WS, editor. Delamination and debonding of materials; 1985. p. 168e88.
12. Scott AE, Mavrogordato M, Wright P, Sinclair I, Spearing SM. In situ fiber fracture measurement in carbon-epoxy laminates using high resolution computed tomography. Compos Sci Technol 2011;71(12):1471e7.
13. Moffat AJ, Wright P, Buffiere JY, Sinclair I, Spearing SM. Micromechanisms of damage in 0 degrees splits in a [90/0](s) composite material using synchrotron radiation computed tomography. Scr Mater 2008;59(10):1043e6.
14. Nixon-Pearson OJ, Hallett S. An investigation into the damage development and residual strengths of open-hole specimens in fatigue. In: Proceedings of the International Conference on composite materials ICCM19. Montreal; 2013.
15. Hexcel 8552, Epoxy matrix product datasheet. Pdf download from: www. hexcelcomposites.com/Markets/Products/Prepreg/PrepregDownld.html.
16. Hallett SR, Green BG, Jiang WG, Wisnom MR. An experimental and numerical investigation into the damage mechanisms in notched composites. Compos Part A 2009;40(5):613e24.
17. Mall S, Weidenaar WA. Tension-compression fatigue behavior of fiberreinforced ceramic matrix composite with circular hole. Composites 1995;26(9):631e6.
18. Withers PJ, Preuss M. Fatigue and damage in structural materials studied by X-ray tomography. Annu Rev Mater Res 2012;42:81e103.
19. Wright P, Moffat A, Sinclair I, Spearing SM. High resolution tomographic imaging and modelling of notch tip damage in a laminated composite. Compos Sci Technol 2010;70(10):1444e52.
20. Sakin R, Ay _ I. Statistical analysis of bending fatigue life data using Weibull distribution in glass-fiber reinforced polyester composites. Mater Des 2008;29(6):1170e81.
21. ASTM. Standard practice for statistical analysis of linear or linearized stresslife (S-N) and strain-life fatigue data. 2010.

CHAPTER 10 HIGH DENSITY POLYETHYLENE COMPOSITES

Runzhou Huang, Xinwu Xu, Sunyoung Lee, Yang Zhang, Birm-June Kim, and Qinglin Wu

Keywords: Glass fiber; talc; HDPE; composites; mechanical; thermal expansion

INTRODUCTION

As a new-generation green composite, co-extruded (core-shell structure) natural fibers reinforced polymer composites (NFPC) has been recently developed and used to enhance performance characteristics of composites. Co-extrusion technology has become one of the most advanced polymer processing technologies due to its unique capacity in creating a multi-layer composite with different complementary layer characteristics, and in making the properties of the final products highly "tunable". In a core-shell structure wood polymer composites (WPC) system, the shell layer, made of thermoplastics unfilled or filled with minerals or natural fibers and other additives, plays a critical role in enhancing overall composite properties [1]. The shell layers with different material combinations, which have quite different properties, are, however, needed to achieve desired product performance.

Filled thermoplastics have been used widely as independent shell layer for co-extruded NFPC with core-shell structure. For example, it was demonstrated that a pure high-density polyethylene (HDPE) or pure polypropylene (PP) shell over a wood polymer composites (WPC) core reduced moisture uptake compare tonon-coextruded NFPC. However, the addition of a pure plastic shell with a relatively low modulus and large thermal expansion over a WPC core negatively affected overall composite modulus and thermal stability [1,2]. Investigations have also been done to develop a stabilized shell layer by blending HDPE and additives including a compatibilizer, a photostabilizer, and a nanosized TiO_2 on the coextruded WPC, by using combined wood and mineral fillers [2], by using carbon nano-tube (CNT) in a shell layer, by using precipitated calcium carbonate (PCC) [3]. Further developments of filled materials as a more cost-effective shell layer for co-extruded WPC is still necessary.

Polymer composites reinforced with glass fibers (GFs) and talc may be achieved in the form of higher modulus and reduced material costs, yet accompanied with decreased strength and impact toughness [4]. Huang et al. [5,6], demonstrated the influence of varying shell moduli and thermal expansion coefficients of GF filled HDPE shells on the overall thermal expansion of co-extruded composites using a finite element model and described effect of the talc content for shell layers on mechanical and thermal expansion properties of core-shell structures WPC.

Hybrid filler reinforced composites form a complex system, and there is inadequate data available about phenomena behind the property changes due to the addition of particulate fillers to the fiber reinforced thermoplastic composites. Although individual classes of fillers or fibers can contribute some desirable properties as reinforcement filler for core-shell structure WPC, hybrid fillers have attracted much attention as the reinforced agents in the shell layer of core-shell structure WPC. The real interest in composites is in optimizing the different contributions from different types of fillers.

The objective of this study described in this paper was to investigate the effect of individual fillers (GF vs. talc) and combined fillers (talc and GF) on morphological, mechanical, and thermal expansion properties of the filled composites as potential shell material for coextrude NFPC/WPC. The result of this study can help provide a fundamental base for developing new functional applications of core-shell structure NFPC/WPC with hybrid fillers reinforced shells.

Citation: Runzhou Huang, Xinwu Xu, Sunyoung Lee, Yang Zhang, Birm-June Kim, and Qinglin Wu. High Density Polyethylene Composites, Materials 2013, 6(9), 4122-4138; DOI: 10.3390/ma6094122

RESULTS AND DISCUSSION

Composites with Individual Glass Fibers

Morphology

Scanning electron microscopy (SEM) was used to characterize the morphology of GF filled HDPE composites. Typical SEM micrographs for GF filled composites are shown in Figure 1 at 10 wt % and 30 wt % loading levels. Certain fiber pullout happened during the fracture process as indicated by the circular voids on the fracture plane. Fiber pullouts were observed on the surface of GF-filled HDPE composites (Figure 1) due to poor bonding of fiber to matrix leads to easy fiber pullout during the impact. Composites showed lower impact strength than neat HPDE due to insufficient fiber to matrix contact. This is consistent with low impact strength at 10 wt % and 30 wt % GF loaded compared with the pure HDPE.

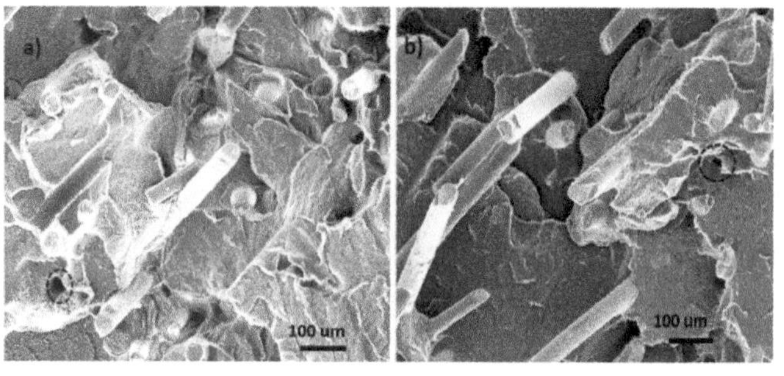

Figure 1. Scanning electron microscopy (SEM) micrographs of fractured surfaces of glass fiber (GF) filled high density polyethylene (HDPE) AD60 composites. (**a**) 10 wt % GF; (**b**) 30 wt % GF.

Most glass fibers were aligned perpendicular to the fracture plane (i.e., along the injection molding flow direction). Fiber breakage can contribute much less to energy than that the fiber pullouts in the net fractured energy [7]. A greater number of fiber pullouts can be observed on the fractured surface of a specimen with 30 wt % GF content than that of 10 wt % GF loaded. This was thought to be due fiber aggregation at the higher loading level in the composite matrix, which reduced their effective bonding.

Tensile Properties

Tensile properties of neat HDPE and its composites with different glass fiber loading levels are summarized in Table 1 and plotted in Figure 2 (HDPE AD60 only). Tensile modulus for both resin systems increased with increase of the filler loading levels. For the neat HDPE AD60 and HDPE 6706, the tensile modulus was found to be 0.86 ± 0.09 GPa and 0.26 ± 0.02 GPa, respectively. The modulus increased to 8.87 ± 0.5 GPa at the 40% GF loading for HDPE AD60, and 5.37 ± 0.24 GPa at the 30% GF filling HDPE 6706. Apparently, GFs showed a better influence on tensile modulus of filled composites than that of pure HDPE matrix. This was due to a larger modulus value, a larger aspect ratio, and surface coupling treatment of the GFs to enhance their bonding to the matrix.

Table 1. Summary of mechanical properties of neat high density polyethylene (HDPE) and filled HDPE composites by individual glass fiber.

System	Filler Content (wt %) [a]	Strength			Modulus	
		Tensile (MPa) [b,c]	Flexural (MPa)	Impact (kJ/m²)	Tensile (GPa)	Flexural (GPa)
HDPE-6706/GF	0	18.9(0.2)A	19.76(0.39)A	8.12(0.21)A	0.26(0.02)A	0.73(0.05)A
	10	27.1(0.38)C	25.3(1.0)C	7.37(0.21)A	1.86(0.21)C	1.2(0.06)BC
	20	35.42(0.15)E	37.7(0.3)E	9.72(0.82)B	3.43(0.41)E	1.6(0.8)DE
	30	46.74(0.88)G	56.3(0.7)G	11.81(0.88)C	5.37(0.24)G	3.4(0.1)G
HDPE-AD60/GF	0	23.8(1.4)B	21.8(1.0)B	28.57(2.0)E	0.86(0.09)B	0.85(0.06)AB
	10	31.54(0.57)D	29.2(0.2)D	9.62(0.37)B	2.46(0.22)D	1.3(0.02)CD
	20	39.34(0.93)F	40.9(0.6)F	10.37(0.37)B	5.17(0.94)F	2.3(0.3)F
	30	48.80(0.50)H	57.8(0.7)H	11.94(0.16)C	6.22(0.97)G	3.6(0.08)G
	40 [d]	64.92(0.34)I	85.9(1.6)I	14.55(0.34)D	8.87(0.51)H	5.8(0.2)H

[a] The filler content was based on the total composite weight; [b] Mean values with the same capital letter for each property are not significantly different at the 5% significance level; [c] Numbers in the parenthesis are standard deviation based on five specimens; [d]Master batch of HDPE-GF blend.

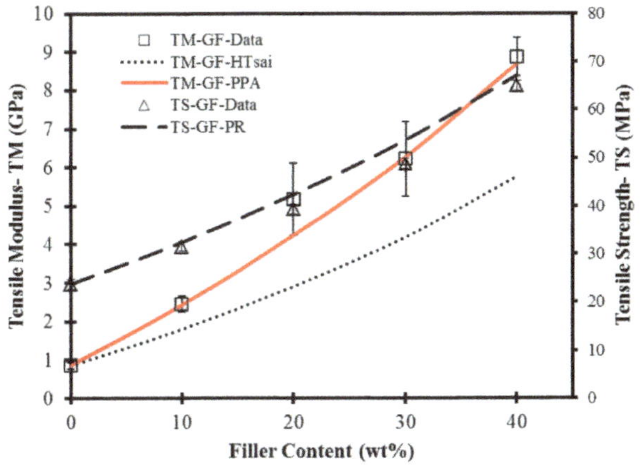

Figure 2. Tensile properties of GF-filled HDPE AD60 composites. Line showing predicted (PR) values with various models.

There is a near linearly increasing trend of tensile strength for composite. The strength increased to 64.92 ± 0.34 MPa for GF-filled composites at the 40% GF loading level. Based on the statistical data analysis, GF filled HDPE composites showed the significantly enhanced behaviors of tensile strength. The bonding at the interface is a crucial parameter in determining the tensile strength.

The increasing trend of GF filled composites was due to surface coupling treatment of GFs helped develop a strong bond between the fiber and matrix, which avoid the stress concentration formed around the GF fiber in the stressed composites and allows the composite to bear more applied load.

For GF-filled composites, several theoretical models can be used to predict tensile properties of fiber composites in terms of the properties of the constituent materials [8]. The Halpi-Pagano micromechanical model [9], which has been utilized to predict the modulus of composites with randomly oriented short fibers [10,11], is given below:

$$E_C^{ran} = \frac{3}{8}E_L + \frac{5}{8}E_T \quad (1)$$

where E_C is Young's modulus of random fiber composites, E_L and E_T are longitudinal and transverse Young's modulus, respectively, of corresponding uniaxial oriented discontinuous fiber composites. This equation is an averaging procedure for estimating elastic moduli of quasi-isotropic laminates. E_L and E_T can be estimated from the following Halpin-Tsai Equations [12,13].

$$E_L = E_m\left[\frac{1+(2l/d)\eta_L\phi_f}{1-\eta_L\phi_f}\right] \text{ and } E_T = E_m\left[\frac{1+2\eta_T\phi_f}{1-\eta_T\phi_f}\right] \quad (2)$$

Among which the constant η_L and η_T are defined as:

$$\eta_L = \frac{(E_f/E_m)-1}{(E_f/E_m)+(2l/d)} \text{ and } \eta_T = \frac{(E_f/E_m)-1}{(E_f/E_m)+2} \quad (3)$$

where E_f and E_m are Young's moduli of the fiber and polymeric matrix, respectively; ϕ_f and l/d is volume fraction and aspect ratio of the fiber in the composites, respectively. The elastic modulus of short fiber composite can also be predicted based on paper physics approach-PPA [14,15]:

$$E_C^{PPA} = \chi_1\chi_2 E_f\phi_f + E_m(1-\phi_f) \quad (4)$$

where χ_1 and χ_2 are, respectively, the fiber length and orientation factors for the composite elastic modulus. To assess predictability of the tensile modulus of GF reinforced HDPE blends, experimental results were compared with calculated data from the above models (i.e., Equations 1 and 4). The properties of the constituent materials of the composite used in the model prediction are E_f = 80 GPa, E_m = 0.86 GPa, density = 2.56 g/cm^3 and l/d = 150 for glass fiber. Figure 2 shows comparisons of model predicted results with the experimental ones obtained from tensile tests as a function of filler loading level. As shown, the theoretical curves predicted by Equation 1 are far below the experimental data points for both filler systems. The behavior could be due to several reasons. First, the model assumes a perfect filler dispersion, while the presence of filler agglomeration may lead to underestimate of filler volume. Second, particle orientation may strongly influence the calculated Young's modulus, while the model assumes a random fiber orientation. SEM micrographs (Figure 1) show significant alignment of the filler, especially for glass fibers. Equation 4 based on paper physics approach, PPA, was also used to calculate the tensile modulus. The calculated values based on a value of lumped parameter, $X_1 * X_2$, of 0.5 for glass fiber are also plotted in Figure 2. The predicted data are in a fairly good agreement with the experimental data.

The Nicolais and Nicodemo's model [16] did not fit the GF-filled composite data at all due to the strong interfacial bonding between fiber and plastic matrix, which led to an increase of the composite tensile strength as fiber content increased.

Prediction of the tensile strength for randomly oriented short-fiber composites is a difficult task, and no universally accepted theory exists on this subject. The difficulty arises because the material's ultimate strength in the case of composites is determined by the onset of fracture, and not via a yielding mechanism. Predictions are particularly difficult in the case of randomly oriented fibers, as cracks tend to propagate by fiber avoidance process as opposed to fiber pullout or fracture [17]. When predictions are attempted, they most frequently take the form:

$$\sigma_C^{PPA} = X_3X_4\sigma_{fu}\phi_f + \sigma_{mu}(1-\phi_f) \quad (5)$$

where σ_{fu} is the fiber tensile strength; σ_{mu} is the tensile stress in the matrix at composite failure strain; X_3 and X_4 are the fiber-length and orientation-correction factors for tensile strength, respectively. The strength properties of the constituent materials of the composite used in the model prediction are σ_{fu} = 2400 MPa, σ_{mu} = 23.8 MPa, and a lumped parameter ($X_3 \cdot X_4$) of 0.1. Figure 2 shows a comparison of the calculated data from Equation 5 and experimental data for tensile strength of GF-based composites, which shows a reasonable agreement between the two data.

Flexural Properties

Flexural properties of GF filled composites with varying amounts of each filler are shown in Table 1. Flexural modulus for GF filled composites exhibited an increasing trend with increased filler content. The neat HDPE AD60 had a flexural modulus of 0.85 ± 0.06 GPa while it was 5.8 ± 0.2 GPa for GF filled HDPE composites having 40 wt % of GF .The increase of flexural modulus was attributed to the enhanced interfacial interaction existed between the filler and matrix, which allowed the transmission of stress from HDPE to GF thereby improving the stiffness of the GF filled HDPE composite.

Based on the statistical data analysis, GF-filled composites showed a significant strength increase with increased GF loading level. At the 40% GF level, the strength was 2.29 times higher than that of the neat resin. Fiber alignment as shown in the SEM micrographs (Figure 1) played an important role in determining the flexural strength. The increased strength benefited from the uniaxially aligned GFs.

Impact Strength

Table 1 shows test data of the notched Izod impact strength of GF filled composites at various leading levels. The neat AD60 resin had an impact strength of 28.57 ± 2.0 kJ/m^2. The strength decreased significantly when the filler was added to the system. At the 10% filler loading level, the impact strength for GF filled composites are 9.62 ± 0.37 kJ/m^2 and 6.03 ± 0.13 kJ/m^2, respectively. For GF-filled composites, the impact strength increased with further increase of GF content beyond the 10 wt % level. As GFs are well bonded to the plastic matrix, the partially aligned GFs in the direction perpendicular to the impact force could help absorb the impact force imparted to the test samples. The enhanced impact strength at higher GF filling beyond 10 wt % due to GF fiber clustering in composite can be ruled out.

Dynamic Mechanical Properties

Storage Modulus (E′)—The effect of temperature on the storage modulus of GF-filled composites having 0%, 10%, and 30% loading levels is shown in Figure 3, respectively. A general trend of increase of the storage modulus with increased filler content in the composites was observed. E′ is more associated with the molecular elastic response of the composites, indicating the stiffness of the material. The increase of E′ with increased filler content was due to mechanical limitation posed by increasing filler concentration embedded in the viscoelastic matrix. The E′ decreased with temperature increase and converged to a narrow range at higher temperatures. The reduction of E′ with increasing temperature was due to the softening of the matrix and initiation of the relaxation process [18].

Loss Modulus (E″)—The loss modulus is a measure of the absorbed energy due to the relaxation and is associated with viscous response or the damping effect of the material. Figure 3 shows the effect of varying filler contents and temperature on loss modulus of the composites. E′ increased with the increased filler concentration and had a peak in the transition region around 50 °C. This relaxation peak was known as α-relaxation of HDPE [19], which was related to a complex multi-relaxation process associated with the molecular motion of the PE crystalline region [20,21]. The E′ at this relaxation temperature was markedly increased with the increase of filler loading level. The presence of a filler in the plastic resin reduced the flexibility of the material by introducing constraints on the segmental mobility of the polymer molecules. The α-relaxation peaks of GF-filled composites shifted to the higher temperature region as compared to neat HDPE resin.

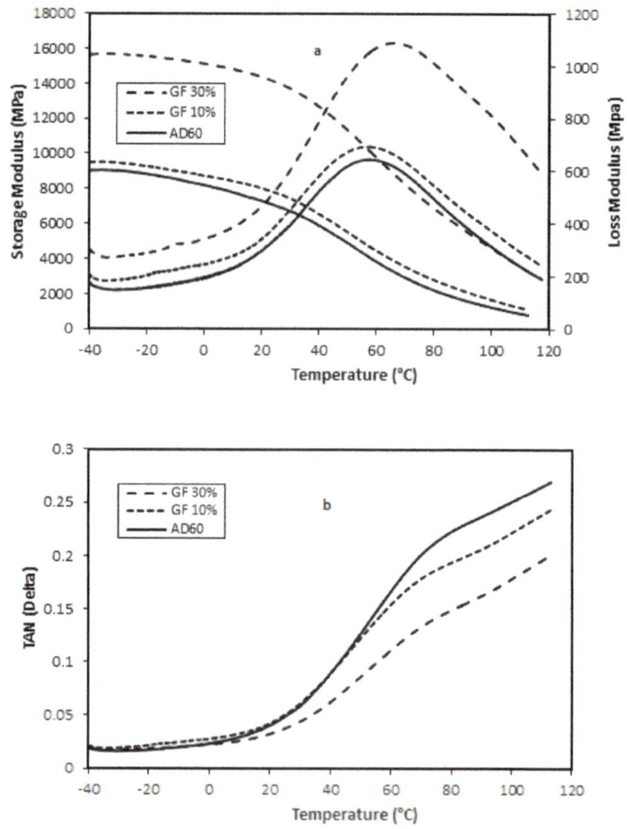

Figure 3. Effects of temperature level on storage modulus and loss modulus (**a**) and damping factor (**b**) of glass fiber-filled HDPE AD60 composites.

Tan δ—The tan δ, damping factor, is a ratio of the loss modulus to the storage modulus. The parameter is independent of the material's stiffness and is widely used to study viscoelastic response of the materials. Figure 3 shows tan δ curves of neat AD60 and filled composites with GF. For GF filled HDPE composite systems, tan δ curves had less distinctive α-relaxation process compared with the E′ data. For GF-filled composites, the damping curves shifted toward lower values as the filler content level increased. This indicates that the damping effect reduced with the increased filler content in the matrix. The result suggested that certain degree of interfacial bonding existed between the fillers and matrix in GF-filled composites. The higher GF levels induced a better fiber packing in the matrix and resulted in more efficient stress transfer from the resin matrix to the fibers, leading to decreased damping effect. Composites showed interfacial adhesion, leading to change in the damping effect.

Thermal Expansion Properties

Typical dimension change data in relation with temperature for the composite system is shown in Figure 4 for neat AD60 resin and GF filled systems. The sample dimension increased as the temperature increased, and decreased as temperature decreased. The linear coefficient of thermal expansion, LCTE, is represented by the slope of the linear portion of the curve. Neat plastic had an obvious larger dimension change than that from filled composites for a given temperature change. Residual deformation is seen for the neat plastic (AD60) at the end of the heating cycle. For the GF-system, the residual deformation was significantly lowered at the two higher GF loading levels (i.e., 30%, and 40%).

The measured LCTE values over the three temperature ranges (i.e., heating: 20–>60 °C, cooling: 60–>–30 °C); and heating: −30–>20 °C) are summarized in Table 2 as a function of filler content levels for GF filled composites. The first heating cycle led to the largest LCTE values at each given filler content levels. LCTE decreased significantly as the GF content was added to the system. At the 40% GF level, the LCTE values are respectively, 11.1, 17.1, and 19.7 × 10^{-6}/°C from 199.1, 162.7, 137.8 × 10^{-6}/°C for the three temperature ranges due to GF causes a mechanical restrain on HDPE chain opening during the temperature change. The values are much smaller than the reported LCTE values of commercial WPC. Thus, the use of chemically modified GFs is more effective in controlling thermal expansion behavior of the filled plastics composite. The system can be used as an effective shell layer in co-extruded NFPC/WPC to control overall thermal expansion properties of the composites [5].

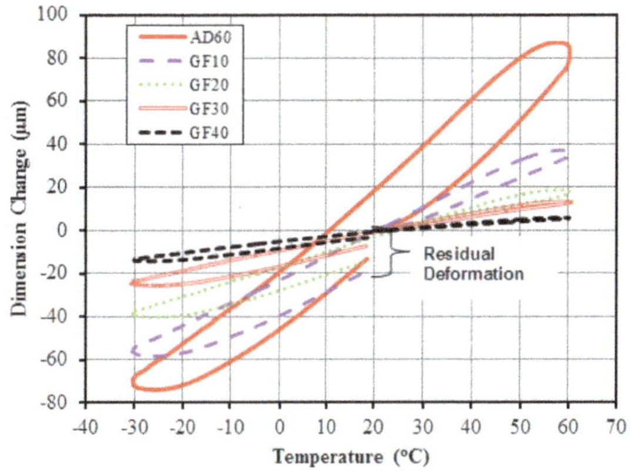

Figure 4. Typical dimension change-temperature history for glass fiber filled HDPE AD60 composite systems.

Table 2. Summary of thermal expansion properties of virgin HDPE and filled HDPE composites with individual glass fibers.

System	Filler content (wt %)	Linear Coefficient of Thermal Expansion (LCTE) (10^{-6}/°C) [a,b,c]		
		20→60 °C	60→−30 °C	−30→20 °C
HDPE 6706/GF	0	203.9(3.0)G	164.8(6.3)G	143.8(4.8)I
	10	71.1(0.4)E	79.7(1.4)E	82.5(1.1)F
	20	32.0(1.1)C	41.6(3.8)C	50.9(0.8)D
	30	19.8(0.7)B	28.6(1.5)B	34.1(1.1)C
HDPE AD60/GF	0	199.1(2.8)G	162.7(2.5)G	137.8(1.0)H
	10	75.1(0.5)F	87.6(3.1)F	87.8(0.7)G
	20	39.6(2.2)D	55.1(0.1)D	57.8(0.7)E
	30	22.4(0.3)B	33.1(1.8)B	33.9(0.3)B
	40	11.1(0.2)A	17.1(0.2)A	19.7(0.2)A

[a] The content of each filler was based on the total composite weight; [b] Mean values with the same capital letter for each property are not significantly different at the 5% significance level; [c] Numbers in the parenthesis are standard deviation based on five specimens.

Composites with Combined Talc and Glass Fiber Fillers

Morphology

Typical SEM micrographs for composites filled with the combined talc and glass fibers (1:2 mixing ratio) are shown in Figure 5 (a and b: 10% loading level; c and d: 30% loading level). Similar to the morphology of the individual filler systems, the plate talc particles were oriented both perpendicular and parallel to the fracture plane. The perpendicular-to-the-fracture-plane orientation of the talc particles is displayed by several line-type pull-out voids in Figure 5d. Most GFs are perpendicular to the fracture plane. The GFs bonded well to the plastic matrix, while there was little bonding between talc and plastic matrix as indicated by the arrows in Figure 5c,d.

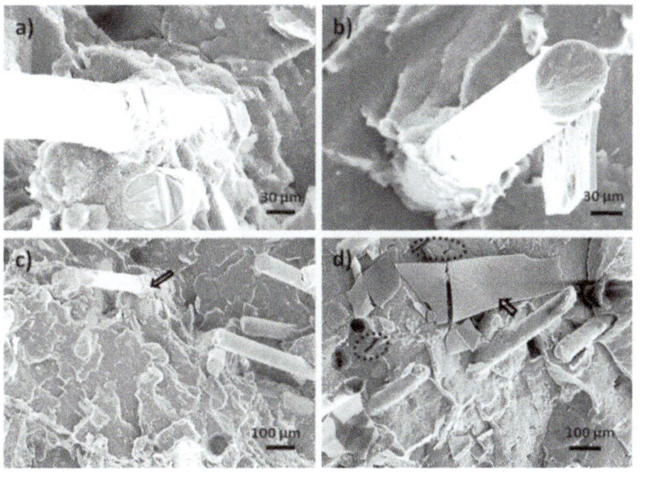

Figure 5. SEM micrograph of fractured surfaces of composites with combined fillers. Upper two charts (**a**,**b**) 10 wt % filer(talc/glass fiber = 1:2); and Lower two charts (**c**,**d**) 30 wt % filler (talc/glass fiber = 1:2).

Mechanical Properties

Mechanical properties of neat HDPE and composites with hybrid talc and GF fillers are shown in Table 3. Tensile modulusvs. talc/glass fiber ratios are plotted in Figure 6. Here, both HDPE 6706 and HDPE AD60 were used to generate a mixed blend (50/50 wt %) as the matrix. For the two neat resins, mechanical properties of HDPE AD60 are higher than those of HDPE 6706. Increase of combined fillers content, from 10% to 30%, led to increased mechanical properties. At a fixed filler loading level, the increase of GF portion of the filler led to increased strength and modulus of composites, especially at 30% loading level.

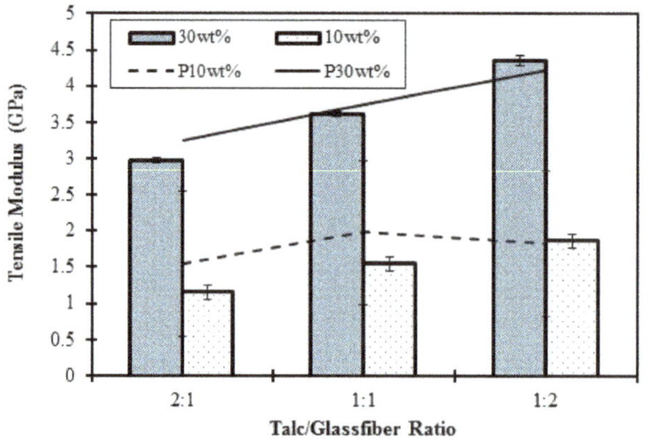

Figure 6. Tensile modulus of composites filled with combined glass fiber and talc fillers. Lines in the upper chart showing predicted values.

Table 3. Summary of mechanical properties of neat HDPE and filled HDPE composites with combined fillers.

An attempt was made to use the paper physics approach to predict tensile modulus of the three component system (i.e., talc, GF, and resin).

System	Filler content (wt%) [a]	Talc/GF Ratio	Strength			Modulus	
			Tensile (MPa) [b,c]	Flexural (MPa)	Impact (kJ/m²)	Tensile (GPa)	Flexural (GPa)
HDPE6706	0	0	18.9(0.2)A	19.8(0.4)A	8.12(0.21)E	0.26(0.02)A	0.73(0.05)A
HDPEAD60	0	0	23.8(1.4)C	21.8(1.0)B	28.57(2.0)F	0.86(0.09)B	0.85(0.06)B
HDPEAD60/HDPE6706/Talc/GF	30	2:1	27.7(0.48)D	33.7(1.4)D	5.77(0.59)A	2.9(0.13)D	2.0(0.20)E
		1:1	31.9(0.41)E	38.8(0.3)E	6.35(0.22)AB	3.6(0.39)E	2.4(0.06)F
		1:2	36.8(0.63)F	44.7(0.6)F	7.80(0.36)DE	4.4(0.33)F	2.8(0.07)G
	10	2:1	21.8(0.22)B	26.4(0.3)C	5.43(0.35)A	1.2(0.15)B	1.2(0.01)C
		1:1	22.6(0.36)B	26.8(0.8)C	6.12(0.30)AB	1.6(0.23)C	1.3(0.05)C
		1:2	24.6(0.23)C	27.1(0.6)C	7.03(0.11)CD	1.9(0.15)C	1.6(0.03)D

[a] The filler content was based on the total composite weight; [b] Mean values with the same capital letter for each property are not significantly different at the 5% significance level; [c] Numbers in the parenthesis are standard deviation based on five specimens.

$$E_C = X_{GF}E_{GF}\Phi_{GF} + X_{Talc}E_{Talc}\Phi_{Talc} + E_m(1 - \Phi_{GF} - \Phi_{Talc}) \quad (6)$$

where X_{GF} and X_{talc} are, respectively, the combined length and orientation factors for GF and talc fillers. Figure 4 showed a comparison between predicted and experimental values of the tensile moduli. With X_{GF} = 0.4 and X_{talc} = 1.0, the predicted values from Equation 6 showed a reasonable agreement with experimental data.

Dynamic mechanical results (Figure 7) supported flexural modulus trend listed in Table 4 base on the E' and E' values. The temperature range of a-relaxation shifted between 50 and 62 °C for both filler weight contents. The comparison betweenTable 3 and Table 4 showed that modulus and strength of composites, modified by combined filler, were sight lower than that of GF alone. The tan δ was decreased with increasing filler content. The 30 wt % combined filler reinforced HDPE composites showed higher tan δ than that of 10 wt % combined filler, and the tan δ values of the combined filler reinforced composite were seen to be lower than that of the pure HDPE matrix. The reason of the above those result was closely similar with that of individual filler.

Table 4. Summary of thermal expansion properties of virgin HDPE and filled HDPE Composites with combined glass fiber and talc filler.

System	Filler content (wt %) [a]	Talc/GF ratio	LCTE (10^{-6}/°C) [a,b,c]		
			20→60 °C	60→30 °C	−30→20 °C
HDPE6706	0	0	203.9(3.0)G	164.8(6.3)G	143.8(4.8)G
HDPEAD60	0	0	199.1(2.8)G	162.7(2.5)F	137.8(1.0)F

HDPE AD60/ HDPE6706/ Talc/GF	10	2:1	144.9(3.4)F	123.2(8.4)E	113.5(2.3)E
		1:1	117.7(0.9)D	111.3(2.1)D	105.1(0.9)D
		1:2	124.0(0.7)E	114.2(3.2)D	107.0(0.5)D
	30	2:1	75.5(2.7)C	67.1(2.5)C	66.3(0.7)C
		1:1	48.2(0.7)A	47.8(1.3)A	52.5(0.3)A
		1:2	54.6(0.5)B	53.8(2.2)B	54.8(0.6)B

[a] The content of each filler was based on the total composite weight; [b] Mean values with the same capital letter for each property are not significantly different at the 5% significance level;[c] Numbers in the parenthesis are standard deviations based on five specimens.

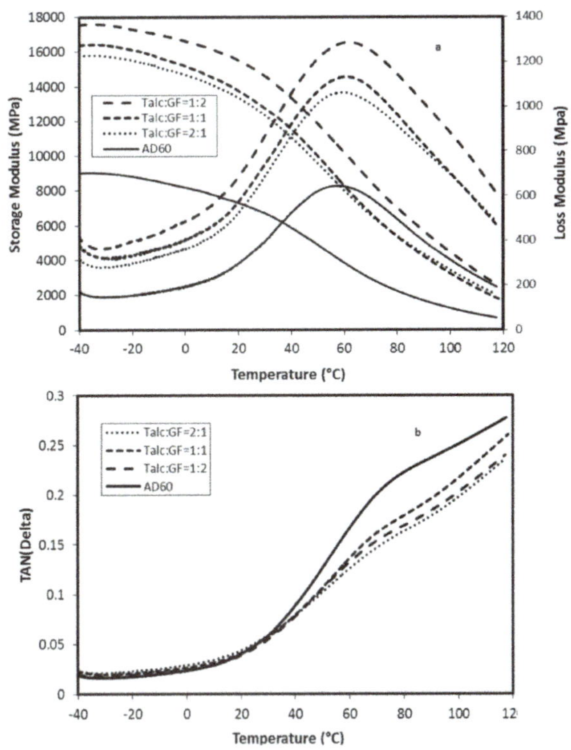

Figure 7. Effects of temperature level on storage modulus and loss modulus (**a**) and damping factor; (**b**) of HDPE AD60 composites filled with combined GF and talc fillers at the 30 wt %.

Thermal Expansion

Typical dimension change and temperature relationship curves for the combined filler systems are shown in Figure 8 in comparison with neat resin data values. LCTE value is represented by the slope of the linear portion of the curve Pure HDPE had an obvious larger dimension change than that from filled composites for a given temperature range. As the GF content increased the curves turned more clockwise toward reduced dimension changes. Residual deformation is seen for the neat HDPE at the end of the heating cycle. The use of combined fillers at the 30% loading level largely reduced residual deformation of the filled composites.

Table 4 lists of measured LCTE data for the composites with combined fillers. LCTE values at the three talc/GF ratios and two total loading levels was showed difference trend. As the total filler level (i.e., 10% vs. 30%) and GF portion of the alc/GF fillers (i.e., 2:1, 1:1, and 1:2) increased, the LETC values decreased. The total of 10% filler loading led to the LCTE values of 124.0, 117.7, and 115.0 × 10^{-6}/°C at talc/GF ratios of 2:1, 1:1, and 1:2, respectively. The use of the 30% combined fillers reduced the LCTE values to 75.5, 48.2, and 54.6 × 10^{-6}/°C at talc/GF ratios of 2:1, 1:1, and 1:2, respectively. The reason what was closely similar with that of individual filler caused the above those results.

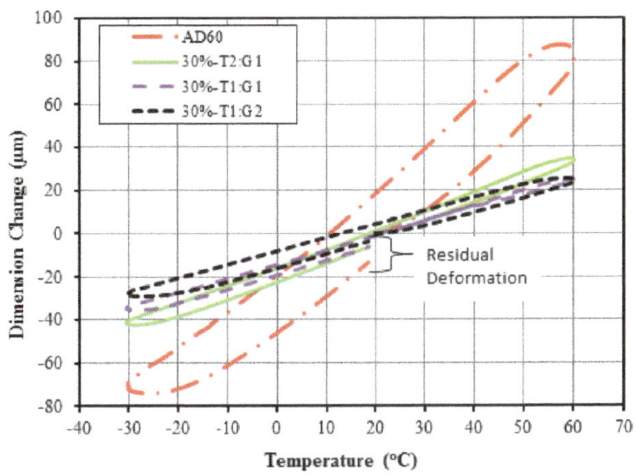

Figure 8. Typical dimension change-temperature history for HDPE AD60 composite filled with combined glassfiber and talc fillers at the 30 wt % loading level.

EXPERIMENTAL SECTION

Raw Material and Experimental Design

SGF reinforced HDPE pellets were provided by RTP Co. (Winona, MN, USA). The material was of the type RTP 707 CC UV Natural, a SGF reinforced HDPE with glass fiber content of 40 wt %. The fiber diameter was 0.014 mm and fiber length was 4 mm prior to compounding. Fibers were sized with a silane-based solution before compounding. UV and coupling agent were added to the system during compounding. Talc was provided by Fiber Glast Development Corp. (Brookville, OH, USA). Two types of neat HDPE were provided by ExxonMobil Chemical Co. (Houston, TX, USA).

Experiment design included two factorial experiments. The first experiment was to investigate the effect of individual filler consisting of six blends covering one filler (glass fiber) and four loading rates (40, 30 , 20, and 10 wt % of total composite weight for glass fiber pallets filler reinforced HDPE AD60 and HDPE 6706, respectively). The second experiment was designed to study the effect of combined polymer and combined fillers system (HDPE AD60/HDPE 6706/Talc/glass fiber pallets), consisting of six blends covering two filler weight contents (10 and 30 wt %) and three talc/glass fiber ratios (2:1, 1:1, and 1:2). The pure HDPE (HDPEAD60 and HDPE 6706) and glass fiber pallets (40 wt % of total composite weight for glass fiber reinforced HDPE AD60) were used as a control.

Sample Preparation

Melt compounding was performed using an intermesh, counter-rotating Brabender twin-screw extruder (Brabender Instruments, Hackensack, NJ, USA) with a screw speed of 40 rpm. The temperature profile of barrels ranged from 150 to 175 °C. The extrudates were air-cooled and then pelletized into granules. The granules were injection-molded into standard mechanical test specimens using a Batenfeld Plus 35 injection molding machine (Batenfeld, NJ). The injection temperatures were 190 and 180 °C for HDPE/filler composites and virgin HDPE, respectively. All specimens were then conditioned for 72 h at a temperature of 23 ± 2 °C and a relative humidity of 50% ± 5% for later characterization.

Characterization and Data Analysis

Morphology Analysis. The scanning electron micrographs was taken by Field Emission Scanning Electron Microscopy (FESEM, FEI Quanta TM 3D FEG Dual Beam SEM/FIB, Hillsboro, OR, USA) and used to analyze the morphology of fractured surfaces of select samples. Prior to the microscopy, samples were frozen with liquid nitrogen for 10 min, and then were impact-fractured. The fractured surfaces of the specimens were coated with gold t to avoid electrical charging. The acceleration voltage used was 10.0 KV in SEM analysis.

Static Strength Property. Flexural testing was done with specimens of 80 mm × 13 mm × 3 mm under three-point bending using an Instron 5582 testing machine (Instron, Norwood, MA) following ASTM D790. A crosshead speed of 1.37 mm/min and a span length of 50 mm were used for test. Tensile testing was done with Type-I dumbbell-shape tensile specimens having a dimension of 165 mm × 13 mm × 3 mm using the same Instron machine according to ASTM D638. A crosshead speed of 5 mm/min and an extensometer with a gage length of 25 mm for strain measurement were used for the test. Notched Izod impact strength was determined from specimens of 63.5 mm × 12.7 mm × 3 mm in size using a Tinius Olsen Mode 1892 impact tester (Tinius Olsen, Horsham, PA, USA) according to ASTM D256. The notch angle of 45o and "V"-type notch depth of about 2.5 mm were used for the test. Five specimens were taken for each test and average data along with corresponding standard deviation were reported.

Dynamic Mechanical Analysis (DMA). The storage modulus, loss modulus, and tan δ of neat resin and its composites were evaluated using a DMA Q800 (TA Instruments Inc., New Castle, PA, USA) with samples of 59.8 (length) mm × 12.7 (width) mm × 3.2 (thickness) mm. The specimens were heated from −40 to 120 °C with a heating rate of 5 °C/min. A dual cantilever mode was used to test all specimens at a fixed frequency of 1 Hz.

Thermal Expansion. Thermal expansion samples were machined with a miniature table saw along the long direction of injection molded samples with dimension of 5.1 mm × 12.7 mm × 1.6 mm. Special attention was made to ensure that the sample ends are parallel. The test was done with a TA Q400 ThermoMechanical Analyzer, TMA .The sample was placed on a quartz base and an extension quartz probe was then placed on the top surface of the sample. A loading of 5 g force was applied to the probe to ensure the proper contact of the probe and the sample. The change in the length of the sample with temperature was measured using a linear variable differential transformer (LVDT) with a sensitivity of ±0.02 μm. The length and temperature data were recorded and analyzed with TA's Universal Analysis software. All tests were done with three heating cycles—(1): 20 °C to 60 °C; (2) 60 °C to –30 °C; and (3): –30 °C to 20 °C. The heating and cooling rates were kept constant at 5 °C/min.

Statistical Data Analysis.

Duncan's multiple range tests for pair-wise comparisons were used to test the effect of various treatments on measured properties using Statistical Analysis Software SPSS 20.0 [22]. Statistical ranking at the 5% significance level was provided among the treatments for each property.

CONCLUSIONS

The effect of individual and combined talc and short glass fibers on mechanical and thermal performance of the filled HDPE composites was studied. Several published models were adapted to fit the measured tensile moduli and strength of various composite systems. The following conclusions can be drawn from the study.

- The use of silane-modified short GFs had a much larger effect in improving mechanical properties and in reducing LCTE values of filled composites compared with the use of talc due to enhanced bonding to the matrix, larger aspect ratio and fiber alignment for GFs.
- Mechanical properties and LCTE values of composites with combined talc and GF fillers varied with talc and GF ratio at a given total filler loading level. The use of a larger portion of GFs in the mix can lead to better composite performance, while the use of talc can help lower the composite costs and increase its recyclability. The use of 30 wt % combined filler seems necessary to control LCTE values of filled HDPE in the data value range generally reported for commercial wood plastic composites.
- Tensile modulus for talc-filled composite can be predicted with rule of mixture, while a PPA-based model can be used to predict the modulus and strength of GF-filled composites.
- The material developed can be used as an effective shell layer in co-extruded NFPCs/WPCs to enhance their performance properties.

ACKNOWLEDGMENTS

The authors would like to thank the financial support from Louisiana Board of Regents (Contract Number: LEQSF-EPS (2011)-OPT-IN-04), from 2013 natural science foundation of Jiangsu Province: Youth Foundation (BK20130966), from the Priority Academic Program Development of Jiangsu Higher Education Institutions, and from Korea Forest Research Institute.

REFERENCES

1. Yao, F.; Wu, Q.L. Coextruded polyethylene and wood-flour composite: Effect of shell thickness, wood loading, and core quality. J. Appl. Polym. Sci. 2010, 118, 3594–3601.

2. Jin, S.; Matuana, L.M. Wood/plastic composites co-extruded with multi-walled carbon nanotube-filled rigid poly(vinyl chloride) cap layer. Polym. Int. 2010, 59, 648–657.
3. Kim, B.J.; Yao, F.; Han, G.P.; Wu, Q.L. Performance of bamboo plastic composites with hybrid bamboo and precipitated calcium carbonate fillers. Polym. Compos. 2012, 33, 68–78.
4. Karrad, S.; Cuesta, J.M.L.; Crespy, A. Influence of a fine talc on the properties of composites with high density polyethylene and polyethylene/polystyrene blends. J. Mater. Sci. 1998, 33, 453–461.
5. Huang, R.; Xiong, W.; Xu, X.; Wu, Q. Effect of hybrid mineral and bamboo fillers on thermal expansion behavior of bamboo fiber and recycled polypropylene-polyethylene composites. BioResources 2012, 7, 4563–4574.
6. Huang, R.; Kim, B.J.; Lee, S.Y.; Zhang, Y.; Wu, Q. Co-extruded wood plastic composites with talc filled shells: morphology, mechanical and thermal expansion performance. BioResources 2013, 8, 2283–2299.
7. Singh, S.; Mohanty, A.K. Wood fiber reinforced bacterial bioplastic composites: Fabrication and performance evaluation. Compos. Sci. Technol. 2007, 67, 1753–1763.
8. Nielson, L.E.; Landel, R.F. Mechanical Properties of Polymers and Composites; CRC Press: New York, NY, USA, 1993.
9. Tsai, S.W.; Pagano, N.J. Invariant Properties of Composite Materials; Defense Technical Information: Fort Bellvoir, VA, USA, 1968.
10. Pilla, S.; Gong, S.; O'Neill, E.; Rowell, R.M.; Krzysik, A.M. Polylactide-pine wood flour composites. Polym. Eng. Sci.2008, 48, 578–587.
11. Johnson, R.K.; Zink-Sharp, A.; Renneckar, S.H.; Glasser, W.G. Mechanical properties of wetlaid lyocell and hybrid fiber-reinforced composites with polypropylene. Compos. Part. A Appl Sci. Manuf. 2008, 39, 470–477.
12. Fornes, T.D.; Paul, D.R. Modeling properties of nylon 6/ clay nanocomposites using composite theories. Polymer 2003,44, 4993–5013.
13. Tsai, S. W.; Halpin, J. C.; Pagano, N.J. Composite Materials Workshop; Technomic Publishing Co. Inc.: Stamford, CT, USA, 1968; pp. 233–253.
14. Fu, S.Y.; Lauke, B. The elastic modulus of misaligned short-fiber-reinforced polymers. Compos. Sci. Technol. 1998, 58, 389–400.
15. Jayaraman, K.; Kortschot, M.T. Correction to the Fukuda-Kawata Young's modulus theory and the Fukuda-Chou strength theory for short fibre-reinforced composite materials. J. Mater. Sci. 1996, 31, 2059–2064.
16. Nicodemo, L.; Nicolais, L. Mechanical-properties of metal polymer composites. J. Mater. Sci. Lett. 1983, 2, 201–203.
17. Piggott, M.R. Short fibre polymer composites: A fracture-based theory of fibre reinforcement. J. Compos. Mater. 1994,28, 588–606.
18. Huda, M.S.; Mohanty, A.K.; Drzal, L.T.; Misra, M.; Schut, E. "Green" composites from recycled cellulose and poly(lactic acid): Physico-mechanical and morphological properties evaluation. J. Mater. Sci. 2005, 16, 4221–4229.
19. Matuana, L.M.; Balatinecz, J.J.; Park, C.B. Effect of surface properties on the adhesion between PVC and wood veneer laminates. Polym. Eng. Sci. 1998, 38, 765–773.
20. Danyadi, L.; Renner, K.; Moczo, J.; Pukanszky, B. Wood flour filled polypropylene composites: Interfacial adhesion and micromechanical deformations. Polym. Eng. Sci. 2007, 47, 1246–1255.
21. Mohanty, S.; Nayak, S.K. Interfacial, dynamic mechanical, and thermal fiber reinforced behavior of MAPE treated sisal fiber reinforced HDPE composites. J. Appl. Polym. Sci. 2006, 102, 3306–3315.
22. IBM. IBM SPSS Advanced statistics 20; IBM Corporation: Armonk, NY, USA, 2011.

CHAPTER 11

FRICTION STIR PROCESSING OF PARTICLE REINFORCED COMPOSITE MATERIALS

Yong X. Gan, Daniel Solomon, and Michael Reinbolt

Keywords: Friction stir processing; particle reinforcements; composite materials

INTRODUCTION

Friction stir processing (FSP) is a solid state process known for its ability to modify microstructures and provide improved properties over conventional processing technologies [1,2,3,4,5,6,7,8,9,10,11,12]. The development of friction stir processing (FSP) is based on the friction stir welding (FSW) technology. Both FSW and FSP have the same process principle and share the same facilities as shown in Figure 1(a). FSW works by plunging a spinning tool into the joint of two materials and then traversing the rotating tool along the interface. The friction caused by the tool heats up the materials around the pin to a temperature below the melting point. The rotation of the tool "stirs" the material together and results in a mixture of the two materials. In FSP, a specially designed rotating pin, as shown in Figure 1(b), is first inserted into the material to be processed with a proper tool tilt angle and then move along the designed paths. The pin produces frictional and plastic deformation heating within the processing zone. As the tool pin moves, materials are forced to flow around the pin. Material flows to the back of the pin, where it is extruded and forged behind the tool, consolidated and cooled under hydrostatic pressure conditions. It is evident that FSW and FSP share the same mechanism as schematically illustrated in Figure 1(c). That is why in some literatures, the uses of the two terms are interchangeable. However, they do have different purposes in practical applications. The goal of FSW is to join two plates together, whereas FSP aims at modifying the microstructure of a single workpiece or multiple workpieces.

Citation: Yong X. Gan, Daniel Solomon, and Michael Reinbolt. Friction stir processing of particle reinforced composite materials, Materials 2010, 3(1), 329-350; DOI:10.3390/ma3010329.

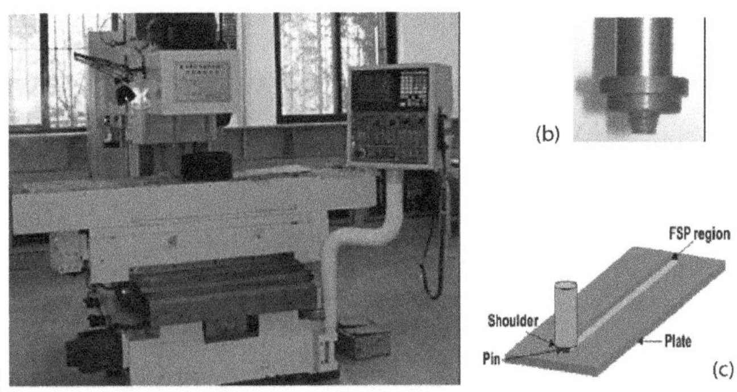

Figure 1. Equipment for friction stir processing: (a) picture of friction stir processing machine (after Wang *et al.* [3]); (b) tool pin (after Ceschini *et al.* [11]); (c) schematic showing the principle of friction stir processing (after Mishra *et al.* [12]).

Friction stir processing is most commonly used with aluminum [13,14,15,16,17,18,19,20,21,22], but it is also used for processing other alloys, for example, Ni-based intermetallic compound modified bronzes [23], Zr alloys [24], tool steels [25,26] and Mg alloys [27,28,29,30,31,32,33]. The advantages offered by friction stir processing method make it suitable for many applications. Recently, friction stir processing technology has been used in aerospace, automotive, marine, and railroad industries along with various other applications. Experimental [34,35,36,37,38,39,40,41,42,43], analytical [44,45] and computational studies of friction stir processing nanocomposites [46,47] have been performed. Specifically, the effects of processing parameters on microstructure evolution, deformation behaviors and mechanical properties are investigated [48,49,50,51,52].

FRICTION STIR PROCESSING OF ALUMINUM MATRIX COMPOSITES

Friction stir processing has been used in various type of particle reinforced aluminum composite materials. SiC particle reinforced aluminum alloys have been studied the most [25,43,53,54,55,56]. For example, the feasibility of making bulk SiCpSiCpreinforced aluminum metal matrix composites (MMCs) with the dimension of 150 mm in length, 60 mm in width and 6 mm in depth via friction stir processing (FSP) was demonstrated by Wang *et al.* [3]. Good interface bonding between particles and the base metal can be obtained. The volume percentage of SiCpSiCp is about 1.5% in the reinforced region. The microhardness of the reinforced MMCs is 10% higher than that of the base metal, Al-6Mg-Mn.

Tewari *et al.* [53] studied the effect of SiC particle orientation change due to friction stir processing. High-resolution, large-area images were obtained to show the detailed microstructural data for SiC/A6061 composite materials, both before and after single friction stir processing step. The SiC reinforcement particles were found to have anisotropic shape with an average aspect ratio around 1.6 to 1.8. The scanning electron microscopic image of the composite in Figure 2(a) shows such a morphological feature. From particle orientation statistics, it was found that there are preferred orientations for the nonequiaxed SiC particles after extrusion processing. As shown in Figure 2(b), the alignment of the SiC particles is parallel to the extrusion direction. The extrusion axis is vertical. The preferred orientation can be modified during passage of the friction stir tool. In Figure 2(c), the redistributed particles at 45o45o to the extrusion and transverse directions can be seen. The FSP tool motion is horizontal, left to right. The microstructural data consistently indicate that significant microstructural modifications occur during FSP, including re-orientation of the reinforcement particles, and a significant reduction in the levels of microstructural heterogeneity and microstructural anisotropy.

The grain structure change in friction stir processed materials has caught considerable attention. Prangnell and Heason [57] investigated the microstructure development in selected zones. What they found is that the deformation field surrounding the rotating pin caused the formation of the ultrafine grained nugget material. In the friction stir processing zone far away from the tool pin, the grains can be seen splitting into deformation bands due to the flow of the Al4.8Cu0.9Li alloy as marked by **i** in Figure 3(a). In the zone closer to the tool pin, increase in misorientation and reduce in spacing of the grains were found as marked by **ii** at the upper left corner in Figure 3(b). Finely spaced, parallel bands were also found and they broke up along their length by thermally and mechanically assisted boundary migration to form fine nugget-scale grains marked by **iii** at the lower right corner in Figure 3(b). In Figure 3(c), the enlarged fine nugget-scale grains are shown. The material is immediately next to the pin surface.

Friction stir processing alumina particle reinforced aluminum alloys has also been studied [9,58,59,60,61]. Shafiei-Zarghani *et al.* [9] used friction stir processing (FSP) to incorporate nano-sized Al_2O_3 into AA6082 aluminum alloy to form particulate composite surface layer as shown in Figure 4(a). The Al_2O_3 particles have an average size of about 50 nm. Perfect bonding between the surface composite and the aluminum alloy substrate was achieved as shown by the defect-free interface in Figure 4(b). Mechanical properties include microhardness and wear resistance were tested. The results show that the increasing in number of FSP passes causes more uniform distribution of nano-sized alumina particles. The microhardness of the surface improves by three times as compared to that of the as-received Al alloy. A significant improvement in wear resistance in the nano-composite surfaced Al was observed as compared to the as-received Al alloy. The wear rate is reduced to one third of that of the as-received Al alloy when friction stir processed composite layer was incorporated as shown by the curves in Figure 4(c).

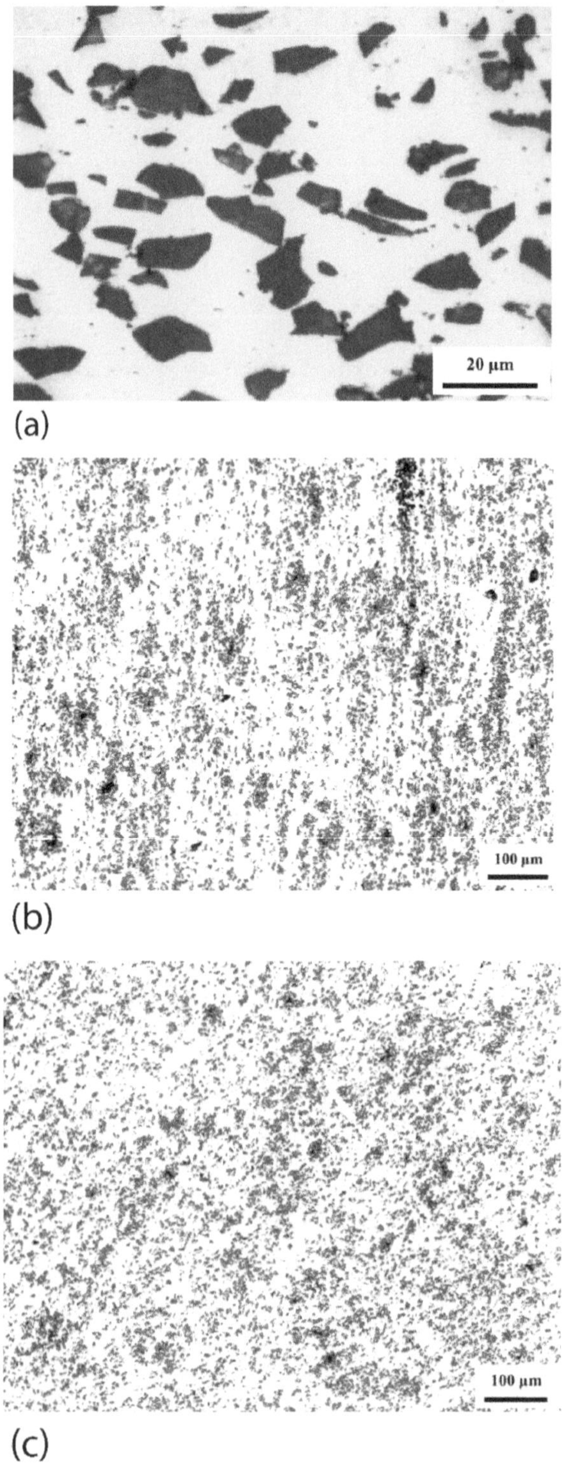

Figure 2. Microstructure change in SiC particle reinforced A6061 due to friction stir processing (after Tewari *et al.* [53]): (a) scanning electron microscopic image of the composite showing the anisotropic shape of SiC particles; (b) as-extruded SiC/Al; (c) after FSP.

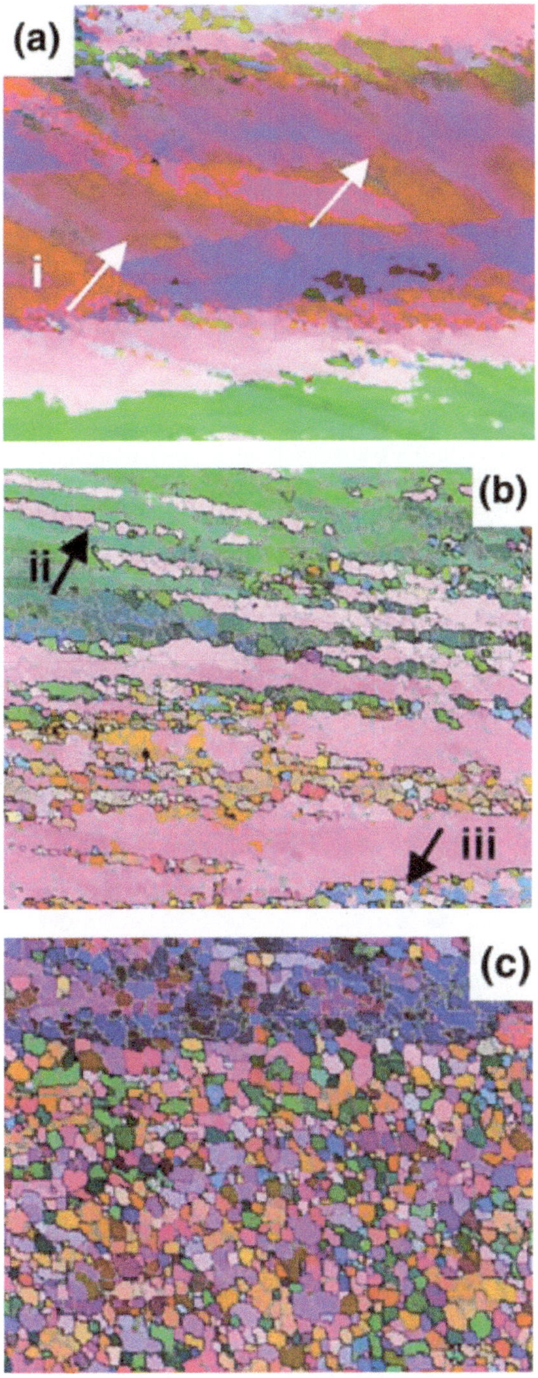

Figure 3. Electron backscatter diffraction (EBSD) maps showing the development of grain structures in different zones associated with the strain and temperature change in friction stir processing (after Prangnell and Heason [57]): (a) materials far away from the tool pin showing split grains into deformation bands; (b) materials close to the tool pin showing parallel bands separated by fine grains; (c) very fine nugget grains in the material immediately next to the tool pin surface.

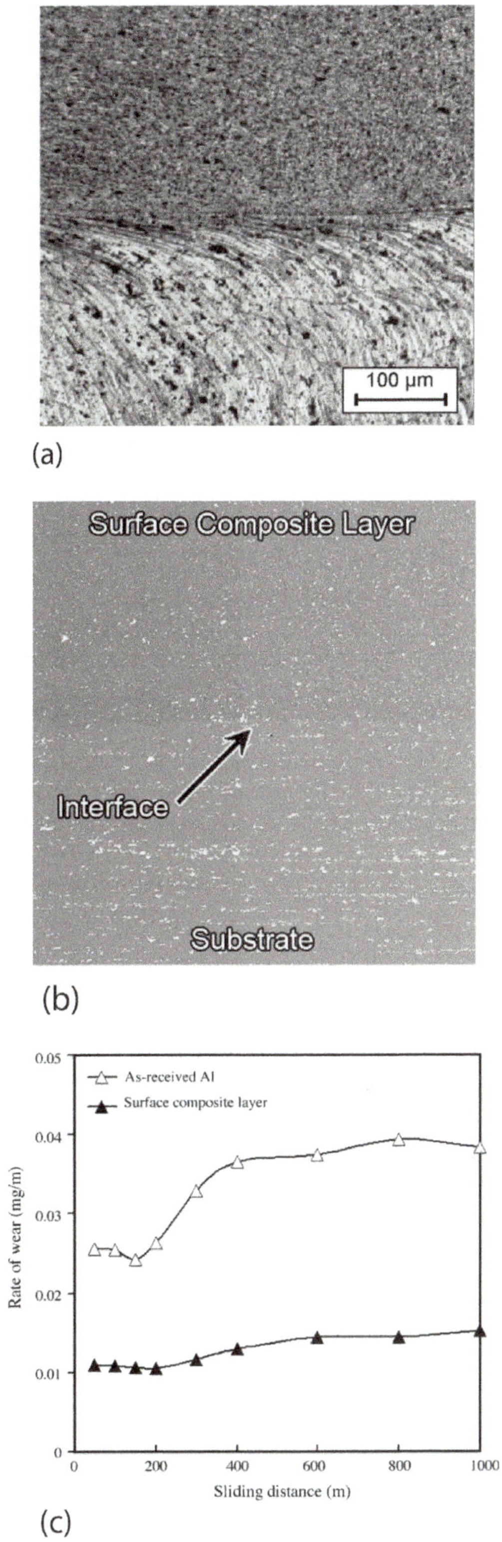

Figure 4. Microstructure and wear property of friction stir processing alumina particle reinforced A6082 aluminum alloy (after Shafiei-Zarghani *et al.* [9]): (a) optical micrograph showing the composite top layer and the bottom layer of matrix alloy; (b) SEM image showing the well bonded interface between the two layers; (c) comparison of wear rate of the as-received alloy and the material with a surface nano-composite layer produced by four FSP passes.

Similar to SiC reinforced aluminum composites, the stirring of the FSP tool has a substantial influence on the distribution of alumina particles in Al_2O_3/Al composites [60]. The stirring of the tool can also change the shape of the reinforcement particles. It breaks off the sharp edges of the bigger particles, rounding them up at the same time [61]. This action results in smaller, round particles in the nugget, which can be seen from the micrographs in Figure 5. Figure 5(a) represents the morphology of the Al_2O_3 particle reinforced AA6061 before FSP. Sharp corners of the particles are very clearly shown. After FSP, scattering in the size distribution of the Al_2O_3 particles within the composite material were more pronounced, which can be easily found from the morphological features shown in Figure 5(b). Many small particles are around those big-sized ones. This is due to the fracture of the original Al_2O_3 particles. In addition, some sharp Al_2O_3 particles became rounded. Structure transition was also observed in the area close to the nugget as shown in Figure 5(c).

Macrostructure observations on friction stir processed Al_2O_3 particle reinforced aluminum composites revealed three distinct zones, i.e. the heat affected zone (HAZ), thermomechanically affected zone (TMAZ) and dynamically recrystallized nugget zone. The microstructure change in HAZ is less obvious. However, particle fracture was found in this zone as indicated by Cavaliere [61]. In the thermomechanically affected zone (TMAZ), deformation bands was observed as shown inFigure 6(a). In the nugget zone, recrystallized microstructure and fractured particles are the major morphological features as illustrated by Figure 6(b). There exists difference in microhardness of these zones. As shown in Figure 6(c), the microhardness for the material in the nugget zone is the highest because the grain size is the smallest in this zone.

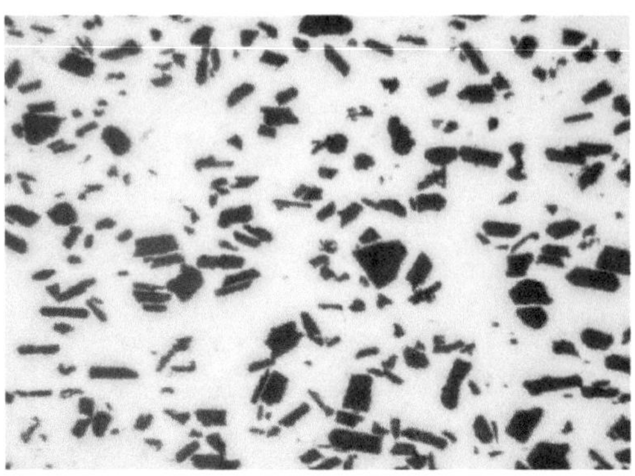

(a)

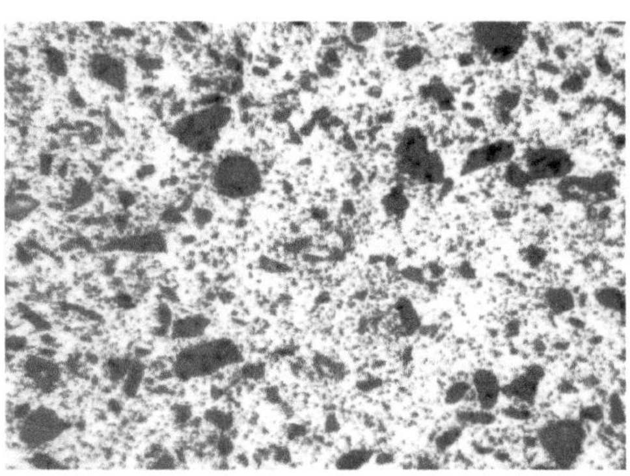

(b)

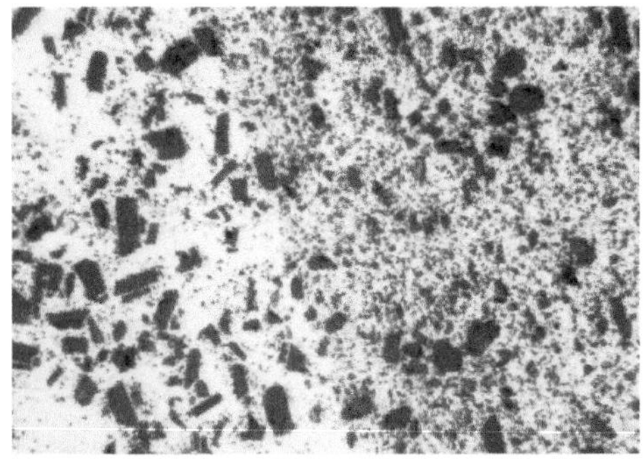

(c)

Figure 5. Effect of friction stir processing on the distribution of alumina particles in 6061 aluminum alloy (after Marzoli *et al.* [60]): (a) micrograph showing the composite before FSP; (b) micrograph showing the composite after FSP; (c) microstructure transition near the nugget zone.

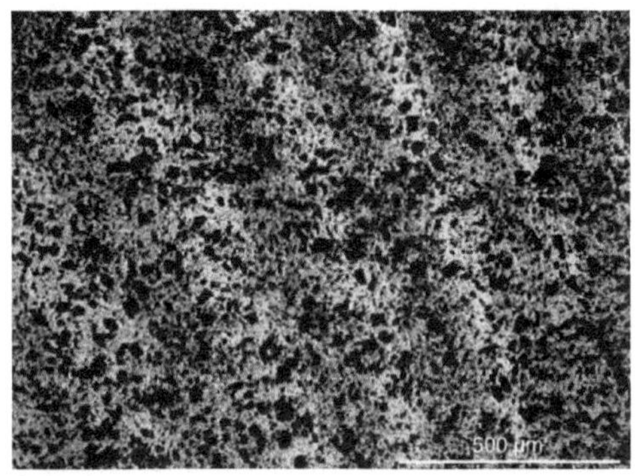

(a)

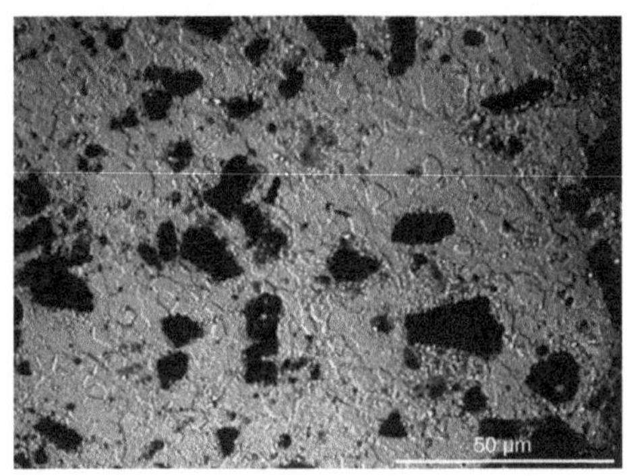

(b)

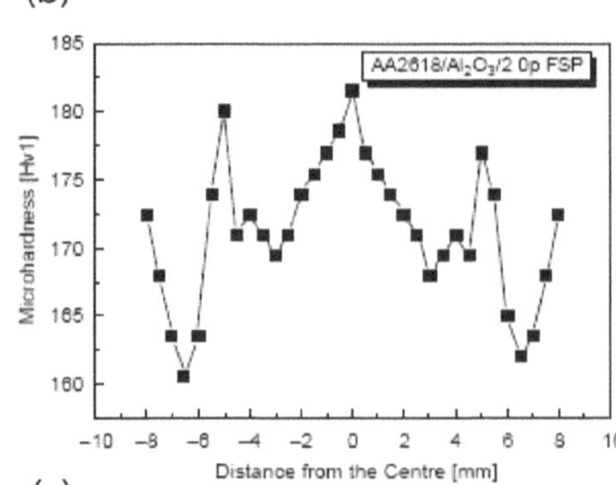

(c)

Figure 6. Different zones generated by friction stir processing as revealed by morphology and microhardness profile (after Cavaliere [61]): (a) micrograph showing the deformation bands in TMAZ; (b) recrystallization in nugget zone; (c) microhardness profile in the FSP Al_2O_3/AA2618 composite.

In addition to SiC and Al_2O_3 particles, B_4C particles were used to make discontinuously reinforced aluminum composite. The structures of the composite with AA6063 matrix reinforced by 6 and 10.5 vol % B_4C particles were modified by friction stir processing [62]. The EBSD orientation map, Figure 7(a), shows the grain structure of the AA6063 aluminium alloy. The presence of B_4C particles and the volume content did not change the grain structure much. The composite with 6% of B_4C particles has almost the same equiaxed grain structure as the matrix alloy. In Figure 7(b), the morphology of the composite is shown. After friction stir processing, the refined grains were observed within the composite as revealed by Figure 7(c). This indicates that the grain size of the AA6063 aluminium matrix is reduced considerably in the stirred zone. However, the size and shape of B_4C particles were not significantly changed by FSP. Rather, the particle distribution in the matrix was more uniform and the number and size of reinforcement particle clusters were reduced in the stirred zone. It was also found that during FSP, dynamic recrystallization happened in the nugget zone. The continuous evolution of aluminium grain structure from coarse grains of the AA6063 matrix alloy via subgrain development occurred within coarse grains to the refined, equiaxed grains in the center of FSP zone.

(a)

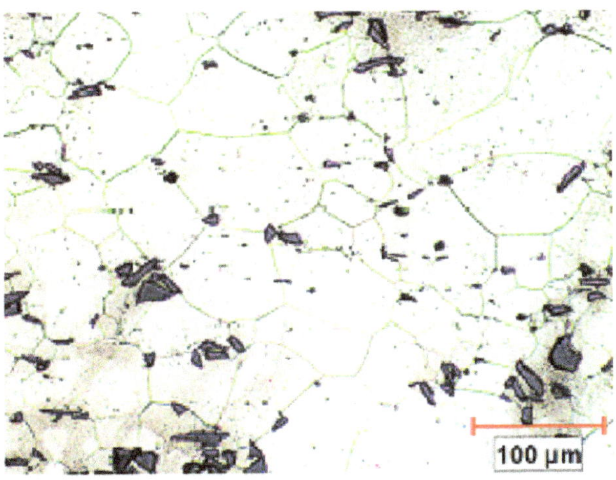

(b)

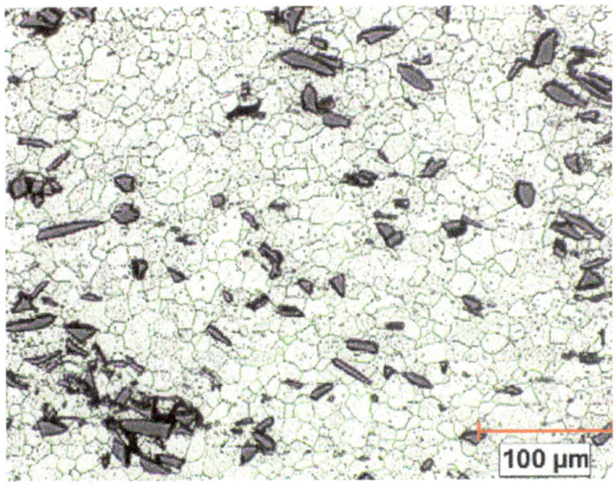

(c)

Figure 7. AA6063 and B_4C particle reinforced composite (after Chen *et al.* [62]): (a) EBSD orientation map showing the grains of AA6063; (b) B_4C particle reinforced composite before FSP; (c) B_4C particle reinforced composite after FSP.

Nanoscale reinforcements such as carbon nanotubes were used to make reinforced composite materials through FSP [14]. The effects of processing parameters on multi-walled carbon nanotube dispersion, and hardness of the composite were investigated. Increasing the tool rotation speed and the tool shoulder penetration depth improved the distribution of nanotubes in the Al-alloy matrices, 6111-T4 and 7075-T6. A completely uniform distribution could not be achieved when regularly tangled nanotubes as shown in Figure 8(a) were used as the starting material. Fracture of carbon nanotubes was found during friction stir processing as revealed by Figure 8(b), the surface morphology of an etched specimen. The etching helps to expose the carbon nanotubes. Evidently, the long and entangled CNTs are not uniformly dispersed throughout the processing zone. Although it is suggested that multiple tool pin passes may be useful to further improve the uniform dispersion of nanotubes in the composites, fracture of carbon nanotubes could be more serious. Nevertheless, carbon nanotube reinforced composites showed the significant strengthening effect, as evidenced by the drastic increase in the hardness in the friction stir processing zone. That the hardness of the processing zone is higher than other regions is shown in Figure 8(c).

Other types of micro- or nanoscale phases, such as *in-situ* formed particles including Al_2Cu, and Al_3Ti [63], NiTi [64], and AiFe [65] have been taken as the reinforcements for aluminum based composites and FSP has been used to modify the microstructures of these *in-situ* composites. Under high temperature processing conditions such as casting and powder metallurgical formation, considerable interfacial reactions lead to formation of a number of intermetallic phases, such as Al_3Ti, Al_3Ni, Ni_3Ti, Ti_2Ni, etc. depending on the compositions of the raw materials. The presence of these intermetallic phases, if they are physically or mechanically incompatible with the matrix alloys, they become the key sites for fracture initiation and failure. FSP could help to alleviate this problem. The advantage of *in-situ* composites is that particles are dispersed very uniformly within matrices as shown by the backscattered electron images (BEI) in Figure 9. Figure 9(a) represents the

Al_2Cu particles in Al-10Cu alloy. The spots with white contrast are the Al_2Cu particles. The distribution of the particles are very uniform. In Figure 9(b), NiTi particles (white contrast) in the nugget zone are shown. Figure 9(c) reveals the Al_3Fe particles in Al matrix after 2 passes of FSP followed by annealing. The brightest phase is identified as Fe through EDS, the dark gray matrix is Al, and the light gray phase, which surround Fe particle or dispersed in the Al matrix, is AlFe compound.

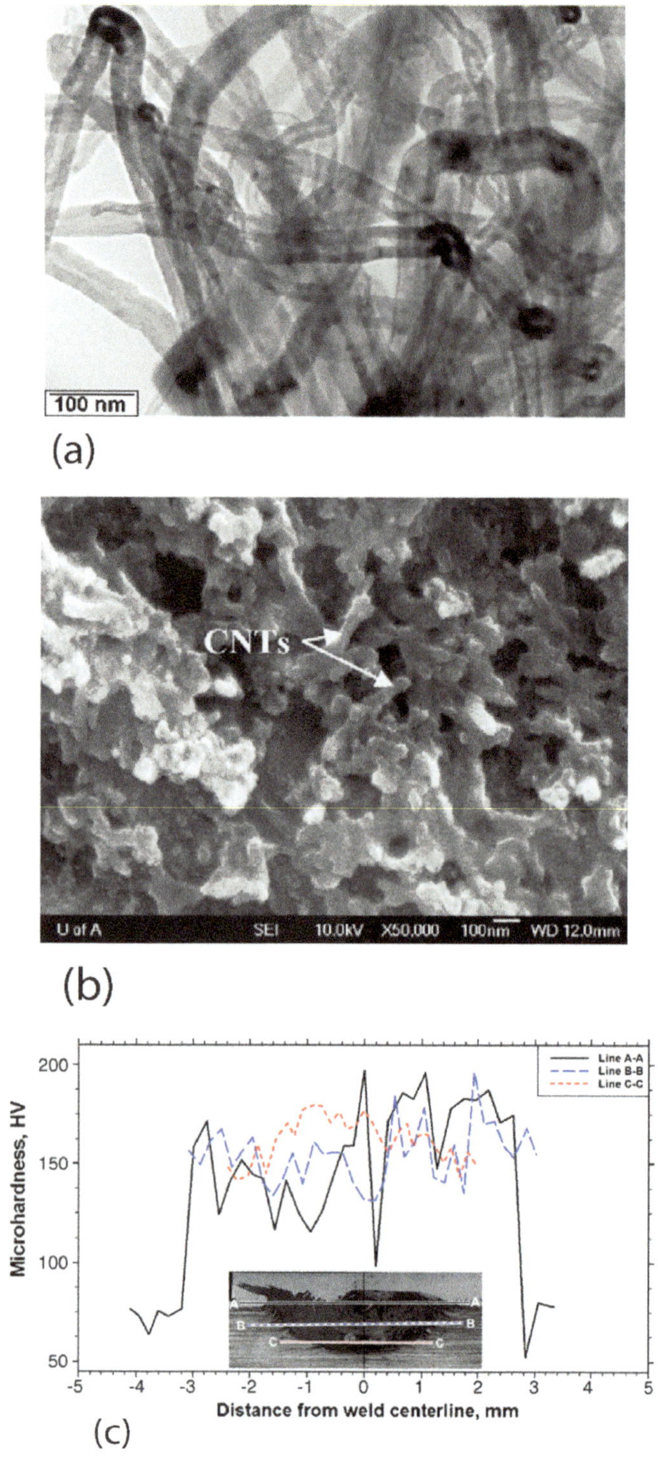

Figure 8. Morphology and hardness profile of friction stir processed carbon nanotube reinforced composite (after Lim *et al.* [14]): (a) TEM image of multi-walled carbon nanotubes; (b) friction stir processed composite; (c) hardness measurement results.

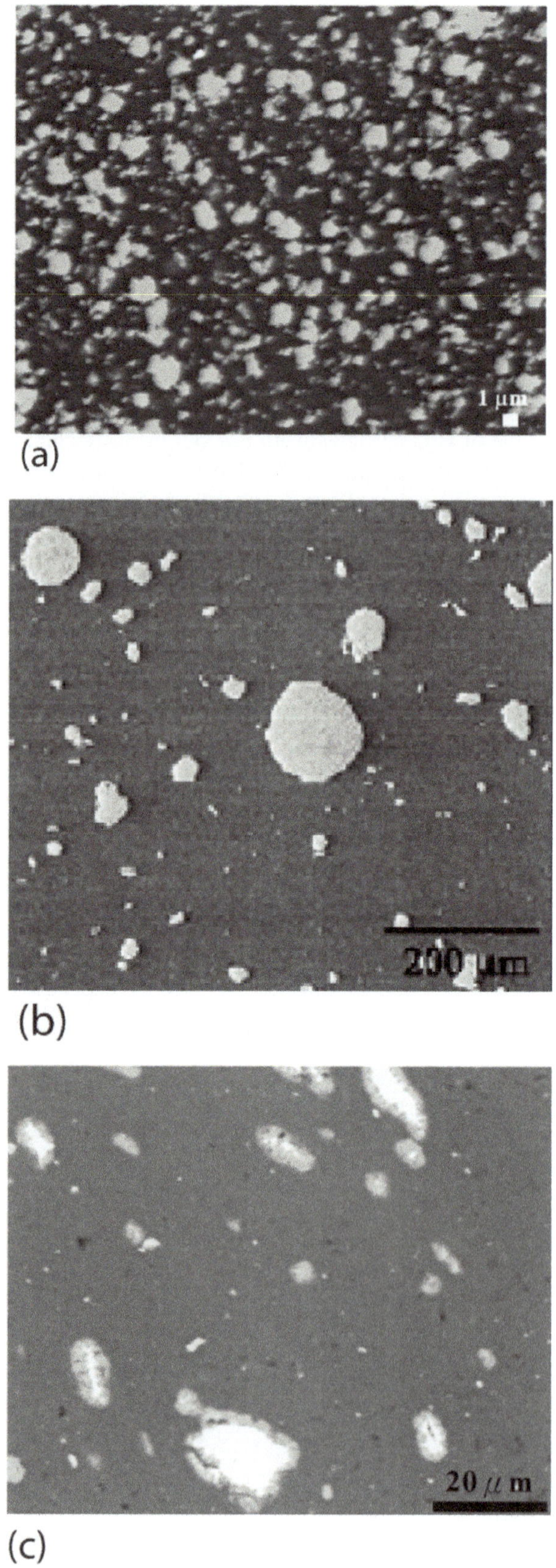

Figure 9. Backscattered electron images showing friction stir processed *in-situ* particle reinforced composite materials: (a) the dispersion of fine Al_2Cu particles (white contrast) in Al10Cu alloy (after Hsu *et al.* [63]); (b) uniformly distributed NiTi particles in the nugget region of FSP composite after annealing (after Dixit *et al.* [64]); (c) Al_3Fe particles in Al matrix (after Lee *et al.* [65]).

FRICTION STIR PROCESSING OF MAGNESIUM MATRIX COMPOSITES

Magnesium has the density as small as 1.7 g/cm^3, which is the lowest among all the structural metals or alloys [66]. Such an outstanding property makes it extremely important for manufacturing light weight structures. However, magnesium is very soft. It is necessary to add either alloy elements or reinforcements to make it stronger. Recently, friction stir processing particle reinforced magnesium has been studied [29,30,32,67,68,69,70]. In the work performed by Lee *et al.* [67], Nanoscale SiO_2 with the dimension of 20 nm was added into an AZ61 magnesium alloy. Friction stir processing (FSP) was applied to incorporate 5-10 vol % nanoparticles into the matrix. The distribution of the SiO_2 particles was uniform after four FSP passes. The grain size of the FSP composites was effectively refined to 0.8 μm, as compared with the 78 μm in the FSP AZ61 alloy processed under the same FSP condition. There is reaction between SiO_2 particles and the Mg alloy during FSP. The products from the reaction include Mg_2Si and MgO. The size of these products is in the range from 5 to 200 nm, which is very fine and good for improving the mechanical properties. The hardness and mechanical strength at room temperature of the AZ61Mg composite with nano-particles were increased, as compared with the AZ61 cast billet. Also found is the high strain rare superplasticity over 400% in the FSP composite.

In the work performed by Morisada *et al.* [68], multi-walled carbon nanotubes (MWCNTs) with the outer diameter of 20 to 50 nm, and length of 250 nm were successfully dispersed into AZ31 magnesium alloy via friction stir processing (FSP). The tool rotating speed was kept at 1500 rpm. Three travel speeds of 100, 50 and 25 mm/min, were used. It was found that the travel speed of 100 mm/min was too fast to produce enough heat flow to produce a suitable viscosity in the AZ31 matrix for the dispersion of the MWCNTs. The sample which was processed at 50 mm/min showed a better dispersion of the MWCNTs. However, there were some regions where aggregation of MWCNTs occurred. A much better dispersion of the MWCNTs was achieved for the sample that was processed at 25 mm/min. The addition of the MWCNTs promoted grain refinement during the FSP. The size of the grains can be controlled as small as 500 nm. The Vickers hardness of the friction stir processed composite material containing MWCNTs is 78 HV. For the sample without MWCNTs treated by the FSP, the measured hardness is 55 HV. The as-received AZ31 sample has the lowest hardness of 41 HV.

Lee *et al.* [69] studied the microstructure and wear properties of friction stir processed SiC particle reinforced AZ91 magnesium alloy composite material. The composite contains 10 vol % SiC particles. In the thermomechanically affected zone (TMAZ), there are deformation bands and the distribution of SiC particles aligned along the bands as can be seen fromFigure 10(a). In the stir zone (SZ), SiC particles are uniformly distributed as shown in Figure 10(b). Concerning the grain structure evolution of magnesium alloy without reinforcement, Suhuddin *et al.* [70] provided detailed results of grain size and orientation to address the twinning, texture formation and recrystallization due to FSP.

At higher magnification as shown in Figure 10(c), it is observed that the sharp corners of SiC particles were rounded up. Before taking SEM images, the surface of the composite was etched for 30 s with a solution containing ethyl alcohol, acetic acid, picric acid and distilled water to reveal the SiC particles. The wear tests for the AZ91, AZ91/SiC/10p and the FSP zone of the AZ91/SiC/10p were performed under wet sliding conditions for 1000 s against a rotating Austenite cast iron disc (hardness 200 HV) at the constant load of 50 N and sliding speed of 1 m/s. The lubricant was distilled water. The specific wear loss, W, was used to measure the wear resistance of the materials, and W is calculated by the following equation

$$W = \frac{Bb^3}{8rPl} \qquad (1)$$

where B is the thickness of the rotating disk, r is the radius of the disk, b is the length of the wear trace, P is the applied load and l is the sliding distance.

Based on the wear test data and the calculation using the above Equation (1), the specific wear loss of the base alloy is about three times higher than that of the SiC reinforced composite before FSP. The wear loss of the composite can be further reduced after FSP as shown in Figure 11.

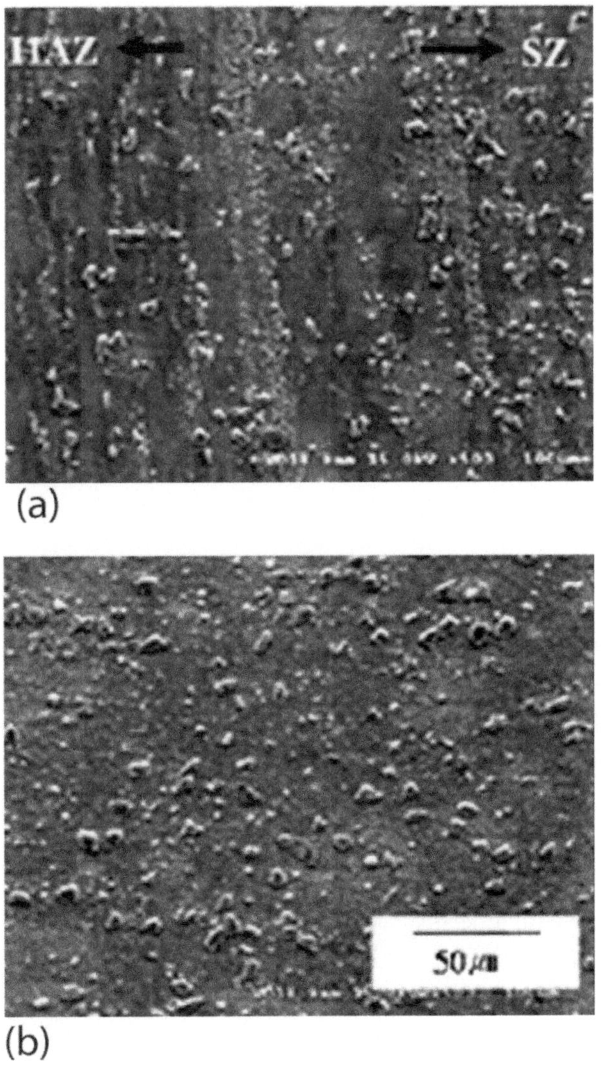

(a)

(b)

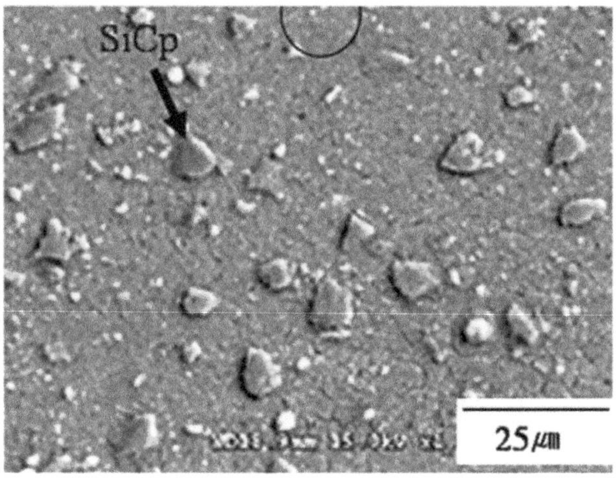

(c)

Figure 10. Microstructure of SiC particle reinforced AZ91 magnesium alloy composite material (after Lee *et al.* [69]): (a) the area showing the transition from thermomechanical affected zone (TMAZ) to stir zone (SZ); (b) uniformly distributed SiC particles in the stir zone; (c) higher magnification image of stir zone showing the SiC particles on the etched surface of the composite.

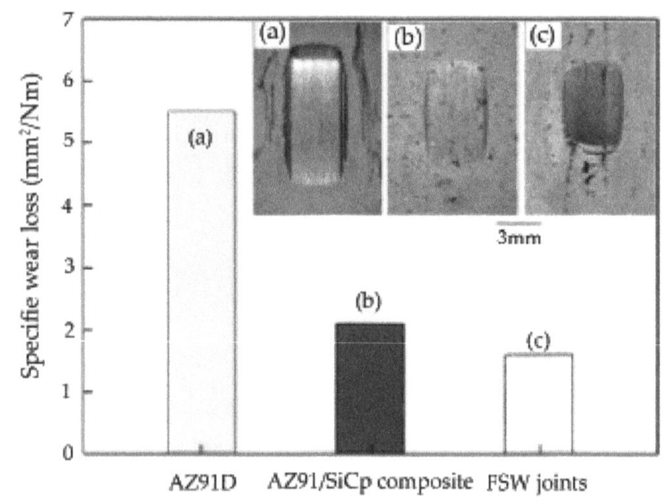

Figure 11. Wear loss of AZ91, SiC/AZ91 composite without and with FSP (after Lee *et al.* [69]).

PERSPECTIVES ON FRICTION STIR PROCESSING

Application for Processing New Metallic Alloys/Composites

Friction stir processing is expected to find more applications, for example, in energy field. Perhaps to process or join vanadium alloys and nanoparticle reinforced copper composite materials for fusion reactor applications is such a good example. V-Ti-Cr alloys and nanocomposites made by electrocodeposition of SiC nanoparticles with copper can be joined by optimizing the parameters such as tool rotating speed, nutting angle and feed rate to obtain effective butt joints. Development of fusion reactor technology for power generation is an attractive proposition because of the abundant availability of hydrogen isotope and the fusion process is much safer than the currently used fission technology. The property requirements for fusion structures such as the first wall and divertor are very stringent. The fusion structural materials should have good mechanical properties at high temperatures, good corrosion resistance with the coolant, be able to accommodate high heat loads, and have low levels of induced radioactivity. Vanadium alloys are under consideration for use in fusion energy systems as structural materials because they offer significant advantages over other iron based alloys in several of the areas mentioned above [71,72]. Extensive research in the past two decades has lead to a basic understanding of the physical, mechanical and metallurgical aspects of vanadium alloys including the effects of radiation. Presently, V-Cr-Ti alloys are considered as the promising ones with improved performance [71,72,73,74,75].

The vanadium alloys must be joined to either the vanadium alloys with different configuration or to lower cost dissimilar metallic materials such as copper alloys to construct an effective power generating system. In structural fabrication, such as the first wall and divertor components of a nuclear reactor, kilometers of welds are expected. Harnessing the excellent mechanical properties and environmental safety features offered by vanadium alloys depends upon the successful development of processing technology for vanadium alloys. There exists a technical challenge in conventional welding or brazing vanadium to vanadium or to copper alloys due to their affinity to interstitial elements such as oxygen, hydrogen and nitrogen [76]. At present time, vanadium alloys can be brazed under high vacuum and high temperature conditions. The high vacuum environments can only be created in a laboratory setup, which can be very expensive to maintain. This poses engineering limitations on the size of the structures that can be processed. In addition, the temperature in brazing process is so high that undesired coarse grains formed in both vanadium alloys and the joining counterpart alloys.

Apart from brazing, other joining methods have also been studied to obtain sound vanadium joints. Arc welding such as gas tungsten arc (GTA) welding, laser welding and electron beam welding were tried in fabrication of joints containing vanadium alloys [77,78,79]. However, the rapid heating and cooling created complex non-equilibrium microstructure in the fusion and heat affected zones (HAZ), close to the fusion line. This problem got further worsened by the affinity of vanadium alloys to interstitial elements such as O, H, and N. Safety and reliability of welded structures are major concerns in critical application areas such as nuclear reactors. The consequences of catastrophic fracture in a critical structure are very severe, because it may result in loss of human life, environmental degradation and loss of revenue. Unfortunately, welded structures are more prone to fracture because of the presence of residual stresses, non-equilibrium microstructure, cracks and other metallurgical defects. Vanadium alloy joins produced using gas tungsten arc, laser and electron beam welding have been found to have low fracture toughness due to the embrittlement of heat affected zone [80,81]. It is obvious that current practices used to process vanadium alloy joints are not good enough for engineering applications. Therefore, developing innovative processing technology to achieve high quality vanadium alloy joints for industrial applications is one of the most significant problems to be solved.

Friction stir welding (FSW) [82] and friction stir processing (FSP) have been applied to various materials including plastics [83] stainless steels [84,85], magnesium alloys [86], titanium alloys [87], pure copper [88], copper-zinc alloys [89] and metal matrix composite materials [90]. FSW or FSP has many advantages such as elimination of the defects named crack and porosity often associated with fusion welding processes, reduced distortion. In addition, joint edge preparation is much less strict. It can be carried out in various positions and can join conventionally non-fusion weldable alloys and improve mechanical properties of weldable alloys. FSP has also been used to refine the grains of casting alloys [91], and to homogenize the microstructure of reinforced metal matrix composite materials [92]. FSP has found applications in transportation industries [93,94]. The application of FSP in manufacturing of fusion reactor structures has also been considered [95]. Therefore, the application of FSP/FSW in preparation and structure modification of energy materials is promising.

Application for Processing Polymeric Materials and Composites

Polymers have much lower processing temperatures than metals. The facilities for implementing FSP could be much simpler and less expensive. As shown in Figure 12, the experimental installation for the friction stir processing just includes the CNC machine, the tool pin connected to the spindle, and the specimen stage. Materials to be processed are fixed by the clamper. Friction stir processing of polymer composites was performed successfully. The polymeric materials used are particle filled plastics such as polystyrene (PS) and nylon. Figure 13(a) shows the nylon-nylon composite laminate, while nylon-polystyrene laminate is illustrated in Figure 13(b). Thermoplastics such as Polypropylene have been treated by friction stir processing [96,97]. It is noted that FSP polymers and polymeric composites are still much less studied than FSP metals and metallic composites [98]. There are some concerns related to friction stir processing polymers because the rotation of tool pin and the severe plastic flow in the nugget zone could break macromolecular chains and change the properties of the materials.

Figure 12. Equipment for friction stir processing polymer composites.

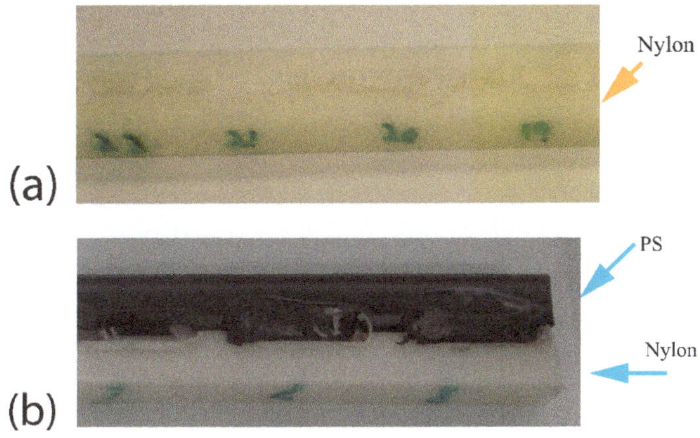

Figure 13. Friction stir processed composites; (a) nylon-nylon laminate, (b) nylon-polystyrene laminate.

Issues on the Challenges and Improvements Needed

There are some challenges associated with friction stir processing approach. Although, friction stir processing has many advantages over various other materials processing technologies, the fact is that the joint efficiency defined as the ratio between the ultimate tensile strength of the welded joint and the one of the base material has hardly reached 80% [62]. This is an indication that the weld strength is still low. In addition to the static tensile strength, the fatigue property of FSP composites should also be improved. Minak *et al.* [99] have recently studied the fatigue properties of friction stir processed particulate reinforced aluminum matrix composites and made extensive statistical analysis. Failure probability curves were generated as shown in Figure 14. The composite material specimens with the set numbers from 1 to 6 have the same composition, but they were processed under different conditions. The processing parameters are given in Table 1. The statistical analysis results in Figure 14 showed that FSP specimens with different processing parameters had worse fatigue behavior than that of the base composite material. The different microstructural homogeneity in the transition from the base to the FSP zone is considered as the major reason for the degradation of fatigue property. As shown in Figure 15(a), the transition in microstructural homogeneity of the composite material generated by FSP can obviously be seen. The transition zone (TZ) seems the weakest part of the composite material because fatigue fracture always occurs in this zone for the specimens with the set numbers from 2 to 6. The macroscopic view of a typical fatigue-fractured specimen is shown in Figure 15(b). Friction stir tool wear is another challenge problem remaining to be solved. This problem may not be so severe in FSP pure alloys without reinforcements. But it is serious during the processing of particle reinforced metal matrix composites (MMCs). In FSP soft metals without reinforcement particles such as aluminum alloys, though frictional heat is generated [100], no obvious tool consumption was created [101]. That is, no significant friction-induced wear occurs. The reason is that the stir zone consists of dynamically recrystallized (DRX) material which creates a solid-state flow regime where friction is minimal [102,103]. Therefore, the rotating pin tool is surrounded by a quasi-hydrodynamic lubricating layer composed of the DRX material [104,105,106].

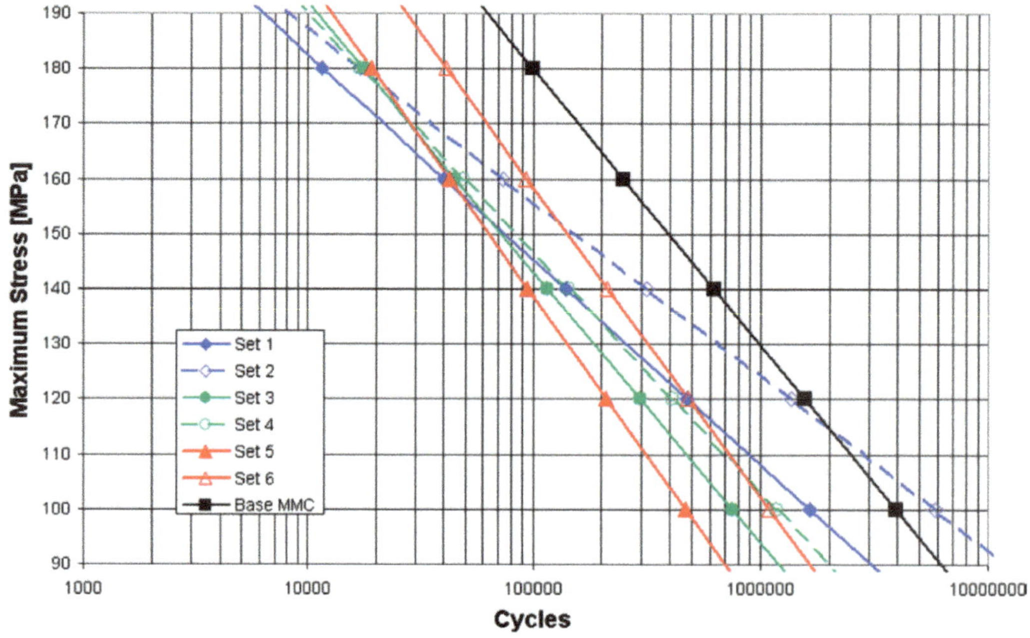

Figure 14. 50% failure probability curves of friction stir processed 22 vol % Al_2O_3particle reinforced AA6061 composites under different processing conditions (after Minak *et al.* [99]).

Table 1. Parameters used for the FSP (data source: [99]).

Specimen set	1	2	3	4	5	6
Rotation rate (rpm)	630	630	630	880	880	880
Tool speed (mm/min)	115	170	260	115	170	260

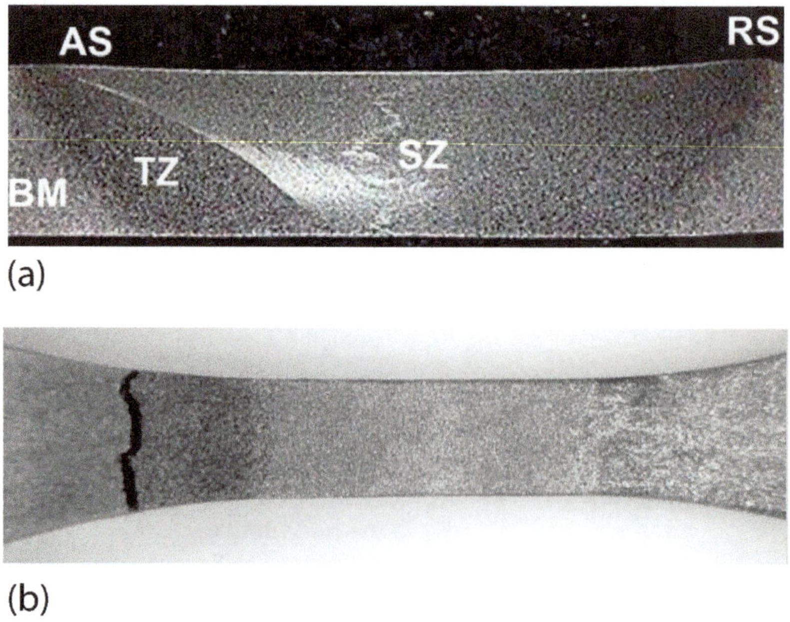

Figure 15. Macroscopic images of the friction stir processed 22 vol % Al_2O_3 particle reinforced AA6061 composite showing: (a) the base material (BM), the transition zone (TZ), the stir zone (SZ), the advancing side (AS), and the retreating side (RS); (b) typical location of fatigue fracture in the transition zone (TZ) (after Minak *et al.* [99]).

Friction stir processing metal matrix composites (MMCs) can exhibit considerable tool consumption as shown in Figure 16[58]. The tool wear was examined in friction stir processing 20 vol % Al_2O_3/AA6061 at the rotation rate of 1000 rpm and the traveling speed of 9 mm/s. Initially, the tool was in the screw shape as shown in Figure 16(a). After being used for FSP, the screw part disappeared, which means that the consumption of the tool occurred. When the tool was in service and traveled 1.96 m, the tool end was rounded up as shown in Figure 16(b). If the tool was further used, the surface wear continued. Figure 16(c) shows when the traveling distance reached 3.66 m, necking due to the increased wear was observed. In order to reduced the wear, cubic boron nitride and tungsten rhenium alloys may be used as tool materials, but they are extremely expensive and difficult to shape.

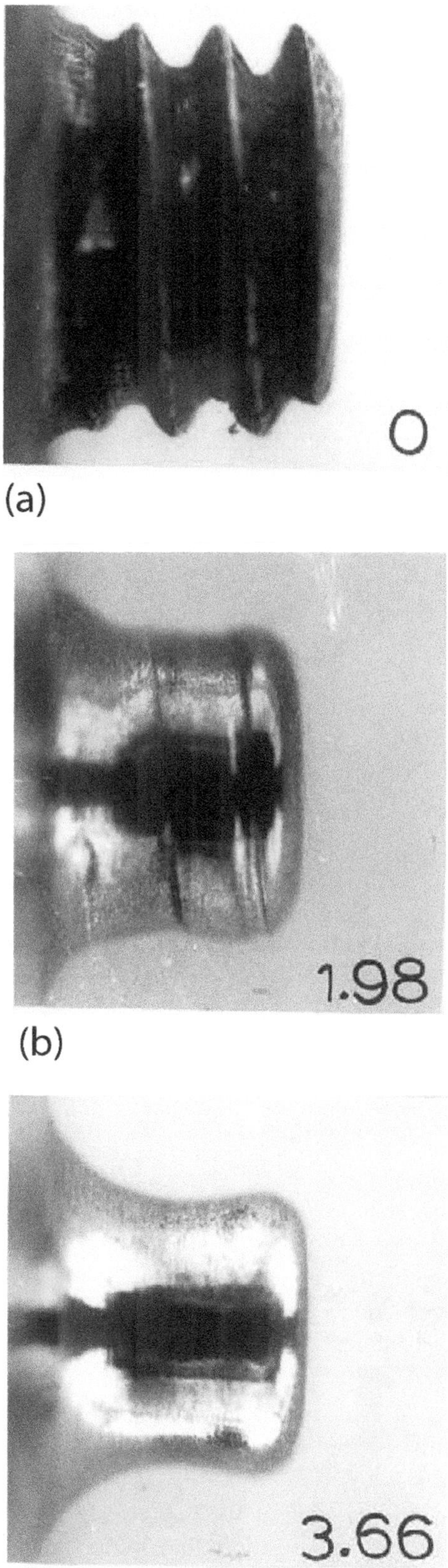

Figure 16. Macroscopic images showing the wear of friction stir processing tool (after Prado *et al.* [58]); (a) the screw tool pin before FSP, (b) after traveling 1.98 m, (c) after traveling 3.66 m.

SUMMARY AND CONCLUSIONS

Friction stir processing (FSP) forces materials flow at the temperature lower than the melting temperature. Materials are extruded, forged, consolidated and cooled under hydrostatic pressure conditions. Thus some of the properties, for example, hardness is high for FSP materials. The primary research on friction stir processing focuses on aluminum alloys. Applying this technology for processing other alloys and materials including stainless steels, magnesium, titanium, and copper are also studied. This technology has found applications in modifying the microstructure of reinforced metal matrix composite materials. It is also used in processing polymeric composite materials. It is possible to transfer FSP technology into energy materials field for example, for fusion reactor materials including vanadium alloy joining and processing. There are some major challenges for improving the FSP technology. One is the tool wear in processing reinforced composite materials. The other challenge is how to increase the joining strength and improve the fatigue property of FSP composite materials.

ACKNOWLEDGEMENTS

YXG acknowledges the support from The Center for Teaching and Learning through a Summer Teaching Fellowship and the research start-up fund provided by The University of Toledo. This work was supported in part by the Durbin Faculty Development Grants at The Cooper Union for the Advancement of Science and Art. We also acknowledge the permissions from Elsevier B.V. Ltd through Copyright Clearance Center's Rightslink service to using Figure 1, Figure 2, Figure 3, Figure 4, Figure 5, Figure 6, Figure 7, Figure 8, Figure 9, Figure 10 and Figure 11, and Figure 14, Figure 15 and Figure 16 in this paper.

REFERENCES

1. Liechty, B.C.; Webb, B.W. Flow field characterization of friction stir processing using a particle-grid method. *J. Mater. Process. Technol.* 2008, *208*, 431–443.
2. Darras, B.M.; Khraisheh, M.K.; Abu-Farha, F.K.; Omar, M.A. Friction stir processing of commercial AZ31 magnesium alloy. *J. Mater. Process. Technol.* 2007, *191*, 77–81.
3. Wang, W.; Shi, Q.; Liu, P.; Li, H.; Li, T. A novel way to produce bulk SiC_p reinforced aluminum metal matrix composites by friction stir processing. *J. Mater. Process. Technol.* 2009, *209*, 2099–2103.
4. Karthikeyan, L.; Senthilkumar, V.S.; Padmanabhan, K.A. On the role of process variables in the friction stir processing of cast aluminum A319 alloy. *Mater. Des.* 2010, *31*, 761–771.
5. Sinha, P.; Muthukumaran, S.; Mukherjee, S.K. Analysis of first mode of metal transfer in friction stir welded plates by image processing technique. *J. Mater. Process. Technol.* 2008, *197*, 17–21.
6. Karthikeyan, L.; Senthilkumar, V.S.; Balasubramanian, V.; Natarajan, S. Mechanical property and microstructural changes during friction stir processing of cast aluminum 2285 alloy. *Mater. Des.* 2009, *30*, 2237–2242.

7. Ma, Z.Y.; Pilchak, A.L.; Juhas, M.C.; Williams, J.C. Microstructural refinement and property enhancement of cast light alloys *via* friction stir processing. *Scr. Mater.* 2008, *58*, 361–366.

8. Du, X.; Wu, B. Using friction stir processing to produce ultrafine-grained microstructure in AZ61 magnesium alloy.*Trans. Nonferrous Met. Soc. China* 2008, *19*, 562–565.

9. Shafiei-Zarghani, A.; Kashani-Bozorg, S.F.; Zarei-Hanzaki, A. Microstructures and mechanical properties of Al/Al2O3 surface nano-composite layer produced by friction stir processing. *Mater. Sci. Eng., A* 2009, *500*, 84–91.

10. Mironov, S.; Sato, Y.S.; Kokawa, H. Microstructural evolution during friction stir-processing of pure iron. *Acta Mater.*2008, *56*, 2602–2614.

11. Ceschini, L.; Boromei, I.; Minak, G.; Morri, A.; Tarterini, F. Microstructure, tensile and fatigue properties of AA6061/20 vol %Al_2O_{3p} friction stir welded joints. *Composites: Part A* 2007, *38*, 1200–1210.

12. Mishra, R.S.; Ma, Z.Y.; Charit, I. Friction stir processing: a novel technique for fabrication of surface composite. *Mater. Sci. Eng., A* 2003, *341*, 307–310.

13. Ma, Z.Y.; Liu, F.C. Achieving exceptionally high superplasticity at high strain rates in a micrograined Al-Mg-Sc alloy produced by friction stir processing. *Scr. Mater.* 2008, *59*, 882–885.

14. Lim, D.K.; Shibayanagi, T.; Gerlich, A.P. Synthesis of multi-walled CNT reinforced aluminium alloy composite *via*friction stir processing. *Mater. Sci. Eng., A* 2009, *507*, 194–199.

15. Surekha, K.; Murty, B.S.; Rao, K.P. Effect of processing parameters on the corrosion behaviour of friction stir processed AA 2219 aluminum alloy. *Solid State Sci.* 2009, *11*, 907–917.

16. Ma, Z.Y.; Sharma, S.R.; Mishra, R.S. Effect of multiple-pass friction stir processing on microstructure and tensile properties of a cast aluminum-silicon alloy. *Scr. Mater.* 2006, *54*, 1623–1626.

17. McNelley, T.R.; Swaminathan, S.; Su, J.Q. Recrystallization mechanisms during friction stir welding/processing of aluminum alloys. *Scr. Mater.* 2008, *58*, 349–354.

18. Johannes, L.B.; Mishra, R.S. Multiple passes of friction stir processing for the creation of superplastic 7075 aluminum.*Mater. Sci. Eng., A* 2007, *464*, 255–260.

19. Cavaliere, P.; De Marco, P.P. Friction stir processing of a Zr-modified 2014 aluminium alloy. *Mater. Sci. Eng., A* 2007,*462*, 206–210.

20. Yazdipour, A.; Shafiei, A.; Dehghani, M.K. Modeling the microstructural evolution and effect of cooling rate on the nanograins formed during the friction stir processing of Al5083. *Mater. Sci. Eng., A* 2009, *527*, 192–197.

21. Liu, F.C.; Ma, Z.Y.; Chen, L.Q. Low-temperature superplasticity of Al-Mg-Sc alloy produced by friction stir processing. *Scr. Mater.* 2009, *60*, 968–971.

22. Johannes, L.B.; Charit, I.; Mishra, R.S.; Verma, R. Enhanced superplasticity through friction stir processing in continuous cast AA5083 aluminum. *Mater. Lett.* 2007, *464*, 351–357.

23. Fuller, M.D.; Swaminathan, S.; Zhilyaev, A.P.; McNelley, T.R. Microstructural transformations and mechanical properties of cast NiAl bronze: Effects of fusion welding and friction stir processing. *Mater. Sci. Eng., A* 2007, *463*, 128–137.

24. Kobata, J.; Takigawa, Y.; Chung, S.W.; Tsuda, H.; Kimura, H.; Higashi, K. Nanoscale amorphous "band-like" structure induced by friction stir processing in $Zr_55Cu_30Al_10Ni_5$ bulk metallic glass. *Mater. Lett.* 2007, *61*, 3771–3773.

25. Morisada, Y.; Fujii, H.; Mizuno, T.; Abe, G.; Nagaoka, T.; Fukusumi, M. Nanostructured tool steel fabricated by combination of laser melting and friction stir processing. *Mater. Sci. Eng., A* 2009, *505*, 157–162.

26. Morisada, Y.; Fujii, H.; Mizuno, T.; Abe, G.; Nagaoka, T.; Fukusumi, M. Modification of nitride layer on cold-work tool steel by laser melting and friction stir processing. *Surf. Coat. Technol.* 2009, *204*, 368–390.

27. Feng, A.H.; Ma, Z.Y. Microstructural evolution of cast Mg-Al-Zn during friction stir processing and subsequent aging.*Acta Mater.* 2009, *57*, 4248–4260.

28. Ni, D.R.; Wang, D.; Feng, A.H.; Yao, G.; Ma, Z.Y. Enhancing the high-cycle fatigue strength of Mg-9Al-1Zn casting by friction stir processing. *Scr. Mater.* 2009, *61*, 568–571.

29. Chang, C.I.; Du, X.H.; Huang, J.C. Producing nanograined microstructure in Mg-Al-Zn alloy by two-step friction stir processing. *Scr. Mater.* 2008, *59*, 356–359.

30. Chang, C.I.; Lee, C.J.; Huang, J.C. Relationship between grain size and Zener-Holloman parameter during friction stir processing in AZ31 Mg alloys. *Scr. Mater.* 2004, *51*, 509–514.

31. Cavaliere, P.; De Marco, P.P. Friction stir processing of AM60B magnesium alloy sheets. *Mater. Sci. Eng., A* 2007, *462*, 393–397.

32. Chang, C.I.; Du, X.H.; Huang, J.C. Achieving ultrafine grain size in Mg-Al-Zn alloy by friction stir processing. *Scr. Mater.* 2007, *57*, 209–212.

33. Commin, L.; Dumont, M.; Masse, J.E.; Barrallier, L. Friction stir welding of AZ31 magnesium alloy rolled sheets: Influence of processing parameters. *Acta Mater.* 2009, *57*, 326–334.

34. Ma, Z.Y.; Mishra, R.S.; Liu, F.C. Superplastic behavior of micro-regions in two-pass friction stir processed 7075Al alloy. *Mater. Sci. Eng., A* 2009, *505*, 70–78.

35. Lakshminarayanan, A.K.; Balasubramanian, V. Process parameters optimization for friction stir welding of RDE-40 aluminium alloy using Taguchi technique. *Trans. Nonferrous Met. Soc. China* 2008, *18*, 548–554.

36. Moreira, P.M.G.P.; de Oliveira, F.M.F.; de Castro, P.M.S.T. Fatigue behavior of notched specimens of friction stir welded aluminium alloy 6063-T6. *J. Mater. Process. Technol.* 2008, *207*, 283–292.

37. Elangovan, K.; Balasubramanian, V. Influences of pin profile and rotational speed of the tool on the formation

of friction stir processing zone in AA2219 aluminium alloy. *Mater. Sci. Eng., A* 2007, *459*, 7–18.

38. Hofmann, D.C.; Vecchio, K.S. Thermal history analysis of friction stir processed and submerged friction stir processed aluminum. *Mater. Sci. Eng., A* 2009, *465*, 165–175.
39. Cui, G.R.; Ma, Z.Y.; Li, S.X. Periodical plastic flow pattern in friction stir processed Al-Mg alloy. *Scr. Mater.* 2008, *58*, 1082–1085.
40. Elangovan, K.; Balasubramanian, V. Influences of tool pin profile and tool shoulder diameter on the formation of friction stir processing zone in AA6061 aluminium alloy. *Mater. Des.* 2008, *29*, 362–373.
41. Elangovan, K.; Balasubramanian, V. Influences of tool pin profile and welding speed on the formation of friction stir processing zone in AA2219 aluminium alloy. *J. Mater. Process. Technol.* 2008, *200*, 163–175.
42. Buffa, G.; Fratini, L.; Pasta, S.; Shivpuri, R. On the thermo-mechanical loads and the resultant residual stresses in friction stir processing operations. *CIRP Ann. - Manuf. Technol.* 2008, *57*, 287–290.
43. Morisada, Y.; Fujii, H.; Nagaoka, T.; Fukusumi, M. Effect of friction stir processing with SiC particles on microstructure and hardness of AZ31. *Mater. Sci. Eng., A* 2006, *433*, 50–54.
44. Liechty, B.C.; Webb, B.W. Modeling the frictional boundary condition in friction stir welding. *Int. J. Mach. Tool. Manuf.*2008, *48*, 1474–1485.
45. Elangovan, K.; Balasubramanian, V.; Babu, S. Predicting tensile strength of friction stir welded AA6061 aluminium alloy joints by a mathematical model. *Mater. Des.* 2009, *30*, 188–193.
46. Zhang, Z.; Zhang, H.W. Numerical studies on controlling of process parameters in friction stir welding. *J. Mater. Process. Technol.* 2009, *209*, 241–270.
47. Wang, Y.; Mishra, R.S. Finite element simulation of selective superplastic forming of friction stir processed 7075 Al alloy. *Mater. Sci. Eng., A* 2007, *463*, 245–248.
48. Cavaliere, P.; De Santis, A.; Panella, F.; Squillace, A. Effect of welding parameters on mechanical and microstructural properties of dissimilar AA6082-AA2024 joints produced by friction stir welding. *Mater. Des.* 2009, *30*, 609–616.
49. Cavaliere, P.; Cabibbo, M.; Panella, F.; Squillace, A. 2198 Al-Li plates joined by Friction Stir Welding: Mechanical and microstructural behavior. *Mater. Des.* 2009, *30*, 3622–3631.
50. Cui, G.R.; Ma, Z.Y.; Li, S.X. The origin of non-uniform microstructure and its effects on the mechanical properties of a friction stir processed Al-Mg alloy. *Acta Mater.* 2009, *57*, 5718–5729.
51. Hirata, T.; Tanaka, T.; Chung, S.W.; Takigawa, Y.; Higashi, K. Relationship between deformation behavior and microstructural evolution of friction stir processed Zn-22 wt.% Al alloy. *Scr. Mater.* 2007, *56*, 477–480.
52. Cavaliere, P.; Squillace, A.; Panella, F. Effect of welding parameters on mechanical and microstructural properties of AA6082 joints produced by friction stir welding. *J. Mater. Process. Technol.* 2008, *200*, 364–372.
53. Tewari, A.; Spowart, J.E.; Gokhale, A.M.; Mishra, R.S.; Miracle, D.B. Characterization of the effects of friction stir processing on microstructural changes in DRA composites. *Mater. Sci. Eng., A* 2006, *428*, 80–90.
54. Fernandez, G.J.; Murr, L.E. Characterization of tool wear and weld optimization in the friction-stir welding of cast aluminum 359+20% SiC metal-matrix composite. *Mater. Charact.* 2004, *52*, 65–75.
55. Uzun, H. Friction stir welding of SiC particulate reinforced AA2124 aluminium alloy matrix composite. *Mater. Des.*2007, *28*, 1440–1446.
56. Feng, A.H.; Xiao, B.L.; Ma, Z.Y. Effect of microstructural evolution on mechanical properties of friction stir welded AA2009/SiC_p composite. *Compos. Sci. Technol.* 2008, *68*, 2141–2148.
57. Prangnell, P.B.; Heason, C.P. Grain structure formation during friction stir welding observed by the 'stop action technique'. *Acta Mater.* 2005, *53*, 3179–3192.
58. Prado, R.A.; Murr, L.E.; Soto, K.F.; McClure, J.C. Self-optimization in tool wear for friction-stir welding of Al 6061+20% Al2O3 MMC. *Mater. Sci. Eng. A* 2003, *349*, 156–165.
59. Pirondi, A.; Collini, L. Analysis of crack propagation resistance of $AlAl_2O_3$ particulate-reinforced composite friction stir welded butt joints. *Int. J. Fatigue* 2009, *31*, 111–121.
60. Marzoli, L.M.; Strombeck, A.V.; Dos Santos, J.F.; Gambaro, C.; Volpone, L.M. Friction stir welding of an AA6061/Al_2O_3/20p reinforced alloy. *Compos. Sci. Technol.* 2006, *66*, 363–371.
61. Cavaliere, P. Mechanical properties of Friction Stir Processed 2618/Al_2O_3/20p metal matrix composite. *Composites: Part A* 2005, *36*, 1657–1665.
62. Chen, X.-G.; da Silvaa, M.; Gougeonb, P.; St-Georges, L. Microstructure and mechanical properties of friction stir welded AA6063B4C metal matrix composites. *Mater. Sci. Eng. A* 2009, *518*, 174–184.
63. Hsu, C.J.; Kao, P.W.; Ho, N.J. Intermetallic-reinforced aluminum matrix composites produced in situ by friction stir processing. *Mater. Lett.* 2007, *61*, 1315–1318.
64. Dixit, M.; Newkirk, J.W.; Mishra, R.S. Properties of friction stir-processed Al 1100NiTi composite. *Scr. Mater.* 2007, *56*, 541–544.
65. Lee, I.S.; Kao, P.W.; Ho, N.J. Microstructure and mechanical properties of AlFe in situ nanocomposite produced by friction stir processing. *Intermetallics* 2008, *16*, 1104–1108.
66. Callister, W.D. *Materials Science and Engineering An Introduction*, 6th ed.; John Wiley: New York, NY, USA, 2005; pp. 349–350.
67. Lee, C.J.; Huang, J.C.; Hsieh, P.J. Mg based nano-composites fabricated by friction stir processing. *Scr. Mater.* 2006, *54*, 1415–1420.
68. Morisada, Y.; Fujii, H.; Nagaoka, T.; Fukusumi, M. MWCNTs/AZ31 surface composites fabricated by friction stir processing. *Mater. Sci. Eng., A* 2006, *419*, 344–348.

69. Lee, W.B.; Lee, C.Y.; Kim, M.K.; Yoon, J.I.; Kim, Y.J.; Yoen, Y.M.; Jung, S.B. Microstructures and wear property of friction stir welded AZ91 Mg/SiC particle reinforced composite. *Compos. Sci. Technol.* 2006, *66*, 1513–1520.
70. Suhuddin, U.F.H.R.; Mironov, S.; Sato, Y.S.; Kokawa, H.; Lee, C.-W. Grain structure evolution during friction-stir welding of AZ31 magnesium alloy. *Acta Mater.* 2009, *57*, 5406–5418.
71. Zinkle, S.J. Overview of the US fusion materials sciences program. *Fusion Eng. Des.* 2005, *47*, 821–828.
72. Kohyama, A. Current status of fusion reactor structural materials R & D. *Mater. Trans.* 2005, *46*, 384–393.
73. Muroga, T. Vanadium alloys for fusion blanket applications. *Mater. Trans.* 2005, *46*, 405–411.
74. Kohyama, A.; Abe, K.; Kimura, A.; Muroga, T.; Jitsukawa, S. Recent accomplishments and future prospects of materials R & D in Japan. *Fusion Eng. Des.* 2005, *47*, 836–843.
75. Muroga, T.; Nagasaka, T.; Nishimura, A.; Chen, J.M. Improvement of vanadium alloys by precipitate control for structural components of fusion reactors. *Mater. Sci. Forum* 2005, *475~479*, 1449–1453.
76. Yao, Z.Y.; Suzuki, A.; Nakasaka, T.; Muroga, T. Behavior of oxygen in fusion candidate vanadium alloysduring oxidation and annealing. *Mater. Sci. Forum* 2005, *475–479*, 1445–1448.
77. Watanabe, H.; Nagamine, M.; Yamasaki, K.; Yoshida, N.; Heo, N.J.; Nagasaka, T.; Muroga, T. The microstructure of laser welded V-4Cr-4Ti alloy after ion irradiation. *Mater. Sci. Forum* 2005, *475–479*, 1491–1495.
78. Chung, H.M.; Strain, R.V.; Tsai, H.C.; Park, J.H.; Smith, D.L. Impact properties and hardening behaviour of laser and electron-beam welds of V-4Cr-4Ti. *Fusion Mater., Semiann. Prog. Rep.* 1996, 55–58.
79. Loomis, B.A.; Koicek, C.F.; Nowicki, L.J.; Smith, D.L. Tensile properties of vanadium-base alloys with a tungsten/inert-gas weld zone. *Fusion Mater., Semiann. Prog. Rep.* 1993, 187–191.
80. Fukumoto, K.; Takahashi, K.; Anma, Y.; Matsui, H. Effects of impurities on microstructural evolution and deformation process of ion-irradiated V-Cr-Ti alloys. *Mater. Trans.* 2005, *46*, 503–510.
81. Hatano, Y.; Hayakawa, R.; Nishino, K.; Ikeno, S.; Nagasaka, T.; Muroga, T.; Watanabe, K. Surface segregation of Ti in a V-4Cr-4Ti alloy and its influence on the surface reaction rates of hydrogen isotopes. *Mater. Trans.* 2005, *46*, 511–516.
82. Thomas, W.M.; Nicholas, E.D.; Needham, J.C.; Murch, M.G.; Templesmith, P.; Dawes, C.J. Friction Stir Butt Welding. International Patent Application No. PCT/GB92/02203 and GB Patent Application No. 9125978, 1991.
83. Arici, A.; Sinmazcelik, T. Effects of double passes of the tool on friction stir welding of polyethylene. *J. Mater. Sci.*2005, *40*, 3313–3316.
84. Cho, J.H.; Boyce, D.E.; Dawson, P.R. Effects of impurities on microstructural evolution and deformation process of ion-irradiated V-Cr-Ti alloys. *Mater. Sci. Eng., A* 2005, *398*, 146–163.
85. Sato, Y.; Nelson, T.W.; Sterling, C.J.; Steel, R.J.; Pettersson, C.O. Microstructure and mechanical properties of friction stir welded SAF 2507 super duplex stainless steel. *Mater. Sci. Eng., A* 2005, *397*, 376–384.
86. Zhang, H.; Lin, S.B.; Wu, L.; Feng, J.C.; Ma, S.L. Defects formation procedure and mathematic model for defect free friction stir welding of magnesium alloy. *Mater. Des.* 2006, *27*, 805–809.
87. Reynolds, A.P.; Hood, E.; Tang, W. Texture in friction stir welds of Timetal 21S. *Scr. Mater.* 2005, *52*, 491–494.
88. Lee, W.B.; Jung, S.B. The joint properties of copper by friction stir welding. *Mater. Lett.* 2004, *58*, 1041–1046.
89. Park, H.S.; Kimura, T.; Murakami, T.; Nagano, Y.; Nakata, K.; Ushio, M. Microstructures and mechanical properties of friction stir welds of 60% Cu-40% Zn copper alloy. *Mater. Sci. Eng., A* 2004, *371*, 160–169.
90. Prado, R.A.; Murr, L.E.; Shindo, D.J.; Soto, K.F. Tool wear in the friction-stir welding of aluminum alloy 6061+20% Al_2O_3: A preliminary study. *Scr. Mater.* 2001, *45*, 75–80.
91. Santella, M.L.; Engstrom, T.; Storjohann, D.; Pan, T.Y. Effects of friction stir processing on mechanical properties of the cast aluminum alloys A319 and A356. *Scr. Mater.* 2005, *45*, 201–206.
92. Hsu, C.J.; Kao, P.W.; Ho, N.J. Ultrafine-grained Al–Al_2Cu composite produced in situ by friction stir processing. *Scr. Mater.* 2005, *53*, 341–345.
93. Thomas, W.M.; Nicholas, E.D. Friction stir welding for the transportation industries. *Mater. Des.* 1997, *18*, 269–273.
94. Williams, S.W. Welding of airframes using friction stir. *Air Space Eur.* 2001, *3*, 64–66.
95. Sanderson, A.; Punshon, C.S.; Russell, J.D. Advanced welding processes for fusion reactor fabrication. *Fusion Eng. Des.*2000, *49–50*, 77–87.
96. Arici, A.; Mert, S. Friction stir spot welding of polypropylene. *J. Reinf. Plast. Compos.* 2008, *27*, 2001–2004.
97. Arici, A.; Selale, S. Effects of tool tilt angle on tensile strength and fracture locations of friction stir welding of polyethylene. *Sci. Technol. Weld. Joining* 2007, *12*, 536–539.
98. Nandan, R.; DebRoy, T.; Bhadeshia, H.K.D.H. Recent advances in friction-stir welding—Process, weldment structure and properties. *Prog. Mater. Sci.* 2008, *53*, 980–1023.
99. Minak, G.; Ceschini, L.; Boromei, I.; Ponte, M. Fatigue properties of friction stir welded particulate reinforced aluminium matrix composites. *Int. J. Fatigue* 2010, *32*, 218–226.
100. McClure, J.C.; Tang, W.; Murr, L.E.; Guo, X.; Feng, Z.; Gould, J.E. A thermal model of friction-stir welding. In *Trends in Welding Research*; Vitak, J.M., David, S.A., Johnson, J.A., Smartt, H.B., Deb Roy, T., Eds.; Am Welding Soc/ASM Int: Materials Park, OH, USA, 1999; pp. 590–595.

101. Mishra, R.S.; Mahoney, M.W. Friction stir processing: A new grain refinement technique to achieve high strain rate superplasticity in commercial alloys. *Mater. Sci. Forum* 2001, *357–359*, 507–512.

102. Rhodes, C.G.; Mahoney, M.W.; Bingel, W.H.; Spurling, R.A.; Bampton, C.C. Effect of friction stir welding on microstructure of 7075 aluminum. *Scr. Mater.* 1997, *36*, 69–75.

103. Liu, G.; Murr, L.E.; Niou, C.-S.; McClure, J.C.; Vega, F.R. Microstructural aspect of the friction stir welding of 6061-T6 aluminum. *Scr. Mater.* 1997, *37*, 355–361.

104. Murr, L.E.; Li, Y.; Flores, R.D.; Trillo, E.A.; McClure, J.C. Intercalation vortices and related microstructural features in the friction-stir welding of dissimilar metals. *Mater. Res. Innovations* 1998, *293*, 150–163.

105. Li, Y.; Murr, L.E.; McClure, J.C. Flow visualization and residual microstructures associated with the friction-stir welding of 2024 aluminum to 6061 aluminum. *Mater. Sci. Eng., A* 1999, *271*, 213–223.

106. Guerra, M.; McClure, J.C.; Murr, L.E.; Nunes, A.C. Metal flow during friction stir welding. In *Friction Stir Welding and Processing*; Jata, K.V., Mahoney, M.W., Mishra, R.S., Semiatin, S.L., Field, D.P., Eds.; TMS (The Minerals, Metals and Materials Society): Warrendale, PA, USA, 2001; pp. 25–34.

CHAPTER 12

THERMAL CONDUCTIVITY OF COMPOSITE MATERIALS

René Salgado-Delgado, Alfredo Olarte-Paredes, Areli Marlen Salgado-Delgado, Zully Vargas-Galarza, Teresa Lopez-Lara, Juan Bosco Hernández-Zaragoza, Israel Rico-Rodríguez, and Gonzalo Martínez-Barrera

INTRODUCTION

The increasing concern about industrial pollution has forced the construction and manufacturing industries to seek innovative materials that are trustworthy and sustainable to replace the conventional synthetic fibers used as reinforcements in structural materials. Natural fibers such as sisal, jute, cotton, linen, hemp, and kenaf have been considered as possible alternatives because of their environmental friendliness and their availability in fibrous form and because they can be obtained from plant leaves at a low cost [1, 2].

Charcoal from sugarcane bagasse (CSB) is a residue generated by the bagasse burning process that has low pozzolanic reactivity due to its high porosity. This material is frequently used as a fertilizer even though it is inadequate for this use due to its high silica content (SiO_2) of more than 60% of the total metallic oxide mass. CSB has a broad and variable particle size distribution in addition to a low nutrient content. Therefore, it is important to promote other uses of CSB that generate an added value and synergistic properties for the material.

The locality of Zacatepec, in the Mexican state of Morelos, has a sugarcane mill. The residues generated by this industry accumulate every year, generating a series of negative consequences for both human health and the environment. In addition, this is a waste of a potential source of products. In this study, a composite material was obtained from the agroindustrial residue CSB and used as a reinforcement and matrix for portland cement (CPC-30R). Various studies on the characterization of composite materials based on cement and various inorganic loads that establish synergistic properties due to the volume and characteristics of the loads are reported. The fillers confer properties such as thermal insulation, light weight, and impact resistance [3, 4].

Other studies involve the use of sugarcane bagasse ash, but these differ from the charcoal from sugarcane bagasse. The use of ash does not improve the mechanical strength of the material and water permeability decreases [5].

Our material intends to use the charcoal from sugarcane bagasse to achieve thermal insulation. Bagasse coal has completely different properties to the ash. In ash the organic part is removed leaving only metal oxides.

This method is commonly employed to explain the properties of compounds even though more precise models that include the impact of the shape, size, and distribution of the particles on the elastic properties of the material have been developed [6, 7].

The thermal conductivity (λ) is a measure of a material's ability to conduct heat. It is most commonly used in Fourier's law of heat conduction. This property is very sensitive to the material's microstructure, and there are thermally conducting and insulating materials.

Fourier's law of heat transfer by conduction states that heat conduction occurs due to a temperature difference between two bodies and obeys the following equation [8]:

$$\lambda \equiv -\frac{q_x \Delta x}{A \Delta T}, \qquad (1)$$

where λ is the thermal conductivity in watts per meter-degree Celsius (W/m∘ C), q_x is the heat flow in the x direction in watts, Δx is the thickness of the sample in meters, ΔT is the temperature difference in ∘ C, and A is the area of the sample in m^2. The method is valid within the linear regime and allows for several material properties, including the thermal conductivity, to be calculated.

Citation: René Salgado-Delgado, Alfredo Olarte-Paredes, Areli Marlen Salgado-Delgado, et al., "Thermal Conductivity of Composite Materials" Advances in Materials Science and Engineering, vol. 2016, Article ID 4950576, 5 pages, 2016. doi:10.1155/2016/4950576

EXPERIMENTAL

Materials

We used charcoal from sugarcane bagasse (CSB) collected from the "Emiliano Zapata Salazar" sugarcane mill in Zacatepec, Morelos, fast-drying composite portland cement (CPC-30R, "Tolteca" brand), and distilled water from the company "Químicos Farmacéuticos Industriales."

Equipment

The following devices were used: a Denver Instrument digital scale with a limit of 320 g, certified to meet ISO quality guideline 9001, a Barnant Series 10 mixer, model number 700-5400, granulometry equipment from FIIC, SA de CV A CRODE SCH 0001 97 drying oven, an insulated hot plate with K-type thermocouples and an Arduino open-source hardware platform, a Thermo Haake K20 water circulation system, a PID temperature controller with a temperature range of 20 to 300°C, and JEOL/JSM-6010LA scanning electron microscope (SEM) and energy-dispersive spectroscope (EDS).

Methods

Preparation of the Charcoal from Sugarcane Bagasse (CSB)

To eliminate any traces of organic fiber, 250 g of CSB was placed in the drying oven for 2 hours at 400°C and then allowed to cool down. Subsequently, different sizes of CSB particles were separated using the granulometry equipment (mesh 100: 0.150 mm and mesh 200: 0.075 mm of hole opening. It is relative with ASTM E11-87).

Preparation of the Composite Materials

The composite materials were prepared by mixing different weight concentrations of CSB (5, 10, and 15 g) with cement (100 g) at a rate of 140 ± 5 rpm for 5 minutes. Then, water was added (40 g) and mixed at a rate of 285 ± 10 rpm for 5 minutes. The mixtures obtained were poured into 10 × 2.5 × 2.5 cm metallic molds. The molds containing the mixtures were left to dry at room temperature for 72 hours. Then, the composite materials were extracted from the molds for characterization.

Characterization

Measurement of Thermal Conductivity

For this study, a hot plate system based on ASTM guideline C177-97 (Standard Test Method for Steady State Heat Flux Measurements and Thermal Transmission Properties by Means of the Guarded Hot Plate Apparatus) was used. Figures 1 and 2 show the experimental setup, which consists of the following elements: a hot plate, samples of the materials, two cold plates, one on each side of the system, a mounting system, an isolation chamber, and the necessary mechanisms for tuning and measurement [9].

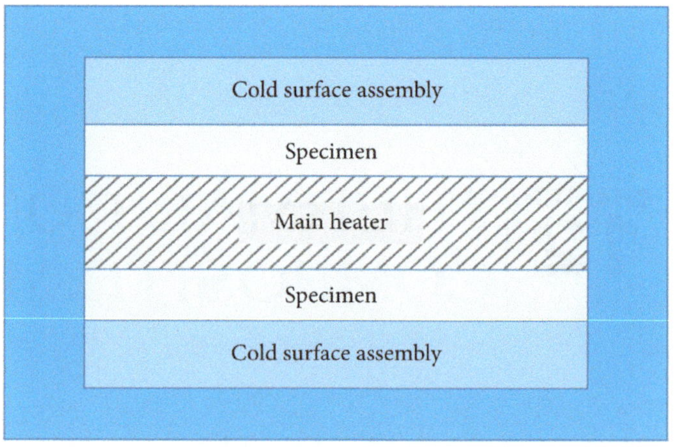

Figure 1: System mounting hot isolated plate by ASTM C177-97 standard.

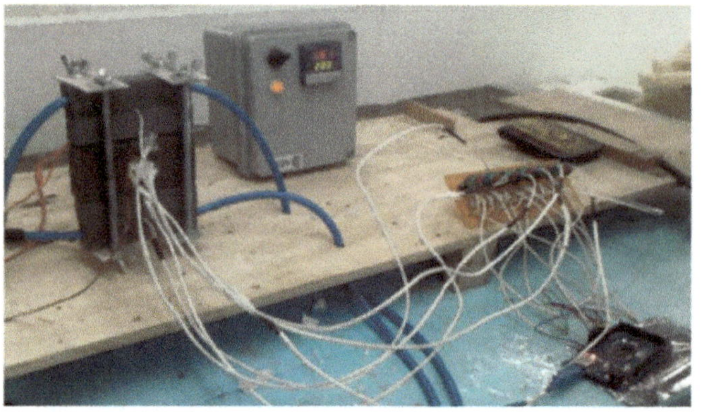

Figure 2: System hot insolated plate technique.

Some of the most important specifications in the guideline are as follows: the method must be applied to materials with thermal conductivities of up to 0.62 kcal/mh°C; the temperature during the test must be kept between −45 and 760°C; the external insulation surrounding the guard ring must have a thermal resistance of at least twice that of the material being tested; the thermocouples must be made from wire with a diameter no greater than 0.57 mm; and finally, the temperature difference between the hot and cold faces of the sample must be at least 22°C [10].

Scanning Electron Microscopy (SEM) and Energy-Dispersive Spectroscopy (EDS) Analyses

The morphology and load distribution of the composite materials were studied using the desktop SEM described in the equipment section. It was set to 3 kV and to a magnification of 1000x. The EDS analysis was performed simultaneously under the same conditions to determine the elemental content and chemical composition of the CSB.

RESULTS AND DISCUSSION

The composite materials were used to create a sample matrix (2 sizes × 3 weight concentrations), as shown in Table 1.

Table 1: Composition of composite materials (CBC/CPC-30R/H_2O).

Mesh number	CBC (g)	CPC-30R (g)	H_2O (g)
M-100	15	100	40
M-100	10	100	40
M-100	5	100	40
M-200	15	100	40
M-200	10	100	40
M-200	5	100	40

To measure the thermal conductivity, the samples were placed on the hot plate apparatus, which maintained a temperature between 30 and 230°C for 4 hours, with a flow temperature of 10°C at the cold plate. The thermal conductivities of the composite are shown in Figure 3.

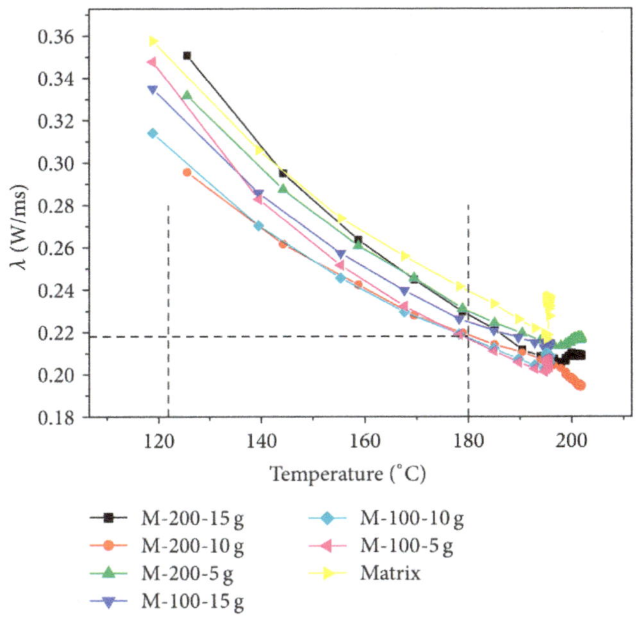

Figure 3: Thermal conductivity of composite materials with two sizes (M-100 and M-200) and three concentrations (15, 10, and 5 g) in the range of 120–180°C.

The results show the behavior of the thermal conductivity as a function of temperature and the influence of the particle size and CSB concentration. At 122°C, the thermal conductivity of the pure cement matrix (CPC-30R) was high at 0.358 W/m°C.

At this temperature, all of the composite materials tested had lower thermal conductivities than the pure matrix; the values were between 0.350 and 0.300 W/m°C. Between 122 and 180°C, a decrease in the conductivity due to the CSB load was observed. For one load size (M-200 or M-100), the 10 g weight concentration marked the threshold for a minimal response because the thermal conductivity increased when the load was decreased (to 5 g) and increased (to 15 g). This may have been due to its distribution within the cement matrix. When the results of the M-200-15 g and M-100-15 g samples were compared, they showed that a larger particle size results in lower conductivity values when the load is constant. The same result was found when comparing the M-200-5 g and M-100-5 g samples. From this, we concluded that a smaller load leads to more efficient heat conduction, a finding that explains the higher conductivities obtained when the CSB particle size was increased [8]. Figure 4 shows that at 195.5°C a maximum conductivity of 0.235 W/m°C for the sample was obtained (matrix: portland cement). Samples containing charge (CSB) showed a lower thermal conductivity in the range of 0.215 to 0.205 W/m°C at this temperature (the latter being shown for M-200-10 g).

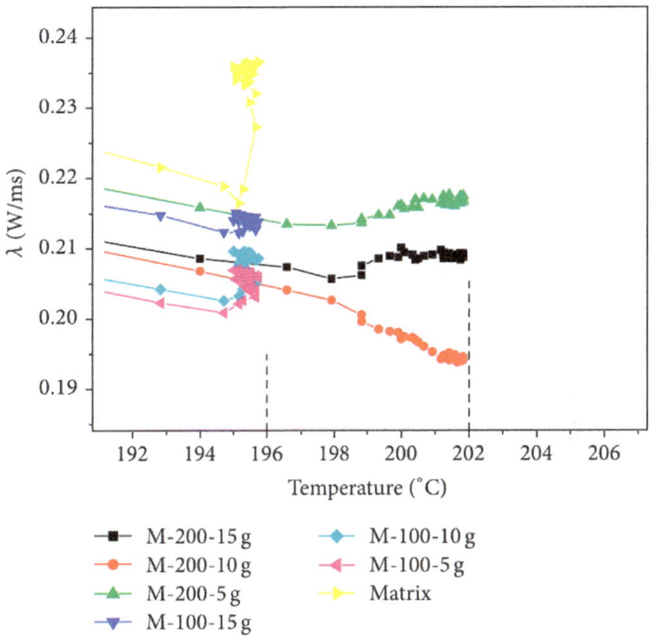

Figure 4: Thermal conductivity of composite materials with two sizes (M-100 and M-200) and three concentrations (15, 10, and 5 g) in the range of 194–202°C.

In the temperature range of 196 to 199°C the thermal conductivity decreases for all samples with CSB.

The sample M-200-10 g shows the lowest thermal conductivity of 0.194 W/m°C at a temperature of 202°C. The CSB is contributing to having a low thermal conductivity. With a concentration of 10 g CBS has the lowest income considering that this concentration is critical because by using 200-mesh distributes the load better generating sufficient space for the thermal conductivity decreases. This demonstrates that a good load distribution and a small particle size favor insulating material properties, as explained by Peña-Rodríguez et al., 2008 [8], where the thermal resistivity obtained polystyrene thermal clay loads and the same behavior is observed in the distribution and particle size.

SEM micrographs of the analysis showed that CPC-30R/H_2O ratio allows a hardening material without leaving a broken surface (Figure 5(a)). Likewise in the sample M-200-10 g good wetting of the matrix on the load can be demonstrated (see Figure 5(b)).

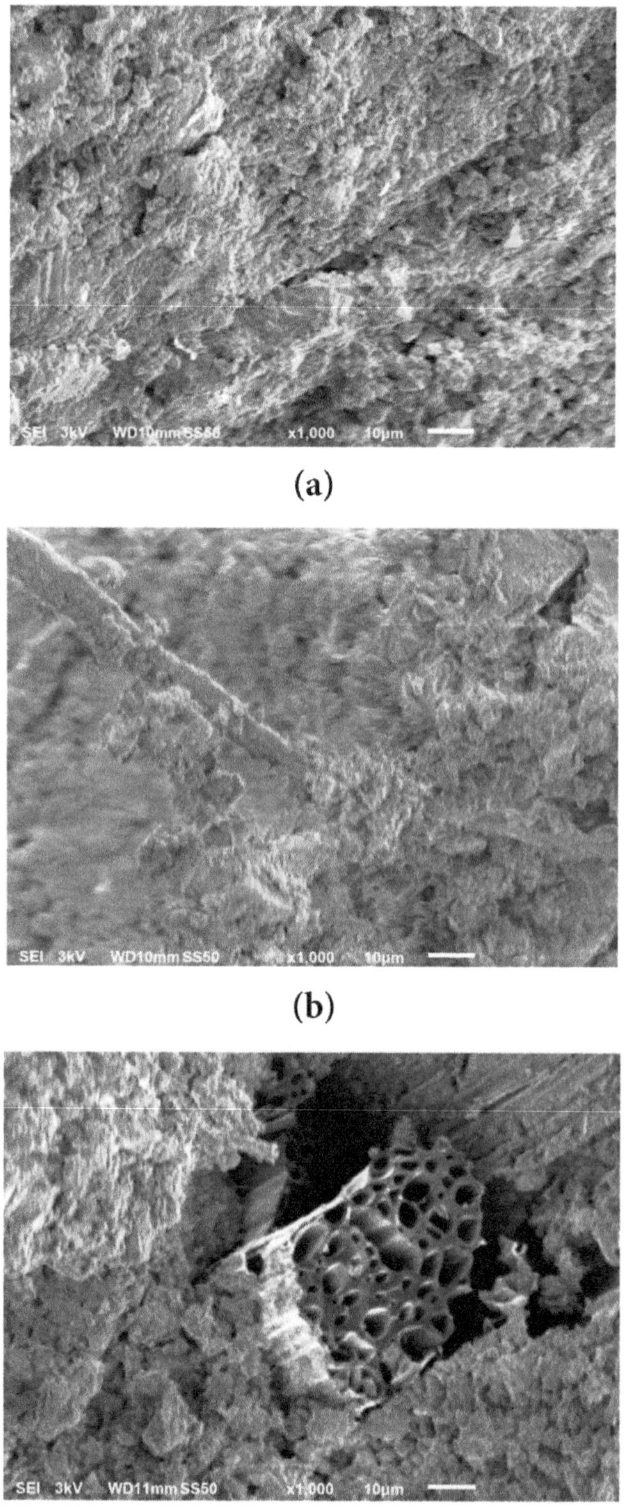

(a)

(b)

(c)

Figure 5: SEM micrographs (1000x) of composite materials: (a) matrix (CPC-30R/H_2O), (b) M-200-10 g, and (c) M-100-10 g.

Figure 5(c) shows a particle load in the fracture of the composite material after undergoing strain. It can be seen that part of the load fell off the matrix suggesting a weak interaction between the filler and the matrix. It notes that the porosity of the charge also contributes to the thermal conductivity of composites decreases.

It is important to establish the composition of the CSB. We used EDS analysis, which provided evidence for the amount of carbon and oxides present in the material for that purpose. Table 2 shows the chemical composition of the CSB surface. Oxides contributed to the decrease in the thermal conductivity of the material, and here, SiO_2 was present in a greater proportion.

Table 2: EDS: elemental analysis of the CBC.

Chemical formula	C	Na_2O	MgO	Al_2O_3	SiO_2	P_2O_5	SO_3	K_2O	CaO	PbO
% mol	70	0.32	1.58	2.31	15.89	0.66	1.53	1.93	5.57	0.21

CONCLUSIONS

Composite materials were created using portland cement (CPC-30R) and charcoal from sugarcane bagasse with varying load particle sizes and concentrations. From studying these materials, we concluded that loading diminishes a composite material's thermal conductivity and that increasing the particle size reduces the thermal conductivity. This is because an increase in the particle size affects the heat conduction by decreasing the contact area. Similarly, an increase in the particle size increases the pore size, which may increase the volume of air inside the load and, therefore, reduce the thermal conductivity. The EDS analysis revealed the presence of oxides as an additional factor contributing to the diminished thermal response. The composite material with the lowest conductivity was the M-200-10 g sample (0.194 W/m°C), which behaved as a thermal insulator; it had a reduced heat transfer rate in comparison to the rest of the samples. The conductivity of the M-200-10 g sample was less than the conductivities reported by other authors for ceramic bricks and asbestos plates (0.87 and 0.36 W/m°C, resp.). The micrographs show both the porosity of the load and the load-matrix interaction at the point of fracture. In future studies, it is recommended that the effects of thermal shock on the mechanical properties of the materials should be investigated.

REFERENCES

1. J. Wei and C. Meyer, "Degradation mechanisms of natural fiber in the matrix of cement composites,"Cement and Concrete Research, vol. 73, pp. 1–16, 2015.
2. F. D. A. Silva, N. Chawla, and R. D. D. T. Filho, "Tensile behavior of high performance natural (sisal) fibers," Composites Science and Technology, vol. 68, no. 15-16, pp. 3438–3443, 2008.
3. M. Salas, R. Luco, and R. Ojeda, "Análisis Estructural de un Camarán de materiales compuestos," Anales de Mecánica de la Fractura, vol. 20, 2003.
4. C. S. Smith, Design of Marine Structures in Composite Materials, Elsevier Science, Amsterdam, The Netherlands, 1990.
5. Marcos De Paula Oliveira, et al., "La caña de azúcar bagazo ash como material parcial- portland-cemento-reemplazo," Dyna, Revista de la Facultad de Minas, vol. 77, no. 163, pp. 47–54, 2010.
6. N. Martínez-Mateos, D. Busquets-Mataix, M. D. Salvador-Moya, and V. Amigó-Borrás, "Propiedades mecánicas de compuestos de matriz de aluminio reforzados con partículas cerámicas, obtenidos por extrusión de polvos," in Proceedings of the VIII Congreso Nacional de Propiedades Mecánicas de Sólidos, pp.

215–224, 2002.

7. G. Bao, J. W. Hutchinson, and R. M. McMeeking, “Particle reinforcement of ductile matrices against plastic flow and creep,” Acta Metallurgica et Materialia, vol. 39, no. 8, pp. 1871–1882, 1991.
8. G. Peña-Rodríguez, J. Sánchez-Molina, and R. Monroy, “Efecto de la concentración de poliestireno expandido en la conductividad térmica efectiva de bloques de termoarcilla,” Revista Colombiana de Física, vol. 40, no. 2, pp. 285–288, 2008.
9. M. M. Pérez-Sánchez, R. Centeno-Lara, and F. Lazcano-Serrano, “Desarrollo de un prototipo para la caracterización térmica de los materiales de construcción regionales,” Ingeniería, vol. 6, no. 2, pp. 13–22, 2002.
10. ASTM C 177-97 (Standard Test Method for Steady State Heat Flux Measurements and Thermal Transmission Properties by Means of the Guarded Hot plate Apparatus).

CHAPTER 13

SIZE EFFECTS OF PT NANOPARTICLE/GRAPHENE COMPOSITE MATERIALS

Chia-Liang Sun, Jheng-Sin Su, Shun-Yi Lai, and Yu-Jen Lu

INTRODUCTION

Since reliable and fast determination of biomolecules is important in many areas such as biotechnology, clinical diagnostics, and food industry, the development of biosensor has attracted extensive attention recently. In general, there are enzymatic and nonenzymatic biosensors in the literature because the enzyme is easily being affected by the environmental factors such as temperature, humidity, and pH values. In addition, the immobilization of enzyme is a complicated and expensive process. Therefore, the nonenzymatic biosensors start to catch the scientists' eyes and attention. The large surface area and excellent electrical properties of graphene allow it to be used in many applications [1–6]. For example, it can connect between the redox centers of an enzyme or protein and an electrode surface. Rapid electron transfer facilitates accurate and selective detection of biomolecules. Its unique structure and properties, such as high specific surface area, high mechanical strength and conductivity, and its relatively low price make graphene suitable for potential applications. In our lab, we used graphene, Pt/graphene, CuO/graphene, graphene oxide nanoribbons, and multiwalled carbon nanotube/graphene oxide nanoribbon core-shell heterostructures to detect biomolecules in the past five years [7–10]. Here we want to utilize the same graphene-based materials to further monitor other biomolecules.

H_2O_2 is a chemical used widely in the food, pharmaceutical, paper, and chemical industries. H_2O_2 is also one of the products of reactions catalyzed by enzymes in many biological and environmental processes. Therefore, the development of a biosensor for detecting H_2O_2 is important [11–23]. The electrode materials for H_2O_2biosensor in the literature can be categorized as polymers, carbon nanotubes, graphene, nanoparticles, their composite materials, and others. After immobilizing horseradish peroxidase (HRP) onto the composite, the H_2O_2 biosensor could be used as a component for investigating bioelectrochemical activity [11–13]. Fan et al. developed a new kind of enzymatic biosensor using biomimetic graphene capsules (GRCAPS) in 2015 [14]. Polybenzimidazole (PBI), polyamic acids (PAAs), benzothiazole (BT), benzoxazole (BO), and their composites as polymers were used to modify gold electrode to determine H_2O_2 in 2011 [15, 16]. Li et al. fabricated a nonenzymatic H_2O_2 sensor by utilizing MWCNTs as the matrix for electrodepositing of Pt nanoparticles [17]. Karuppiah et al. have constructed a novel glucose and H_2O_2 biosensor based on graphene/Co_3O_4 NPs composite modified electrode [18]. H_2O_2 showed a better electrochemical response at the nitrogen and boron codoped graphene modified GC electrode (GCE), much higher than that of graphene solely doped with N atoms (N-G) or with B atoms (B-G) [19]. The Prussian blue nanocubes-nitrobenzene-reduced graphene oxide nanocomposites/GCE showed good electrocatalytic ability for the reduction of H_2O_2 with good stability and selectivity [20]. 3D graphene foam supported PtRu on Ni foam exhibited an excellent electrocatalytic activity toward the H_2O_2 detection [21]. The activity of MoS_2 NPs toward the reduction of H_2O_2 released by cells was demonstrated in 2013 [22]. The as-produced AuCu nanowires have been explored toward the detection of H_2O_2 [23].

There are many important factors that influence the catalytic activity of catalysts. One of the very important parameters is the size of the nanoparticles. Although there are studies using nanoparticles on nanocarbons for sensing H_2O_2, the effects of particle size remain unclear. Therefore, in this study, we try to investigate the size effects of Pt nanoparticle supported on graphene. The electrochemical detection of

Citation: Chia-Liang Sun, Jheng-Sin Su, Shun-Yi Lai, and Yu-Jen Lu, "Size Effects of Pt Nanoparticle/Graphene Composite Materials" Journal of Nanomaterials, vol. 2015, Article ID 861061, 7 pages, 2015. doi:10.1155/2015/861061

H_2O_2 was used to evaluate the properties of these graphene-supported Pt catalysts. The size-dependent electrochemical properties will be displayed and discussed in this study.

MATERIAL AND METHODS

Chemicals

Platinum(IV) chloride ($PtCl_4$, 99%) was purchased from Acros Organics. Ethyl glycol was purchased from J.T. Baker. Nafion (DuPont, 5 wt.%) was used to generate the ink. NaOH and H_2O_2 were obtained from Sigma. All solutions were prepared with deionized water with a resistivity of 18 MΩ/cm.

Preparation of Pt Colloidal Solution

Pt nanoparticles were synthesized using the polyol method reported in detail elsewhere [24–27]. In short, 0.4652 g $PtCl_4$ was dissolved in 50 mL ethylene glycol. In order to control the size of the particles, the appropriate amount of sodium hydroxide (NaOH) was added to the $PtCl_4$ solutions. The mixture was then stirred at room temperature for 30 min with rotational speed of 600 rpm, heated to 160°C for 3 hr, and finally allowed to cool down to room temperature, forming a Pt colloidal solution (1.3, 1.7, 2.9, and 4.3 nm). The NaOH concentrations for Pt colloids of 1.3, 1.7, 2.9, and 4.3 nm are 0.6, 0.4, 0.3, and 0.1 M, respectively.

Deposition of Pt Nanoparticles on Graphene

Graphene oxide powders were prepared following Staudenmaier's method and reduced to graphene powders by annealing at 1050°C under an argon atmosphere. 20 mg of graphene powders was mixed with the Pt colloidal solutions in a 40 mL solution containing 2 M sulfuric acid and ethylene glycol [25–27]. The volume ration between sulfuric acid and ethylene glycol is 1 to 1. The Pt ratio is controlled to be around 20 wt.% for Pt-G catalyst. The solution was then stirred for 24 h and then sonicated using an ultrasonic processor (Part number Q700) for 15 min. The resulting solution was filtered to recuperate the catalyst. Four Pt-G catalysts with different average particle sizes were obtained in this manner.

Material Characterization

Transmission electron microscopy (JEOL JEM-1230, 100 kV) was used to characterize sample morphologies.

Electrode Preparation and Electrochemical Measurements

The catalyst ink for electrochemical measurement was prepared with the Pt-graphene powders. 3 mL deionized water, 2 mL ethanol, 60 µL Nafion, and 6 mg Pt-graphene powders were sonicated to make the ink [25–27]. Potentiostat/galvanostat (CHI 405A) was used for electrochemical measurements. The working electrode was 3 mm-diameter glassy carbon (GC) disc electrode on which 10 µL of the catalyst ink was deposited and dried at room temperature. A silver/silver chloride (Ag/AgCl) electrode and a large surface area platinum electrode were used as the reference and counterelectrode, respectively. All potentials in this study are reported with respect to the Ag/AgCl electrode.

RESULTS AND DISCUSSION

Controlled Synthesis of Size-Selected Pt Colloids

Figure 1 displays TEM images of Pt nanoparticles with different average sizes varying between 1.7 and 4.3 nm and their histograms. The particle sizes were controlled by changing the pH of the $PtCl_4$ solution dissolved in ethylene glycol. The histograms show the size distribution of the particles with an average diameter which was taken over 300 individual particles from the TEM pictures. The NaOH concentrations of ethylene glycol with dissolved $PtCl_4$ are 0.1, 0.3, 0.4, and 0.6 M for making 4.3, 2.9, 1.7, and 1.3 nm Pt colloids. When NaOH concentrations become large, the mean diameters of Pt colloids get small. It is worthwhile to mention that the sequence for mixing $PtCl_4$ solution is very important. Before adding any NaOH, $PtCl_4$ needs to be dissolved in ethylene glycol completely. If $PtCl_4$ was added to the ethylene glycol already with NaOH, there will be no size control effect though Pt nanoparticles can still be formed.

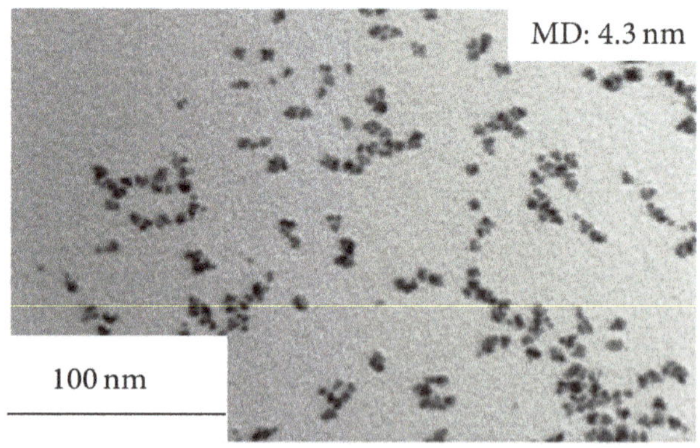

(a)

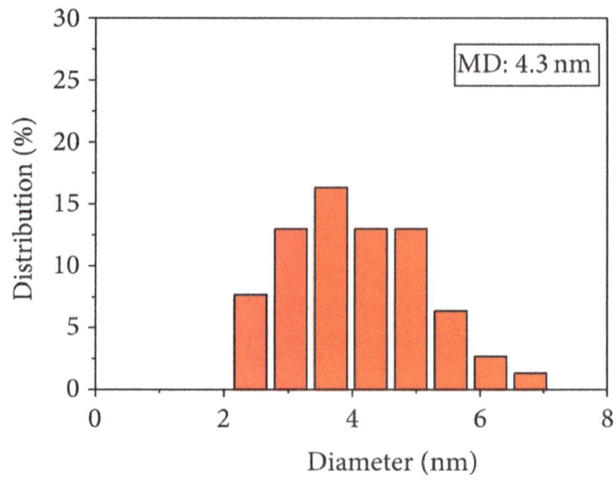

(b)

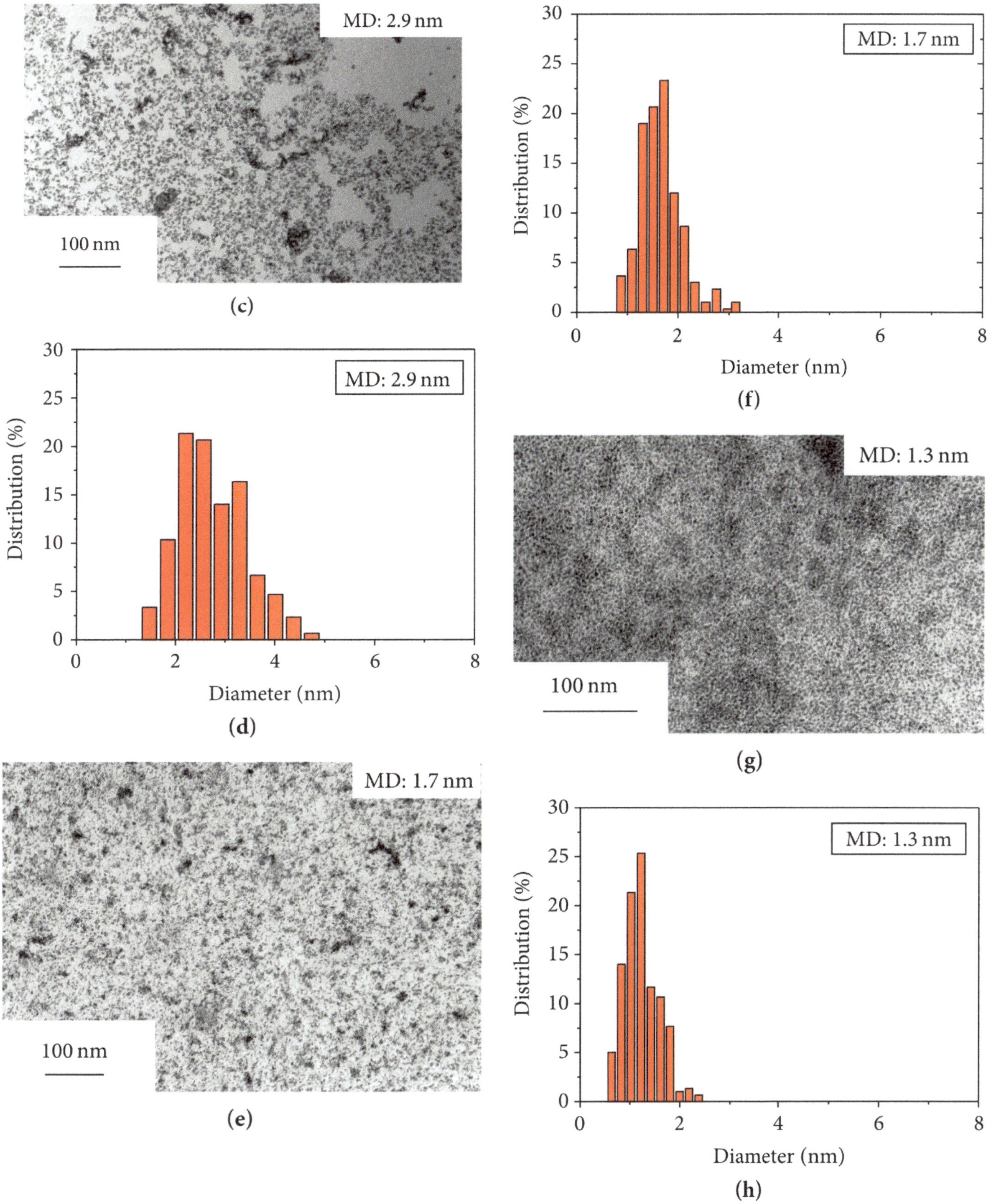

Figure 1: Transmission electron microscope images of (a) 4.3 nm-diameter, (c) 2.9 nm-diameter, (e) 1.7 nm-diameter, and (g) 1.3 nm-diameter Pt colloidal nanoparticles. Histograms show the particle size distribution of (b) 4.3 nm-diameter, (d) 2.9 nm-diameter, (f) 1.7 nm-diameter, and (h) 1.3 nm-diameter Pt nanoparticles.

Material Characterization of Graphene-Supported Pt Nanoparticles

Figure 2 shows the TEM images of the 4.3, 2.9, 1.7, and 1.3 nm particles supported on single graphene sheets. The small dark spots are the Pt nanoparticles adsorbed on multilayered graphene as the background that is relatively gray compared to the white holey in other regions of the Cu grid. The wrinkles on single graphene pieces randomly appear in the pictures. Although most of the time the nanoparticles are uniformly distributed on graphene surfaces, sometimes the aggregates could be formed like the two areas in Figure 2(d). This may be owing to the very high surface area of small particles that tend to reduce the total surface energy in the system. It is suggested that the functional groups on graphene surfaces will help further disperse the small Pt nanoparticles.

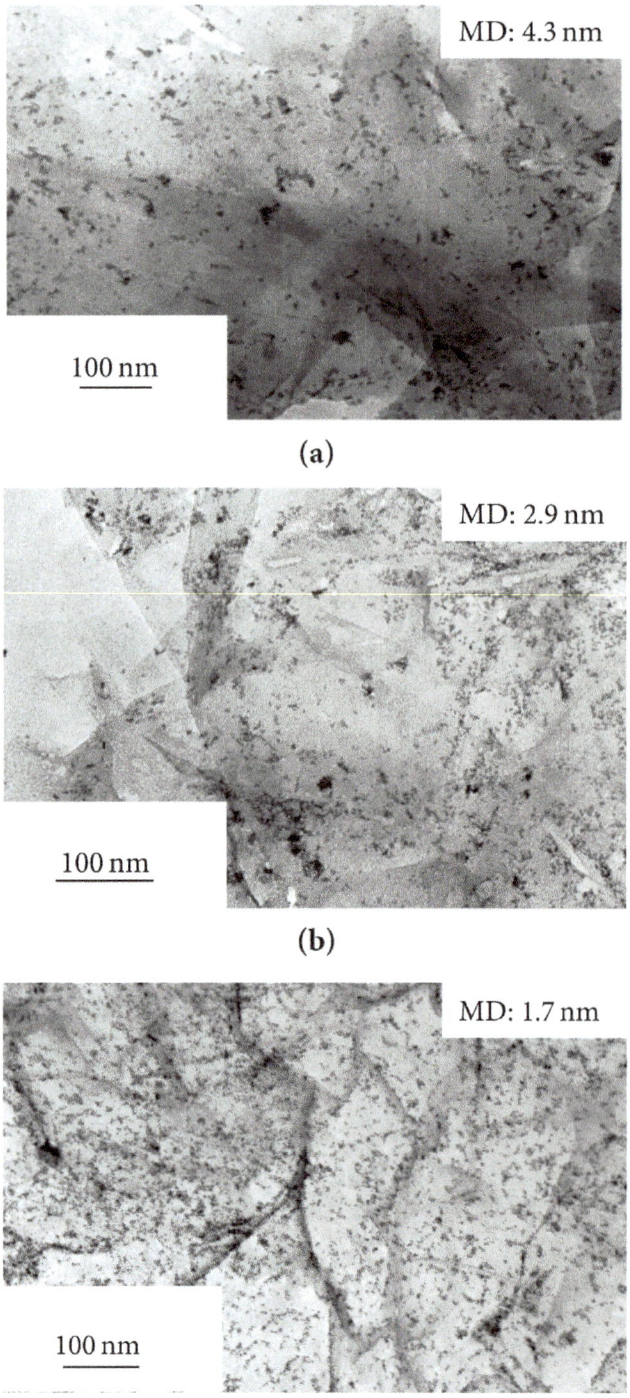
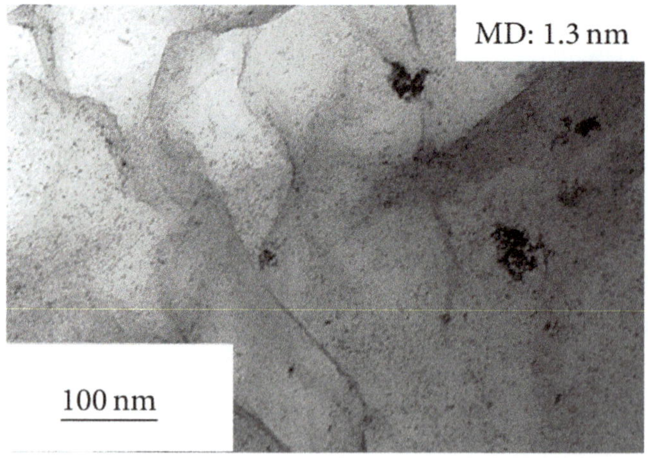

Figure 2: Transmission electron microscope images of (a) Pt-4.3 nm/G catalyst, (b) Pt-2.9 nm/G catalyst, (c) Pt-1.7 nm/G catalyst, and (d) Pt-1.3 nm/G catalyst.

Cyclic Voltammetric Detection of H_2O_2

Cyclic voltammograms in Figure 3 illustrate the reduction of H_2O_2 for each catalyst. In general, the reduction currents gradually increase when lowering the potential after 0.2 V and there is a main reduction peak for each catalyst. The reduction peaks are located around −0.5 V for four catalysts. This is similar to the oxygen reduction reaction (ORR) for the cathode of a fuel cell. The reduction of H_2O_2 is a two-electron process that has low electron transfer number compared to ORR. For the size lower than 10 nm but higher than 1.5 nm, there are three catalysts named Pt-4.3 nm/G, Pt-2.9 nm/G, and Pt-1.7 nm/G. Among these three catalysts, the reduction current will increase along the decrease of particle sizes. Hence the Pt-1.7 nm/G catalyst has the highest reduction current that is 2.5 times higher than that of the Pt-2.5 nm/G one. For the particle sizes more than 10.0 nm and smaller than 1.5 nm, the reduction current becomes smaller than that of Pt-1.7 nm/G catalyst.

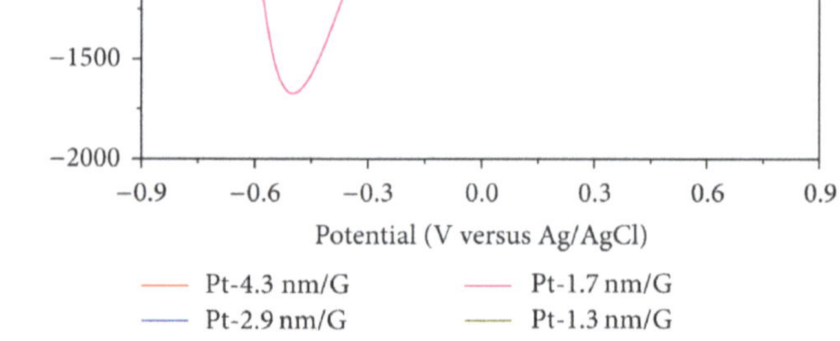

Figure 3: Cyclic voltammograms of four Pt/G-modified GC electrodes in 0.1 M PBS (pH 7.0) 10 mM H_2O_2. Scan rate: 100 mVs^{-1}.

The Amperometric Response of H_2O_2

The amperometric responses of the modified GC electrode to H_2O_2 are depicted in Figures 4 and 5. After adding analyte solutions with different concentrations, the reduction currents were monitored at a fixed potential of −0.5 V. In Figure 4, the same trend as in Figure 3 can be observed. With the same concentration of analyte, the Pt-1.7 nm/G catalyst has the highest reduction current. The linear regression equation is given by $I_{H_2O_2} = -0.089\, C_{H_2O_2} + 1.392$, with a correlation coefficient of r = 0.996. The sensitivity of Pt-1.7 nm/G catalyst is 1264.6 μA mM^{-1} cm^{-2}. This number is higher than 254.8 (Pt-1.3 nm/G), 565.6 (Pt-2.9 nm/G), and 560.8 (Pt-4.3 nm/G) μA mM^{-1} cm^{-2}. The sensitivity numbers among all catalysts are size-dependent.

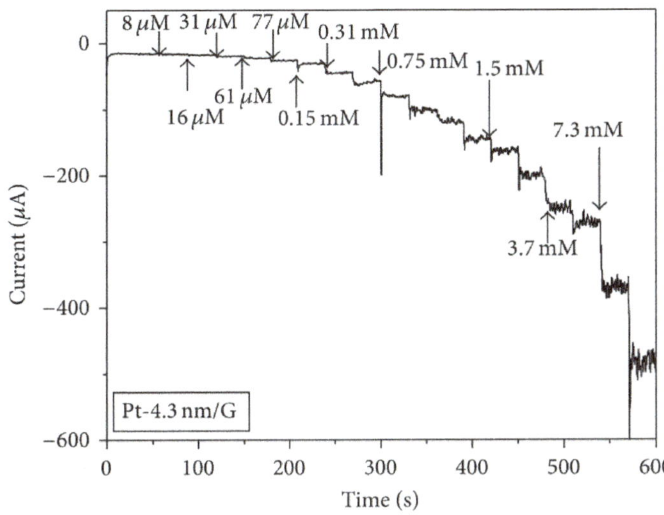

(a)

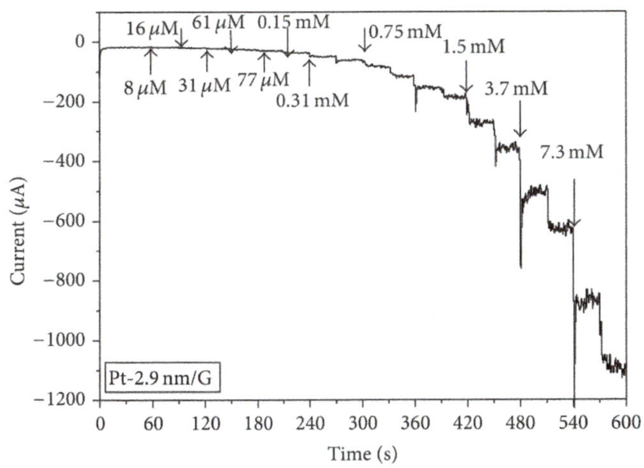

(b)

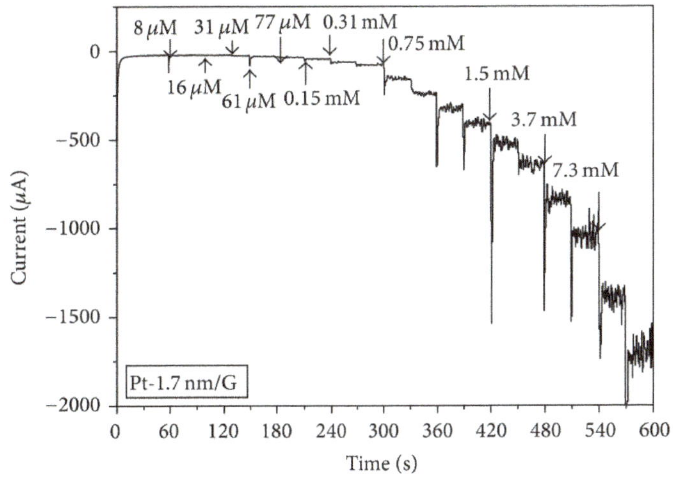

(c)

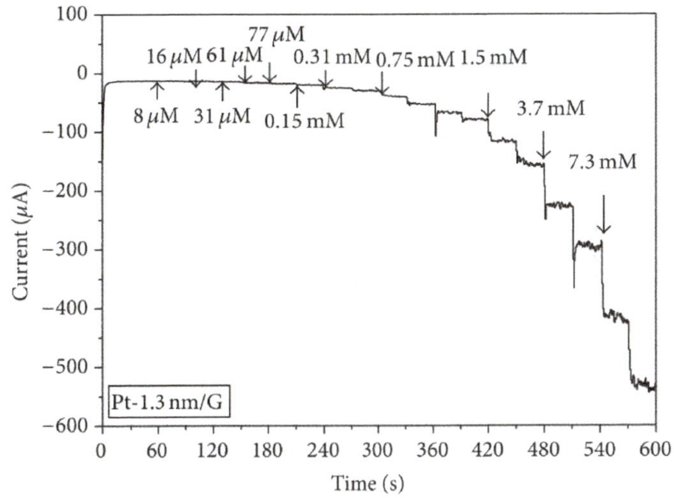

(d)

Figure 4: Amperometric responses of four Pt/G-modified GC electrodes after the subsequent addition of H_2O_2 in a 0.1 M PBS solution at the potential of −0.5 V.

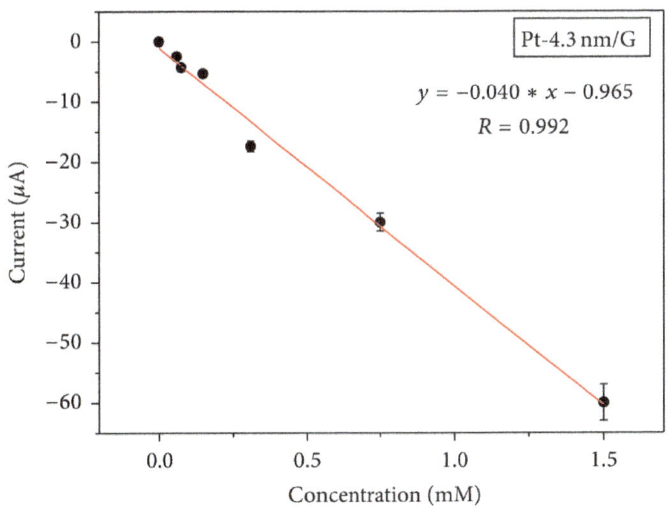

(a)

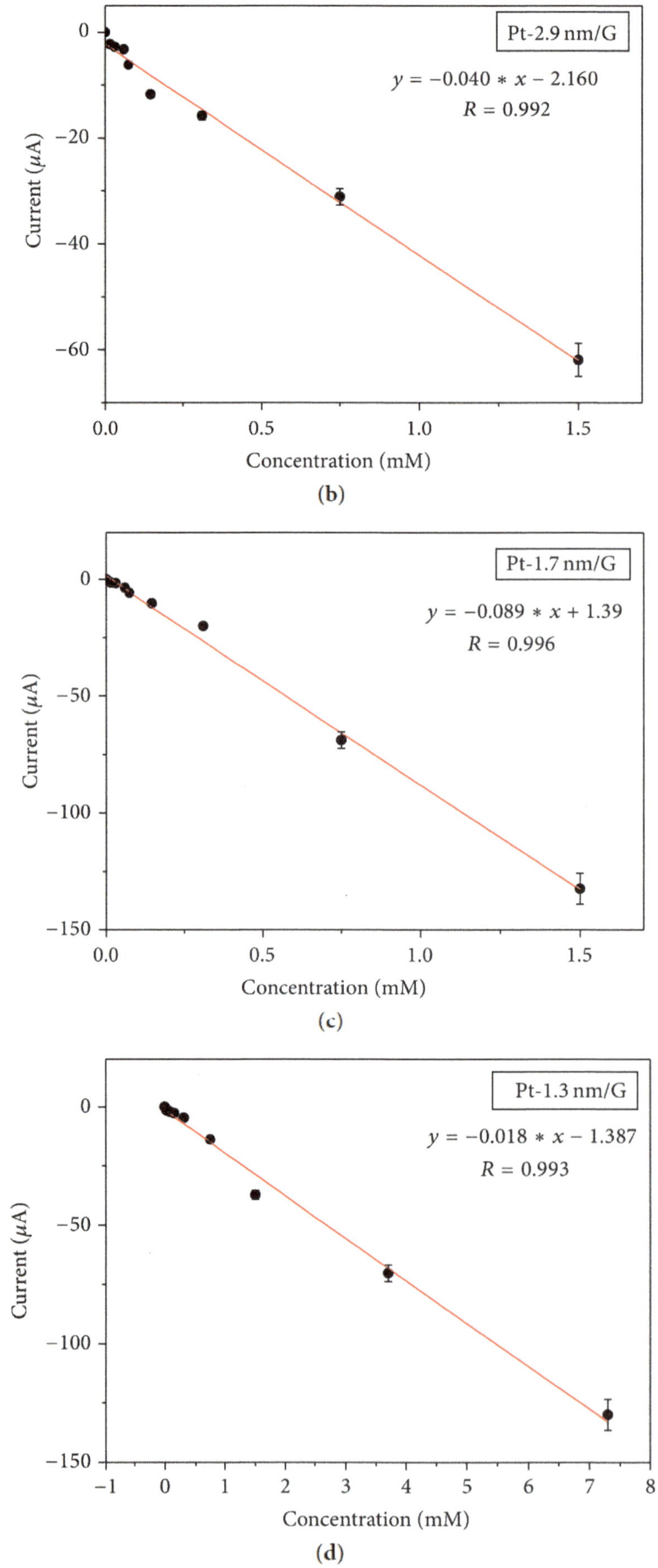

Figure 5: The corresponding plots of the reduction current of H_2O_2 ($I_{H_2O_2}$) versus the H_2O_2 concentration ($C_{H_2O_2}$) using four Pt/G-modified GC electrodes at the potential of −0.5 V.

CONCLUSIONS

In summary, the particles' size of Pt colloids can influence the electrochemical properties of Pt/G catalysts for electrochemical sensing of H_2O_2. The Pt-1.7 nm/G catalyst has the highest sensitivity up to 1264.6 $\mu A mM^{-1} cm^{-2}$, rapid response time of 1.69 s, low detection limit, and good ECSA in the linear range of 8 μM to 1.5 mM. Among all catalysts, it was found that the smaller Pt particles on graphene would give the higher sensitivity and wider linear range. Thus, the sensitivity of Pt-1.7 nm/G catalyst is highest among all catalysts. However, after the size is down to 1.3 nm, the linear range of Pt-1.3 nm/G shifts to the high concentration with the much lower sensitivity.

Conflict of Interests

The authors declare that there is no conflict of interests regarding the publication of this paper.

ACKNOWLEDGMENTS

The authors thank the Ministry of Science and Technology, Chang Gung University, and Chang Gung Memorial Hospital (CMRPD2C0011) for financially supporting this research.

REFERENCES

1. K. S. Novoselov, A. K. Geim, S. V. Morozov et al., "Electric field effect in atomically thin carbon films,"Science, vol. 306, no. 5696, pp. 666–669, 2004.
2. K. S. Novoselov, A. K. Geim, S. V. Morozov et al., "Two-dimensional gas of massless Dirac fermions in graphene," Nature, vol. 438, no. 7065, pp. 197–200, 2005.
3. C.-C. Lin and Y.-W. Lin, "Synthesis of carbon nanotube/graphene composites by one-step chemical vapor deposition for electrodes of electrochemical capacitors," Journal of Nanomaterials, vol. 2015, Article ID 741928, 8 pages, 2015.
4. X. Ma, G. Chen, Q. Liu, G. Zeng, and T. Wu, "Synthesis of $LiFePO_4$/graphene nanocomposite and its electrochemical properties as cathode material for Li-ion batteries," Journal of Nanomaterials, vol. 2015, Article ID 301731, 6 pages, 2015.
5. H.-Y. Tsai, W.-H. Hsu, and Y.-C. Huang, "Characterization of carbon nanotube/graphene on carbon cloth as an electrode for air-cathode microbial fuel cells," Journal of Nanomaterials, vol. 2015, Article ID 686891, 7 pages, 2015.
6. X. Wang, Y. Liu, J. Xu et al., "Molecular dynamics study of stability and diffusion of graphene-based drug delivery systems," Journal of Nanomaterials, vol. 2015, Article ID 872079, 14 pages, 2015.
7. Y.-W. Hsu, T.-K. Hsu, C.-L. Sun, Y.-T. Nien, N.-W. Pu, and M.-D. Ger, "Synthesis of CuO/graphene nanocomposites for nonenzymatic electrochemical glucose biosensor applications," Electrochimica Acta, vol. 82, pp. 152–157, 2012.
8. C.-L. Sun, H.-H. Lee, J.-M. Yang, and C.-C. Wu, "The simultaneous electrochemical detection of ascorbic acid, dopamine, and uric acid using graphene/size-selected Pt nanocomposites," Biosensors and Bioelectronics, vol. 26, no. 8, pp. 3450–3455, 2011.
9. C.-L. Sun, C.-T. Chang, H.-H. Lee et al., "Microwave-assisted synthesis of a core-shell MWCNT/GONR heterostructure for the electrochemical detection of ascorbic acid, dopamine, and uric acid," ACS Nano, vol. 5, no. 10, pp. 7788–7795, 2011.
10. C.-L. Sun, C.-H. Su, and J.-J. Wu, "Synthesis of short graphene oxide nanoribbons for improved biomarker detection of Parkinson's disease," Biosensors and Bioelectronics, vol. 67, pp. 327–333, 2015.
11. J. Shen, X. Yang, Y. Zhu, H. Kang, H. Cao, and C. Li, "Gold-coated silica-fiber hybrid materials for application in a novel hydrogen peroxide biosensor," Biosensors and Bioelectronics, vol. 34, no. 1, pp. 132–136, 2012.
12. L. Zheng, L. Xiong, D. Zheng et al., "Bilayer lipid membrane biosensor with enhanced stability for amperometric determination of hydrogen peroxide," Talanta, vol. 85, no. 1, pp. 43–48, 2011.
13. S. Chen, P. Fu, B. Yin, R. Yuan, Y. Chai, and Y. Xiang, "Immobilizing Pt nanoparticles and chitosan hybrid film on polyaniline naofibers membrane for an amperometric hydrogen peroxide biosensor,"Bioprocess and Biosystems Engineering, vol. 34, no. 6, pp. 711–719, 2011.
14. Z. Fan, Q. Lin, P. Gong, B. Liu, J. Wang, and S. Yang, "A new enzymatic immobilization carrier based on graphene capsule for hydrogen peroxide biosensors," Electrochimica Acta, vol. 151, pp. 186–194, 2015.
15. M.-Y. Hua, H.-C. Chen, R.-Y. Tsai, and C.-S. Lai, "A novel polybenzimidazole-modified gold electrode for the analytical determination of hydrogen peroxide," Talanta, vol. 85, no. 1, pp. 631–637, 2011.
16. M.-Y. Hua, H.-C. Chen, C.-K. Chuang et al., "The intrinsic redox reactions of polyamic acid derivatives and their application in hydrogen peroxide sensor," Biomaterials, vol. 32, no. 21, pp. 4885–4895, 2011.
17. X. Li, X. Liu, W. Wang, L. Li, and X. Lu, "High loading Pt nanoparticles on functionalization of carbon nanotubes for fabricating nonenzyme hydrogen peroxide sensor," Biosensors and Bioelectronics, vol. 59, pp. 221–226, 2014.
18. C. Karuppiah, S. Palanisamy, S.-M. Chen, V. Veeramani, and P. Periakaruppan, "A novel enzymatic glucose biosensor and sensitive non-enzymatic hydrogen peroxide sensor based on graphene and cobalt oxide nanoparticles composite modified glassy carbon electrode," Sensors and Actuators B: Chemical, vol. 196, pp. 450–456, 2014.
19. G.-H. Yang, Y.-H. Zhou, J.-J. Wu et al., "Microwave-assisted synthesis of nitrogen and boron co-doped graphene and its application for enhanced electrochemical detection of hydrogen peroxide," RSC Advances, vol. 3, no. 44, pp. 22597–22604, 2013.
20. L. Wang, Y. Ye, X. Lu et al., "Prussian blue nanocubes on nitrobenzene-functionalized reduced graphene oxide and its application for H_2O_2 biosensing," Electrochimica Acta, vol. 114, pp. 223–232, 2013.

21. C.-C. Kung, P.-Y. Lin, F. J. Buse et al., "Preparation and characterization of three dimensional graphene foam supported platinum-ruthenium bimetallic nanocatalysts for hydrogen peroxide based electrochemical biosensors," Biosensors and Bioelectronics, vol. 52, pp. 1–7, 2014.

22. T. Wang, H. Zhu, J. Zhuo et al., "Biosensor based on ultrasmall MoS_2 nanoparticles for electrochemical detection of H_2O_2 released by cells at the nanomolar level," Analytical Chemistry, vol. 85, no. 21, pp. 10289–10295, 2013.

23. N. Wang, Y. Han, Y. Xu, C. Gao, and X. Cao, "Detection of H_2O_2 at the nanomolar level by electrode modified with ultrathin AuCu nanowires," Analytical Chemistry, vol. 87, no. 1, pp. 457–463, 2015.

24. C. Bock, C. Paquet, M. Couillard, G. A. Botton, and B. R. MacDougall, "Size-selected synthesis of PtRu nano-catalysts: reaction and size control mechanism," Journal of the American Chemical Society, vol. 126, no. 25, pp. 8028–8037, 2004.

25. C.-L. Sun, J.-S. Tang, N. Brazeau et al., "Particle size effects of sulfonated graphene supported Pt nanoparticles on ethanol electrooxidation," Electrochimica Acta, vol. 162, pp. 282–289, 2015.

26. J.-M. Yang, S.-A. Wang, C.-L. Sun, and M.-D. Ger, "Synthesis of size-selected Pt nanoparticles supported on sulfonated graphene with polyvinyl alcohol for methanol oxidation in alkaline solutions," Journal of Power Sources, vol. 254, pp. 298–305, 2014.

27. C.-L. Sun, J.-S. Su, J.-H. Tang et al., "Investigation of the adsorption of size-selected Pt colloidal nanoparticles on high-surface-area graphene powders for methanol oxidation reaction," Journal of the Taiwan Institute of Chemical Engineers, vol. 45, no. 3, pp. 1025–1030, 2014.

CHAPTER 14

COLLAGEN/HYDROXYAPATITE COMPOSITE MATERIALS

Laura Cristina Rusu, Ioan Avram Nedelcu, Mădălina Georgiana Albu, Maria Sonmez, Georgeta Voicu, Marius Radulescu, Denisa Ficai, Anton Ficai, Meda-Lavinia Negrutiu, and Cosmin Sinescu

INTRODUCTION

Collagen (COLL) and hydroxyapatite (HA) are the main components of the bone [1]. This is why many materials designed for bone grafting and repair are based on collagen, hydroxyapatite, or composite materials containing the aforementioned components. The performances of these synthesized materials are strongly dependent on the composition and processing induced morphology [1–4].

Due to the compositional similitude with the bone, collagen/hydroxyapatite (COLL/HA) composite materials seem to be the most suitable material for bone engineering. Unfortunately, the lower properties of these materials comparing with natural bones made it necessary to improve the composition of these materials by adding third components [5] or to change the processing route in order to obtain certain morphology [2, 6, 7].

The collagen-carboxymethylcellulose/hydroxyapatite (COLL-CMC/HA) was previously obtained in our group and seems to be proper material for bone grafting and repair [5]. The three components seem to be highly compatible, the FTIR maps obtained at wavelength corresponding to CMC (carboxymethylcellulose) ($715\,cm^{-1}$), COLL ($1655\,cm^{-1}$), and HA ($1033\,cm^{-1}$) revealing a very homogeneous material. COLL-CMC/HA composite material containing both fibres (CMC) and particles (HA) as reinforcing agent is a true hybrid material, expecting superior mechanical properties compared to COLL/HA composite materials.

One of the most important shortcomings of the surgical intervention is related to the possibility of infections. This is why, in many surgical interventions, systemic administration of antibiotic is prescribed in both pre- and postoperative steps [8–10]. Unfortunately, systemic administration of antibiotics also involves systemic toxicity and high incidence of antibiotic resistance [11, 12]. To counter the increasing antibiotic resistance there are two solutions meaning to produce new and new antibiotics or, much better, to use antibiotics in a rational way in order to avoid the development of antibiotic resistance. The most important ways to avoid antibiotic resistance development are to avoid the irrational use of antibiotics, to avoid the excessive use of antibiotics, to use local instead of systemic therapies, the rational choice of the used antibiotic, and so forth [13–17]. Tetracycline, being a broad-spectrum antibiotic indicated against many bacterial infections, was chosen to be used in our study.

The aim of this work is to reduce the amount of antibiotics (tetracycline) by a more rational way of use. For this purpose, tetracycline is embedded in the COLL-CMC/HA composite material, the final tetracycline loaded composite materials being able to assure the osteoregeneration and to assure the anti-infective properties without the need of systemic administration of antibiotic. The locoregional delivery of tetracycline is very important because the overall systemic side effects are limited or even removed.

MATERIALS AND METHODS

Materials and Reagents

Type I fibrillar collagen gel with a molecular weight of about 300.000 Da and acidic pH (2.6) was obtained in the Collagen Department of National Research & Development Institute for Textiles and Leather as we previously described [18] and purified by dialysis, against water free of endotoxins. Carboxymethylcellulose sodium salt (low viscosity) and tetracycline hydrochloride (BioReagent, suitable for cell

Citation: Laura Cristina Rusu, Ioan Avram Nedelcu, Mădălina Georgiana Albu, et al., "Collagen/Hydroxyapatite Composite Materials" Journal of Nanomaterials, vol. 2015, Article ID 361969, 5 pages, 2015. doi:10.1155/2015/361969

culture) were purchased from Sigma-Aldrich (Saint Louis, USA) and were used without any further purification.

Hydroxyapatite was obtained in situ, in the presence of the carboxymethylcellulose-collagen gel. The precursors used for HA synthesis were calcium hydroxide (puriss. p.a.) and sodium phosphate monobasic monohydrate (ACS reagent), both purchased from Sigma-Aldrich (Steinheim, Germany).

Synthesis of COLL/HA-Tetracycline Composite Material

COLL/HA-tetracycline was obtained as schematically presented in Figure 1.

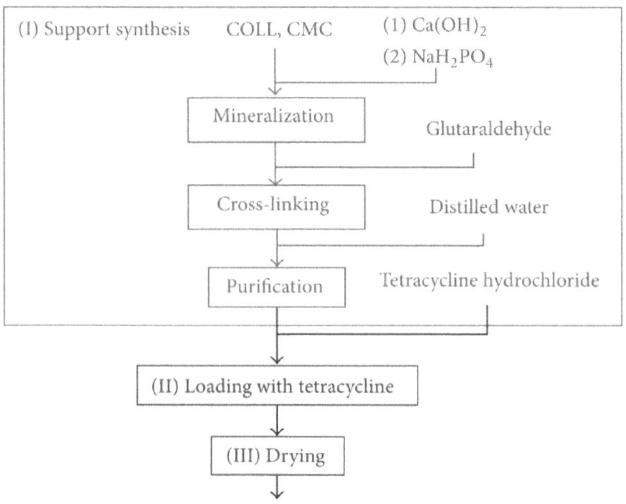

Figure 1: Schematic representation of preparation of COLL-CMC/HA-tetracycline.

The first step is devoted to the preparation of the support material, the second step to the loading with tetracycline, and the last step to the drying. The support preparation consists in homogenisation of the collagen gel and carboxymethylcellulose (COLL : CMC weight ratio is 2 : 1) followed by mineralization with $Ca(OH)_2$and $NaH_2PO_4 \cdot H_2O$ as previously presented [5, 7, 19], cross-linked with glutaraldehyde 1% (reported to the dry collagen), and finally washing with distilled water (three times, each washing step taking 30 min) to remove the soluble salts. The mineralization was performed in order to obtain COLL : CMC : HA composite support material with a mass ratio of 2 : 1 : 8. The loading with tetracycline was done by absorbing the corresponding amount of tetracycline hydrochloride (0.5 g tetracycline/10 g composite) as aqueous solution. The drying can be done by controlled air drying or by freeze-drying, depending on the desired properties, as previously described by Andronescu et al. [1].

Characterisation and Testing

After synthesis and freeze-drying, the composite material was analyzed by using Fourier Transform–Infrared (FTIR) spectroscopy and microscopy and scanning electron microscopy (SEM). The release of tetracycline was monitored by using UV-VIS spectroscopy.

IR microscopy/spectroscopy was performed by using a Thermo FTIR Nicolet iN10 MX microscope; the spectra were recorded in ATR mode over the wave number range of 675–4000 cm^{-1}, with a resolution of 4 cm^{-1}. For a better identification of the peaks, the obtained spectra were resolved using a Gaussian-Lorentzian peak resolve procedure, with no baseline (previously for all the spectra baseline correction was done). For IR microscopy, the data were recorded using an imaging detector (MCT array detector) in reflection mode over the 715–4000 cm^{-1} range, the collection time being 3 s.

SEM analyses were performed on a HITACHI S2600N electron microscope with EDAX, on samples covered with silver layer.

The drug released was evaluated by ultraviolet-visible spectrophotometry based on the peak from 270 nm using a Thermo Evolution 300 spectrophotometer, in quartz cuvettes of 10 mm with a scan speed of 240 nm/min and data interval of 1 nm while the bandwidth was set at 1 nm. For this purpose, 1 g of sample was immersed in 500 mL phosphate buffer, 0.1 M (7.4 pH), maintained at 37°C. At fixed time intervals, sample of 5 mL was extracted and replaced with the same volume of fresh, preheated acceptor phase. All samples were measured in triplicate.

RESULTS AND DISCUSSION

Scanning electron microscopy is a useful tool for characterizing materials, the morphology of the materials being essential both for bone grafts and for drug delivery systems. At low magnification (Figure 2 (right-down side)), spherical and ovaloidal pores can be identified, their size being optimal for bone cell colonization (20–150 μm) [20–23].

Figure 2: Representative SEM images of COLL-CMC/HA-tetracycline.

At higher magnification, micronic and submicronic structures can be identified which, based on our previous study, can be the ends of CMC microfibres [24].

Figure 3 presents the experimental spectra of COLL/HA, COLL-CMC/HA, COLL-CMC/HA-tetracycline, tetracycline hydrochloride, and carboxymethyl cellulose.

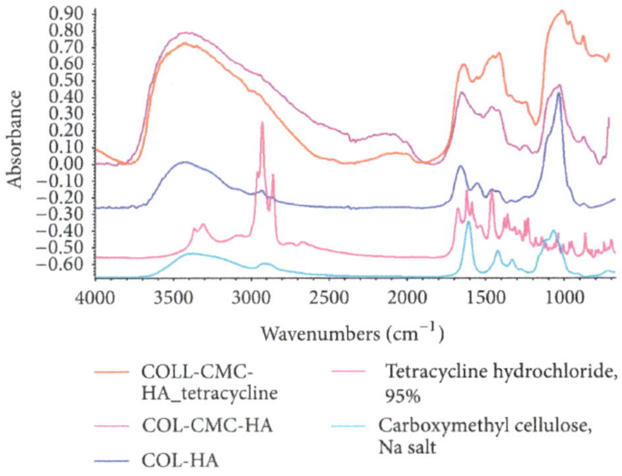

Figure 3: FTIR spectra of tetracycline, COLL/HA, and COLL/HA-tetracycline.

FTIR spectrum of the COLL-CMC/HA-tetracycline sample exhibits the main absorption bands of collagen, carboxymethyl cellulose, and hydroxyapatite as well as the bands of tetracycline. The intensities of these bands are strongly influenced by composition and molar absorptivity of each characteristic vibration. The low content of tetracycline (5% tetracycline) as well as the moderate intensity of its characteristic peaks makes it difficult to clearly identify these peaks. For this reason the deconvolution of the spectrum over the 1200–1800 cm^{-1}spectral range is necessary. In this range, the deconvolution procedure permits the identification of over 30 absorption bands, some of them being assigned to the main bands of tetracycline (1235, 1283, 1384, 1406, 1468, 1515, 1551, 1585, 1616, 1645, and 1660 cm^{-1}). It is also worth mentioning that the relative intensity of these bands is similar to that from tetracycline and consequently proves the presence of the drug. Carboxymethyl cellulose can be also identified both in support material (COLL-CMC/HA) and in the antimicrobial DDS (COLL-CMC/HA-tetracycline). Based on the three experimental spectra, differences can be observed once the number of components increases. For instance, the spectrum of the ternary COLL-CMC/HA composite material differs comparing with the COLL/HA, especially in the region of the absorption bands of CMC (1200–1600 cm^{-1}).

The ternary composite material was also analyzed by FTIR microscopy. The maps were obtained based on the three main independent absorption bands of collagen (1650 cm^{-1}), hydroxyapatite (1022 cm^{-1}), and cellulose (715 cm^{-1}). The three maps are presented in Figure 4.

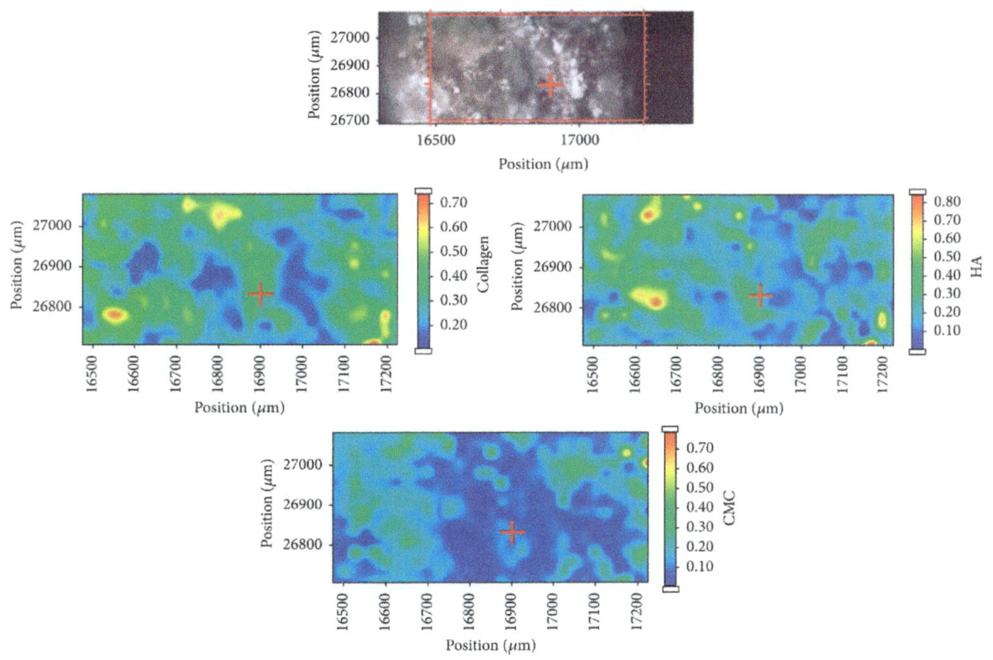

Figure 4: FTIR microscopy recorded on COLL-Cell/HA-tetracycline.

The FTIR maps, colored from blue (low intensity) to red (high intensity), reveal the distribution and relative intensity of the monitored bands (components). Based on the correlation with the video image, as well as the relative intensities of the three monitored bands, it can be seen that collagen and hydroxyapatite are intimately associated but also carboxymethylcellulose microfibers are well dispersed into the COLL/HA hybrid composite material, the blue/red distribution being especially a result of the hill/valley topology of the analyzed section.

The release process was monitored by recording the UV-VIS spectra of the solutions at different times (Figure5). It can be seen that the COLL/HA samples are stable in time, the spectra recorded after three days being identical to that recorded after some minutes of immersion which means that only soluble salts are delivered into the solution. The delivery of tetracycline depends on time. The tetracycline spectra recorded after 1, 3, and 6 days clearly show that the absorbance of both peaks increases. Based on this we can affirm that these systems exhibit prolonged delivery and could be proper candidates as bone grafts for infected bone defects. For quantification, a calibration curve was plotted between 0.1 µg/mL and 10 µL, the linearity being better than $R^2 = 99.98$. Based on the calibration curve, the correlation between absorbance and concentration is given by the equation n $A = 0.0896 \cdot c + 0.0031$. Comparing with the theoretical concentration, after 6 days, the recovered tetracycline was 79.8% which is good enough to assure antimicrobial activity during the healing of the wound. In the case of infections, over 6 days of delivery could be enough to treat the infections, and, if not, supplementary antibiotic can be administered in classical way.

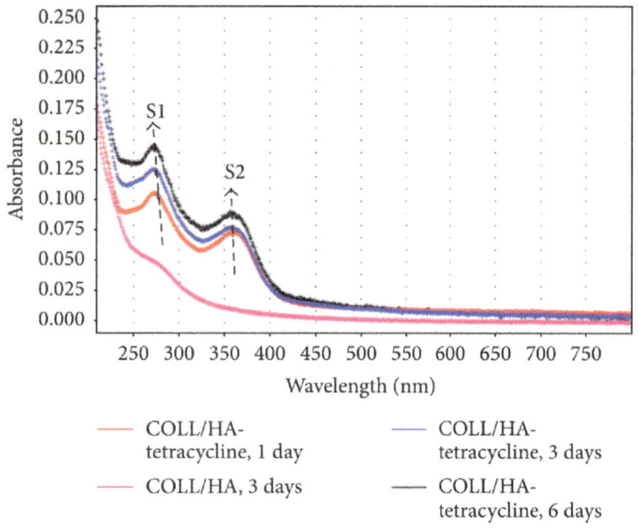

Figure 5: UV-VIS spectra of the solution recorded during the delivery process.

CONCLUSIONS

New antimicrobial bone grafts with prolonged delivery of tetracycline were obtained by coprecipitation of hydroxyapatite in the presence of the carboxymethylcellulose-collagen gel followed by loading with tetracycline. This material was designed to be used only for infected bone defects. This locoregional delivery of tetracycline is essential to reduce the systemic side effects associated with the use of antibiotics. These materials can be considered with prolonged delivery because even after 6 days the recovery of tetracycline is lower than 80%. This result is consistent with the clinical protocols of treating infections but reduces the amount of tetracycline and reduces the systemic side effects.

ACKNOWLEDGMENTS

This work was supported by Romanian National Authority for Scientific Research, MPNS COST Action MP1301: New Generation Biomimetic and Customized Implants for Bone Engineering and Project no. PIII-C2-PCFI (2015-2016), DENTALOCT.

REFERENCES

1. E. Andronescu, A. Ficai, M. G. Albu et al., "Collagen-hydroxyapatite/cisplatin drug delivery systems for locoregional treatment of bone cancer," Technology in Cancer Research and Treatment, vol. 12, no. 4, pp. 275–284, 2013.
2. A. Ficai, E. Andronescu, D. Ficai, M. Sonmez, M. G. Albu, and G. Voicu, "Mimicking the morphology of long bone," Central European Journal of Chemistry, vol. 10, no. 6, pp. 1949–1953, 2012.
3. S. V. Dorozhkin, "Calcium orthophosphate-based bioceramics," Materials, vol. 6, no. 9, pp. 3840–3942, 2013.
4. P. Fratzl and R. Weinkamer, "Nature's hierarchical materials," Progress in Materials Science, vol. 52, no. 8, pp. 1263–1334, 2007.
5. I.-A. Nedelcu, A. Ficai, D. Ficai, G. Voicu, M. G. Albu, and E. Andronescu, "Hybrid COLL-CMC/HA composite materials for bone tissue regeneration," University Politehnica of Bucharest Scientific Bulletin Series B, vol. 77, no. 1, pp. 3–14, 2014.
6. A. Ficai, M. G. Albu, M. Sonmez, D. Ficai, and E. Andronescu, "Synthesis of collagen/hydroxyapatite composite materials with oriented microstructure induced by the interaction of the flowing mineralised collagen gel with the support," ChemXpress, vol. 3, pp. 107–110, 2014.
7. A. Ficai, E. Andronescu, G. Voicu et al., "Self-assembled collagen/hydroxyapatite composite materials,"Chemical Engineering Journal, vol. 160, no. 2, pp. 794–800, 2010.
8. M. Solignac, "New perspectives in antibiotic prophylaxis," Presse Medicale, vol. 27, no. 3, pp. 27–29, 1998.
9. D. Lebeaux, N. Fernández-Hidalgo, A. Chauhan et al., "Management of infections related to totally implantable venous-access ports: challenges and perspectives," The Lancet Infectious Diseases, vol. 14, no. 2, pp. 146–159, 2014.
10. A. Lipp and G. Lusardi, "Systemic antimicrobial prophylaxis for percutaneous endoscopic gastrostomy,"The Cochrane Database of Systematic Reviews, no. 11, Article ID CD005571, 2013.
11. M. Zilberman and J. J. Elsner, "Antibiotic-eluting medical devices for various applications," Journal of Controlled Release, vol. 130, no. 3, pp. 202–215, 2008.
12. T. Y. Wu, Q. C. Zhang, W. P. Ren et al., "Controlled release of gentamicin from gelatin/genipin reinforced beta-tricalcium phosphate scaffold for the treatment of osteomyelitis," Journal of Materials Chemistry B, vol. 1, no. 26, pp. 3304–3313, 2013.
13. L. Zhang, Y. Huang, Y. Zhou, T. Buckley, and H. H. Wang, "Antibiotic administration routes significantly influence the levels of antibiotic resistance in gut microbiota," Antimicrobial Agents and Chemotherapy, vol. 57, no. 8, pp. 3659–3666, 2013.
14. C. García-Rey, J. E. Martín-Herrero, and F. Baquero, "Antibiotic consumption and generation of resistance in Streptococcus pneumoniae: the paradoxical impact of quinolones in a complex selective landscape," Clinical Microbiology and Infection, vol. 12, no. 3, pp. 55–66, 2006.
15. A. Rodríguez-Rojas, J. Rodríguez-Beltrán, A. Couce, and J. Blázquez, "Antibiotics and antibiotic resistance: a bitter fight against evolution," International Journal of Medical Microbiology, vol. 303, no. 6-7, pp. 293–297, 2013.
16. A. Dashti, D. Ready, V. Salih et al., "In vitro antibacterial efficacy of tetracycline hydrochloride adsorbed onto Bio-Oss bone graft," Journal of Biomedical Materials Research Part B: Applied Biomaterials, vol. 93, no. 2,

pp. 394–400, 2010.

17. S. Praveen and M. Rohaizak, "Local antibiotics are equivalent to intravenous antibiotics in the prevention of superficial wound infection in inguinal hernioplasty," Asian Journal of Surgery, vol. 32, no. 1, pp. 59–63, 2009.
18. M. G. Albu, Collagen Gels and Matrices for Biomedical Applications, Lambert Academic Publishing, Saarbrücken, Germany, 2011.
19. E. Andronescu, M. Ficai, G. Voicu, D. Ficai, M. Maganu, and A. Ficai, "Synthesis and characterization of collagen/hydroxyapatite: magnetite composite material for bone cancer treatment," Journal of Materials Science: Materials in Medicine, vol. 21, no. 7, pp. 2237–2242, 2010.
20. A. Ficai, E. Andronescu, G. Voicu, and D. Ficai, "Advances in collagen/hydroxyapatite composite materials," in Advances in Composite Materials for Medicine and Nanotechnology, B. Attaf, Ed., chapter 1, InTech, Rijeka, Croatia, 2011.
21. B.-S. Chang, C.-K. Lee, K.-S. Hong et al., "Osteoconduction at porous hydroxyapatite with various pore configurations," Biomaterials, vol. 21, no. 12, pp. 1291–1298, 2000.
22. H. Develioglu, E. Koptagel, R. Gedik, and L. Dupoirieux, "The effect of a biphasic ceramic on calvarial bone regeneration in rats," Journal of Oral Implantology, vol. 31, no. 6, pp. 309–312, 2005.
23. O. Gauthier, J.-M. Bouler, E. Aguado, P. Pilet, and G. Daculsi, "Macroporous biphasic calcium phosphate ceramics: influence of macropore diameter and macroporosity percentage on bone ingrowth,"Biomaterials, vol. 19, no. 1–3, pp. 133–139, 1998.
24. D. E. Fetterolf and R. J. Snyder, "Scientific and clinical support for the use of dehydrated amniotic membrane in wound management," Wounds, vol. 24, no. 10, pp. 299–307, 2012.

CHAPTER 15

A NOVEL COMPOSITE MATERIAL DESIGNED FROM FESI POWDER AND $MN_{0.8}ZN_{0.2}FE_2O_4$ FERRITE

Magdalena Streckova, Radovan Bures, Maria Faberova, Pavel Kurek, Pavla Roupcova, Hynek Hadraba, Vladimir Girman, and Jozef Strecka

INTRODUCTION

Soft magnetic composites (SMCs), which are of technological relevance in electromagnetic applications, can be described as ferromagnetic powder particles surrounded by an electrical insulating film [1]. It is well known that the core/shell particles can be processed by powder metallurgy (PM) methods. PM technologies are indeed able to produce SMCs with a high enough density and sufficiently stable mechanical properties, whereas an insulating layer between magnetic powder particles ensures a high electric resistivity that minimizes the overall magnetic losses [2]. The main advantage of SMCs prepared by PM methods with respect to laminated soft magnetic materials lies in their unique magnetic properties such as an isotropic three-dimensional ferromagnetic behavior, low eddy current losses, and relatively lower total core losses at medium and high frequencies [3, 4]. SMCs with superior magnetic properties are guaranteed by a choice of suitable soft magnetic material serving as a core and convenient insulating material serving as a coating. A lot of efforts have been devoted to a preparation of SMCs composed of different ferromagnetic base materials (Fe, Permalloy, Vitroperm, FeSi, etc.) and different hybrid organic resins (mostly of phenol-formaldehyde type) used as an insulating coating [5, 6]. However, it is worthwhile to remark that a thermal treatment of these materials is strictly limited by a thermal resistance of the insulating organic material, which is inadequate in order to remove plastic deformations created during a preparation of green compacts [3]. In this regard, another promising way in order to produce SMCs with excellent properties consists in using spinel ferrites as a coating with an extremely high electric resistivity and almost zero eddy current losses in alternating magnetic fields, which will not basically affect the magnetic characteristics of metal cores (high magnetic flux density and low coercive force) [7]. In general, the ferrites are metal oxides containing divalent and trivalent metal ions. The spinel-type ferrites with the general formula MFe_2O_4 (M is divalent metal ion, e.g., Ni^{2+}, Mn^{2+}, Zn^{2+}, etc.) involve two sites, namely, tetrahedral sites A and octahedral sites B. According to a distribution of the divalent metal ions M^{2+} between tetrahedral and octahedral sites one recognizes two different types of the ferrites: the normal spinels $[M^{2+}]_A[Fe^{3+}Fe^{3+}]_BO_4$ in which the divalent metal ions M^{2+} occupy the tetrahedral sites and the trivalent Fe^{3+} ions occupy the octahedral sites and, respectively, the inverse spinels $[Fe^{3+}]_A[M^{2+}Fe^{3+}]_BO_4$ in which the octahedral sites are occupied in part by the divalent metal ions M^{2+} and in part by the trivalent Fe^{3+}ions in contrast with the tetrahedral sites occupied exclusively by the trivalent Fe^{3+} ions [8]. Note, furthermore, that it is possible to control to a certain extent basic magnetic characteristics of spinel-type ferrites by changing the type of divalent cations [9], the size and shape of nanoparticles [10–12], which also have fundamental impact on a chemical composition and crystal structure [13, 14].

$Mn_xZn_{1-x}Fe_2O_4$ ferrites belong to the most versatile ferrites among the inverse spinel ferrites due to their promising magnetic properties like high saturation magnetization, high magnetic permeability, high electrical resistance, and biocompatibility [15, 16]. Up to date, the MnZn ferrites have been widely used in the field of electronic and electromagnetic devices [17–19]. Iron, zinc, and manganese are not only the main components of the MnZn ferrites, but they are also the main components of spent alkaline batteries. It is noteworthy that Yang et al. have developed a novel way to recycle spent alkaline batteries using a combination of sol-gel autocombustion and microwave digestion methods producing a powder of MnZn ferrite [20]. The microwave-assisted technique is environmentally friendly and economic low-cost method for a synthesis of ferrites

Citation: Magdalena Streckova, Radovan Bures, Maria Faberova, et al., "A Novel Composite Material Designed from FeSi Powder and Mn0.8Zn0.2Fe2O4 Ferrite," Advances in Materials Science and Engineering, vol. 2015, Article ID 924859, 8 pages, 2015. doi:10.1155/2015/924859

[21,22]. To the best of our knowledge, the combination of sol-gel autocombustion method used for a preparation of the MnZn ferrite with the microwave-assisted sintering process of the composite samples has been not reported yet and it is therefore the main subject of the present work.

In this work, our particular attention will be focused on a design of perspective class of new SMCs based on spherical FeSi particles serving as a core and the inverse spinel ferrite $Mn_{0.8}Zn_{0.2}Fe_2O_4$ serving as a coating. The MnZn ferrite was synthesized by the sol-gel autocombustion method. The coating process was carried out immediately after gel creation and the autocombustion process run in a presence of the ferromagnetic FeSi particles. The crystal structure of MnZn ferrite was verified by TEM analysis. The coating deposited on spherical FeSi particles and the distribution of ferrite in the final microcomposite FeSi/$MnZnFe_2O_4$ samples was investigated by SEM and EDX analysis. The coated powder was compressed to a cylindrical shape and the samples were sintered by exactly defined microwave process. The mechanical test and electric resistivity measurement were performed.

EXPERIMENTAL

Materials

The commercial powder of spherical FeSi particles supplied by Höganäs Corporation was used as the base ferromagnetic material, which is available in the granulometric fraction from 45 µm to 150 µm [23]. The analytical pure chemicals $Mn(NO_3)_2{\cdot}4H_2O$ (99%, Acros Organic), $Zn(NO_3)_2{\cdot}4H_2O$ (98%, Acros Organic), $Fe(NO_3)_2$ (99+%, Acros Organic), and $C_6H_8O_7{\cdot}H_2O$ (99.8% CentralChem) were used to synthesize $Mn_{0.8}Zn_{0.2}Fe_2O_4$ ferrite. The equimolar ratio of metal nitrates was dissolved in minimum amount of distilled water. The equimolar ratio of citric acid to the total molar amount of metal nitrates was added to the aqueous solution as a complexing agent during continuous stirring. The pH = 6 was adjusted by NH_4OH (26% aq., Sigma Aldrich) after total dissolution of citric acid. Roughly 100 mL of the prepared aqueous solution was heated at 70°C for 4 hours until the viscous gel was created. 5 g of the prepared gel was mixed with 20 g of FeSi particles and heated at 200°C for 12 hours in laboratory dryer. The exothermic reaction and self-propagating combustion process initiated after the creation of xerogel. The 4 wt% of ferrite was created in the final composite powder agglomerate samples.

Experimental Procedures

The coated powder was compacted at 600 MPa in cylindrical die without any additives. Density of powder and bulk composite was measured by He-pycnometer AccuPyc II 1340 and calculated from the sample dimensions, respectively. The prism-shaped samples ((dimensions 5 × 4 × 20 mm ($w{\times}h{\times}l$)) were used to evaluate elastic properties represented by Young modulus (E)-Impulse excitation method and software Buzz-o-Sonic and transverse rupture strength (TRS) measured by 3-point bending test using universal testing machine Tiratest 2100. The cylindrical shaped samples (dimensions of 10 × 3 mm ($d{\times}h$)) were used to measure the hardness tests using TUKKON 1102 hardness tester. The electric resistivity was measured by 4-point probe method using Mitsubishi Loresta AXMCP-T370. The sintering process was realized in air atmosphere using single mode microwave sintering device (microwave generator 750 W). Cylindrical cavity of diameter 28 mm and of height 80 mm was used. The sintering temperature profile is shown in Figure 1.

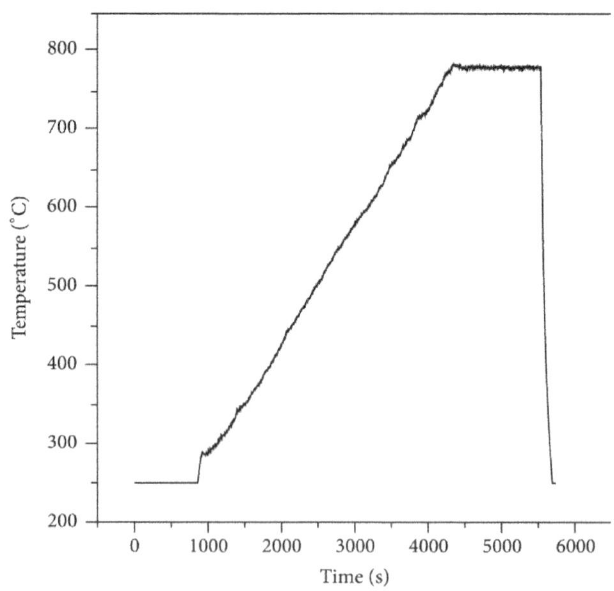

Figure 1: Time-temperature dependence of microwave sintering process.

The microstructure and morphology of all the samples were examined by the scanning electron microscope SEM (JEOL JSM-7000F or TESCAN Lyra3) equipped with the energy dispersive X-ray analyzer (EDX) after the carbon coating. The bright-field TEM or STEM images were explored at 200 kV by the transmission electron microscopy TEM (JEOL 2100F). The structural analysis was carried out using X-ray diffraction XRD (X'Pert Pro PanAnalytical) in Bragg-Brentano geometry, with CoKα radiation, linear detector, and β-filter in incident path.

RESULTS AND DISCUSSION

Synthesis of $Mn_{0.8}Zn_{0.2}Fe_2O_4$

The well-described sol-gel synthesis accompanied with the autocombustion process was used for preparation of $Mn_{0.8}Zn_{0.2}Fe_2O_4$. The exothermic reaction and decomposition of xerogel result in a creation of the ferrite ash, which shows interesting structure resembling coral (see Figure 2). The synthesized powder particles $Mn_{0.8}Zn_{0.2}Fe_2O_4$ were carefully examined with the aim to simplify the latter identification of the MnZn ferrite coating on FeSi substrate. The temperature schedule enabling the complete decomposition process and the ferrite formation was set according to TG-DSC data published by Gabal et al. [24]. The typical thermal decomposition of the prepared precursor with Zn molar content of 0.2 has been reported in [24] in a temperature range up to 600°C. The decomposition proceeds through three well-defined TG steps. The most important step has been observed at 200°C, at which a sharp exothermic peak accompanied with the main weight loss (78%) occurs. According to [25], the complete decomposition of the MnZn ferrite precursor occurs at 340°C, whereas the nitrate-citrate combustion reaction can be presented as follows:

$$0.8Mn(NO_3)_2 \cdot 4H_2O + 0.2Zn(NO_3)_2 \cdot 6H_2O + 2Fe(NO_3)_3 \cdot 9H_2O + C_6H_8O_7 \cdot H_2O \longrightarrow Mn_{0.8}Zn_{0.2}Fe_2O_4 + 4N_2 + 6CO_2 + 28.4H_2O + 5O_2 \quad (1)$$

Figure 2: Photograph of $Mn_{0.8}Zn_{0.2}Fe_2O_4$ ferrite after autocombustion process.

Note that in our synthesis it was not necessary to heat the MnZn ferrite up to 340°C, because the subsequent sintering process of the final composite material FeSi/$Mn_{0.8}Zn_{0.2}Fe_2O_4$ requires the temperature schedule shown in Figure 1 that goes up to 780°C.

A comparison of the X-ray diffraction pattern of the synthesized powder with the PDF-4 database (1-089-7550) demonstrates a presence of the spinel-type structure only (see Figure 3). The computation of crystal parameters from the diffraction pattern by Rietveld method with structural model based on ICSD (28516) determines the lattice parameter a = 0.8445 nm and crystalline size about 20 nm from Scherrer formula [26]. The shape of peaks and broadening of bottom of peaks indicate the presence of highly disordered (amorphous) phase. The crystalline phase in the sample can be indexed as the face centered cubic (FCC) structure corresponding to the inverse spinel structure almost identical to the one reported in [27]. The spinel structure was also verified by the selected area electron diffraction pattern obtained by TEM analysis for the powder sample (see Figure 5(c)).

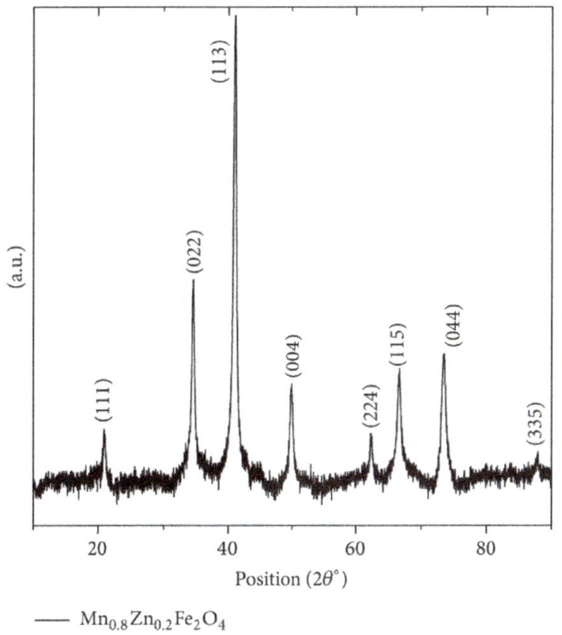

Figure 3: XRD pattern of $Mn_{0.8}Zn_{0.2}Fe_2O_4$.

SEM and TEM of as-burned sample with an extremely fine structure are displayed in Figures 4 and 5. SEM analysis has confirmed sharp edges of the synthesized MnZn ferrite (Figure 4), while TEM analysis has demonstrated the layered structure (Figure 5(a)). The chemical composition of MnZn ferrite powder was verified by EDX analysis shown in Figure 5(d). It is quite evident from Figure 5(a) that one layered particle can be seen as a large agglomerate composed of extremely fine nanoparticles. A more detailed analysis of the substructure (zoom Figure 5(b)) verified the formation of small partially ordered nanocrystalline particles with a diameter around 20 nm, which are separated from each other by the amorphous part. The nanocrystalline particles incorporated in amorphous part were also confirmed by XRD pattern of the MnZn ferrite.

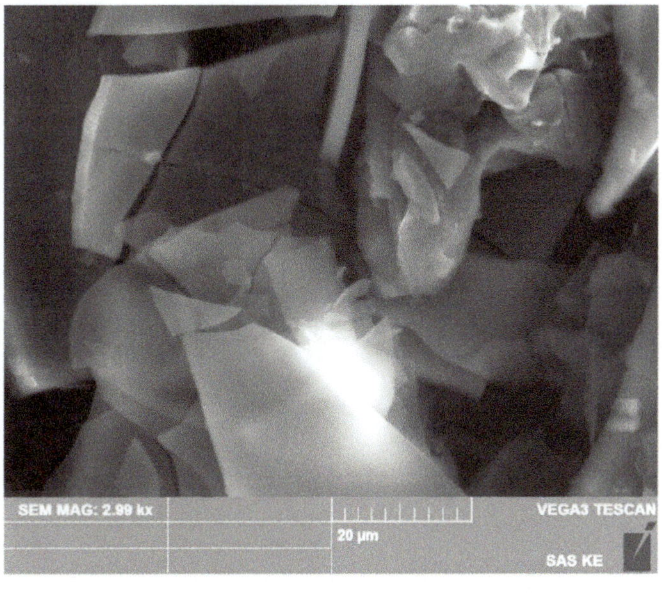

Figure 4: SEM image of $Mn_{0.8}Zn_{0.2}Fe_2O_4$.

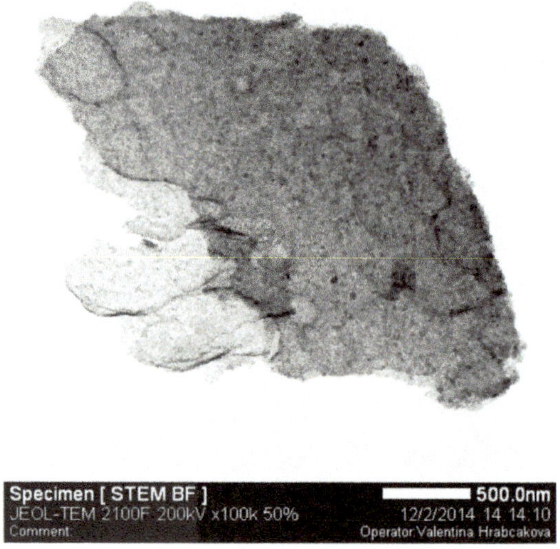

(a)

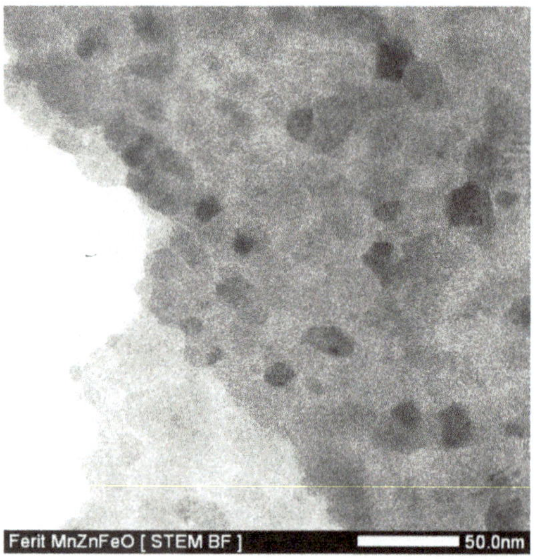

(b)

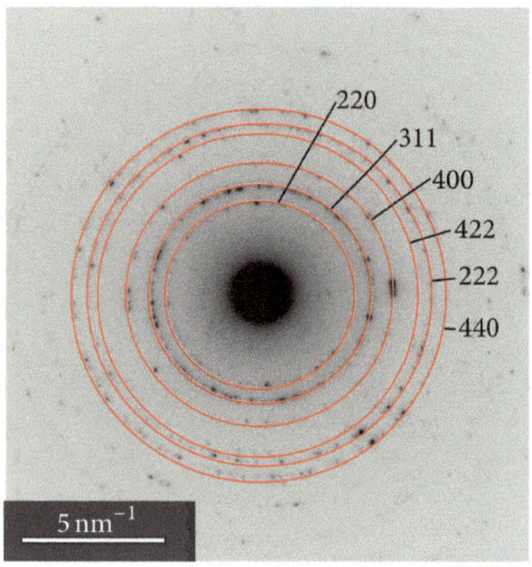

(c)

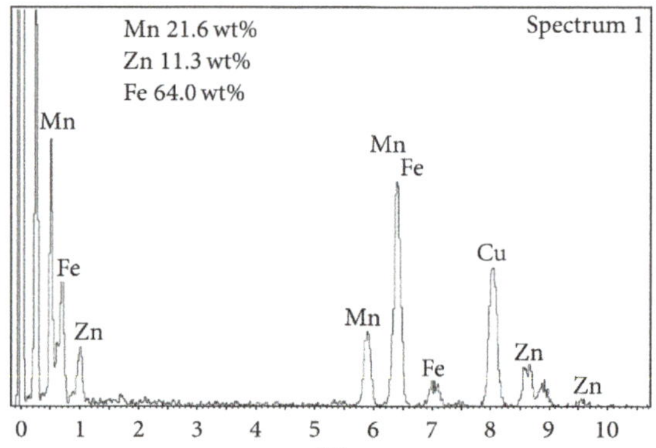

(d)

Figure 5: (a) TEM image of $Mn_{0.8}Zn_{0.2}Fe_2O_4$ powder, (b) zoom of $Mn_{0.8}Zn_{0.2}Fe_2O_4$powder, (c) diffractogram, and (d) EDX analysis.

Preparation of FeSi/$Mn_{0.8}Zn_{0.2}Fe_2O_4$ Composite

Figure 6(a) shows the SEM image of the original spherical FeSi particles and Figure 6(b) shows the SEM image of the as-prepared FeSi/$Mn_{0.8}Zn_{0.2}Fe_2O_4$ composite, in which the spherical FeSi particles are linked together through the very fine MnZn ferrite in order to form large agglomerate. Note that the MnZn ferrite does not instantaneously produce a continuous coating on FeSi particles, but one may instead observe a porous layer of the MnZn ferrite surrounding FeSi particles as illustrated in Figures 7(a) and 7(b). The compaction is responsible for breaking down of the as-prepared ferrite coral structure and, consequently, the continuous coating of the MnZn ferrite surrounding spherical FeSi particles is built up during the compaction process as it is evident from Figures 8(a) and 8(b). It is worthwhile to remark that the fine porous ferrite layers enclosing FeSi particles greatly simplify the compaction process and, thus, the spherical particles may form a dense body. The microwave heating of the final green compact FeSi/$Mn_{0.8}Zn_{0.2}Fe_2O_4$, as described in Section 2.2, was carried out as the final sintering process. It is worthwhile to recall that the very continuous network of the secondary ferrite phase was confirmed from the polished samples of FeSi/$Mn_{0.8}Zn_{0.2}Fe_2O_4$ created after microwave sintering process (see Figures 8(a) and 8(b)). Moreover, the fine structure of the MnZn ferrite leads to a creation of the composite samples with a very tight arrangement without any significant porosity. It is quite apparent that the compaction procedure has not caused any noticeable deformation of the spherical FeSi particles. The EDX line analysis was employed in order to verify the chemical composition of the final sintered samples. The weight percentages of individual components are displayed in Figure 9 along the line crossing an interface between MnZn ferrite and spherical FeSi particle, which corresponds to the scale of zoom on the top of Figure 9. The chemical composition at the distances 0.5 μm and 3 μm of EDX line analysis is specifically mentioned in two charts involved in Figure 9. The weight percentages of individual components at the former distance are in a good accordance with the chemical composition of $Mn_{0.8}Zn_{0.2}Fe_2O_4$ ferrite, while the chemical composition at the latter distance points to almost pure FeSi phase. This finding has unambiguously confirmed a successful preparation of FeSi/$Mn_{0.8}Zn_{0.2}Fe_2O_4$ composite.

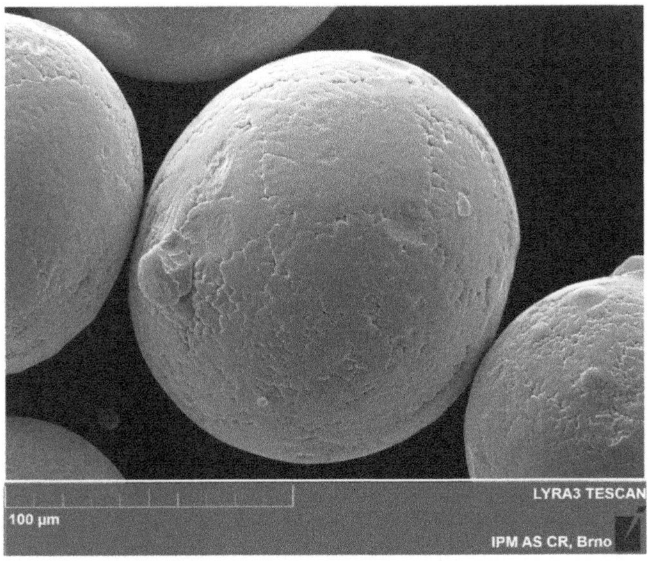

(a)

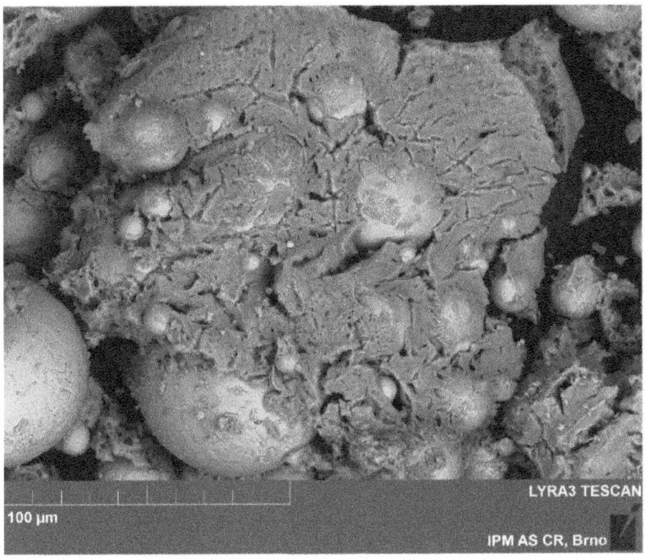

(b)

Figure 6: SEM images from (a) the original FeSi spherical powder (97% Fe and 3% Si); (b) the agglomerate of FeSi particles embedded in $Mn_{0.8}Zn_{0.2}Fe_2O_4$ ferrite.

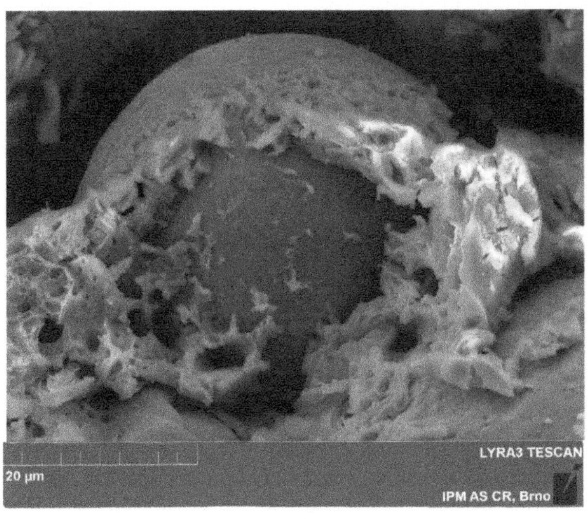

(a)

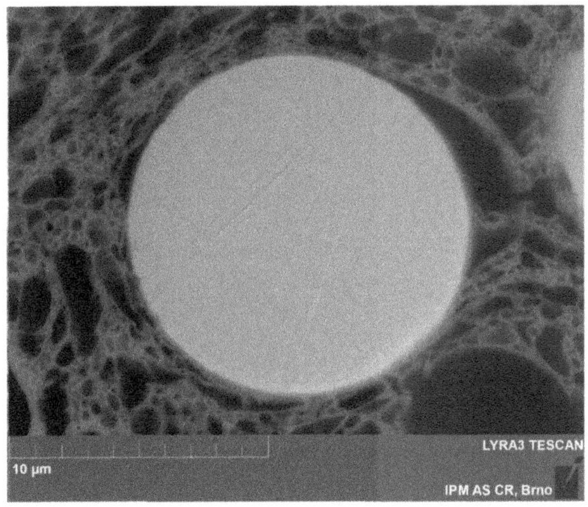

(b)

Figure 7: SEM images from (a) the individual FeSi particle covered by $Mn_{0.8}Zn_{0.2}Fe_2O_4$ferrite; (b) the cross section of FeSi particle surrounded by $Mn_{0.8}Zn_{0.2}Fe_2O_4$ ferrite.

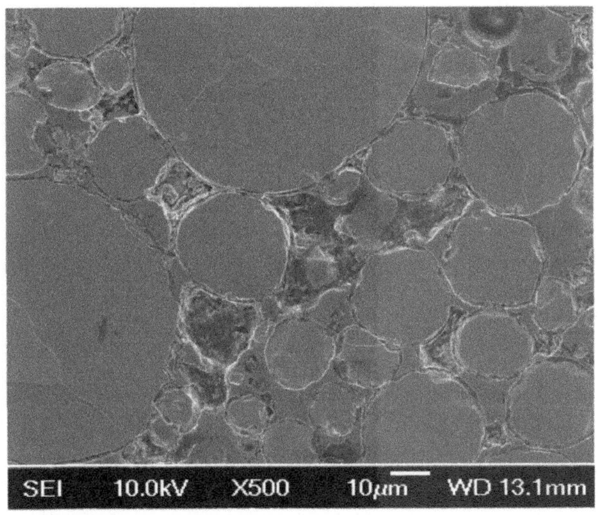

(a)

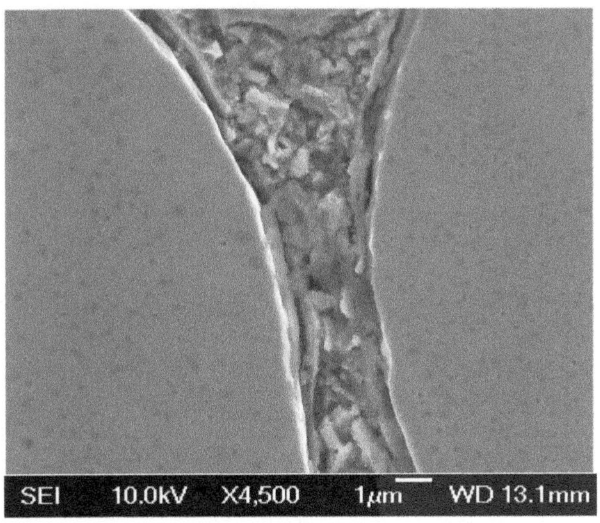

(b)

Figure 8: (a) and (b) SEM images displaying a cross section from the sintered FeSi/$Mn_{0.8}Zn_{0.2}Fe_2O_4$.

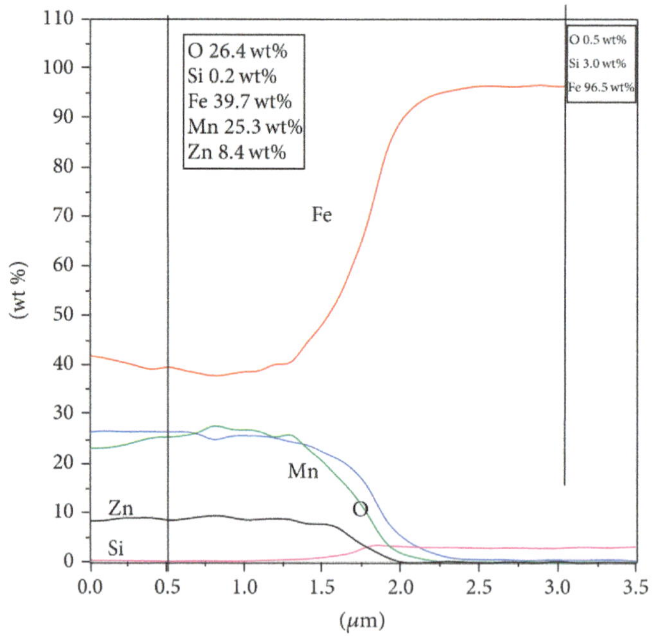

(a)

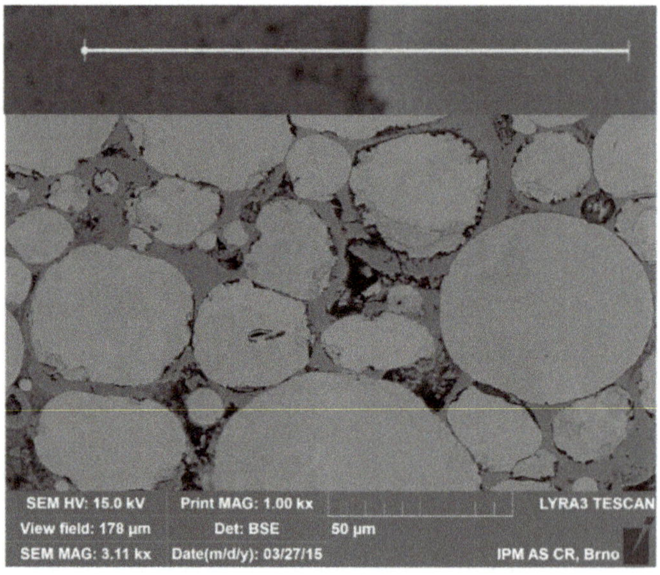

(b)

Figure 9: EDX line analysis of the final sintered sample FeSi/$Mn_{0.8}Zn_{0.2}Fe_2O_4$ supplemented with two charts involving the chemical composition at the distances 0.5 μm and 3 μm according to the scale of zoom located on the top.

Characterization of FeSi/$Mn_{0.8}Zn_{0.2}Fe_2O_4$ Composite

Table 1 shows that the mechanical properties and electric resistivity of the final composite FeSi/$Mn_{0.8}Zn_{0.2}Fe_2O_4$ were significantly affected by the distribution, thickness, and amount of the secondary ferrite phase. The measured and calculated values of the bulk density indicate insignificant porosity in the prepared composite samples. For a comparison, Table 1 enumerates basic characteristics of a few comparative sintered materials. The pure polycrystalline $Mn_{0.8}Zn_{0.2}Fe_2O_4$ sample sintered at 1250 or 1300°C in the air atmosphere at desired partial pressure of oxygen has TRS from 195 MPa to 265 MPa and Young modulus from 175 GPa to 177 GPA, which indicate very hard (HV approximately 6 GPa) and brittle material. It is of technological importance that the measured values of mechanical hardness, transverse rupture strength, and Young modulus for the FeSi/$Mn_{0.8}Zn_{0.2}Fe_2O_4$ composite are quite superior with respect to the ones of the analogous composite involving the same FeSi powder and the phenol-formaldehyde resin as the insulating coating (cf. the first and the last row in Table 1). In addition, it is quite clear from Table 1 that all the mechanical properties of FeSi/$Mn_{0.8}Zn_{0.2}Fe_2O_4$ are comparable (of the same order of magnitude) with the basic characteristics of the pure sintered $Mn_{0.8}Zn_{0.2}Fe_2O_4$ ferrite. The homogeneous distribution of the insulating coating based on the MnZn ferrite manifests itself also through a sufficiently high specific electric resistivity of the FeSi/$Mn_{0.8}Zn_{0.2}Fe_2O_4$ composite, which is almost two orders of magnitude greater than the specific electric resistivity of the FeSi/phenol-formaldehyde resin composite. This result would suggest that the prepared FeSi/$Mn_{0.8}Zn_{0.2}Fe_2O_4$ composite could represent the improved

SMC due to lower eddy current losses at medium and higher frequencies, which will be significantly hindered by a rather high specific electric resistivity.

Table 1: The mechanical properties and electric resistivity of FeSi/$Mn_{0.8}Zn_{0.2}Fe_2O_4$. The columns determine the measured and calculated bulk density, transverse rupture strength (TRS), Young modulus (E), mechanical hardness (HV), and specific electric resistivity.

Sample	ρ(measured bulk) (g/cm³)	ρ(calculated bulk) (g/cm³)	TRS (MPa)	E (GPa)	HV (GPa)	ρ_e (Ohm·cm)
FeSi/$Mn_{0.8}Zn_{0.2}Fe_2O_4$	6.534 ± 0.01	6.069	141 ± 10	121 ± 0.3	1.4 ± 0.02	16 ± 4.5
$Mn_{0.8}Zn_{0.2}Fe_2O_4$ (polycrystalline) [28]	—	—	195–265	175–177	6–6.7	—
$Mn_{0.8}Zn_{0.2}Fe_2O_4$ (size 27–37 nm) [29]	4.69–5.29	—	—	—	—	$3 \cdot 10^6$
Fe_3Si (sintered) [30]	7.15	—	—	—	—	$5 \cdot 10^{-5}$
FeSi/phenol-formaldehyde resin [31]	6.41	—	97	34	0.45	0.2

CONCLUSION

The novel composite material was designed from the spherical FeSi particles and the soft magnetic ferrite $Mn_{0.8}Zn_{0.2}Fe_2O_4$ with the inverse spinel structure. The aforementioned MnZn ferrite was synthesized by the sol-gel autocombustion method from the nitrate-citrate aqueous solution. The composite powder coated by the MnZn ferrite was prepared in the form of large agglomerates, which were used for the subsequent compaction and final sintering of green compacts. The inverse spinel structure of the prepared $Mn_{0.8}Zn_{0.2}Fe_2O_4$ was confirmed by XRD analysis. SEM images have enabled us to identify very fine structure of the prepared powder with sharp edges. TEM analysis has evidenced the layered structure of the pure MnZn ferrite and it has also visualized large agglomerates composed of fine nanoparticles of approximate size 20 nm.

It has been found that the fine porous MnZn ferrite structure provides a very convenient secondary phase during the compaction process, because tight arrangement of spherical FeSi particles onto dense body without any significant porosity is possible after overall breaking down of the porous MnZn ferrite structure. It turns out that the mechanical properties of the final FeSi/$Mn_{0.8}Zn_{0.2}Fe_2O_4$ composite are quite superior with respect to that ones of the analogous composites, in which the organic binder (mostly phenol-formaldehyde resin) is used as the insulating coating instead of the MnZn ferrite. The homogeneous distribution of the insulating coating based on the MnZn ferrite is reflected in a rather high specific electric resistivity of the FeSi/$Mn_{0.8}Zn_{0.2}Fe_2O_4$ composite, which indicates perspective application of this SMC material at medium and high frequencies due to a substantial reduction of relevant eddy current losses. According to this, the investigation of basic magnetic characteristics and their optimization represents a challenging task for our future work.

ACKNOWLEDGMENTS

This work was supported in part by the projects "the progressive technology for preparation of microcomposite materials for electrotechnology," ITMS 26220220105, and by the Scientific Grant Agency of the Ministry of Education of Slovak Republic and the Slovak Academy of Sciences, Project VEGA 2/0185/15. This work was also realized in CEITEC (Central European Institute of Technology) with research infrastructure supported by the Project CZ.1.05/1.1.00/02.0068 financed from European Regional Development Fund and Czech Science Foundation P108/11/1350.

REFERENCES

1. H. Shokrollahi and K. Janghorban, "Soft magnetic composite materials (SMCs)," Journal of Materials Processing Technology, vol. 189, no. 1–3, pp. 1–12, 2007.
2. B. Ślusarek, B. Jankowski, K. Sokalski, and J. Szczygłowski, "Characteristics of power loss in soft magnetic composites a key for designing the best values of technological parameters," Journal of Alloys and Compounds, vol. 581, pp. 699–704, 2013.
3. S. Wu, A. Sun, W. Xu et al., "Iron-based soft magnetic composites with Mn-Zn ferrite nanoparticles coating obtained by sol-gel method," Journal of Magnetism and Magnetic Materials, vol. 324, no. 22, pp. 3899–3905, 2012.
4. A. H. Taghvaei, H. Shokrollahi, and K. Janghorban, "Magnetic and structural properties of iron phosphate-phenolic soft magnetic composites," Journal of Magnetism and Magnetic Materials, vol. 321, no. 23, pp. 3926–3932, 2009.
5. M. Streckova, R. Bures, M. Faberova, L. Medvecky, J. Fuzer, and P. Kollar, "A comparison of soft magnetic composites designed from different ferromagnetic powders and phenolic resins," Chinese Journal of Chemical Engineering, 2014.
6. B. V. Neamţu, O. Geoffroy, I. Chicinaş, and O. Isnard, "AC magnetic properties of the soft magnetic composites based on Supermalloy nanocrystalline powder prepared by mechanical alloying," Materials Science and Engineering B: Solid-State Materials for Advanced Technology, vol. 177, no. 9, pp. 661–665, 2012.
7. H. Kronmuller, Handbook of Magnetism and Advanced

Magnetic Materials, Wiley, 2007.

8. Y. Mastai, "Crystalization in spinel ferrite nanoparticles," in Advances in Crystallization Processes, chapter 14, InTech, 2012, http://cdn.intechopen.com/pdfs/36363/InTech-Crystalization_in_spinel_ferrite_nanoparticles.pdf.
9. M. P. Gonzalez-Sandoval, A. M. Beesley, M. Miki-Yoshida, L. Fuentes-Cobas, and J. A. Matutes-Aquino, "Comparative study of the microstructural and magnetic properties of spinel ferrites obtained by co-precipitation," Journal of Alloys and Compounds, vol. 369, no. 1-2, pp. 190–194, 2004.
10. J. Wang, F. Ren, B. Jia, and X. Liu, "Solvothermal synthesis and characterization of $NiFe_2O_4$ nanospheres with adjustable sizes," Solid State Communications, vol. 150, no. 25-26, pp. 1141–1144, 2010.
11. Z. Zhang, Y. Liu, G. Yao, G. Zu, X. Zhang, and J. Ma, "Solid-state reaction synthesis of $NiFe_2O_4$nanoparticles by optimizing the synthetic conditions," Physica E: Low-Dimensional Systems and Nanostructures, vol. 45, pp. 122–129, 2012.
12. P. Sivakumar, R. Ramesh, A. Ramanand, S. Ponnusamy, and C. Muthamizhchelvan, "Synthesis, studies and growth mechanism of ferromagnetic $NiFe_2O_4$ nanosheet," Applied Surface Science, vol. 258, no. 17, pp. 6648–6652, 2012.
13. E. Darezereshki, "One-step synthesis of hematite (α-Fe_2O_3) nano-particles by direct thermal-decomposition of maghemite," Materials Letters, vol. 65, no. 4, pp. 642–645, 2011.
14. J. L. Gunjakar, A. M. More, K. V. Gurav, and C. D. Lokhande, "Chemical synthesis of spinel nickel ferrite ($NiFe_2O_4$) nano-sheets," Applied Surface Science, vol. 254, pp. 5844–5848, 2008.
15. R. Raeisi Shahraki, M. Ebrahimi, S. A. Seyyed Ebrahimi, and S. M. Masoudpanah, "Structural characterization and magnetic properties of superparamagnetic zinc ferrite nanoparticles synthesized by the coprecipitation method," Journal of Magnetism and Magnetic Materials, vol. 324, no. 22, pp. 3762–3765, 2012.
16. S. A. Seyyed Ebrahimi, S. M. Masoudpanah, H. Amiri, and M. Yousefzadeh, "Magnetic properties of MnZn ferrite nanoparticles obtained by SHS and sol-gel autocombustion techniques," Ceramics International, vol. 40, no. 5, pp. 6713–6718, 2014.
17. W. J. Wang, C. G. Zang, and Q. J. Jiao, "Synthesis, structure and electromagnetic properties of Mn-Zn ferrite by sol-gel combustion technique," Journal of Magnetism and Magnetic Materials, vol. 349, pp. 116–120, 2014.
18. B. S. Zlatkov, N. S. Mitrović, M. V. Nikolić et al., "Properties of MnZn ferrites prepared by powder injection molding technology," Materials Science and Engineering B, vol. 175, no. 3, pp. 217–222, 2010.
19. G. Kogias, V. Tsakaloudi, P. van der Valk, and V. Zaspalis, "Improvement of the properties of MnZn ferrite power cores through improvements on the microstructure of the compacts," Journal of Magnetism and Magnetic Materials, vol. 324, no. 2, pp. 235–251, 2012.
20. L. Yang, G. Xi, and J. Liu, "MnZn ferrite synthesized by sol-gel auto-combustion and microwave digestion routes using spent alkaline batteries," Ceramics International, vol. 41, no. 3, pp. 3555–3560, 2015.
21. S. Kumar, T. J. Shinde, and P. N. Vasambekar, "Microwave synthesis and characterization of nanocrystalline Mn-Zn ferrites," Advanced Materials Letters, vol. 4, no. 5, pp. 373–377, 2013.
22. L. Zhenyu, X. Guangliang, and Z. Yalin, "Microwave assisted low temperature synthesis of MnZn ferrite nanoparticles," Nanoscale Research Letters, vol. 2, no. 1, pp. 40–43, 2007.
23. Hoganas datasheet, https://www.hoganas.com/.
24. M. A. Gabal, R. S. Al-luhaibi, and Y. M. Al Angari, "Recycling spent zinc-carbon batteries through synthesizing nano-crystalline Mn–Zn ferrites," Powder Technology, vol. 258, pp. 32–37, 2014.
25. M. A. Gabal and Y. M. Al Angari, "Low-temperature synthesis of nanocrystalline NiCuZn ferrite and the effect of Cr substitution on its electrical properties," Journal of Magnetism and Magnetic Materials, vol. 322, no. 20, pp. 3159–3165, 2010.
26. H. P. Klug and L. E. Alexander, X-ray Diffraction Procedures for Polycrystalline and Amorphous Materials, Wiley, New York, NY, USA, 1997.
27. S. Kumar, S. N. Dolia, and R. K. Singhal, "Application of Rietveld method to the structural characteristics of some bulk and nanocrystalline materials," AIP Conference Proceedings, vol. 1391, pp. 65–67, 2011.
28. T. Tanaka, "Young's and shear moduli, hardness and bending strength of polycrystalline Mn-Zn ferrites,"Japanese Journal of Applied Physics, vol. 14, no. 12, pp. 1897–1901, 1975.
29. M.-R. Syue, F.-J. Wei, C.-S. Chou, and C.-M. Fu, "Magnetic, dielectric, and complex impedance properties of nanocrystalline Mn–Zn ferrites prepared by novel combustion method," Thin Solid Films, vol. 519, no. 23, pp. 8303–8306, 2011.
30. http://www.pramiz.com/softm.html.
31. M. Strečková, J. Füzer, L. Kobera et al., "A comprehensive study of soft magnetic materials based on FeSi spheres and polymeric resin modified by silica nanorods," Materials Chemistry and Physics, vol. 147, no. 3, pp. 649–660, 2014.

CHAPTER 16

EXPERIMENTAL INVESTIGATION OF DAMPING FLEXURAL VIBRATIONS IN GLASS FIBRE COMPOSITE PLATES CONTAINING ONE- AND TWO-DIMENSIONAL ACOUSTIC BLACK HOLES

E.P. Bowyer and V.V. Krylov

Keywords: Vibration damping; Composite plates; Composite panels; Acoustic black hole effect; Wedges of power-law profile; Circular indentations

INTRODUCTION

Composite materials and structures are found in an increasing variety of applications in aeronautical, automotive, and marine industries, where they replace many parts and components traditionally made of metals and metallic alloys. The main advantages of composite structures over their metallic counterparts are their higher stiffness-to-weight ratio and better resistance to corrosion. Another useful feature of composite materials is higher values of their intrinsic energy loss factors, which results in lower levels of undesirable structural vibrations under the same operational conditions. Nevertheless, structural vibrations in composite materials and structures do occur and are a key factor in delamination and crack propagation, therefore their reduction remains an important engineering problem that continues to be investigated (see e.g. [1], [2] and [3]).

Citation: E.P. Bowyer, V.V. Krylov, Experimental investigation of damping flexural vibrations in glass fibre composite plates containing one- and two-dimensional acoustic black holes, Composite Structures, Volume 107, January 2014, Pages 406-415, ISSN 0263-8223, http://dx.doi.org/10.1016/j.compstruct.2013.08.011.

Usually, passive damping of structural vibrations is achieved by adding layers of highly absorbing materials to the structure in order to increase elastic energy dissipation of propagating (mostly flexural) waves [4], [5] and [6]. The main disadvantage of this approach is the necessity to attach rather thick layers of absorbing materials to structural surfaces in order to achieve desirable levels of damping.

An alternative approach, which is applicable to damping of resonant structural vibrations, is based on reduction of reflections of structural waves from free edges of structures, rather than on increasing energy losses in the process of wave propagation (see e.g. [4]). To implement this approach in a more efficient way, a new method of damping flexural vibrations based on the so-called 'acoustic black hole effect' has been recently developed and investigated [7], [8] and [9]. This method has been initially applied to one-dimensional metal plates of power-law profile (wedges) that had to be covered by narrow strips of absorbing layers near sharp edges [9]. Ideally, if the power-law exponent is equal or larger than two, the flexural wave never reaches the sharp edge and therefore never reflects back [7], [8], [9] and [10]. Thus, such a sharp edge of power-law profile materialises the 'acoustic black hole' (ABH) for flexural waves, with no reflection and with total energy absorption in the attached piece of absorbing layer. It has been established theoretically [7] and [8] and confirmed experimentally [9] that this method of damping structural vibrations is much more efficient than traditional damping methods.

In addition to the above-mentioned wedges of power-law profile, which materialise one-dimensional (1D) acoustic black holes, a serious attention has been paid also to circular indentations of power-law profile made inside plate-like structures [11], [12], [13], [14],[15] and [16]. Such circular indentations materialise two-dimensional (2D) acoustic black holes. Like 1D black holes, such 2D black holes have also proven to be efficient means of damping structural vibrations. The advantage of two-dimensional acoustic black holes over their one-dimensional equivalents is that such holes, as well as their combinations (arrays), can be situated inside plate-like structures, which eliminates sharp edges from constructions and thus makes them safer, while reducing the risk of damage to the indentation tips.

The main practical advantage of the method of vibration damping using the acoustic black hole effect is that it requires very small amounts of absorbing materials to be attaches at the sharp edges, which is especially important for light-weight structures. In the same time, the acoustic black hole effect is robust enough to remain effective even for structures with geometrical and material imperfections resulting from manufacturing [17].

The aim of the present paper is to investigate vibration damping in glass fibre composite plates and panels using 1D and 2D acoustic black holes. Note in this connection that a composite panel with smooth outer edges is one of the most commonly found composite structures. The construction of this type of a composite panel that can incorporate the damping abilities of an acoustic black hole forms the ultimate aim of this paper. Five types of composite structures containing acoustic black holes and their effect on vibration damping have been investigated. These structures are: a composite strip with a wedge of power-law profile (1D acoustic black hole), a composite plate with a circular indentation of power-law profile (2D acoustic black hole), combinations of composite plates containing circular indentations of power-law profile, smooth surface composite panel configurations with enclosed circular indentations of power-law profile, and a honeycomb structure with added plates containing circular indentations of power-law profile.

MANUFACTURING OF EXPERIMENTAL SAMPLES

Fifteen glass fibre composite samples were manufactured for this investigation; three strips of dimensions 250 × 50 mm and a thickness of 6 mm; the additional wedge being 50 mm long and of power-law profile with $m = 2.2$. A wedge of power-law profile with$m = 2.2$ was also produced in order to be attached to a steel strip, dimensions being the same as for the composite strip. The eleven glass fibre composite plates were of dimensions 310 × 185 mm and consisted of two 3 mm thick plates and nine 6 mm thick plates. The circular indentations of power-law profile with $m = 4$ had a diameter of 110 mm with a central hole of 10 mm, leaving a profile length of 50 mm.

The glass fibre composite used for these samples was SE84LV- Low Temperature Cure Epoxy Prepreg System. This composite has a high compressive strength and it is widely used in large heavily loaded components, such as yacht hulls, spars, aviation panels, and also in non-structural applications. SE 84LV is also widely used in sandwich structures with honeycombs. Each sheet had a thickness of 0.2 mm. The composite plates and strips were layed-up to the required thickness and then cured using the vacuum bagging process.

Fourteen of the profiles were created in the traditional way: A CNC (Computer Numerically Controlled) milling machine operating at a cutter speed of 1200 rpm with a carbide cutter was used to produce the wedges and circular indentations. The main problem encountered when utilizing this method of manufacturing was that it is not possible to construct directly a panel with internal cavities while utilizing a 'vacuum only processing' method. The outer layer needs to be cured separately and attached using epoxy resin. This can lead to a more lengthy manufacturing time.

These fourteen samples consisted of a reference strip and strip with an additional wedge (Fig. 1(a)); examples of the other three types of plates can be seen in Fig. 1(b–d). Fig. 2displays the cross-sectional view of the plate samples when viewed from the narrow end. The average profile tip thickness for each of the samples is given in Table 1.

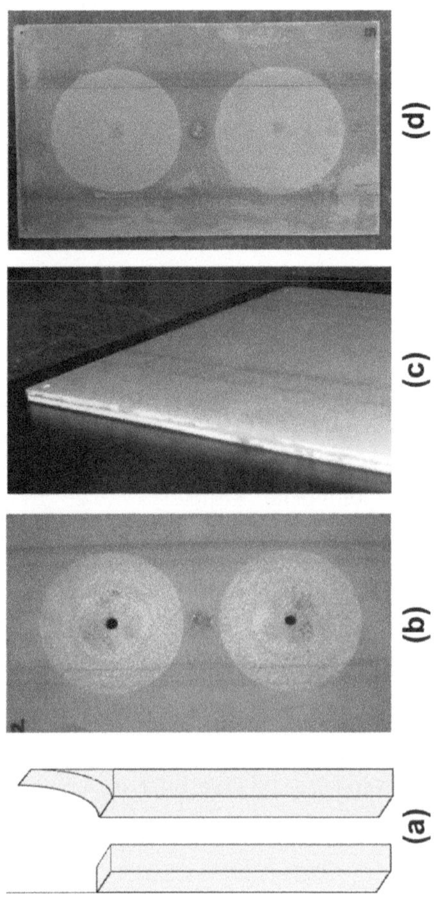

Figure 1. (a) Strip with and without a wedge, (b) plate (6 mm) containing two 2D acoustic black holes, (c) combined composite plate (6 mm), and (d) composite panel (6 mm).

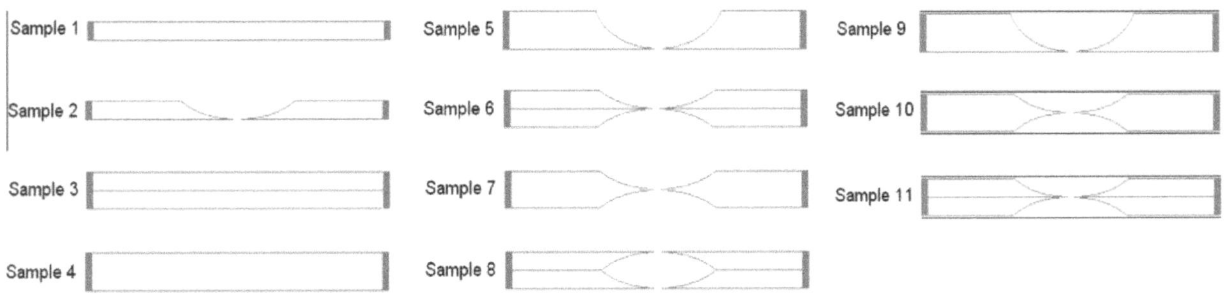

Figure 2. Cross-section views of sample plates 1–11.

Table 1. Average profile tip thickness for each sample (in mm).

Sample	5	6	7	8	9	10	11
Tip thickness (mm)	0.10	0.11	0.10	0.10	0.10	0.10	0.11

Two types of glass fibre composite honeycomb sandwich panels were created for this investigation, see Fig. 3: a reference composite honeycomb sandwich panel and a composite honeycomb sandwich panel containing two acoustic black holes in each of the composite plates.

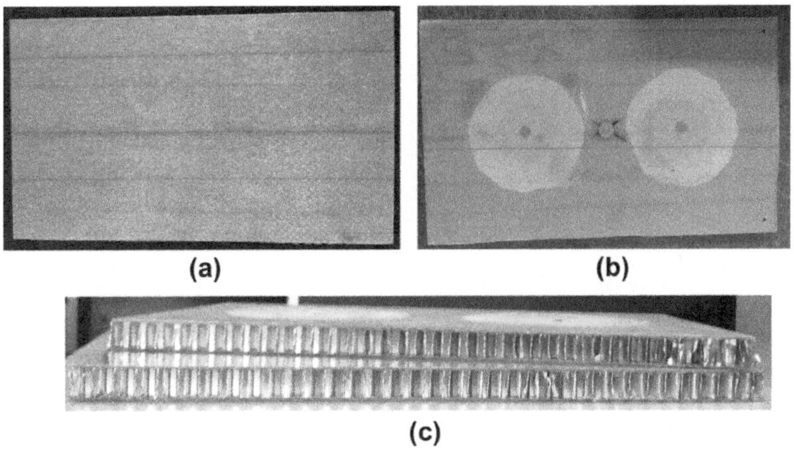

Figure 3. (a) Composite honeycomb sandwich panel – reference plate, (b) composite honeycomb sandwich panel with two acoustic black holes, and (c) side view of the panels.

There are several methods that could have been employed in the manufacture of these honeycomb sandwich structures: For this investigation, a method that was specific to composites and that would allow the integration of acoustic black holes into the structure was required. The standard method has four stages, the first being the curing of the outer composite plates. This was done using the same vacuum curing technique described above. The composite plates, aluminium honeycomb and the adhesive sheets are then cut to size. Each composite plate is then cured in turn onto the aluminium honeycomb,Fig. 4(a).

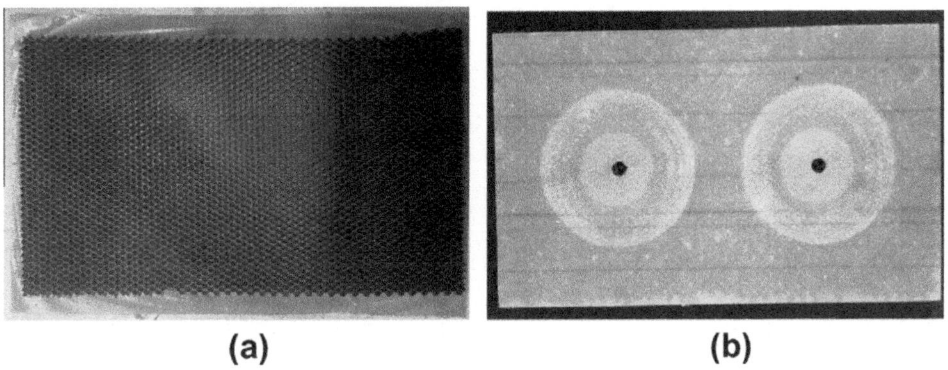

Figure 4. (a) Composite plate cured onto the aluminium honeycomb and (b) honeycomb sandwich panel with two exposed indentations.

In addition to the method used above, two further stages were added for the construction of the sandwich panel incorporating the acoustic black holes (ABHs). The first being the manufacture of the ABHs, a CNC (Computer Numerically Controlled) milling machine operating at a cutter speed of 1200 rpm with a carbide cutter was used to produce the circular indentations in the composite. This stage was introduced prior to the bonding of the composite sheets to the honeycomb, resulting in a honeycomb sandwich panel with exposed indentations; Fig. 4(b). The second being the fixing of the single sheet of composite to the outer surfaces of the panel to enclose the ABHs, leaving a smooth surfaced panel. This stage was added to end of the manufacturing procedure. A cross-section view displaying the composition of the sandwich panel as seen from the narrow end of the samples can be seen in Fig. 5.

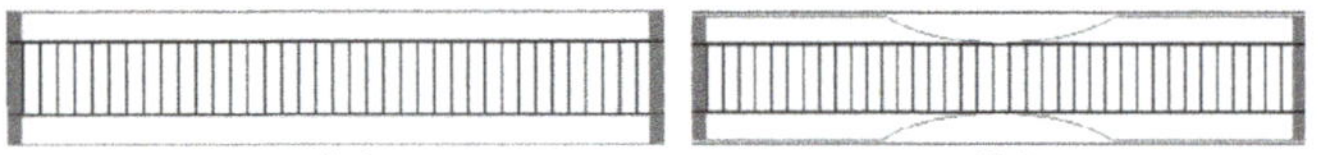

Figure 5. Cross-section view of (a) composite honeycomb sandwich panel – reference plate and (b) composite honeycomb sandwich panel with acoustic black holes (not to scale).

The dimensions of the final samples were 310 × 185 mm and a thickness of 20 mm; the circular indentations of power-law profile with $m = 4$ had a diameter of 110 mm with a central hole of 10 mm, leaving a profile length of 50 mm. The composite plates had a thickness of 1.2 mm each. The 5052 aluminium honeycomb centre had a thickness of 12.7 mm, a cell size of 3 1/16 in, a foil thickness of 15 μm and a mass of 70 kg per m^2. A single profile was chosen for this investigation rather than a double profile. There was only a small difference in the damping performance of the two profiles, so the above profile was selected to ensure accurate results.

EXPERIMENTAL SET UP

The experimental set-up has been designed to allow nearly free vibration of the sample plates (i.e. to eliminate clamping of edges), take the weight off the plate edges and introduce minimal damping to the system, see Fig. 6(a). The excitation force was applied centrally on the plate via an electromagnetic shaker attached to the plate using glue and fed via a broadband signal amplifier. The response was recorded by an accelerometer (B&K Type 4371) that was attached to the one surface, directly in line with the force transducer (B&K Type 8200), also attached using glue, Fig. 6(b). The acquisition of the point accelerance was utilised using a Bruel & Kjaer 2035 analyser and amplifier. A frequency range of 0–6 kHz was investigated; above this range no discernable response could be detected.

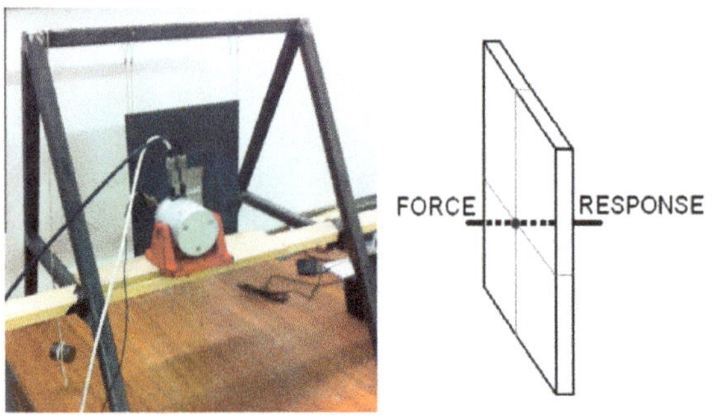

Figure 6. (a) Experimental set up and (b) locations of the shaker (force) and of the accelerometer (response) on an experimental sample.

RESULTS AND DISCUSSION

Introduction of a Wedge of Power-law Profile to a Composite Strip

In the first instance it seemed prudent to first ascertain whether the introduction of a wedge of power-law profile to a composite strip could produce the acoustic black hole effect as seen in previously tested steel samples [9], [13] and [17]. In this section, two types of sample were tested: a reference strip and a strip with a machined wedge of power-law profile (1D acoustic black hole). It was found during initial testing that a composite sample, unlike the steel samples, required no additional damping layer to be attached to the wedge tip to produce the acoustic black hole effect. This is primarily due to the increased loss factor of the composite material itself (~0.1 to 0.2). For this reason, all of the following results show the samples without any additional damping layers.

A comparison of the frequency responses for a strip with and without a power-law profiled wedge is shown in Fig. 7. There is no difference between the two samples below 250 Hz. After this point, an increase in the reduction of the resonant peaks is seen up until a maximum reduction of 3.5 dB from the reference sample, which is seen at 2.2 kHz. At 2.6 kHz the two peaks match and no reductions are seen. Beyond this point the sample with the wedge has damped all remaining peaks.

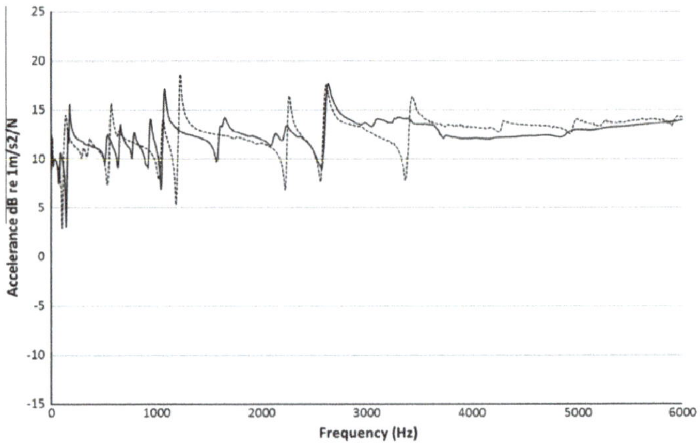

Figure 7. Measured accelerance for a strip with a wedge of power-law profile (solid line) compared to the case of a reference strip (dashed line).

Tapered Circular Indentation of Power-law Profile in Glass Fibre Composites

The next step, as with the previously tested steel samples [13], was to introduce circular indentations of power-law profile (2D ABHs) into glass fibre composite plates. This section describes the effect of the addition of two 2D ABH's into both a 2.5 mm (Sample 2) and 5 mm thick plate (Sample 5) when compared to a respective thickness reference plate (Samples 1 and 4). This section also looks at the effect of a double profiled indentation of the same power-law profile (Sample 7) compared to the single profiled 5 mm thick sample (Sample 5) and the reference plate. The cross-sections of these profiles can be seen in Fig. 8.

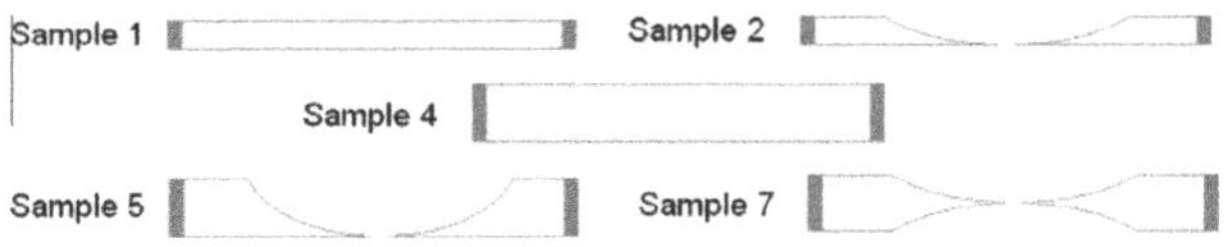

Figure 8. 2.5 mm reference plate (Sample 1), 5 mm reference plate (Sample 4), cross-sections of samples containing tapered circular indentations (Samples 2, 5 and 7).

Fig. 9 shows the results for the 2.5 mm samples: Sample 2 when compared to a plain reference plate, Sample 1. One can see that the effect of adding two circular indentations of power-law profile is immediately obvious, with considerable damping of resonant peaks easily observed. Below 500 Hz little to no damping is seen. An increase in the reduction of the peak responses of the reference plate is seen in the profiled sample until a maximum reduction from the reference plate of 7.5 dB can be observed at 1.2 kHz. After 2.7 kHz the response is smoothed, with all resonant peaks seen in the reference sample heavily damped if not completely removed.

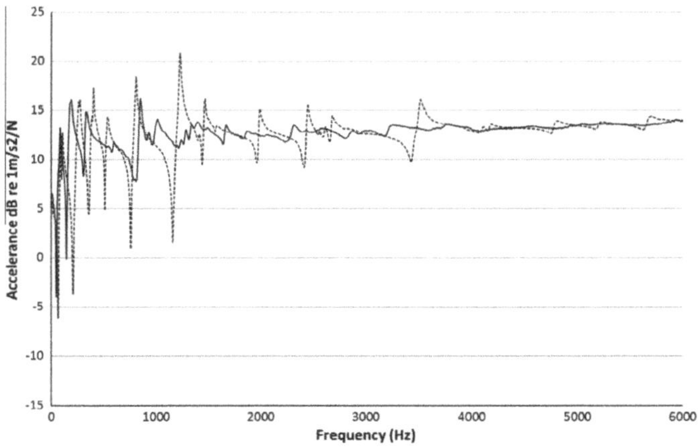

Figure 9. Measured accelerance for Sample 1 – 2.5 mm reference plate (dashed line) compared to that for Sample 2 – 2.5 mm plate with two 2D acoustic black holes (solid line).

The same test as above was repeated for 5 mm thick plate with the addition of a third sample for comparison, Sample 7, a double profiled sample. From Fig. 10 it can be seen that, when the double profiled sample (Sample 7) is compared to a plain reference plate (Sample 4) and a single profile plate (Sample 5), the damping values achieved for Samples 5 and 7 are very similar. Sample 7 performs slightly (1–1.5 dB) more effectively than Sample 5 below 1 kHz. Above this frequency this trend is reversed, with Sample 5 behaving slightly (again 1–1.5 dB) more effectively than Sample 7. As expected, below 450 Hz there is little to no damping. A maximum reduction from the reference plate by both Samples 5 and 7 of 8.5 dB can be seen at 2.4 kHz. After 2.7 kHz both Samples 5 and 7 show a smoothing of the reference plate resonant peaks. Samples 5 and 7 can therefore be classified as interchangeable. Whether a sample has a double profile, as with Sample 7, or a single profile, as with Sample 5, the damping performance obtained is the same.

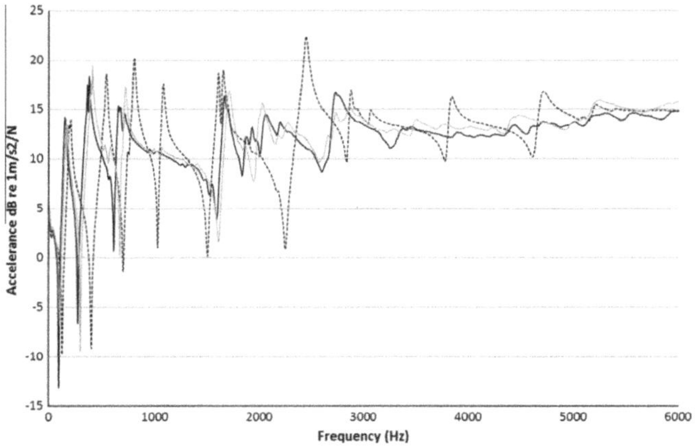

Figure 10. Measured accelerance for Sample 7 (solid black line) compared to those for Sample 5 – 5 mm plate with two 2D acoustic black holes (solid grey line) and Sample 4 – 5 mm reference plate (dashed line).

Combined Plates Containing Circular Indentations of Power-law Profile

In this section, different ways of combining 2.5 mm plates were considered in order to maximise the damping performance of the composite plates. The initial motivation for this combining of plates (Sample 6) was due to two main drivers: to produce a profile combination that could easily be converted into a panel with a continuous outer surface and also to increase the ease of manufacture of the double profiled indentations. The cross-sections of these profiles can be seen in Fig. 11. The plates were combined using a long cure epoxy resin.

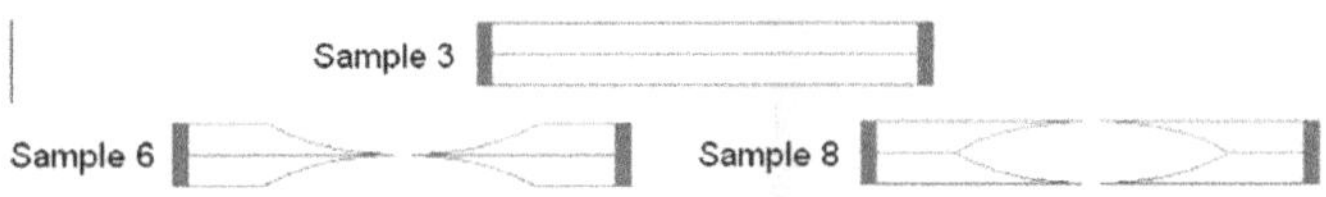

Figure 11. Cross-sections of combined composite samples, Sample 3 – reference plate, Samples 6 and 8.

Sample 7 from Section 4.2 was also used for comparison in this section, as it could show a direct comparison between a sample manufactured from two separate composite sheets and then combined to a sample manufactured from a single sheet of composite. Two different styles of reference plate were considered in order to gauge the effect of combining the plates and to determine which should be used for comparison. A plain plate (Sample 4) and a combined plate (Sample 3), see Fig. 10, were considered as both had merit as a reference. Both reference plates were of the same dimensions as the profiled plates, the combined plate consisted of two 2.5 mm plates combined with epoxy resin.

The results showed that the effect of the combination of the two plates and the addition of the epoxy resin had no quantifiable effect on the response of the plates by either increasing, decreasing or shifting the resonant peaks of the plain reference plate. Both samples therefore had merit as a reference plate. However, as the main comparison considered in this paper was the effect of these changes on existing complete panels, Sample 4, the plain reference was chosen to be used for comparison both in this and the following sections. A comparison of the two combined double profiles, Samples 6 and 8, and Sample 7 is shown in Fig. 12. Below 500 Hz little difference in the responses of the three samples is seen. In general, it can be said that the amplitudes of the resonant peaks for Sample 8 are generally lower than for the other two samples, and therefore it performs more efficiently at damping resonant peaks. Sample 8 shows a maximum reduction from Sample 6 of 3 dB at 640 Hz and the same reduction from both the other two samples at 1.65 and 2.7 kHz. Above 2.9 kHz the response of all three samples is smoothed with resonant peaks heavily damped if not completely removed.

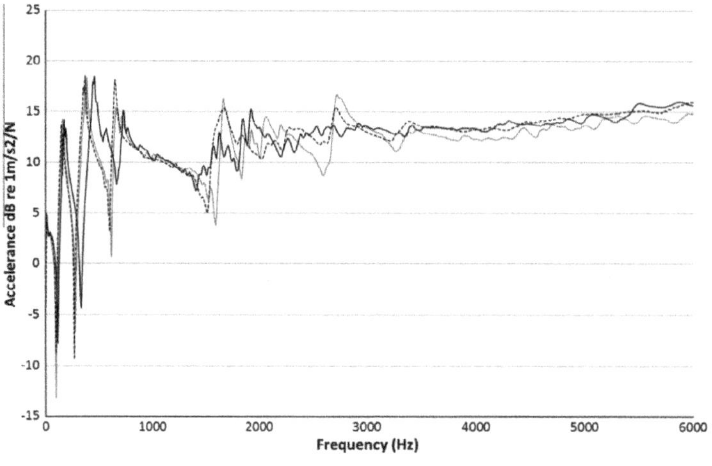

Figure 12. Measured accelerance for Sample 8 (solid black line) compared to Sample 7 (solid grey line) and to Sample 6 (dashed line).

Fig. 13 shows the results for Sample 8 when compared to a plain reference plate. Below 450 Hz little to no damping is seen. The reduction in the amplitude again increases until a maximum reduction from the reference plate of 10 dB can be seen at 2.4 kHz. After this point the response is smoothed with resonant peaks heavily damped. It should be noted that the resonant peak seen at 1 kHz in the reference sample has been completely damped in Sample 8.

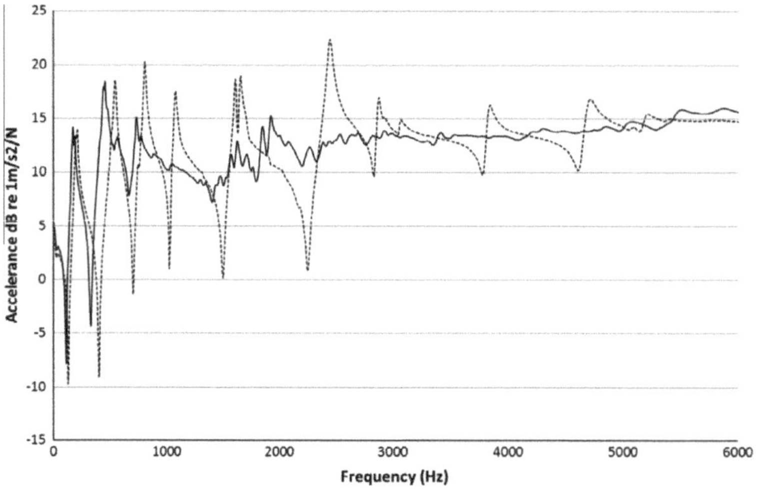

Figure 13. Measured accelerance for Sample 8 (solid line) compared to Sample 4 – reference plate (dashed line).

Comparison of Composite Panel Constructions

The present section deals with encasing Samples 5–7 with a single sheet of pre-preg glass fibre composite in order to create a panel with a continuous outer surface, as shown in Fig. 14. Despite the good damping performance of Sample 8, the two 10 mm diameter holes on the either side of the outer surface of the panel meant that this configuration would be impractical to convert to a smooth surfaced panel without interfering with the circular indentations at the tip. Therefore, in this section the responses of the three enclosed panels will be investigated when compared to each other, to a reference plate, and finally when compared to the equivalent sample without the additional casing. This final comparison will show if there are any adverse effects to encasing the sample and whether such an effect, if present, can be off set against the advances gained by encasing the panels.

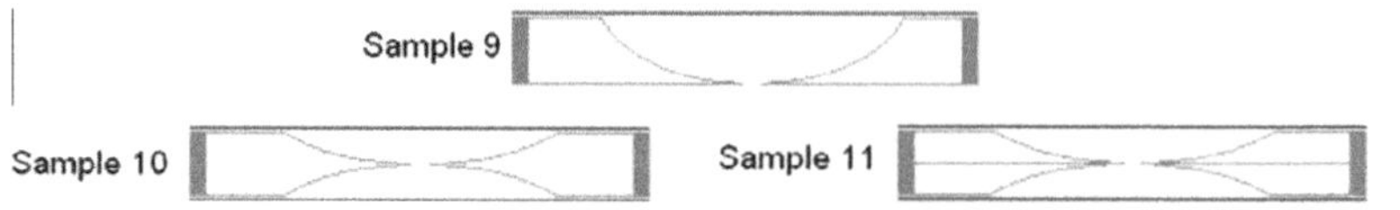

Figure 14. Cross-sections of Samples 9–11.

From Fig. 15, it can be seen that, when the composite panels, Samples 9–11, are compared, there is very little difference between the three samples. For Samples 9 and 10, there is no quantifiable difference between the resonant peaks. However, as Sample 9 has two 10 mm diameter holes on one side of the plate, it does not have two smooth surfaces, therefore Sample 10 would best fit the specification. Sample 11 shows a reduction of 0.5–1 dB at 460 and 700 Hz and 2 kHz from the other two samples. Below 460 Hz there is no difference in the response of the three samples, and above 3 kHz the response is smoothed with resonant peaks heavily damped if not completely removed, thus producing a similar response.

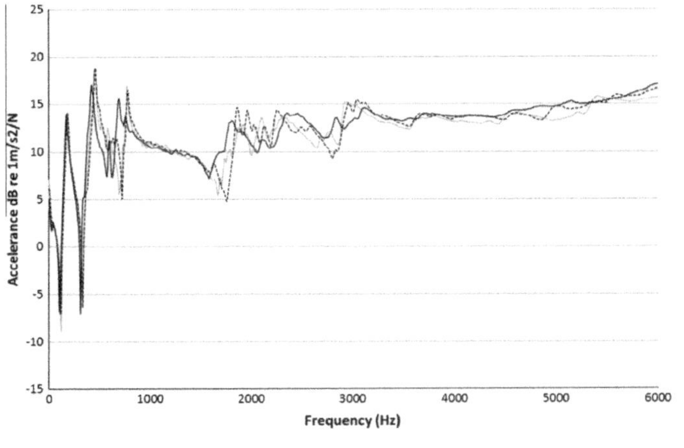

Figure 15. Measured accelerance for Sample 11 (solid black line) compared to Sample 9 (solid grey line) and to Sample 10 (dashed line).

A comparison of most effective damping panel; Sample 11, when compared to a reference plate, is shown in Fig. 16. A peak shift to the left from reference is observed, however, in the case of the combined composite plates, this effect occurs at a much lower frequency than previously observed, with the peak shift occurring as low as at 500 Hz. There is no difference between the two samples below 450 Hz. After this frequency, the dB reduction of the peak amplitudes increases until a maximum reduction of 10 dB from the reference plate is achieved at 2.4 kHz. It can also be seen that the reference plate resonant peak at 1 kHz has been damped completely in Sample 11. Thus, the combination of the composite plates and sheets results in an effective method of damping flexural waves in smooth surfaced composite panels.

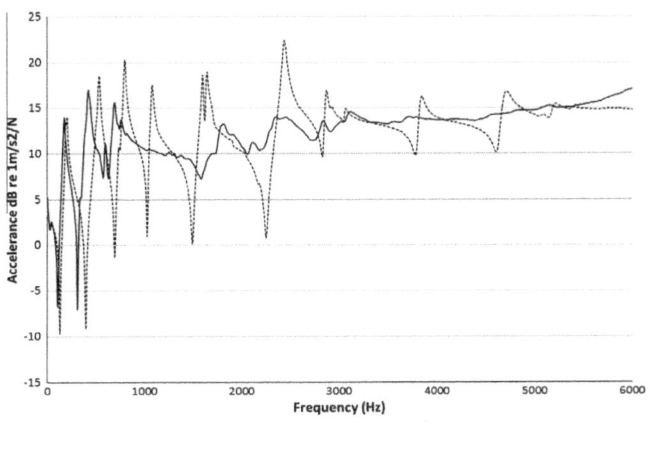

Figure 16. Measured accelerance for Sample 11 (solid line) compared to Sample 4 – reference plate (dashed line).

Finally, from Fig. 17 it can be seen that, when Sample 11 is compared to its exposed indentation equivalent; Sample 6, the damping achieved for Sample 11 is in fact greater (1–2 dB) or equal to that of Sample 6. Below 450 Hz there is little to no difference in the samples. A maximum reduction of 3 dB can be seen at 650 Hz and 1.6 kHz. The most likely explanation for this increase is the thin layer of epoxy resin between the sample and the composite sheets. It is enclosed within, providing a slight increase in the loss factor of the sample. This result shows that circular indentations of power-law profile can be successfully enclosed in a smooth surface panel with a positive effect on the damping performance of the plate. There also appears to be little to no significant 'drum skin effect' over the area of the enclosed indentations. This result was replicated in the other samples tested.

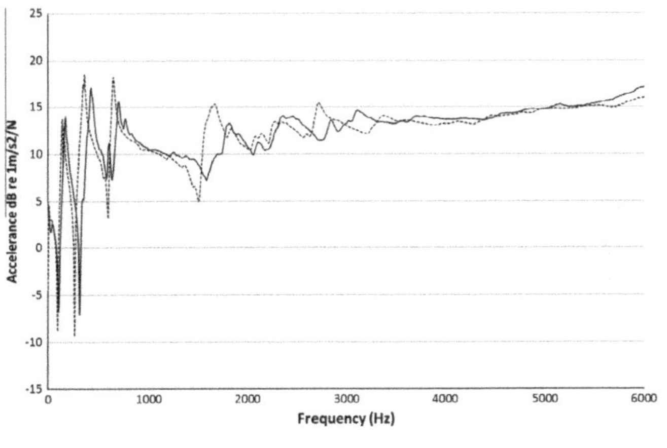

Figure 17. Measured accelerance for Sample 11 (solid line) compared to Sample 6 (dashed line).

Introduction of Indentations of Power-law Profile to a Honeycomb Sandwich Panel

This section describes the investigation of a composite honeycomb structure incorporating the smooth surfaced composite panels containing circular indentations of power-law profile designed above.

Fig. 18 shows the measured accelerance for a composite honeycomb sandwich reference panel compared to a composite honeycomb sandwich panel with enclosed indentations of power-law profile. As expected, there is a peak shift to the left as a result of reduced mass and stiffness. Above 1 kHz, the resonance peaks of the acoustic black hole plate show increasing reductions in amplitude compared to the reference sample. Above 2.4 kHz, the response is smoothed, with all resonant peaks seen in the reference sample heavily damped if not completely removed. A maximum reduction of 6 dB is seen at 2.5 kHz and at 3.4 kHz. One of the configurations discussed in this paper was a composite honeycomb sandwich panel with indentations on both sides of the panel. There may be applications though where this is not suitable, for example on boat hulls. In cases such as these indentations can be placed on only one side of the panel, see Fig. 19. As a result, a reduction in damping performance will occur, but a substantial amount of damping can still be achieved while increasing the strength of the outer surface of the panel.

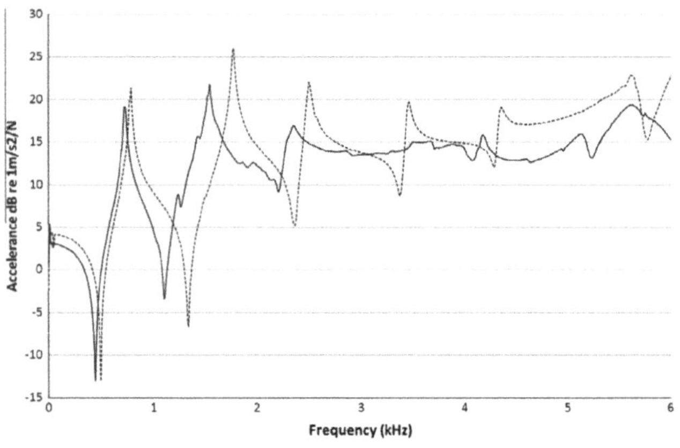

Figure 18. Measured accelerance for a composite honeycomb sandwich reference panel (dashed line) compared to a composite honeycomb sandwich panel with enclosed indentations of power-law profile (solid line).

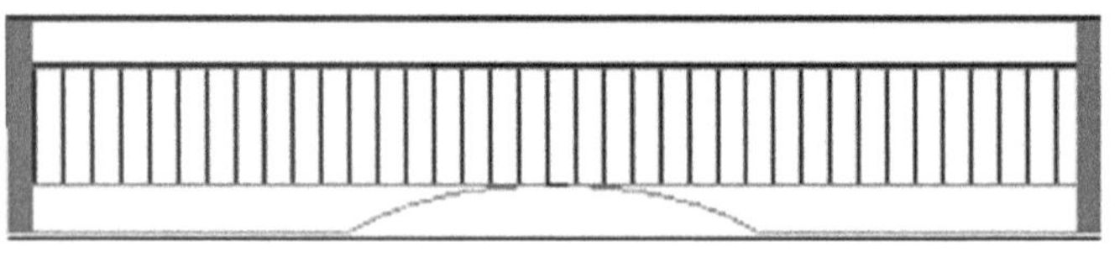

Figure 19. Composite honeycomb sandwich panel with one 'normal side' and one containing enclosed indentations of power-law profile.

Theoretically the double profile, Fig. 20, would be the most effective profile to use, not only due to the damping performance, but also because of a greater contact area to be attached to the honeycomb to reduce the implications of the indentation on the structural performance of the panel.

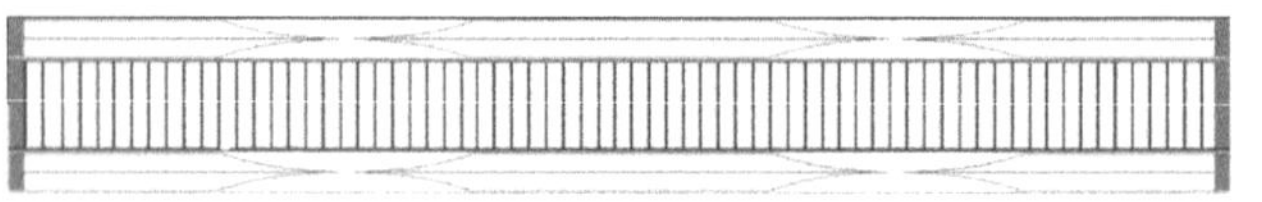

Figure 20. Composite honeycomb sandwich panel with a different configuration of enclosed indentations of power-law profile.

It is important to note that further testing is required into the structural effects of the indentations on the composite and honeycomb panels. One would expect that making power-law profiled indentations in panels will reduce their structural strength in the same way as the introduction of non-profiled holes does as a result of increased mechanical stresses due to the reduction of the effective cross-sections of the panels, e.g. in the case of tensile strength of composite plates with through holes [18]. Generally, the requirement to introduce power-law profiled indentations into structures to be damped may compromise their structural strength. This represents the main disadvantage of the method of damping structural vibrations based on the acoustic black hole effect. Therefore, this method is limited to some specific structures and applications where structural strength is either not so important or can be improved.

CONCLUSIONS

Glass fibre composite strips containing wedges of power-law profile and plates containing circular indentations of power-law profile, materialising 1D and 2D acoustic black holes (ABHs) respectively, are effective configurations to be used for damping flexural vibrations. A glass fibre composite strip behaves in much the same way as steel when a wedge of power-law profile is added to the end of the strip. A 1D ABH in a strip produces a maximum reduction of 3.5 dB at 2.2 kHz.

A glass fibre wedge can be successfully attached to a steel strip in place of a steel wedge to produce increased reduction of resonant peaks. In theory this method can be also used to create 2D acoustic black hole inserts that can be integrated into existing structures with existing holes/slots.

There is little to no difference in the amplitudes of the responses when compared a machined wedge to a cured wedge. There is however, a difference in the frequency at which the resonances occur. This observation could be useful when a shift and reduction in resonance is required. The 2D indentation can also be easily manufactured via the curing method of manufacture.

As in the case of wedges of power-law profile, 2D ABHs perform in composite plates in much the same way as in steel plates, with a maximum reduction of 7.5 dB at 1.2 kHz in a 3 mm thick plate and with a maximum reduction of 8.5 dB at 2.4 kHz in a 6 mm plate, when compared to a reference sample.

The composite plates can be combined to attain more effective damping combination and to achieve a more time/cost and skill effective method of production. A combined plate provides a similar if not better reduction of the resonant peaks as a sample machined out of a single plate of composite. A maximum reduction of 10 dB at 2.4 kHz was achieved by Sample 8 when compared to a reference sample.

When a composite plate is enclosed by a composite sheet to form a smooth surfaced panel, there is relatively little reduction in damping performance when compared to an unenclosed plate. A maximum reduction of 10 dB at 2.4 kHz was achieved by Sample 11 when compared to a reference sample. Little to no drum skin effect is seen over the enclosed ABHs. Therefore, an enclosed smooth surfaced composite panel can be manufactured to give the same level of damping of flexural waves that can be achieved by a plate with exposed indentations.

It can be shown that curing is an acceptable method of manufacture to produce a wedge of power-law profile and to materialise the acoustic black hole effect. This method can be used to mass produce panels incorporating acoustic black holes with relative ease.

A smooth surfaced composite panel with ABHs is an effective method of damping flexural vibrations in composite panels. These panels can be substituted into applications where a non-structural composite panel is in use and a flexural vibration problem is present as an effective damping method. Such applications would include aircraft internal panels, internal panels in boats and formula one vehicles.

As in the case of composite panels, a composite honeycomb panel with enclosed indentations of power-law profile is an effective configuration for damping flexural vibrations within the structure. A maximum reduction of 6 dB is seen at 2.5 and 3.4 kHz, with heavy damping or elimination of resonant peaks above 2.4 kHz. The circular indentations, as it was seen before, provide substantial reductions in resonant peaks over a large frequency range.

Enclosing the circular indentations of power-law profile within the composite panel results in a surface texture similar to that of an un-machined conventional honeycomb sandwich panel, it also increases the damping performance of the panels. Honeycomb sandwich panels can be made with either both sides or a single side machined, depending on the application and damping requirements.

ACKNOWLEDGEMENT

The research reported here has been partly supported by EPSRC Grant EP/F009232/1.

REFERENCES

1. Alam N, Asnani NT. Vibration and damping analysis of fibre reinforced composite material plates. J Compos Mater 1986;20:2–18.
2. Chow ST, Liew KM, Lam KY. Transverse vibration of symmetrically laminated rectangular composite plates. Compos Struct 1992;20:213–26.
3. Kramer MR, Liu Z, Young YL. Free vibration of cantilevered composite plates in air and in water. Compos Struct 2013;95:254–63.
4. Heckl M, Cremer L, Ungar E. Structure borne sound. 2nd ed. Springer-Verlag; 1988.
5. Mead DJ. Passive vibration control. Chichester: Wiley; 1998.
6. Ross D, Kerwin E, Ungar E. Damping of plate flexural vibrations by means of viscoelastic laminae. In: Ruzicka JE, editor. Structural damping, Pergamon Press, Oxford. 1960. p. 49-87.
7. Krylov VV, Tilman FJBS. Acoustic black holes for flexural waves as effective vibration dampers. J Sound Vib 2004;274:605–19.
8. Krylov VV. New type of vibration dampers utilising the effect of acoustic 'black holes'. Acta Acust United Acust 2004;90:830–7.
9. Krylov VV, Winward RETB. Experimental investigation of the acoustic black hole effect for flexural waves in tapered plates. J Sound Vib 2007;300:43–9.
10. Mironov MA. Propagation of a flexural wave in a plate whose thickness decreases smoothly to zero in a finite interval. Soviet Phys Acoust 1988;34: 318–9.
11. Krylov VV. Propagation of plate bending waves in the vicinity of one- and twodimensional acoustic 'black holes'. In: Proceedings of the ECCOMAS international conference on computational methods in structural dynamics and earthquake engineering (COMPDYN 2007), Rethymno, Crete, Greece, 13– 16 June, 2007 [CD-ROM].
12. Georgiev V, Cuenca J, Molerón-Bermúdez MA, Gautier F, Simon L, Krylov VV. Numerical and experimental investigation of the acoustic black hole effect for vibration damping in beams and elliptical plates. In: Proceedings of the European conference on noise control "Euronoise 2009", Edinburgh, UK, 26–28 October, 2009 [CD-ROM].
13. Bowyer EP, O'Boy DJ, Krylov VV, Gautier F. Experimental investigation of damping flexural vibrations using two-dimensional acoustic 'black holes'. In: Sas P, Bergen B, editors, Proceedings of the international conference on noise and vibration engineering (ISMA 2010), Leuven, Belgium, 20–22 September, 2010. p. 1181–92.
14. Georgiev VB, Cuenca J, Gautier F, Simon L, Krylov VV. Damping of structural vibrations in beams and elliptical plates using the acoustic black hole effect. J Sound Vib 2011;330:2497–508.
15. O'Boy DJ, Krylov VV. Damping of flexural vibrations in circular plates with tapered central holes. J Sound Vib 2011;330:2220–36.
16. O'Boy DJ, Bowyer EP, Krylov VV. Point mobility of a cylindrical plate incorporating a tapered hole of power-law profile. J Acoust Soc Am 2011; 129:3475–82.
17. Bowyer EP, O'Boy DJ, Krylov VV, Horner JL. Effect of geometrical and material imperfections on damping flexural vibrations in plates with attached wedges of power law profile. Appl Acoust 2012;73:514–23.
18. Mallick PK. Effect of hole stress concentration and its mitigation on the tensile strength of sheet moulding compound (SMC-R50) composites. Composites 1988;19:283–7.

CHAPTER 17

MECHANICAL PROPERTIES OF REINFORCED POLYMER COMPOSITES

Stefano Del Rosso, Lorenzo Iannucci, and Paul T. Curtis

Keywords: Microbraids; Robotised filament winding; Polymer-matrix composites

INTRODUCTION

High performance polymeric fibres are extensively used to make personal protective textiles and as reinforcing phase in polymer composite materials. Thanks to their high tenacity and toughness, low elongation at break as well as the ability to dissipate shock waves over large areas in a short amount of time, they are very suitable for applications where impact resistance and energy absorption capabilities are of vital importance.

Braiding is the process of interlacing three or more threads in such a way that they cross one other and are laid together at a bias angle. In theory, any material, in the form of strips or filaments, can be braided to produce linear, flat, tubular or solid forms. Braids can be produced as 2D, in flat or tubular form, and as 3D structures. The former contains only two sets of strands through the thickness, and axial yarns in case of triaxial braids, whereas the latter have several strands through the thickness. Over the past decades, braided reinforced polymer composites (BRPC) have been increasingly used in high performance structures due to their outstanding properties such as damage and impact resistance, high delamination resistance, greater through-the-thickness reinforcement and lesser notch sensitivity with respect to unidirectional (UD) and woven reinforced composites. Moreover, the investment and labour costs can be minimised due to the inexpensive machinery, high production rate and level of automation which the braiding technique offers [1], [2] and [3].

Brunnschweiler [4] and [5] and subsequently Ko and Pastore [1] discussed in details the principles of braid manufacture and the use of braided fabrics as reinforcing phase within engineering structures. Recently, Carey and Ayranci [6] reviewed the published studies on 2D braided composites outlining advantages and disadvantages of this technique, different characterisation methods currently employed and applications of BRPC in the composite industry. Omeroglu [7] investigated the properties of dry 2D polypropylene (PP) braided ropes by varying the braid architecture, fibre linear density and take-up speed. Regular braids showed higher tenacity, modulus and yield strength with respect to diamond braids. The higher the take-up speed, the higher the aforementioned properties. Moreover, the Young's modulus and tenacity were noted to be higher for braids made of finer PP strands. Harte and Fleck [8] studied the tensile behaviour of glass–epoxy braided tubes having different braid angles. Although they noted a lower Young's modulus and tensile strength with increasing fibre bias angle, the strain to failure and the energy absorption increased for the same angles. Moreover, the failure mechanism of the tubular composites changed from brittle to ductile with increasing the fibre bias angle from 23° to 55°.

Usually, braided reinforced composites are produced by stacking many braided slit sleeves or flattened tubes in order to create layers of desired orientation and thickness. For instance, Kelkar et al. [9] investigated the tensile and fatigue properties of epoxy-reinforced laminates made from 2/2 carbon braid slit sleeves and flattened braided tubes. As the fibre bias angle increased, the ultimate tensile strength and Young's modulus of the composites decreased whilst the endurance increased with respect to increasing braid angle. Fouinneteau and Pickett [10] studied the properties of carbon and glass braided composites made from flattened braided tubes and thermoset epoxy resin. For the same braid angle, they noted a higher tensile strength and strain to failure for the carbon braided composites.

Citation: Stefano Del Rosso, Lorenzo Iannucci, Paul T. Curtis, Mechanical properties of reinforced polymer composites, Volume 125, July 2015, Pages 509-519, ISSN 0263-8223, http://dx.doi.org/10.1016/j.compstruct.2015.02.036.

However, premature failure occurred locally in the region close to the specimen tabbed area, regardless of the material. The tensile strength and strain to failure of the carbon braided composites were detrimentally affected by as much as 27.5% and 39.1%, respectively, when the specimens had cut edges. Falzon and Herszberg [11] found a reduction of 20% in the tensile strength of braided composite laminates with respect to UD ones. They attributed this reduction to fibre damage while braiding.

To the authors' best knowledge, there are very few studies in the open literature in which the mechanical behaviour of braids and microbraids made of high performance polymeric fibres has been experimentally assessed (for example, in [7], [12],[13] and [14]). Moreover, there are no existing studies of microbraids directly used as reinforcing phase in composite materials. Sakaguchi et al. [15] and Fujihara et al. [16]claim the manufacture of microbraid reinforced composites. They braided matrix filaments over high performance fibres. The manufactured braids were filament wound over a steel plate and then cured. However, after melting the braided filaments, a composite material reinforced by unidirectional fibres would appear. Moreover, a linear density of the used microbraids above 10,000 dtex (the microbraid's diameter and linear density was not stated in either paper) would not be truly applicable to a "micro" range of dimensions.

The main aim of this work is to investigate the potential use of 2D microbraids as the primary constituent in high performance textiles and as the reinforcing phase within polymer composite systems. In this contest, the present investigation is concerned with the mechanical characterisation of high performance polymeric yarns and 2D microbraids. A comprehensive series of mechanical tests is presented and results reported. Thus, a novel and unique method was developed and used to manufacture microbraid reinforced polymer composites (mBRPC) via robotised filament winding and hot-pressing. The manufactured mBRPC were tested in tension and results herein presented.

MATERIALS, MANUFACTURE AND TESTING METHODS

Materials

Two high performance yarns were investigated in this study: Dyneema®SK75 and Kevlar®49. Fibre diameters were determined by analysis of images from scanning electron microscope (SEM). For the manufacture of microbraid reinforced composites, Rayofix TP, a thermoplastic resin film, was used. Physical properties of the investigated materials are listed in Table 1. The number in brackets indicates the number of single fibres examined.

Table 1. Physical properties of the investigated materials.

Yarns	Density (g/cm^3)	Linear density (dtex)	Single fibre diameter (μm)	No. filaments/yarn
Dyneema®SK75	0.97	220	17.28 ± 0.58 (112)	100
Kevlar®49	1.44	215	12.14 ± 0.41 (108)	130
Matrix	Density (g/cm^3)	Areal density (g/m^2)	Thickness (μm)	
Rayofix TP	0.932	71.63	75	

Manufacture of Dry Microbraids

The manufacture of 2D microbraids was carried out using the Herzog RU2-16/80, a Maypole-type braiding machine having two working heads, 8 horn gears per head and equipped with 16 carriers in the "fully-occupied" setup. In order to determine the influence of the braiding architecture and the number of braided yarns on the mechanical properties of the microbraids, diamond 1/1 and regular 2/2 patterns were created by varying the number of working carriers and the carrier disposition on the braiding path, respectively. The different braid patterns are sketched in Fig. 1. For each braid architecture, microbraids having different braid angle α were manufactured by changing the cogwheel ratio on the braiding machine. The diameter of the microbraids and their bias angles were determined by analysis of SEM images (Fig. 2). The microbraids linear densities were determined according to the ASTM D1577-07 Standard Test Methods for Linear Density of Textile Fibers [17]. Specifications of the manufactured microbraids are presented in Table 2.

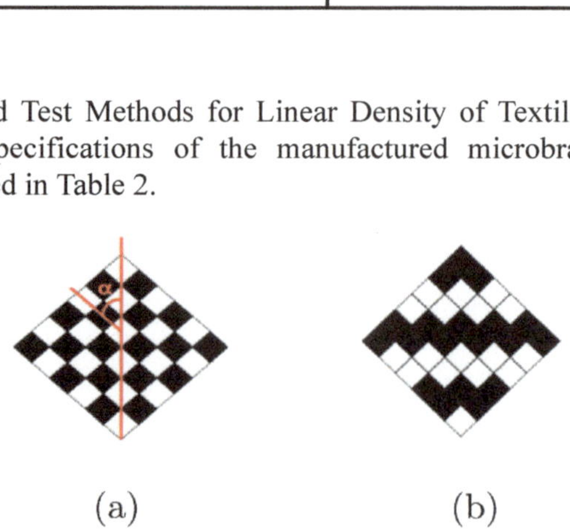

Figure 1. Braid patterns: (a) Diamond 1/1; (b) Regular 2/2.

(a) (b)

Figure 2. SEM images of two different microbraids: (a) bDA1, (b) bKA2.

Table 2. Physical properties of the manufactured microbraids.

bID	Material	Number of braided yarns	Braid pattern	Braid diameter (mm)	Braid angle (°)	Linear density (dtex)
bDA1	Dyneema®SK75	8	1/1	0.86 ± 0.02	15.0 ± 0.8	1816 ± 54
bDB1	Dyneema®SK75	8	1/1	0.85 ± 0.02	19.3 ± 1.3	1996 ± 48
bDC1	Dyneema®SK75	8	1/1	0.67 ± 0.01	28.7 ± 1.1	2238 ± 61
bDA2	Dyneema®SK75	16	2/2	1.34 ± 0.01	22.0 ± 0.6	3878 ± 59
bDB2	Dyneema®SK75	16	2/2	1.2 ± 0.01	31.9 ± 1.5	4419 ± 66
bDC2	Dyneema®SK75	16	2/2	0.97 ± 0.01	43.9 ± 0.7	5066 ± 48
bKA1	Kevlar®49	8	1/1	0.84 ± 0.02	13.1 ± 0.7	1890 ± 57
bKB1	Kevlar®49	8	1/1	0.84 ± 0.01	23.9 ± 0.5	1934 ± 64
bKC1	Kevlar®49	8	1/1	0.69 ± 0.02	39.1 ± 0.9	2026 ± 53
bKA2	Kevlar®49	16	2/2	1.25 ± 0.01	21.7 ± 1.6	3920 ± 62
bKB2	Kevlar®49	16	2/2	1.12 ± 0.01	28.6 ± 1.4	4117 ± 49
bKC2	Kevlar®49	16	2/2	0.98 ± 0.01	40.1 ± 1.1	4478 ± 69

A generic dry microbraid will belong to the class "bXYZ", where:

- b stands for dry microbraid.
- X will be the microbraid's material, in particular D for Dyneema®SK75 and K for Kevlar®49.
- Y will denote the braid angle, where A < B < C.
- Z will represent the braiding architecture, in particular "1" for diamond 1/1 and "2" for regular 2/2.

Manufacture of Microbraid Reinforced Polymer Composites (mBRPC)

The manufactured microbraids were wound in a unidirectional fashion over a spinning aluminium plate using a robotised filament winding system (Fig. 3(a)). The robot was programmed to move across the plate a distance equal to the diameter of the microbraid per each revolution of the plate. The tension of the rewinding process was controlled by a motor-driven creeling machine able keep the tension constant by changing the material supplying rate. After the winding process was completed, the plate was removed from the motor flange, the thermoplastic film was wrapped over the faces of the dry fabric and finally placed in the hot-press for the consolidation stage. When the plate was cold, the resin-impregnated fabric was cut from the edges of the plate to obtain two prepregs (Fig. 3(b)). Hence, the latter were hand laid-up in a cross-ply orientation to create the final composite panels. The temperature profile used for curing the microbraid fabrics was identical for both materials. However, the pressure used to consolidate the Kevlar®49 microbraid fabrics was lower than the one used for consolidating the Dyneema®SK75 ones. The temperature vs. pressure profile is shown in Fig. 3(c). In order to directly compare the properties of the mBRPC with cross-ply laminates made with unidirectional fibres and manufactured via the same route, composites having similar areal density and fibre volume fraction were manufactured from Dyneema ®SK76 1760dtex and Kevlar®49 1580dtex, respectively. The mBRPC fibre volume fraction was determined according to the ASTM D3171–11 Standard Test Methods for Constituent Content of Composite Materials [18], whereas the void content was determined according to ASTM D2734-09 Standard Test Methods for Void Content of Reinforced Plastics [19]. Physical properties of the manufactured composites are listed in Table 3. The different lamination sequence for the UD-fibre composites and mBRPC reinforced with 8 yarn microbraids rose not only to keep the fibre volume fraction as high as possible and fairly constant among different composites, but also to maintain the same cross-ply stacking sequence. The generic microbraid reinforced composite "cXYZ" was manufacture using the microbraid "bXYZ".

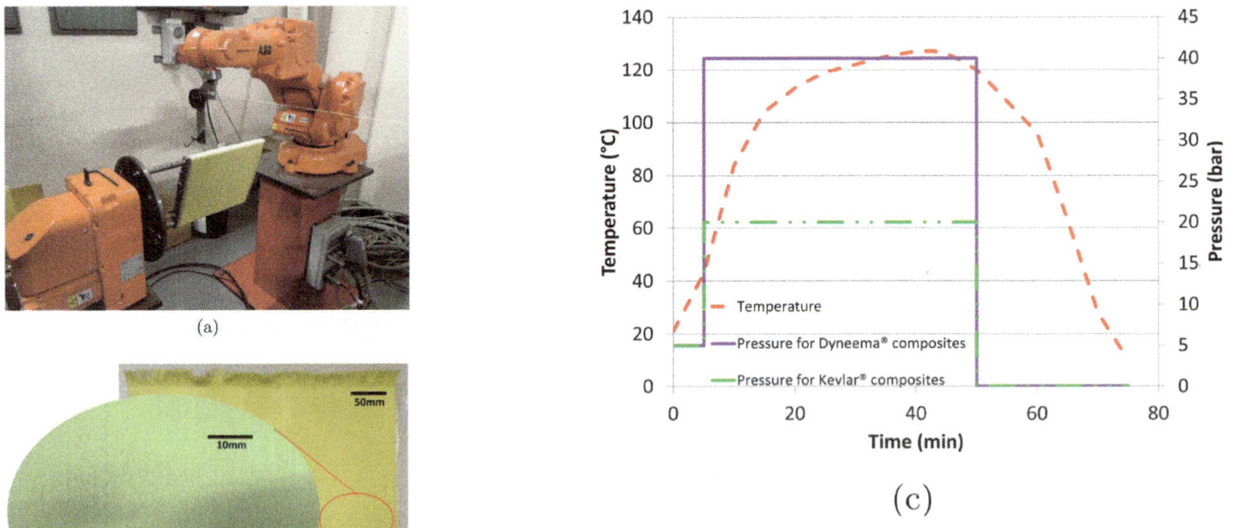

Figure 3. mBRPC manufacture: (a) Robotised filament winding; (b) cKA1 prepreg; (c) Temperature vs. Pressure consolidation profile.

Table 3. Physical properties of the manufactured composites.

cID	Number of layers	Stacking sequence	Laminate thickness (mm)	Areal density (kg/m²)	Fibre volume fraction (%)	Void content (%)
cDUD	10	$[0/(90_2/0_2)_2/90]$	3.99 ± 0.24	2.81 ± 0.02	72.50 ± 1.18	6.54 ± 1.06
cDA1	10	$[0/(90_2/0_2)_2/90]$	3.56 ± 0.16	2.70 ± 0.03	68.24 ± 1.32	7.55 ± 1.27
cDB1	10	$[0/(90_2/0_2)_2/90]$	3.79 ± 0.19	2.83 ± 0.05	78.66 ± 0.83	7.72 ± 0.61
cDC1	10	$[0/(90_2/0_2)_2/90]$	4.51 ± 0.23	3.51 ± 0.03	81.06 ± 1.40	7.66 ± 1.28
cDA2	6	$[0/90]_3$	3.06 ± 0.23	2.21 ± 0.02	73.94 ± 1.82	8.73 ± 1.64
cDB2	6	$[0/90]_3$	3.28 ± 0.18	2.45 ± 0.02	77.46 ± 0.70	7.13 ± 0.59
cDC2	6	$[0/90]_3$	3.76 ± 0.22	2.88 ± 0.02	80.52 ± 2.21	6.03 ± 2.09
cKUD	10	$[0/(90_2/0_2)_2/90]$	3.11 ± 0.15	2.78 ± 0.03	61.08 ± 1.39	10.48 ± 1.15
cKA1	10	$[0/(90_2/0_2)_2/90]$	3.10 ± 0.25	2.80 ± 0.11	67.85 ± 2.18	12.97 ± 1.54
cKB1	10	$[0/(90_2/0_2)_2/90]$	3.45 ± 0.16	2.83 ± 0.01	67.32 ± 0.63	13.68 ± 0.59
cKC1	10	$[0/(90_2/0_2)_2/90]$	4.48 ± 0.22	3.79 ± 0.18	64.43 ± 2.70	16.61 ± 1.95
cKA2	6	$[0/90]_3$	2.64 ± 0.28	2.21 ± 0.01	63.43 ± 0.65	13.46 ± 0.54
cKB2	6	$[0/90]_3$	2.66 ± 0.12	2.44 ± 0.02	61.7 ± 1.13	16.90 ± 0.94
cKC2	6	$[0/90]_3$	3.99 ± 0.23	3.02 ± 0.09	53.87 ± 1.24	28.11 ± 0.77

Testing Methods

Quasi-static tensile tests on yarns were performed at room temperature using an Instron 5969 universal tensile testing machine equipped with a 50kN load cell having an accuracy of ±0.5% of the displayed force. Specimens were clamped using Instron 2714-004 pneumatic capstan grips. Up to 2500 data-points per second were recorded by the acquisition system during each test. The strain was measured by a high speed camera: two points were marked along the gauge length and their relative displacement subsequently measured by motion tracking software developed in house. In order to investigate the effects of the strain rate and gauge length on the aforementioned yarns, tensile tests were performed with three different gauge lengths of 100 mm, 250 mm and 350 mm, and at three different strain rates of 0.01 s^{-1}, 0.001 s^{-1} and 0.0001 s^{-1}, respectively. Only for a gauge length of 100 mm, tensile tests were performed at a strain rate of 0.1 s^{-1}. For each test series, at least 5 valid tests (failure within the gauge length) were performed and collected.

Yarns were also cyclic loaded up to different force levels. This was done to understand to what extent a pre-stress introduced in the yarns prior to be braided would influence the final mechanical properties of the dry microbraids, and also any possible deformation due to stress relaxation of the created architecture. The test rate was kept constant throughout the loading and unloading parts at 0.01 s^{-1}. One and five consecutive loading and unloading cycles were performed on both fibres, respectively. Same equipment and data acquisition settings used for the tensile tests were adopted for cyclic tests. Tensile tests on dry microbraids were performed at only one gauge length and strain rate (250 mm and 0.01 s^{-1}, respectively), using the same tensile testing machine and procedures adopted for testing the raw yarns.

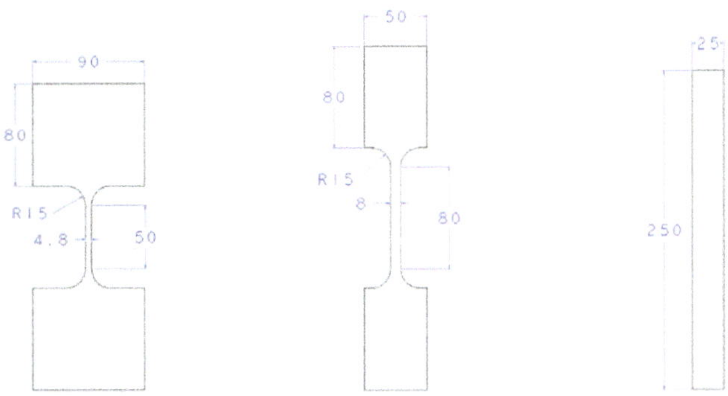

(a) (b) (c)

Figure 4. Geometries of the tensile specimens for: (a) Unidirectional composites; (b) Dyneema®SK75 mBRPC; (c) Kevlar®49 mBRPC. All dimensions in mm.

Quasi-static tensile tests on mBRPC were performed at room temperature using an Instron 5985 universal testing machine equipped with a 250 kN load cell having an accuracy of ±0.5% of the displayed force. Testing specimens, waterjet cut from the manufactured plates, were clamped using hydraulic grips to prevent slippage. Up to 50 data-points per second were recorded by the acquisition system. Strain was measured contactlessly by a camera tracking the relative displacement of points drawn along the gauge length of the specimens. All tests have been performed at cross-head speed of 10 mm/min. Specimen geometries are sketched in Fig. 4 (all dimensions are in mm).

RESULTS

Quasi-static Tensile Test

Fig. 5 shows the engineering stress vs. strain curves for Dyneema®SK75 and Kevlar®49 yarns. Only one curve among the tests performed is shown for clarity purposes. It can be seen that Kevlar®49 yarn had a reasonable linear response up to failure regardless of the strain rate at which the yarn was tested. The Young's modulus, tensile strength and strain were little affected by changing the test speed for a fixed gauge length, meaning a very small dependency of the aforementioned mechanical properties over the investigated gauge lengths. A small decrease in tensile strength and strain to failure was noted with increasing gauge length and strain rate by as much as 9% and 4%, respectively. The consistency of the test results would imply an even distribution of defects and flaws along the length of the yarn although the likelihood of finding weaker points would be higher in longer fibres.

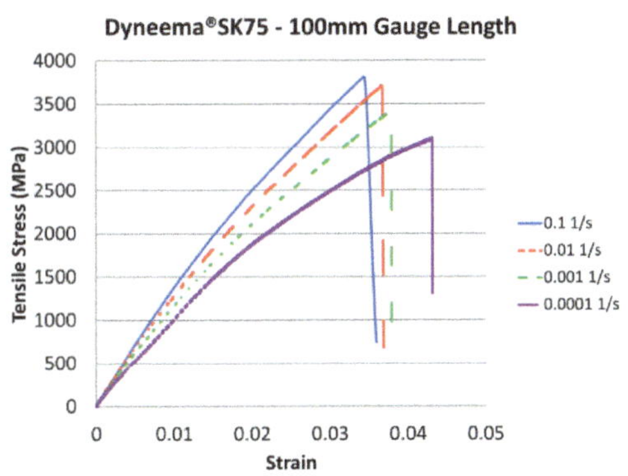

(a)

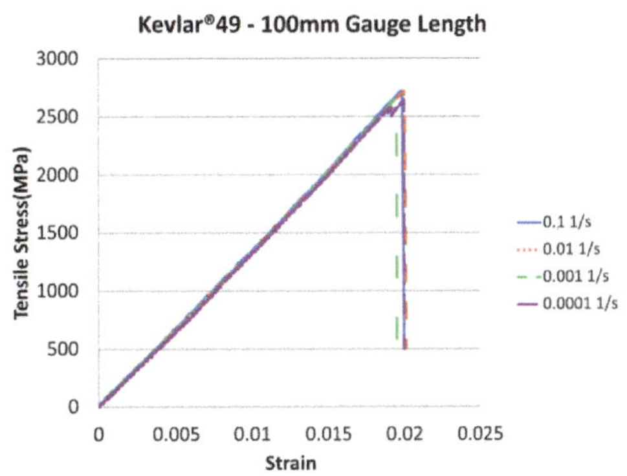

(b)

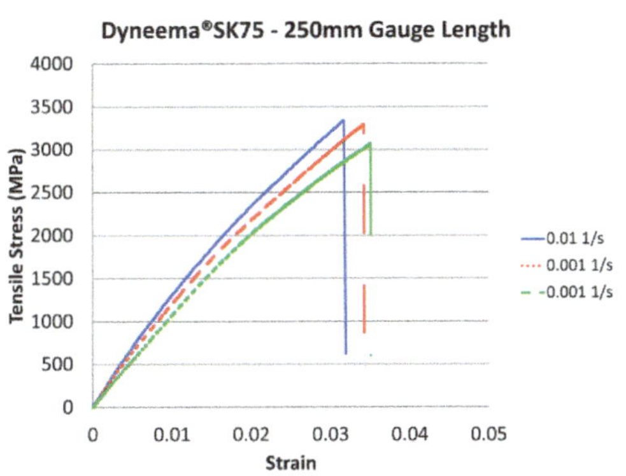

(c) Dyneema250mm

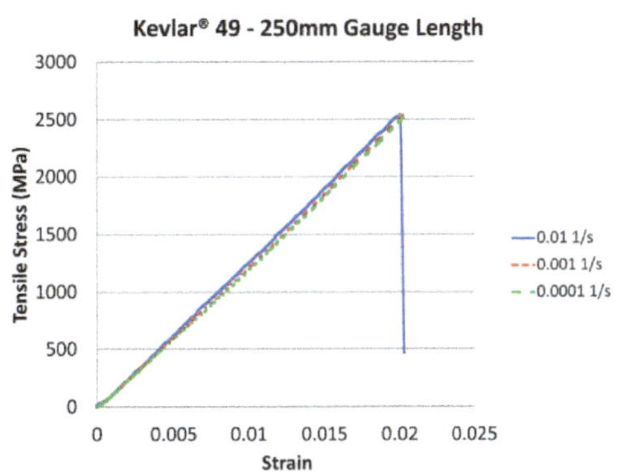

(d)

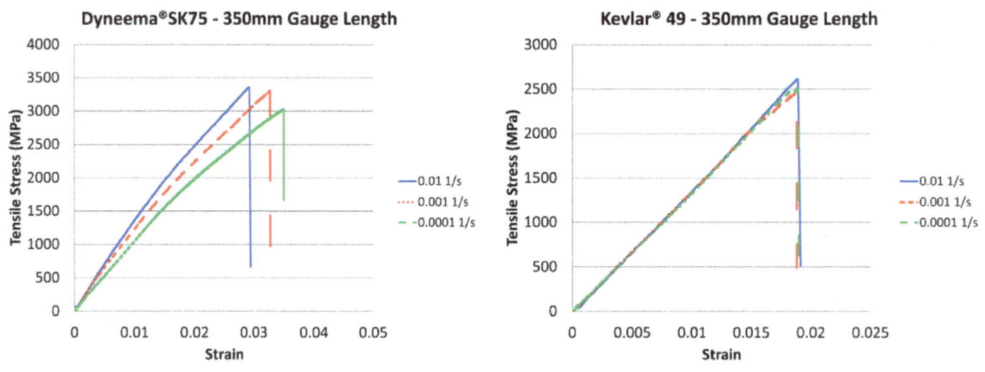

(e) (f)

Figure 5. Engineering stress vs. strain curves for different fibre gauge lengths: (a), (c) and (e): Dyneema®SK75; (b), (d) and (f): Kevlar®49.

On the other hand, the tensile behaviour of Dyneema®SK75 yarn showed a marked dependence with respect to the testing conditions. The Young's modulus increased with increasing strain rate by as much as 17% and 23% over the investigated gauge lengths and strain rates, respectively. The tensile strength remained reasonably constant, within the scatter errors, over the investigated gauge lengths, meaning an even distribution of flaws and defects along the length of the yarn. However, it increased as much as 23% over the investigated strain rates. This is clearly due to the viscoelastic nature of the material itself.

Despite the energy absorption of the two investigated yarns were very similar for the same testing conditions, the toughness and tenacity calculated for Dyneema®SK75 were superior to those noted for Kevlar®49. This is because of the higher strength and strain to failure, as well as lower specific density of the former material with respect to the latter.

Cyclic Tensile Tests

Fig. 6 presents the stress vs. strain curves from cycling tests up to different force levels performed on Dyneema®SK75 and Kevlar®49 yarns, respectively. Only one curve among the performed tests is shown for clarity purposes.

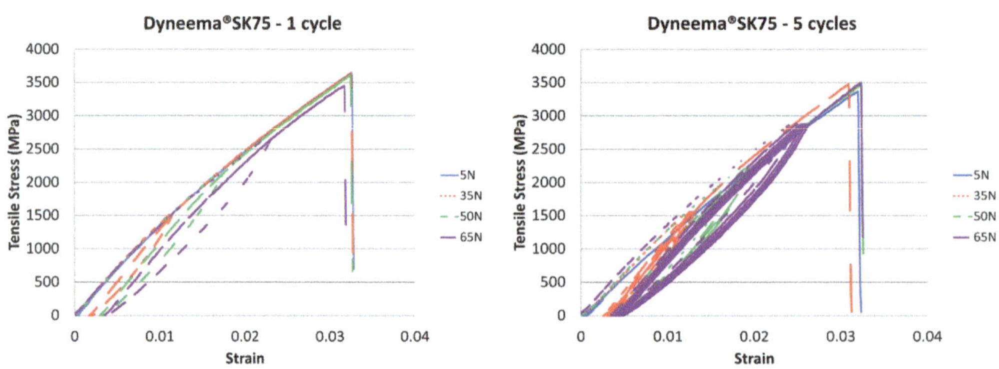

(a) (b)

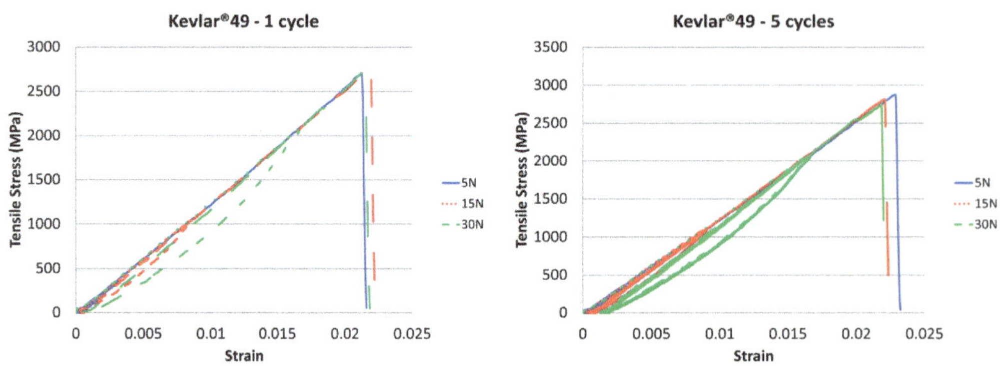

(c) (d)

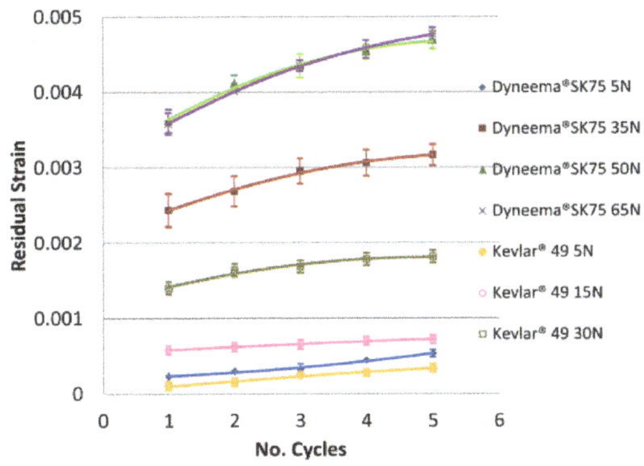

(e)

Figure 6. Cyclic tensile stress vs. strain curves for: (a) Dyneema®SK75 1 cycle; (b) Dyneema®SK75 5 cycles; (c) Kevlar®49 1 cycle; (d) Kevlar®49 5 cycles; (e) Residual strain vs. Number of cycles.

It is evident from Fig. 6(a) and (b) that, after the first cycle, a residual strain remained in the Dyneema®SK75 yarn. When the yarn was stressed during the loading part of the test, the polymeric chains were further aligned to the loading direction and the deformation occurring during this stretching was not fully recovered within the unloading time. The residual strain was dependent on the level of load at which the fibre was pre-stressed prior to being brought to failure. The higher the pre-stress, the bigger the residual strain in the fibre. It was also evident an increase in the slope of the second loading part of the stress vs. strain curve with respect to the monotonic one. This is probably due to the better alignment of the polyethylene chains to the loading direction after being straightened out, as also reported by Berger et al. [20]. Once completing the first cycle, the tensile strength and strain to failure of Dyneema®SK75 were noted to be the same, within the scatter error, as if the yarn was not cyclic loaded, i.e. the fibre were not damaged during the cycle. As the number of loading and unloading cycles increased, the residual strain did so although it tended to level off for higher number of cycles (Fig. 6(e)).

On the other hand, the cyclic loading history had very little influence on the mechanical response of Kevlar®49, which residual strain did not exceed 0.17% when the fibre was pre-loaded at ~75% of the maximum yarn breaking force. As seen for Dyneema®SK75 yarns, the tensile strength and strain of Kevlar®49 was not affected by the number of cycling loadings.

Van der Werff and Pennings [21] described the possible mechanisms occurring in ultra high molecular weight polyethylene (UHMwPE) fibres during tensile and cyclic loads. The proposed flow mechanism assumed an induced flow of the polymeric chains due to thermal activated processes - in this case the cyclic tensile deformation. This effect would be much greater in materials having a low melting temperature (T_m) such as UHMwPE than in para-aramids, which T_m is about three time higher. It should be also noted the differences in the chemical structure between the two materials. While UHMwPE has the simplest monomer and chemical structure among all polymers, its chains are easily prone to deform under external loads, i.e. the C–H bonds and CHC angles along the carbon backbones can be easily stretched, rotated and opened, while the stiffer, benzene ring-rich structure and stronger intramolecular forces present in the aramid fibre make this polymer less prone to deform and faster in recovering the original, more stable, entropy favourable conformation.

Quasi-static Tensile Test on Dry Microbraids

The results obtained from cycling tests of Dyneema®SK75 and Kevlar®49 showed that these yarns experienced a deformation even when stressed at small loads. Although this deformation would be small and possibly time-recoverable [21], the tension in the yarn during the spooling process was controlled to not exceed 2 N tension in order to minimise any possible physical change in the raw materials and in the architecture of the braid after being shaped. On the other hand, the rewinding speed and the carriers revolution speed was kept high at 120 m/min and 300 rpm, respectively, in order to not give to the polymeric chains enough time to respond to the external load. Higher processing speeds, as well as low working tensions, would minimise the residual strain in the fibre after the external stresses are removed.

Fig. 7 shows the tenacity vs. strain curves for Dyneema®SK75 and Kevlar®49 microbraids obtained from quasi-static tensile tests. Engineering properties are graphically presented in Fig. 8. It clearly appears that the braid angle, defined as the angle between the braid axial direction and the bias yarns, played a fundamental role in determining the final properties of the dry microbraids. The strain to failure approached 20% for sample bDC2, i.e. more than five times higher the strain to failure of the relative UD counterpart. The higher the bias angle, the higher the strain to failure. On the other hand, it can be seen that microbraids having smaller braid angles had a stiffer response after jamming occurred with respect to those having bigger bias angles. It also appears from Fig. 7 that the tenacity of the investigated microbraids tended to diminish with increasing braid angle. Although this was always true for Dyneema®SK75 microbraids, however, the tenacity of bKA1 and bKA2 samples was higher than the tenacity of their unidirectional counterpart by as much as 17.51% and 3.17%, respectively, despite their higher linear densities and crimped yarns. It is reasonable to think that the mechanical interlocks created during the braiding process would prevent an early failure of the whole structure, i.e. the microbraid was still able to withstand the external load even though the structure was damaged and some filaments already failed. The reason why this effect appeared only in Kevlar®49 microbraids would be due to the higher coefficient of friction with respect to that of Dyneema®SK75. The fibre–fibre coefficient of friction for Dyneema®SK75 yarn is reported to be 0.05–0.065 [22] and [23] whereas the fibre–fibre coefficient of friction for Kevlar®49 yarns is as high as 0.15–0.22 [23] and [24]. The higher coefficient of friction of Kevlar®49 yarns would make more difficult the sliding of the yarns and the rearrangement of the braid geometry under external load. In fact, the jamming point of Kevlar®49 microbraids would occur at lower strains with respect to Dyneema®SK75 microbraids for the same braid angle and braid diameter. The rubbing of the jammed yarns would give extra strength to the braid structure. However, when normalising the area under the tenacity vs. strain curves with respect to the microbraid linear density, the normalised energy values obtained for Dyneema®SK75 and Kevlar®49 dry microbraids were always lower with respect to those noted for the respective UD counterparts. Although the normalised energy absorption ability of Dyneema®SK75 decreased with increasing linear density, there was no significant difference in normalised energy amongst Kevlar®49 microbraids having the same architecture but different α, meaning that the capacity of absorbing energy of these microbraids is approximately the same regardless of the braid angle and the linear density.

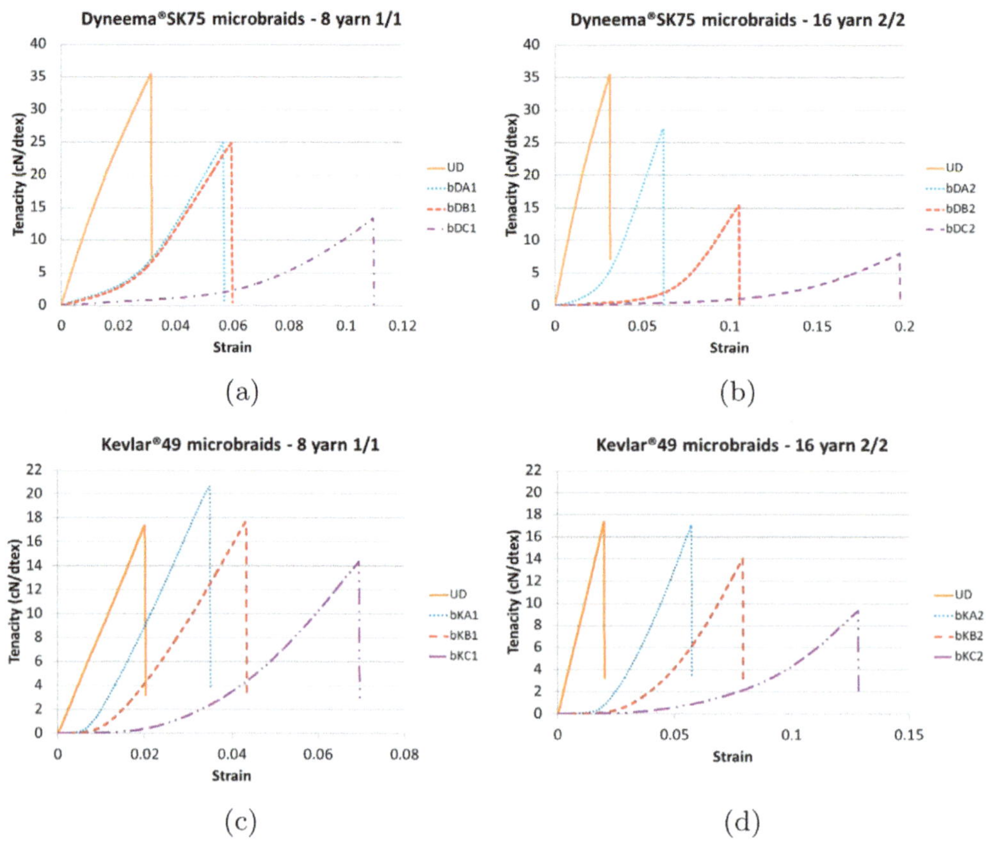

Figure 7. Tenacity vs. strain curves for different microbraids: (a) Dyneema®SK75 8y 1/1; (b) Dyneema®SK75 16y 2/2; (c) Kevlar®49 8y 1/1; Kevlar®49 16y 2/2.

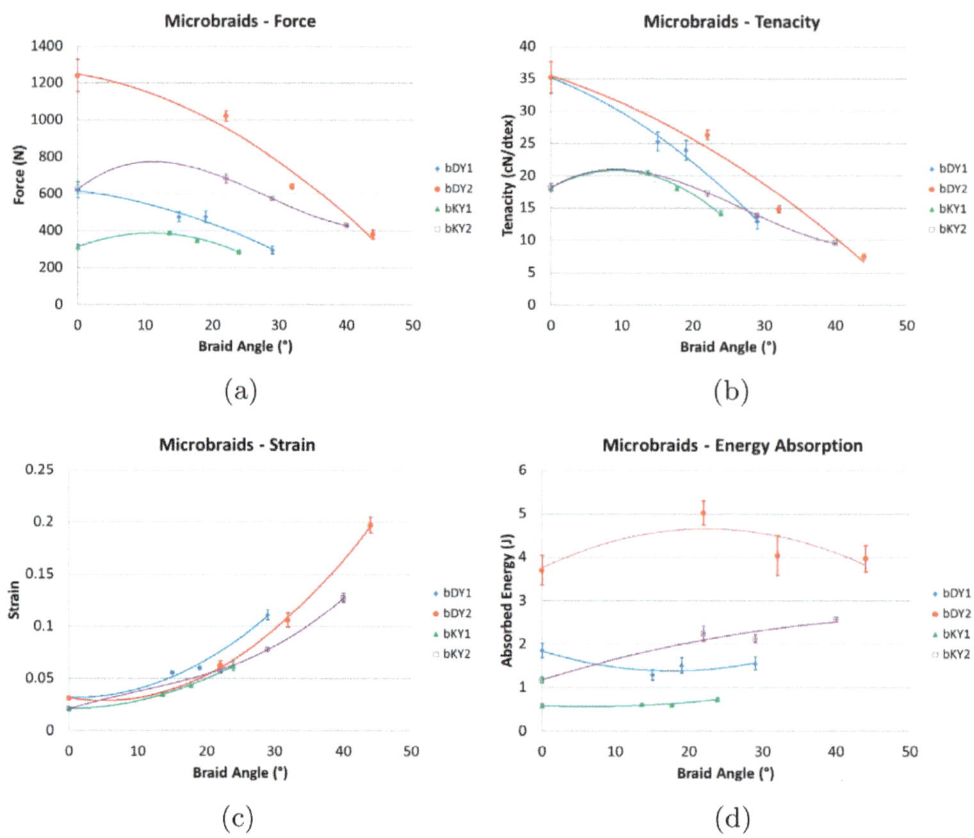

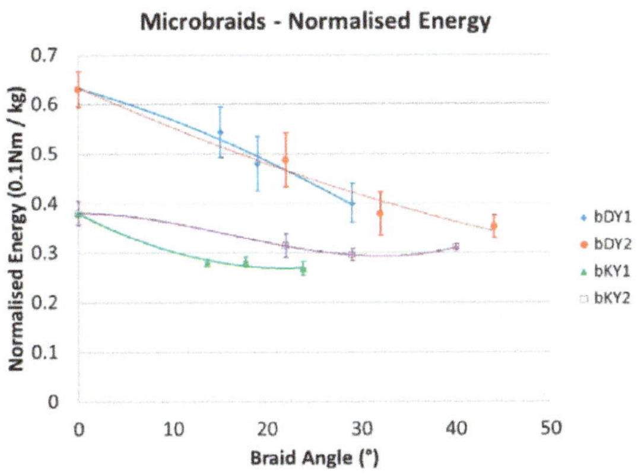

(e)

Figure 8. Tensile properties of dry microbraids: (a) Force vs. braid angle; (b) Tenacity vs. braid angle; (c) Strain vs. braid angle; (d) Energy absorption vs. braid angle; (e) Normalised energy absorption vs. braid angle.

Quasi-static Tensile Test on mBRPC

Tensile tests on mBRPC were performed according to ASTM D3039-08 Standard Test Method for Tensile Properties of Polymer Matrix Composite Materials [25]. Preliminary test results on unidirectional composites and Dyneema®SK75 mBRPC performed using rectangular specimens were unsuccessful (Fig. 4(c)). Specimens failed at the gripped region (Fig. 9) due to the low shear strength of the composites. As pointed out in different papers [26] and [27], it is very difficult to introduce axial stresses from the tabbed regions of the specimen to its gauge length by shear, especially for slippery, low shear strength materials. Therefore, in order to promote failure within the gauge length, specimens having larger dogbones and narrower width of the gauge part were waterjet cut from the manufactured panels (Fig. 4(a) and (b)).

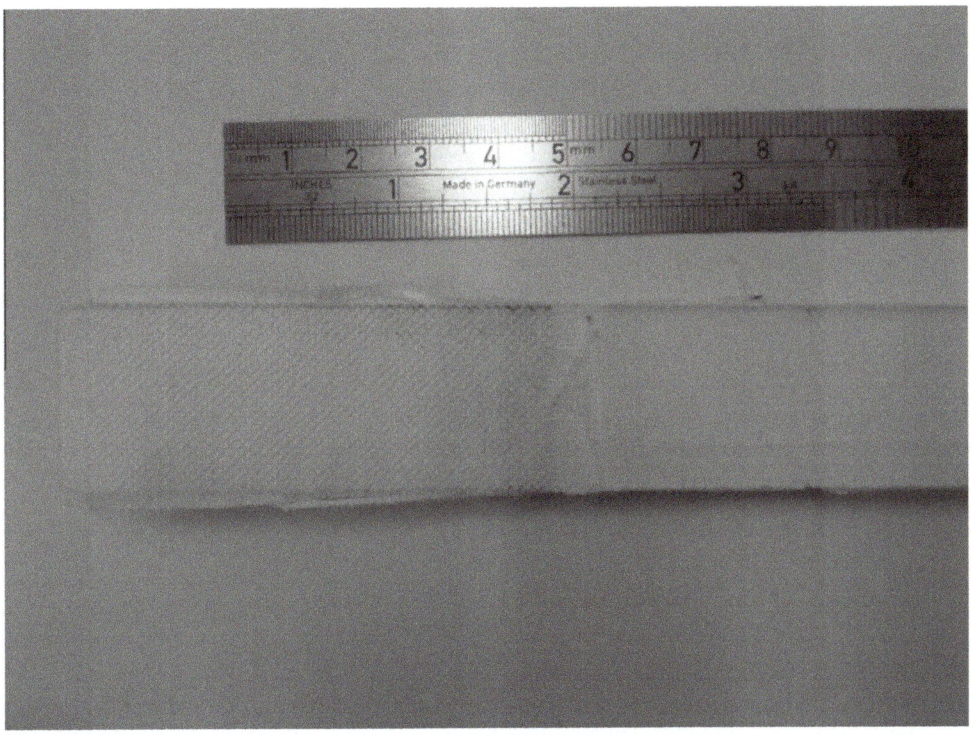

Figure 9. Dyneema®SK75 mBRPC rectangular specimen incorrectly failed at the gripped region.

Fig. 10 shows the engineering stress vs. strain behaviour for Dyneema®SK75 and Kevlar®49 microbraid reinforced composites. Only one curve among the tests performed is shown for clarity purposes. Engineering properties are graphically shown in Fig. 11.

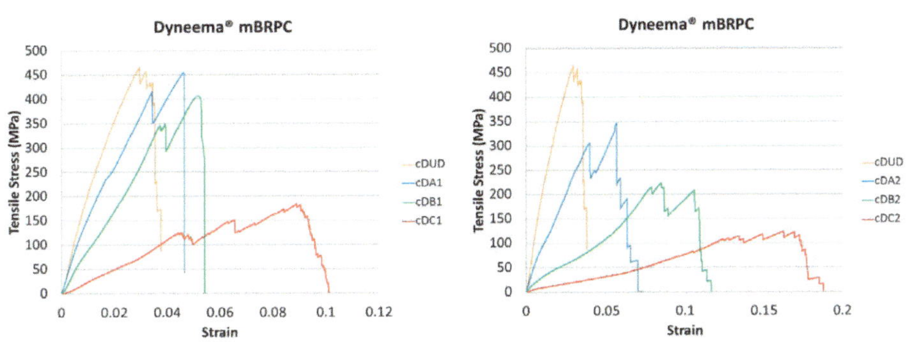

(a) (b)

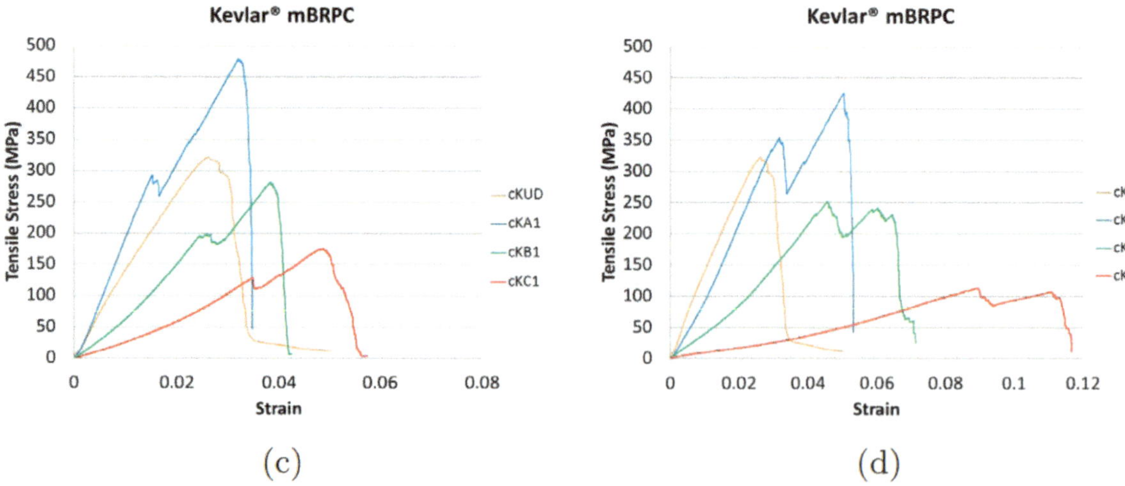

Figure 10. Engineering stress vs. strain curves for different microbraid reinforced composites: (a) cDY1; (b) cDY2; (c) cKY1; cKY2.

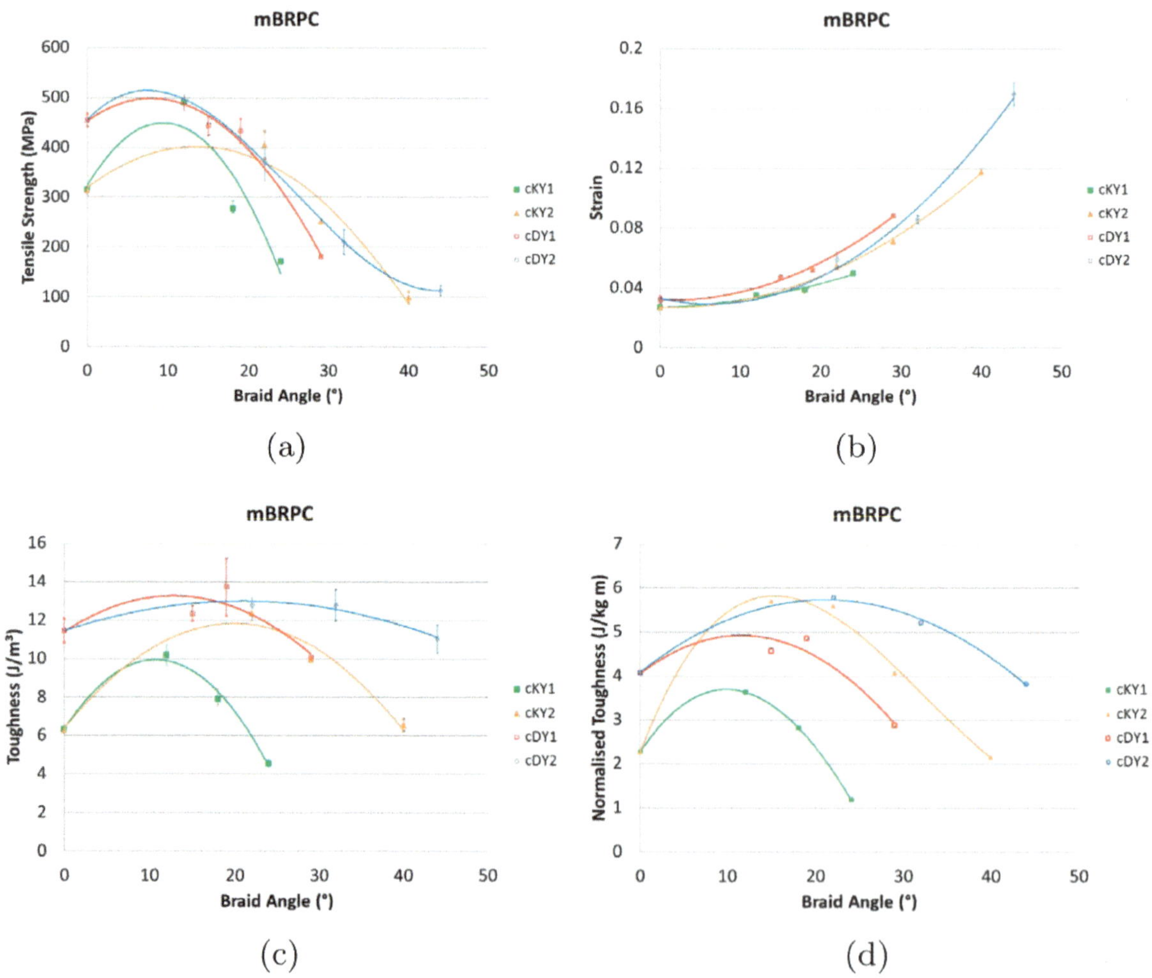

Figure 11. Tensile properties of microbraid reinforced polymer composites: (a) Tensile strength vs. braid angle; (b) Strain vs. braid angle; (c) Toughness vs. braid angle; (d) Normalised energy absorption vs. braid.

It can be observed from Fig. 10 that the stress vs. strain curves of both Dyneema®SK75 and Kevlar®49 microbraid reinforced composites had similar trends observed when testing dry microbraids. The smaller the braid angle, the higher the tensile strength. On the other hand, the higher the braid angle, the higher the strain to failure. However, the failure mode of mBRPCs was different from the brittle-catastrophic mode of failure experienced by the dry microbraids. In proximity of failure, the outermost layers of the microbraid reinforced composites failed, making the load to drop slightly (Fig. 12). Nevertheless, the specimen was still able to carry the external load until complete failure occurred thereafter. Delamination occurred prior to failure in all tested specimens. Moreover, the higher the yarns bias angle, the higher the extent of delamination among the laminate layers, which can be also deducted by the smoother fall of the stress vs. strain curves after ultimate tensile strength, for both materials.

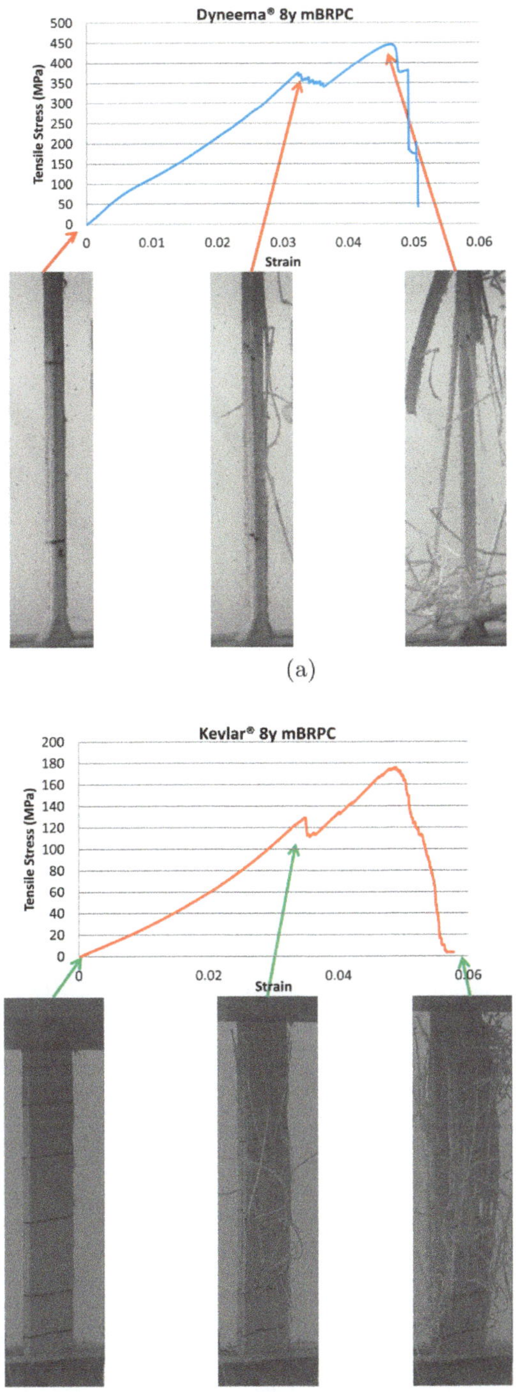

Figure 12. mBRPC failure: (a) cDA1; (b) cKC1.

The strain to failure of each microbraid reinforced material is comparable, within the scatter errors, with the strain to failure of the constituent microbraid by which the panel was manufactured. For Dyneema®SK75 mBRPC, the tensile strength of the laminates decreased with increasing braid angle and no significant differences can be appreciated between composites reinforced with microbraids made of 8 and 16 yarns, as far as the tensile strength is concerned. The toughness, calculated as the area under the stress vs. strain curve, remained fairly constant regardless of the braid angle at the value of the unidirectional composites manufactured using same technique. However, this property tended to diminish with increasing braid angle. On the other hand, the ultimate tensile strength of Kevlar®49 mBRPC manufactured by braids having the smallest braid angle was higher than the ultimate tensile strength of the unidirectional counterpart by as much as 55.7% and 28.9% for composites reinforced with 8 yarn and 16 yarn microbraids, respectively. Coupling the higher strength with higher strain to failure, the toughness of these two particular composites was higher than the toughness calculated for the unidirectionally aligned fibre composites by as much as 61.3% and 96.2%, respectively. Normalising the toughness with respect to the areal density of the manufactured composites, it appears from Fig. 11(d) that laminates reinforced with microbraids having small braid angle had superior ability to absorb energy with respect to laminates reinforced with unidirectional fibres, for both materials, although this property tended to diminish with increasing braid angle. This result can be attributed to the inherent nature of the braid, which structure made of mechanically intertwined threads could help to distribute more uniformly the external load throughout the whole structure. However, these observations must be confirmed with other experimental tests in order to assess to what extent the difference in specimen geometry, thickness, fibre volume fraction and areal density affected the mechanical response of this novel class of composite materials.

CONCLUSION

In this paper, the tensile response of two high performance fibres were experimentally investigated via a comprehensive series of mechanical tests. Experimental results showed a significant difference in the tensile behaviour of the investigated materials as far as the stress vs. strain behaviour is concerned. Different types of microbraids were manufactured from the as supplied yarns. Both Dyneema®SK75 and Kevlar®49 microbraids showed different tensile properties with respect to those observed for the constitutive materials. The final mechanical properties of braids depended not only on the material properties but also on the fibre bias angle and architecture. As the braid angle increased, also the strain to failure did so although whilst the tenacity decreased. However, for some architectures and braid angles, the tenacity of the dry microbraids exceeded the tenacity of the unidirectional yarn.

In order to manufacture microbraid reinforced polymer composites having high fibre volume fraction, a robotised filament winding system and hot-pressing technique were successfully employed. Tensile tests on specimens waterjet cut from the manufactured composites showed always higher strain to failure when compared with unidirectional composites made using the same manufacturing route. Moreover, for certain braid angles, it was also noted a 55.7% higher tensile strength and a 96.2% higher toughness. The progressive failure mode noted when tensile testing the mBRPCs would imply more damage tolerant structures able to absorb more external energy prior to failure.

The results of this study indicate that the braiding process can be used to manipulate and modify, to some extent, the mechanical properties of the precursor materials for the creation of new materials with unique and enhanced mechanical properties. Further research needs to examine the mechanical properties of dry microbraids and mBRPC under dynamic loading conditions in order to demonstrate the applicability of microbraid reinforced systems in high energy absorption applications.

ACKNOWLEDGEMENTS

The authors would like to acknowledge the funding from DSTL MAST STC and EPSRCunder CASE award DSTL-X-1000061561. DSM Dyneema and DuPont® are acknowledged for the provision Dyneema®SK75 and Kevlar®49 yarns, respectively.

REFERENCES

1. Head A, Ko F, Pastore C. Handbook of industrial braiding. Atkins and Pearce; 1989.
2. Lee S. Handbook of composite reinforcements. Wiley; 1992.
3. Cox B, Flanagan G. Handbook of analytical methods for textile composites, NASA contractor report, National aeronautics and space administration; 1997.
4. Brunnschweiler D. Braids and braiding. J Text Inst Proc 1953;44(9):P666–86. http://dx.doi.org/10.1080/19447015308687874.
5. Brunnschweiler D. The structure and tensile properties of braids. J Text Inst Trans 1954;45(1):T55–77. http://dx.doi.org/10.1080/19447025408662631.
6. Ayranci C, Carey J. 2d braided composites: a review for stiffness critical applications. Compos Struct 2008;85(1):43–58. http://dx.doi.org/10.1016/j.compstruct.2007.10.004.
7. Omeroglu S. The effect of braiding parameters on the mechanical properties of braided ropes. Fibres Text East Eur 2006;14(4):53–7.
8. Harte A-M, Fleck NA. On the mechanics of braided composites in tension. Eur J Mech – A/Solids 2000;19(2):259–75. http://dx.doi.org/10.1016/S0997-7538(99)00164-3.
9. Tate JS, Kelkar AD, Whitcomb JD. Effect of braid angle on fatigue performance of biaxial braided composites. Int J Fatigue 2006;28(10):1239–47. http://dx.doi.org/10.1016/j.ijfatigue.2006.02.009 [The third international conference on fatigue of composites the third international conference on fatigue of composite].
10. Fouinneteau M, Pickett A. Shear mechanism modelling of heavy tow braided composites using a meso-mechanical damage model. Compos Part A: Appl Sci Manuf 2007;38(11):2294–306. http://dx.doi.org/10.1016/j.compositesa.2006.12.006 [compTest 200].
11. Falzon PJ, Herszberg I. Mechanical performance of 2-d braided carbon/epoxy composites. Compos Sci Technol 1998;58(2):253–65. http://dx.doi.org/ 10.1016/S0266-3538(97)00133-4 [Australasian special issue on manufacturing processes and mechanical properties characterisation of advanced composites].
12. Viju S, Thilagavathi G. Fabrication and characterization of silk braided sutures. Fibers Polym 2012;13(6):782–9. http://dx.doi.org/10.1007/s12221-012-0782- 8.
13. Davies P, Reaud Y, Dussud L, Woerther P. Mechanical behaviour of hmpe and aramid fibre ropes for deep sea handling operations. Ocean Eng 2011;38(17– 18):2208–14. http://dx.doi.org/10.1016/j.oceaneng.2011.10.010.
14. McKenna H, Hearle J, O'Hear N. 4 – properties of rope. In: McKenna H, Hearle J, O'Hear N, editors. Handbook of fibre rope technology. Woodhead publishing series in textiles. Woodhead Publishing; 2004. p. 101–40. http://dx.doi.org/ 10.1533/9781855739932.101.
15. Sakaguchi M, Nakai A, Hamada H, Takeda N. The mechanical properties of unidirectional thermoplastic composites manufactured by a micro-braiding technique. Compos Sci Technol 2000;60(5):717–22. http://dx.doi.org/10.1016/ S0266-3538(99)00175-X.
16. Fujihara K, Huang Z-M, Ramakrishna S, Hamada H. Influence of processing conditions on bending property of continuous carbon fiber reinforced peek composites. Compos Sci Technol 2004;64(16):2525–34. http://dx.doi.org/ 10.1016/j.compscitech.2004.05.014.
17. ASTM standard D1577-07. Standard test methods for linear density of textile fibers. In: ASTM international, West Conshohocken, PA; 2012. http://dx.doi.org/10.1520/D1577-07R12.
18. ASTM standard D3171. Standard test methods for constituent content of composite materials. In: ASTM international, West Conshohocken, PA; 2011. http://dx.doi.org/10.1520/D3171-11.
19. ASTM standard ASTMD2734-09. Standard test methods for void content of reinforced plastics. In: ASTM international, West Conshohocken, PA; 2009. http://dx.doi.org/10.1520/D2734-09.
20. Berger L, Kausch H, Plummer C. Structure and deformation mechanisms in uhmwpe-fibres. Polymer 2003;44(19):5877–84. http://dx.doi.org/10.1016/ S0032-3861(03)00536-6.
21. Van Der Werff H, Pennings A. Tensile deformation of high strength and high modulus polyethylene fibers. Colloid Polym Sci 1991;269(8):747–63. http:// dx.doi.org/10.1007/BF00657441.
22. Marissen R. Design with ultra strong polyethylene fibers. Mater Sci Appl 2011;2(5):319–30. http://dx.doi.org/10.4236/msa.2011.25042.
23. McKenna H, Hearle J, O'Hear N, Institute T. Handbook of fibre rope technology. Woodhead publishing in textile technology series. CRC Press; 2004.
24. Briscoe B, Motamedi F. The ballistic impact characteristics of aramid fabrics: the influence of interface friction. Wear 1992;158(1-2):229–47. http://dx.doi.org/10.1016/0043-1648(92)90041-6.
25. ASTM standard ASTMD3039-08. Standard test method for tensile properties of polymer matrix composite materials. In: ASTM international, West Conshohocken, PA; 2008.
26. Russell B, Karthikeyan K, Deshpande V, Fleck N. The high strain rate response of ultra high molecular-weight polyethylene: from fibre to laminate. Int J Impact Eng 2013;60(0):1–9.
27. Karthikeyan K, Russell B, Fleck N, Wadley H, Deshpande V. The effect of shear strength on the ballistic response of laminated composite plates. Eur J Mech – A/Solids 2013;42(0):35–53.

CHAPTER 18

MEASUREMENT OF FIBRE WAVINESS IN INDUSTRIAL COMPOSITE COMPONENTS

M.P.F. Sutcliffe, S.L. Lemanski, and A.E. Scott

Keywords: A. Carbon fibres; B. Defects; D. Optical microscopy

INTRODUCTION

Compressive strength is an important property of fibre reinforced polymer matrix composites, but it is difficult to control and predict. There are various competing mechanisms of compressive failure, as reviewed by several authors, e.g. [1], but for standard engineering composites the governing failure mechanism is generally agreed to be plastic microbuckling. Argon [2], observing that fibre composites have regions of fibre misalignment, derived a simple expression for the compressive strength σ_c of unnotched material as $\sigma_c = k/\bar{\phi}$ where k is the in-plane shear yield stress of the composite and $\bar{\phi}$ is the initial misalignment angle of the uniformly misaligned fibres. This formula illustrates the critical role that fibre misalignment plays in determining compressive strength of fibre composites.

The size of the region of fibre misalignment also affects the compressive strength. Fibre bending stiffness causes an increase in compressive strength for small defects, but is negligible once the wavelength is greater than around 200 times the fibre diameter [3]. Lemanski and Sutcliffe [4] have shown that the compressive strength is sensitive to the ratio of the width of the wavy region to the width of a plate of finite size. Hence the spatial distribution of fibre orientation is also important for understanding and modelling compressive strength of fibre composites.

A variety of techniques have been developed to measure fibre waviness. Yurgatis [5]described a sectioning method, using the elliptical cross section of fibres cut on an oblique plane, to estimate fibre orientation. An automated version of this method has been used to measure large numbers of fibres with misalignments greater than 40° [6]. This method has the advantage of being able to measure orientation in two planes simultaneously, but relies on accurate cross-sectional geometry. More recently X-ray computed tomography has been used to measure fibre orientation. Salaberger et al. [7] measured the orientation of glass fibres in a polypropylene matrix, showing that the quality of the results depends on the algorithms used to derive the fibre orientation. Requena et al. [8] use synchrotron micro-tomography to image carbon fibre reinforced metal and polymer matrix composites. A series of image analysis steps is used to track the trajectory of individual fibres. The method is effective at measuring the three-dimensional fibre orientation field, although limited by the resources needed for such imaging and the relatively small volumes that can be measured. Creighton et al. [9]proposed the multiple field image analysis (MFIA) method to analyse relatively low-resolution micrographs of planes sectioned parallel to the nominal fibre direction. The standard deviation in the misalignment angle of prepreg, in a preliminary study, was found to be 0.7° [10]. An alternative image analysis algorithm based on Fourier analysis was developed by Kratmann et al. [11] and found to give an accuracy for the mean fibre angle of better than 0.2°. The technique has been used to analyse prepreg and pultruded samples [12]. Changes in pultrusion details reduced the standard deviation in the fibre misalignment angle from 1.3° to 0.9°. While the above techniques provide methodologies to measure fibre waviness, the methods have only been systematically applied to pultruded samples or a limited range of laboratory pre-preg samples. Wang et al. [13]present measurements of tow-level misalignment in an industrial component. Their measurements relate to a specific defect associated with a nearby ply drop and a lay-up defect in an adjacent ply, so is representative of

Citation: M.P.F. Sutcliffe, S.L. Lemanski, A.E. Scott, Measurement of fibre waviness in industrial composite components, Composites Science and Technology, Volume 72, Issue 16, 16 November 2012, Pages 2016-2023, ISSN 0266-3538, http://dx.doi.org/10.1016/j.compscitech.2012.09.001.

defects rather than a normal part of the sample. The paucity of data for real components has hindered model development and our understanding of the role of fibre waviness in compressive strength in industrial structures.

While knowledge of spatial variations in fibre misalignment are critically important, there is a lack of systematic measurement of the size of misaligned regions. Kyriakides et al.[14] observed using an optical microscope that the wavelength of fibre misalignment in APC-2/AS4 composites ranged between 2.1 and 5.6 mm. Clarke et al. [15] and [16]applied confocal laser-scanning microscopy to measure fibre waviness in a glass fibre-reinforced epoxy. They found, in preliminary results, wavelengths in the fibre direction typically within the range of 0.5–4 mm. This range of wavelengths was in good agreement with that observed by Wisnom [17] for carbon fibre composites. Potter et al.[18] observed a regular and systematic waviness in carbon fibre prepreg with a mean misalignment angle of 3.8° and a mean wavelength of 3 mm. Finally Liu et al. [10] found from spectral analysis that wavelengths of the order of a few 100 μm were predominant in CFRP material made from prepreg.

The theoretical work described above demonstrates the importance of both the angle and the spatial variation of fibre misalignment for compressive strength of long fibre composites. While there have been a number of studies measuring the misalignment angle of laboratory samples, relatively little research has characterised the spatial variation of fibre misalignment. Moreover the measurements have been almost entirely for well controlled laboratory samples which may not be typical of actual components. This study adopts the MFIA method for measurement of fibre orientation, with relatively minor extensions to increase efficiency. The method is used in two novel ways; as applied to industrial components and as a means to characterise both the misalignment angle and the spatial distribution of misalignment in fibre composite samples. Moreover we demonstrate that the analysis method, up to now applied to polished sections, can also be used with X-ray images.

SAMPLE DETAILS

Two industrial components with standard lay-ups and manufacturing routes were provided by our collaborators. Both components had a predominantly 0° lay-up of unidirectional carbon fibres reinforced with epoxy, although other layers were present. One component was made from prepreg, consolidated under vacuum. The other was laid up as a dry preform and then resin transfer moulded (RTM). These are referred to as the 'prepreg' and 'RTM' components. Samples were taken from two representative locations on the two components. Both components had a slight curvature with in-plane dimensions significantly larger than the thickness, which was between 5 and 10 mm for the RTM component and between 10 and 30 mm for the prepreg component.

IMAGING METHODOLOGY

Orientation of Sections

Section planes parallel and perpendicular to the surface of the component were used to measure in-plane and out-of-plane fibre waviness, as illustrated in Fig. 1, with corresponding in-plane and out-of-plane waviness misalignment angles φ. In all cases the zero direction for fibre orientation was taken as the mean fibre orientation angle in a section. The fibre angle varied as a function of position, and a prime is used to denote the standard deviation of the angle[1]. A characteristic length l and width w for the waviness regions is also determined independently on the in-plane and out-of-plane sections, as illustrated schematically in Fig. 1.

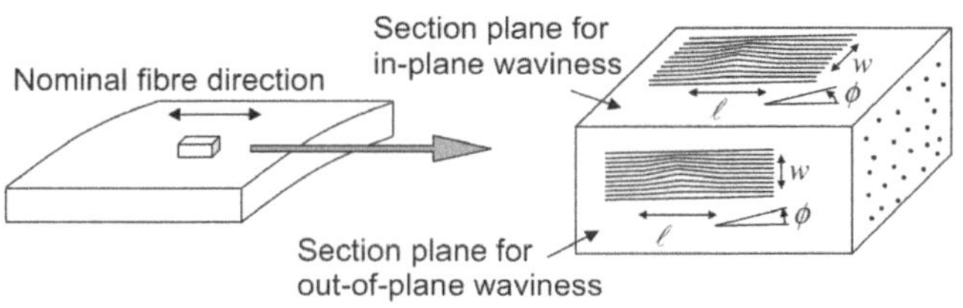

Figure 1. Orientation of sections within samples.

Polished Sections

Samples cut from the components were progressively polished, with a final polish using 3200 grit paper. Images were obtained using an Olympus BX51 optical microscope; the magnification was such that the pixel side length was 1.51 μm, giving adequate resolution to identify individual fibres. The section planes were nominally parallel to the fibres but there was inevitably some angular offset between these planes and the fibres, so that the fibres appeared as very elongated ellipses. Multiple images were taken using a motorised stage and stitched together using post-processing software [20], to give areas of typically around 5 × 5 mm. In total 16 analysis areas were measured, spanning the two specimens and both orientations. For the thicker prepreg sample the analysis area was predominantly 0° fibres. For the thinner RTM component the ply block thickness varied from a fraction of a millimetre to several millimetres, so that the regions of 0° material in the out-of-plane sections of this part necessarily covered a smaller area.

X-ray Imaging

Micro-CT X-ray imaging was undertaken using a 225 kV Nikon/Metris HMX on a sample cut from the prepreg component. The imaging volume, obtained by concatenating images, was 4 × 7 × 12 mm^3, with the longest dimension running along the fibre direction. A beam energy of 65 kV and current of 770 μA was used with a 2 s exposure. Two-dimensional radiographs were recorded at 2544 angular positions over 360° of rotation via a 2 × 2 k flat panel detector to give a voxel side length of 6.4 μm. Section planes in and out of the plane of the component were extracted from the 3D data at 100 μm intervals, as shown in Fig. 2.

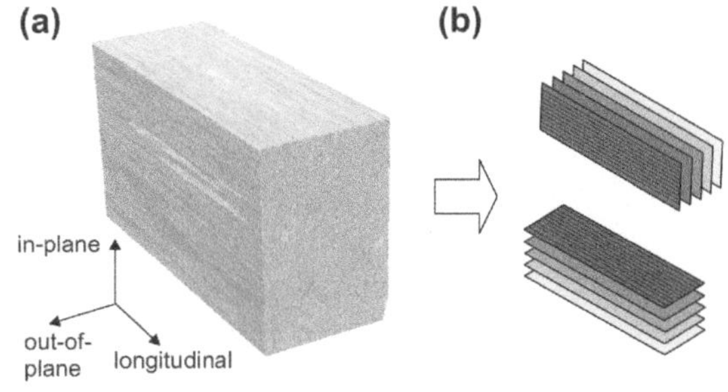

Figure 2. X-ray data for prepreg specimen; (a) 3D data and (b) extraction of perpendicular slices.

IMAGE ANALYSIS METHODOLOGY

This section presents details of the method used to calculate fibre orientation and derive characteristic dimensions for the waviness features. A discussion of errors and sensitivity to analysis parameters is given in Section 4.3.

Estimate of Fibre Orientation Angles

The MFIA method [9] is used to extract the fibre orientation from images of the polished and X-ray samples. There are many methods available for identification of orientation features, for example an FFT approach [11] or Hough transforms [21]. However the MFIA method was chosen as experience showed that this algorithm performs well for the given application. The method works by calculating the variation in correlation between features in the image and a 'trial' fibre, as a function of orientation angle of the trial fibre, averaged over a region of interest. The trial fibre orientation giving the minimum variation in the correlation function identifies the fibre angle within this region, as illustrated in Fig. 3. This bespoke algorithm is particularly effective because the fibre and matrix features defining the orientation are long thin bands of light and dark with a more-or-less uniform intensity. Three relatively minor changes were made to the original algorithm: (i) some optimisation of the code was undertaken to speed up the original algorithm, (ii) problems associated with alignment of fibres with the pixel direction were addressed and (iii) a more accurate method of determining the minimum in correlation was used. Previous developments of the MFIA method were limited by the calculation time [11], but the revised implementation was relatively rapid. The total CPU time required to analyse a typical micrograph was about 10 min on an Intel Core i7 2.8 GHz processor.

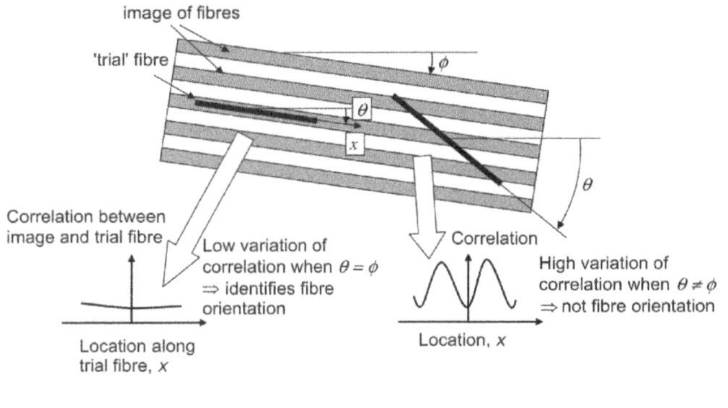

Figure 3. MFIA method used to identify fibre angle. The trial fibre orientation θ giving the minimum variation in the correlation identifies the fibre angle φ.

The circles shown in Fig. 4 give a typical variation of correlation function with trial angle. A quadratic function is fitted to the measured dependence of correlation function on orientation angle, through the data points close to the minimum. The minimum in this fit is used to determine the fibre orientation angle. There is a particular issue where the image edges are aligned with the fibre direction. Spurious peaks were seen in the correlation function when a large number of trial fibre orientations were used to determine the correlation function, as shown in Fig. 4. This perturbation, which was also noted in [9], would cause an error in estimating the correct orientation angle. It was observed empirically that this error does not occur at angles equal to $\sin^{-1}(i/N)$, where i is an integer and N is the number of pixels in the averaging region in the fibre direction. Presumably the error is associated with the discrete nature of the pixel representation of the image, perhaps arising from interpolation of the image at an angle to the edges of the image. By using only angles equal to $\sin^{-1}(i/N)$ for the correlation evaluation, this error can be effectively eliminated, as shown by the circular markers at such 'nodes' plotted in Fig. 4. The effectiveness of this approach was checked by confirming that the resulting values of the correlation function averaged over many images did not contain spurious peaks. However the preferred approach is to ensure that fibres are not aligned along the pixel direction by appropriate alignment of the specimen with respect to the camera pixel orientation, so avoiding this problem.

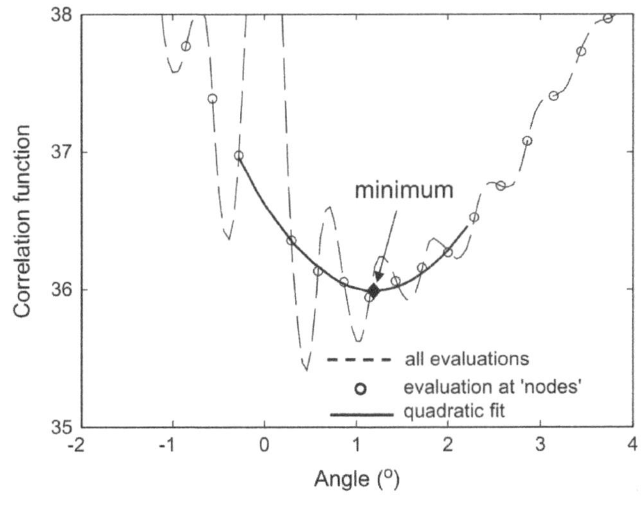

Figure 4. Typical variation of correlation function with trial angle from a polished micrograph section. When a large number of angles are used spurious perturbations are introduced. Evaluation only at selected 'nodes' avoids these perturbations to give a smooth function which is fitted at the minimum by a quadratic.

Fig. 5 shows typical results for polished and X-ray images. The regions of interest over which the orientation is averaged are given by the inclined squares. An averaging area of side length 300 μm was used. The length of the 'trial fibre' was 40 μm for the polished samples and 64 μm for the X-ray image. Results are not sensitive to the choice of this length; these values were chosen as being comparable to the length of the features observed in the images which ensured the robustness of the method. The slightly longer length for the X-ray image, corresponding to 10 pixels, was used on account of the larger pixel size for this data. The estimate of the fibre orientation is given by the colour/greyscale value of the patch within each box, as denoted by the scale bar. Estimates match well with a visual check. The image analysis algorithm was reasonably robust, even when the fibres were not so elongated because the sectioning plane was at a significant angle to the fibres or when the image not so sharp (e.g. because of polishing or focussing difficulties).

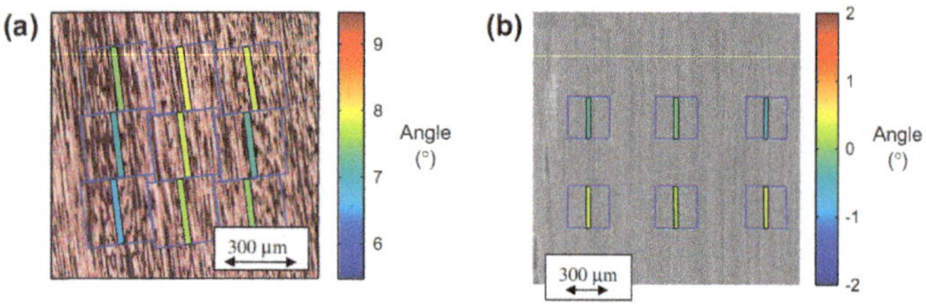

Figure 5. Typical analyses of the fibre orientation angle: (a) polished micrograph sample and (b) X-ray image. The patch colour/grayscale value denotes the orientation angle in degrees, as per the scale bar. The superimposed squares show the extent of the averaging regions, which have side lengths of 300 μm.

Fig. 6b shows the results of the analysis on the polished sample given in Fig. 6a. The orientation angle was evaluated at regions of interest of side-length 300 μm spaced 300 μm apart, to give around 300 measures of the fibre angle. In this case three sets of images have been stitched together, so evaluations have been avoided where images are joined. Regions in which a fibre orientation evaluation was not possible are indicated by grey patches and superimposed crosses. Many of these points correspond to areas outside the 0° plies, though there are some regions where the image quality was not good enough or where the fibre orientation was not well defined.

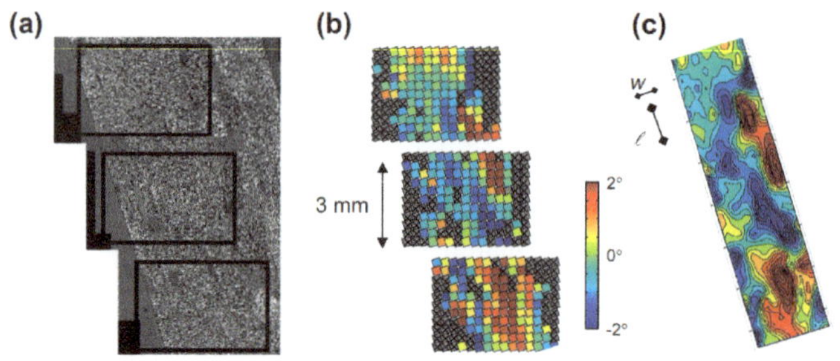

Figure 6. Example of analysis of a polished micrograph section, (a) polished section, (b) fibre orientation map of 0° plies (contour scale bar gives the orientation, relative to the mean value, in degrees) and (c) interpolated gridded data, including correlation length and width of patch as estimated from the corresponding autocorrelation functions shown in Fig. 7. All images are to same scale, as per the scale bar. Grey patches containing crosses mark regions without a valid estimate of the orientation.

Characterisation of Typical Size of Misaligned Region

The method described above can extract the fibre orientation angle as a function of position. In this section we extend the work in [9] to estimate typical dimensions for the patches of misalignment. Previous work has used a spectral approach to estimating wavelengths [10]. Here we simplify the analysis to extract a single measure for the characteristic length l along the fibre direction and the width w transverse to the fibres of regions of misalignment. The analysis was undertaken by first selecting as large a rectangular region as possible which contains continuous 0° material from each section. The measured variation in misalignment angle data over the area was conditioned using the matlab gridfit algorithm [22], which generates a regular interpolating grid of values from a randomly spaced set of data and avoids difficulties with missed data points. Parameters were chosen to minimise smoothing of the dataset. Fig. 6c shows an example of the resulting interpolated data taken from the central region of the polished sample of Fig. 6a. A 2D autocorrelation function was then determined for the gridded data and used to find the autocorrelation functions along and transverse to the fibre direction. The autocorrelation function R(x), scaled to equal one at zero lag x, is defined for a function f(x) with mean μ and variance σ^2 as

$$R(x) = \frac{E[(f(x') - \mu)(f(x' + x) - \mu)]}{\sigma^2} \quad (1)$$

where E is the expected value operator. The correlation length is defined here as the lag distance x where the autocorrelation function equals 0.1, following standard practice in analysis of surface roughness length scales [23]. The autocorrelation function is used in a similar way to define a length scale in other fields, for example in turbulent flows [24]. Autocorrelation functions for the misalignment data given in Fig. 6are plotted in Fig. 7. In this case the correlation length and width of waviness are estimated as 0.58 and 1.21 mm, respectively. These lengths are included in Fig. 6c, illustrating how the correlation length provides a reasonable approximation of the size of the waviness regions.

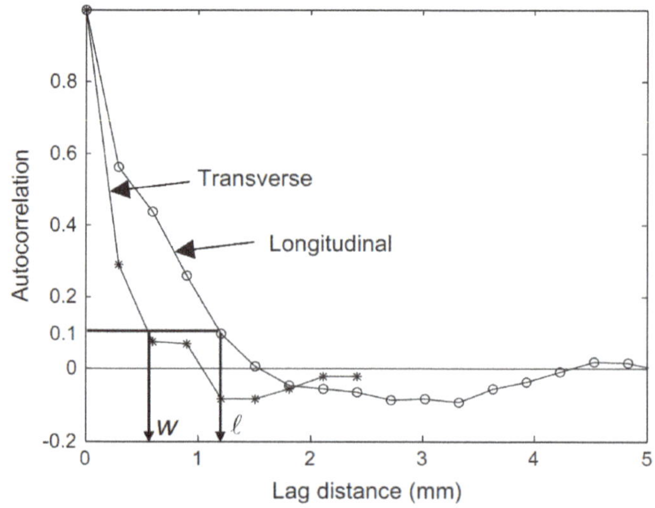

Figure 7. Autocorrelation functions for data from Fig. 6 for a polished micrograph sample used to identify the correlation length and width of waviness regions.

Discussion of Potential Errors and Sensitivity to Analysis Parameters

There are three areas of error or uncertainty present in identification of fibre orientation: errors arising from poor definition of the orientation in the image, the accuracy of the algorithm itself and uncertainty over how to choose an appropriate area over which to average the fibre directions.

Poor Definition of the Orientation in the Image

One of the advantages of the MFIA algorithm is that relatively low resolution microscope images can be used to extract a fibre orientation. Moreover it is not essential that the fibres are precisely sectioned along their length as long as a preferred orientation can be seen. However there comes a point at which the visual image no longer provides sufficient detail to detect a fibre direction. It is important with the automated approach to detect and exclude spurious orientation estimates arising, for example, where there are no 0° fibres, poorly polished regions or large non-fibre features (e.g. voids). On the other hand, where such 'missing data' occur, subsequent interpolation may miss real variations in misalignment near defects, so that it is important to minimise such exclusions. Hence the MFIA algorithm was modified to include three criteria which were used to validate an angle measurement: (i) the reduction in correlation function associated with alignment of the trial fibre and image features had to exceed a threshold value, typically 5% of the background value, (ii) a good fit was required between the calculated variation of correlation function with angle and a quadratic fit through the data near the minimum (typically the standard deviation in the difference between the data and the fit was less than 25% of the range of the correlation function), and (iii) the average intensity within the region had to fall within a specified range. A wide range was allowed for the last criterion, which was used to exclude in a simple way regions of the micrograph beyond the edge of the sample or plies which were not at a 0° orientation. Values of these validation parameters were tuned by comparison with visual examination for the particular type of image used. These criteria provided a robust and automated way of excluding regions which were obviously inappropriate, on closer inspection, as not covering regions with well defined fibre orientations, while not affecting the measurement for good parts of the image. Although the choice of these parameters was not critical, nevertheless it was important to select reasonable values to avoid spurious measurements. Alternative screening of appropriate parts of the sample, for example using manual segmentation, could have been used, although this would have been more cumbersome.

Accuracy of the Algorithm

Visual spot-checks were made on the data to verify that the detected orientation matched a reasonable 'by eye' estimate. To assess the accuracy in an objective way, artificial fibre patterns were created with thin elliptical shapes all orientated at 10° but with random centre locations. Ten realisations were constructed, see Fig. 8a for an example. Using the standard algorithm with a relatively course interval between angle steps of 1°, the estimated angle differed from the nominal value of 10° by at most 0.13° in any of the 10 realisations. Introducing noise into the signal causes a significant reduction in accuracy only when the image itself is seriously degraded. For example the accuracy is still better than 0.17° for the image of Fig. 8b, which has random noise introduced at every pixel, with the level of noise uniformly distributed in the range ±194% of the difference between light and dark intensity levels. With this level of noise the measures for determining whether the angle estimate is valid are near the values used to reject estimates.

(a) (b)

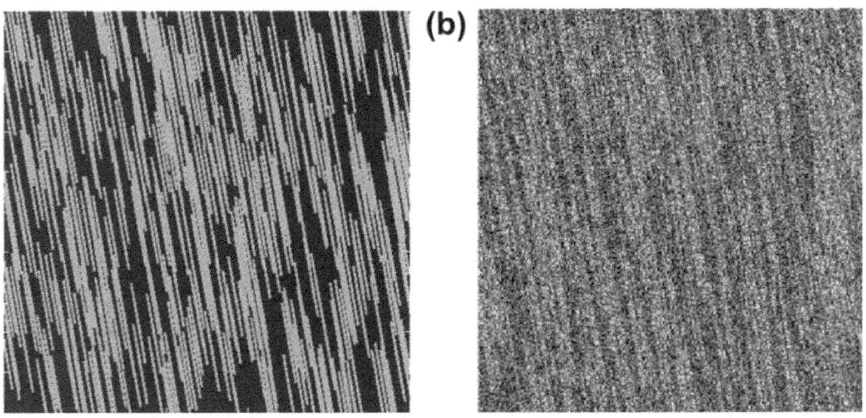

Figure 8. Assessment of accuracy of fibre orientation, (a) typical artificial fibre pattern and (b) artificial pattern with noise which is beginning to degrade the accuracy. As these are artificial images there is no associated length scale.

Choosing the Averaging Area

To guide the choice of averaging area, it is noted that fibre waviness patches of size smaller than about 100 times the fibre diameter d are significantly strengthened by fibre bending and so not critical to the strength. For the samples measured in this work, very local perturbations in fibre waviness appears to be relatively minor, so that it is appropriate to average over an area of side-length 150–600 μm, corresponding to 25–100d. The choice of averaging area is then dictated by the minimum size that gives an accurate estimate. Fig. 9 compares calculations of fibre orientation angle with averaging area side-lengths of 150 and 300 μm. The overall pattern of orientation is similar for the two analyses, but there seems to be some local variations for the smaller averaging area which may be unreliable. Moreover there are more data-points rejected as unreliable for the smaller sampling distance, as it becomes more difficult to extract the orientation over a small area. The standard deviation in the fibre orientation l, taking the mean of three prepreg out-of-plane samples, equals 0.85°, 0.70° and 0.56° for averaging area side-lengths of 150, 300 and 600 μm, respectively. There is a significant decrease in φ' with increasing averaging area, as minor changes within the sample area are averaged out. There is a similar sensitivity of correlation length and width to sample area size. For example the correlation length l, taking the mean of three prepreg out-of-plane samples, equals 1.22,

1.57 and 1.96 mm for averaging area side-lengths of 150, 300 and 600 µm, respectively. To give a compromise of reliability while maintaining adequate spatial sensitivity a side-length of 300 µm is chosen in the analysis.

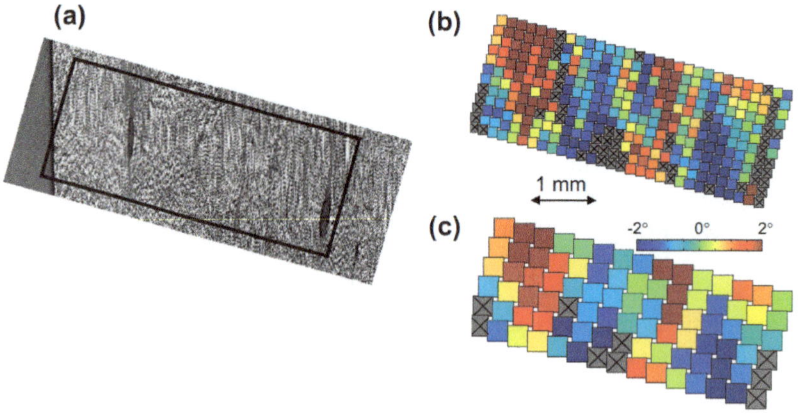

Figure 9. Effect of analysis area on orientation angle for RTM sample. (a) polished micrograph section, (b) 150 µm averaging area side-length and (c) 300 µm side-length. The scale bar shown on the figure applies to all three images.

RESULTS

Polished Sections

Fig. 10a and b shows typical normalised histograms of fibre orientation angles on out-of-plane sections, relative to the mean orientation, for prepreg and RTM samples. There is an approximately normal distribution of angles which can be characterised by the standard deviation φ' in the fibre orientation angle. Fig. 11a plots φ' for the different samples, grouped by section orientation and component. Values are in the range 0.6–1.8°, i.e. similar to waviness data reported previously for carbon fibre composites [5], [9],[10] and [12]. The mean values of the angle standard deviations φ' for the prepreg component are 1.12° and 0.70°, for in-plane and out-of-plane sections, respectively. The corresponding values for the RTM in-plane and out-of-plane sections are 1.37° and 1.63°, respectively. The fibre mis-orientation angles for out-of-plane waviness of the prepreg component are significantly smaller than for the other sample measurements (significance at the 1% level confirmed using a Student's t-test). This probably reflects both differences in the fabric geometry and factors associated with the consolidation process.

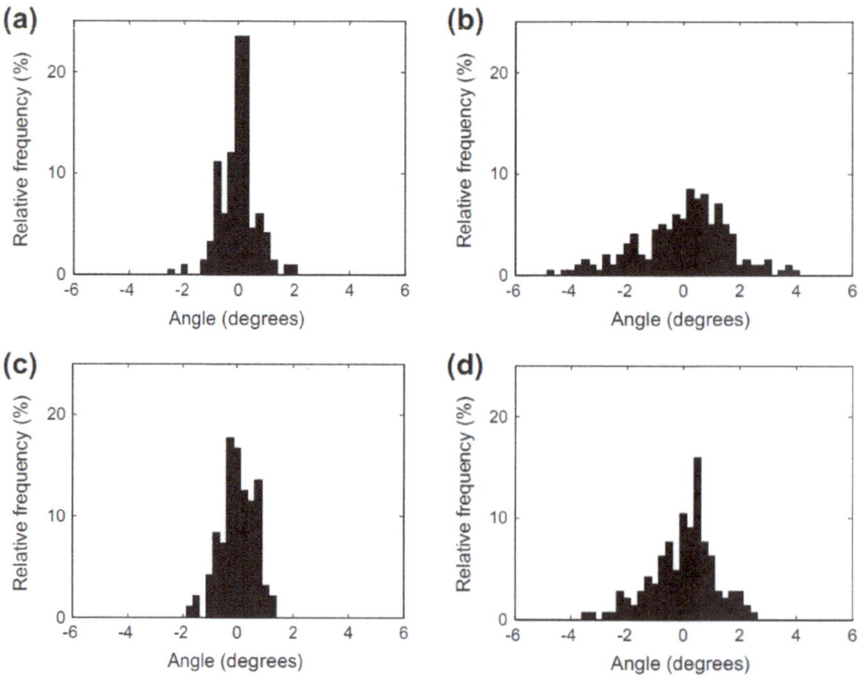

Figure 10. Normalised histogram of fibre orientation angles, relative to the mean orientation, for: (a) polished micrographs of out-of-plane prepreg, (b) polished micrographs of out-of-plane RTM, (c) X-ray out-of-plane prepreg and (d) X-ray in-plane prepreg.

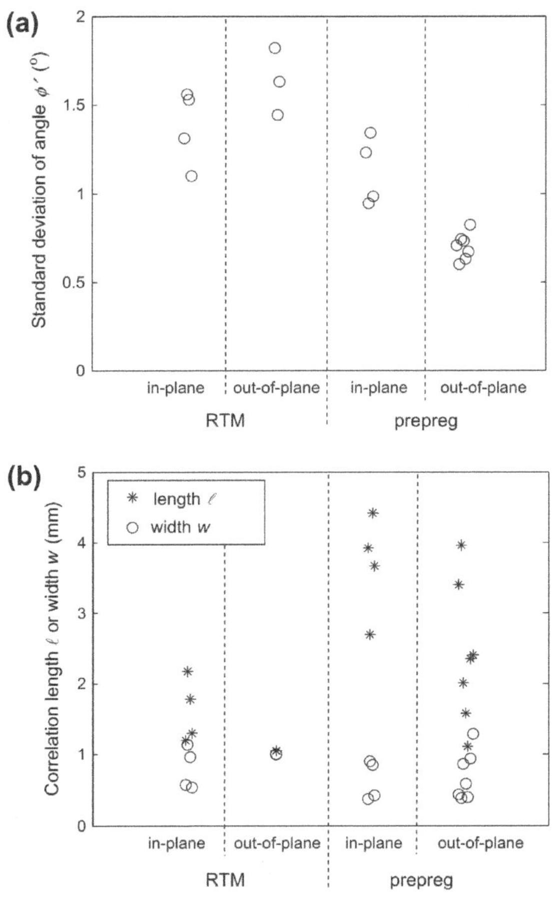

Figure 11. Polished micrograph sample measurements: (a) standard deviation in fibre orientation and (b) correlation length scales.

Fig. 11b plots corresponding values of the correlation length l and width w of the different samples. There is significant scatter within a given category, part of which may be due to the fact that the areas used for characterisation are relatively small compared with these correlation lengths. The mean values of the correlation length l for the prepreg component are 3.7 mm and 2.4 mm, for in-plane and out-of-plane sections, respectively. Values for the RTM in-plane and out-of-plane sections are 1.6 mm and 1.1 mm, though in the later case there is only one data point rather than a mean as ply blocks were too thin to make a reliable evaluation of l and w in most of these cases. These lengths match reported values in the literature for the wavelengths of waviness in carbon fibre composites [10], [14], [15], [16] and [17]. Regions of waviness are significantly longer in the prepreg as compared with the RTM component (significance at the 5% level was confirmed using a Student's t-test, but results were not significant at the 1% level). The nature of the manufacturing process is presumably affecting these length scales. The width w of the patches tends to be significantly smaller than the length (significance at the 1% level confirmed using a Student's t-test), ranging from 0.37 to 1.30 mm, being broadly similar across the sample types. The mean width, averaging over all sample types, is 0.73 mm.

X-ray Images

Fig. 10c and d shows histograms of fibre orientations, relative to the mean, from X-ray images of the prepreg sample. Again fibre angles are approximately normally distributed. The standard deviations of fibre orientation are 1.21° and 0.75° for in-plane and out-of-plane misalignment, respectively. These estimates agree well with values of 1.12° and 0.70° (in-plane and out-of-plane, respectively) for the mean of the standard deviations from all the corresponding polished samples.

An advantage of X-ray imaging is that in and out-of-plane misalignment angles can be measured at the same location, c.f. the results of Requena et al. [8]. It might be expected that factors causing high misalignment in one plane would lead to high misalignment in another. However an insignificant Pearson's correlation coefficient of 0.29 was observed between the misalignment angles in the two fibre orientations at the same location. This lack of correlation suggests that little extra information would be gained for these samples by characterising the waviness in a 3D manner, rather than using 2D slices from the images. However such a 3D quantification might add useful additional information where the quality of the sample imaging was relatively poor.

CONCLUSIONS

The method described by Creighton et al. [9] to measure fibre misalignment in composite materials has been improved and applied to the unidirectional ply blocks of carbon fibre in two industrial components. These were made either using resin transfer moulding or prepreg/vacuum consolidation. Measurement of waviness in the plane of the component and out-of-plane were undertaken by appropriate sectioning, either of polished samples or micro-CT X-ray images. Values for the standard deviation in fibre orientation angle were in the range 0.6–1.8°. The fibre alignment varied significantly less for in-plane waviness of the prepreg component, compared with the other sample measurements. The measurement technique of [9] has been extended using an autocorrelation function approach to characterise the length and width of the waviness region along and transverse to the fibres. The length of the waviness patch is in the range 1.1–4.4 mm, being significantly greater for the prepreg than the RTM component. The width of the patches tend to be significantly smaller than the length, ranging from 0.37 to 1.30 mm, being broadly similar across the sample types. Measurements of waviness from the X-ray images were in good agreement with the data from the polished sections, confirming the effectiveness of the fibre orientation measurement technique with X-ray images. The methods presented in this paper to characterise the spatial distribution of fibre orientation could equally be applied to other measurement techniques, either from polished micrographs [11] or X-ray imaging [7] and [8].

ACKNOWLEDGEMENTS

The authors acknowledge the support of EPSRC (reference EP/G012938/1) and MOD/DSTL, and thank colleagues at the Universities of Cambridge and Bristol and at the industrial collaborators for their help, particularly with supply of the samples.

REFERENCES

1. Fleck NA. Compressive failure of fibre composites. Adv Appl Mech 1997;33:43–119.
2. Argon AS. Fracture of composites, Treatise Mat Sci & Tech, vol. 1. New York: Academic Press; 1972.

3. Shu JY, Fleck NA. Microbuckle initiation in fibre composites under multiaxial loading. Proc Roy Soc 1997;A453:2063–83.
4. Lemanski SL, Sutcliffe MPF. Compressive failure of finite size unidirectional composite laminates with a region of fibre waviness. Composites Part A 2012;43:435–44.
5. Yurgartis SW. Measurement of small angle fiber misalignments in continuous fiber composites. Compos Sci Technol 1987;30:279–93.
6. Barwick SC, Papathanasiou TD. Identification of sample preparation defects in automated topological characterization of composite materials. J Reinf Plast Compos 2003;22:655–69.
7. Salaberger D, Kannappan KA, Kastner J, Reussner J, Auinger T. Evaluation of computed tomography data from fibre reinforced polymers to determine fibre length distribution. Int Polym Process 2011;26:283–91.
8. Requena G, Fiedler G, Seiser B, Degischer P, Di Michiel M, Buslaps T. 3Dquantification of the distribution of continuous fibres in unidirectionally reinforced composites. Composites: Part A 2009;40:152–63.
9. Creighton CJ, Sutcliffe MPF, Clyne TW. A multiple field image analysis procedure for characterisation of fibre alignment in composites. Composites: Part A 2001;32:221–9.
10. Liu D, Fleck NA, Sutcliffe MPF. Compressive strength of fibre composites with random fibre waviness. J Mech Phys Solids 2004;52:1481–505.
11. Kratmann KK, Sutcliffe MPF, Lilleheden LT, Pyrz R, Thomsen OT. A novel image analysis procedure for measuring fibre misalignment in unidirectional fibre composites. Compos Sci Technol 2009;69:228–38.
12. Kratmann KK, Thomsen OT. Experimental characterisation of parameters controlling the compressive failure of pultruded unidirectional carbon fibre composites. Appl Mech Mater 2010;24–25:15–22.
13. Wang J, Potter KD, Hazra K, Wisnom MR. Experimental fabrication and characterization of out-of-plane fibre waviness in continuous fibre-reinforced composites. J Compos Mater 2012;46:2041–53.
14. Kyriakides S, Arseculeratne R, Perry EJ, Liechti KM. On the compressive failure of fibre reinforced composites. Int J Solids Struct 1995;32:689–738.
15. Clarke AR, Archenhold G, Davidson NC. A novel technique for determining 3D spatial distribution of glass fibres in polymer composites. Compos Sci Technol 1995;55:75–91.
16. Clarke AR, Archenhold G, Davidson NC, Fleck NA. Determining the power spectral density of the waviness of unidirectional glass fibres in polymer composites. Appl Compos Mater 1995;2:233–43.
17. Wisnom MR. The effect of fibre waviness on the relationship between compressive strength of unidirectional composites. J Compos Mater 1994;28:66–76.
18. Potter KD, Langer C, Hodgkiss B, Lamb S. Sources of variability in uncured aerospace grade unidirectional carbon fibre epoxy preimpregnate. Composites: Part A 2007;38:905–16.
19. Mardia KV, Jupp P. Directional statistics. 2nd ed. John Wiley and Sons Ltd.; 2000.
20. d'Angelo P. Hugin-2010.2.0. 2010: [accessed 29.06.2012].
21. Illingworth J, Kittler J. A survey of the Hough transform. Comp Vision Graph Image Process 1988;44:87–116.
22. D'Errico J. 2010. Surface fitting using gridfit. 2010: [accessed 29.06.2012].
23. Williams JA. Engineering tribology. Cambridge University Press; 2005.
24. Tennekes H, Lumley JL. A first course in turbulence. MIT Press; 1972.

CHAPTER 19 NANOSTRUCTURED BIOCOMPOSITES

Farhan Ansari, Mikael Skrifvars, and Lars Berglund

Keywords: Nano composites; Fracture toughness; Interface; Fractography; Cellulose nanofiber (CNF)

INTRODUCTION

The use of natural fibers (cellulosic plant fibers) as reinforcement for polymers is of interest due to the favorable mechanical properties of the fibers. Plant fibers are considered as a substitute for the more commonly used glass fibers. The recent breakthrough in isolation of cellulose nanofibrils (CNF, also termed 'cellulose nanofibers') with low energy consumption has opened new possibilities for these 'green' materials [1],[2] and [3]. Cellulose nanofibrils are the primary load carrying unit in plant fibers and have high tensile modulus. During the past decade, the interest in nanocellulosic-based materials has gradually increased, including nanocellulose-reinforced composite materials [4], [5] and [6].

In the present study, CNF is reinforcing unsaturated polyester (UP), one of the most commonly used thermoset resins. UPs are short chain prepolymers formed by condensation reaction between diols and di-functional unsaturated and saturated acids. This oligomer is mixed with up to 40 wt% styrene, which gives a fluid and processable resin and serves to crosslink the resin into a three-dimensional molecular network[7] and [8]. The properties of the resin can be tuned by variation in either the type of acid/diol or their relative amounts. This, combined with low cost, wide availability and flexible processing characteristics, makes UP attractive for a wide variety of applications including adhesives, automotive components, boat hulls, small aircraft structures, etc. Existing literature on CNF/UP composites primarily focuses on low content of CNF (up to 20%). Chirayil et al. [9] and [10] studied UP composites with CNF content up to 5 wt% and observed that the mechanical properties were favorable at 0.5 wt%. The drop in properties at higher CNF content is usually due to agglomeration of the fibrils, and the agglomerates initiate premature failure. A reduction in gel time of UP in the presence of CNF was reported and it was suggested that CNF accelerates the curing reaction. The mechanism for such an effect is unclear.

Coupling agents have been used to improve cellulose-UP interactions at the interface. Hu et al. [11] used silanes to modify bacterial cellulose (BC) fibrils, which were then mixed with UP resin. At 10 vol% of fibrils, the strength increased from 60 MPa for untreated BC composites to 131 MPa for the treated BC counterparts, possibly due to improved dispersion and stress transfer at the interface. Gao et al. also modified BC fibrils with silanes and the strength was optimum at 20 vol% fibrils [12]. More recently, Kargarzadeh et al. used silane compounds to modify cellulose nanocrystals and studied UP composites with up to 6 wt% of nanocrystals [13]. Builes et al. used block copolymers to disperse 1 wt% of CNF in UP resin, and observed marginal increase in fracture toughness [14]. Since coupling agents represent a significant cost when used with high specific surface area nanofibers, no coupling agents are used in the present study.

A significant advantage of CNF is the ability to form a network when filtered from water suspension. This network, often referred to as nanopaper, has a modulus of ca. 10–14 GPa and a strength up to 300 MPa [6], [15], [16] and [17]. After filtration, the wet network of fibrils can be dried to form a porous template [16] and [18], which can serve as mat reinforcement for cellulose biocomposites. The problem with agglomeration is reduced in this approach since the CNF can be preserved in the form of a well-dispersed nanofiber network. Yano et al. [19] impregnated a nanopaper with acrylate monomers followed by hot-pressing and curing. The resulting composites

Citation: Farhan Ansari, Mikael Skrifvars, Lars Berglund, Nanostructured biocomposites, Composites Science and Technology, Volume 117, 29 September 2015, Pages 298-306, ISSN 0266-3538, http://dx.doi.org/10.1016/j.compscitech.2015.07.004.

showed high optical transparency. Formaldehyde, epoxy (EP) and other resins for thermoset polymers have been used to prepare composites with CNF via similar network approach [19], [20], [21],[22] and [23].

In the present work, we report on UP biocomposites with high CNF content. A specific aim is to preserve individual, discrete nanofibers and other nanostructural features of the CNF network also in the biocomposite structure. The strain hardening and ductility of the nanostructured nanopaper network by itself can possibly address the intrinsic brittleness problem of UP thermosets. Nanostructured UP biocomposites with CNF content as high as 45 vol%, much higher than in any previous study, are successfully processed and characterized. The effect of moisture is studied by mechanical testing of the materials at different relative humidities. The roles of the interface and matrix characteristics as well as the importance of nanoscale dispersion of individual cellulose nanofibers are discussed.

MATERIALS AND METHODS

The Polyester used in this work was an isophthalic acid based unsaturated polyester under the trade name Aropol™ S580, kindly supplied by Ashland. Benzoyl Peroxide (BOP) (25% in water) was obtained from Sigma Aldrich.

Cellulose nanofibrils (CNF) were prepared from never-dried pulp containing 13.8% hemicelluloses, 0.7% lignin (kindly supplied by Nordic Paper, Sweden) according to a method reported by Henriksson et al. [1]. Wood pulp fibers are based on fibers in the form of tubes of about 3 mm in length and 25 μm in diameter with a cell wall thickness of 2–5 μm. The pulp fibers (obtained as a wet cake with a dry content of ca. 20%) was suspended in water and the solid content was adjusted to ca. 2 wt%. It was then pretreated via enzymatic action (Novozym® 476) which facilitated the subsequent disintegration process. Thereafter, the pulp fiber suspension was passed through a microfluidizer (Microfluidics Inc, USA), with chambers of 400 μm/200 μm (3 passes) and 200 μm/100 μm (5 passes). High shear forces act on the fibers when passed through the microfluidizer. As a result, individual CNF nanofibers are disintegrated from the fibers, with diameters below 10 nm and lengths in the range of typically 0.8–2 μm. No large cell wall fragments were present in the CNF used. After processing, the CNF was obtained as a stable viscous gel (~2 wt% dry content). The specifics of the enzymatic pretreatment and the fibrillation process have been discussed in detail in Refs. [1] and [24].

Processing

The gel like suspension of CNF was diluted to 0.1% concentration in order to facilitate subsequent filtration. The suspension was homogeneously dispersed by mechanical stirring in an Ultraturrax (model D125 Basic, IKA, Germany) at a speed of 10,000 rpm. The shearing was carried out until the bundle like flocculates of CNF gel (a few mm in size) were well dispersed, which typically took 10 min. A detailed study of CNF colloidal suspensions is carried out in Refs. [25] and [26], where dynamic light scattering data is used to confirm the lack of large aggregated particles. The suspension was filtered over vacuum using a 0.6 μm pore size membrane (Millipore, USA), to ensure that the fibrils were not lost during filtration. The water in the CNF template was exchanged to acetone by placing the network in an acetone bath. The impregnating solution was a combination of UP and acetone, mixed in a predetermined ratio. 3 wt% BOP was added as the initiator. UP curing takes place by styrene monomers reacting to form short chemical crosslinks between double bonds on the polyester molecules [7] and [8]. The CNF-acetone template was soaked in the UP-acetone solution. The amount of UP in the UP-acetone solution was varied to obtain composites with different compositions. The CNF template was impregnated in solutions containing 80%, 40% and 20% of UP (in acetone) to obtain composites with 16, 30 and 45 vol% of CNF respectively. The template was later removed from the bath and kept at room temperature (for a few minutes) to allow the acetone to evaporate. It was then cured at 90 °C for 2 h followed by post curing at 140 °C for 3 h. The evaporation of acetone and curing at high temperature caused the loss of some styrene. However, the weight loss was minimal (less than 5% by gravimetric analysis) since benzoyl peroxide decomposes rapidly at 90 °C to start the curing reaction. The samples were then hot pressed (Fontijne Presses) for 30 min at 120 °C at a pressure of 5 MPa. Composite samples of 100–200 μm thickness were obtained, while the wet CNF-acetone preform had a thickness of ~500 μm. Fig. 1a summarizes the preparation process. The neat CNF sheet was prepared by drying the wet CNF nanopaper between filter papers at 93 °C for 15 min, in vacuum.

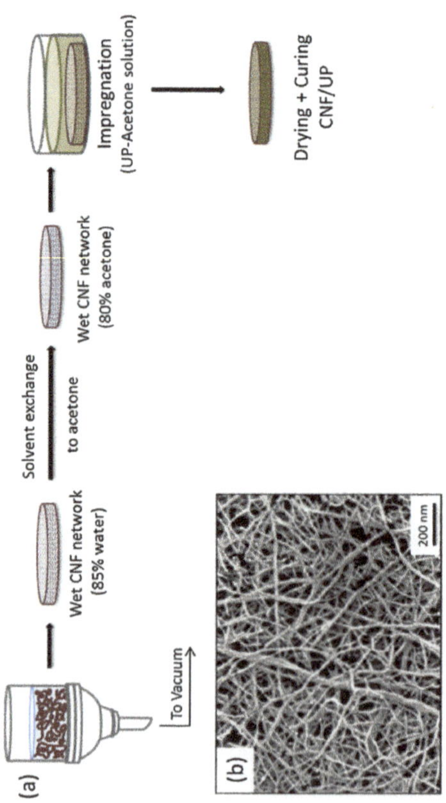

Figure 1. (a) An illustration of the preparation method, note that impregnation occurs by UP component diffusion from UP-acetone solution (b) Surface morphology of CNF aerogel corresponding to the CNF network used in the present study.

The weight fraction of CNF in the final composite was determined based on the initial dry weight of CNF in the suspension and the final weight of the dried composite. Conversion was performed using a density of 1.18 g/cm³ for

UP and 1.5 g/cm³ for CNF [27]. Vol% is used to describe CNF content in composites.

Characterization

Fourier transform infrared spectroscopy (FTIR) analysis was performed on a Perkin Elmer spectrum 2000 using a triglycine sulfate (TGS) detector. The FTIR instrument was equipped with a Golden Gate single reflection diamond ATR-accessory from Graseby Specac Ltd. The spectra were collected in the range of 4000–600 cm^{-1}. The mechanical properties were measured on Instron 5944 (500 N load cell). Rectangular samples (30 mm by 5 mm) were conditioned at 23 °C and Relative Humidity (RH) of 50% (or 90%) for 2 days and tested in tensile mode at a strain rate of 10%/min. The strain was measured using 2D Differential Speckle Photography (DSP) and correlation was done with a LIMESS software. The values reported here are averaged over a minimum of 5 samples. The weight change of the samples during moisture sorption was measured and is presented as % increase based on the initial weight (dry) of the composite. Yield stress for the composites was calculated from the intersection of the tangents at the initial elastic region and the following plastic region; for the neat UP, a line parallel to the elastic region at an offset of 0.2% was drawn and its intersection with the curve taken as yield stress. The strain hardening coefficient was calculated from the Hollomon›s equation. The engineering stress (and strain) were first converted to trues stress/strain, and were thereafter fitted in an exponential relation to obtain the strain hardening coefficient (n) [28].

$$\sigma_t = K\varepsilon_t^n \tag{1}$$

Where σ_t is true stress, ε_t is the true strain, k is a constant and n represents the strain hardening coefficient. The value of n was calculated from the slope of a logarithmic plot of equation (1).

A micromechanics model (equation (2)) based on a combination of the Halpin Tsai model (equations (4) and (5))) and rule of mixtures (equation (3)), originally developed to predict the modulus of random-in-plane composites with cylindrical fibers [29] and [30], was used to back-calculate the effective CNF modulus.

$$E_C = \frac{3}{8}E_L + \frac{5}{8}E_T \tag{2}$$

where,

$$E_L = E_{f,l}\cdot V_f + E_m\cdot\left(1 - V_f\right) \tag{3}$$

$$E_T = E_m\left(\frac{1 + 2\cdot\eta\cdot V_f}{1 - \eta\cdot V_f}\right) \tag{4}$$

$$\eta = \frac{\left(E_{f,t} - E_m\right)}{\left(E_{f,t} + 2E_m\right)} \tag{5}$$

In the above series of equations, V_f and V_m represent the volume fraction of CNF and matrix respectively; E_c, E_m, $E_{f,l}$, $E_{f,t}$ represent the composite modulus, matrix modulus, axial CNF modulus and transverse CNF modulus respectively. Experimental values of E_m and E_c were used in these equations, $E_{f,t}$ was taken as 15 GPa [31] and $E_{f,l}$ was treated as the unknown variable. By substituting the respective values, $E_{f,l}$ was calculated for each volume fraction of the composite. The value of $E_{f,l}$ so calculated is a measure of the effective contribution from the CNF.

Fracture surfaces from tensile tests (at RH 50%) were observed in a Field-Emission Scanning Electron Microscope (FE-SEM) (Hitachi S-4800, Japan) after sputtering the surface with a few nanometer-thin platina-palladium layer.

Dynamic Mechanical Thermal Analysis (DMTA) testing was done on a TA instrument Q800 in tensile mode. Sample size was ca. 3 mm by 10 mm. Measurements were taken at a frequency of 1 Hz and an amplitude of 10 µm in the range of –20 °C to 200 °C at a rate of 5 °C/minute. The glass transition temperature was estimated from the position of the loss modulus peak.

A fracture toughness measure for the materials was obtained from single edge notched specimens. The width (w) of the samples was in the range of 7–10 mm and cracks of lengths a = 3–5 mm were introduced above the glass transition temperature. Samples with a/w in the range of 0.45–0.55 are reported here. Crack growth started at maximum stress and resulted in immediate catastrophic failure. The apparent fracture toughness was calculated from the following equation [32].

$$K_Q = Y\sigma a^{0.5} \tag{6}$$

where; K_Q = apparent fracture toughness, σ = maximum stress at failure, a = crack length, Y is the geometry factor and for single edge notched specimens is expressed by the following equation.

$$Y = 1.99 - 0.41(a/w) + 18.7(a/w)^2 - 38.48(a/w)^3 + 53.85(a/w)^4 \tag{7}$$

All samples tested for fracture toughness were 200 ± 20 µm in thickness. Samples are unlikely to be subjected to plain strain conditions; hence the term K_Q is used when referring to apparent fracture toughness.

RESULTS AND DISCUSSION

Processing

The composites were prepared using a prepreg approach. The final prepreg consists of a random-in-the-plane cellulose nanofiber network in a matrix of soft, unreacted UP mixture. This prepreg can thus be stacked and shaped into 3D structures, to be cured at elevated temperature. The prepregs can also be used as an adhesive film.

A dilute colloidal water suspension of CNF (0.1 wt% solid content) was first filtered using vacuum to obtain a wet CNF network [25], [26] and [33]. In order to preserve the dispersion of individual nanofibers, this network was solvent exchanged to acetone and then impregnated in a UP-acetone solution (see Fig. 1a). After impregnation, the acetone was evaporated and composites were cured at elevated temperature. By diluting the impregnating solution to different extents, the matrix content diffusing into the CNF network could be controlled. This procedure ensures fairly well-dispersed CNF fibrils, compared with UP impregnation of dried and porous CNF mats.

Unsaturated polyesters can have a molar mass up to several kDa and would theoretically diffuse slower than styrene monomers into the CNF network. However, in this study the impregnating solution was diluted with acetone. This facilitated UP diffusion. This was confirmed by spectroscopic analysis where the intensity of carbonyl peaks (from polyesters) and aromatic peaks (from styrene) from FTIR spectra were compared for the neat UP matrix and also for the composites. The ratio of these two peaks was approximately the same, indicating that the relative amount of the two components (styrene and polyester) was the same in the composites as in neat UP.

The starting network with 15 vol% dry content (used in this study) corresponds to a network with 85% porosity. The network structure of a CNF network with 15% dry content, prepared by supercritical drying as described in Ref. [13] is presented in Fig. 1b. This structure is comparable to the network in the present CNF-acetone gel. The pore size is at the 100 nm scale. Note that the scale of the structure is very fine; a typical plant fiber diameter is about 20 times the width of Fig. 1b. One advantage with 'conventional' plant fiber based composites is that there are industrial processing methods available. CNF–UP processing at an industrial scale remains a challenge, but the present process is one possibility with an industrial potential.

Mechanical and Hygro-mechanical Properties

The storage modulus (E′) as a function of temperature (T) is presented in Fig. 2a. The composites have improved modulus, which increases with increasing CNF content. The neat UP as well as the composites are stable up to 60 °C, after which the glass transition begins. The neat UP undergoes a broad transition and the modulus is reduced by 2 orders of magnitude. The broad transition is because the distribution of styrene cross-links is heterogeneous in length and location. The CNF/UP composites preserve high modulus even after the T_g because the CNF network retains mechanical integrity. The reinforcement effect of the CNF is stronger in the rubbery state. The rubbery state modulus of the 45% CNF/UP composite is even higher than the glassy state modulus of the neat UP matrix.

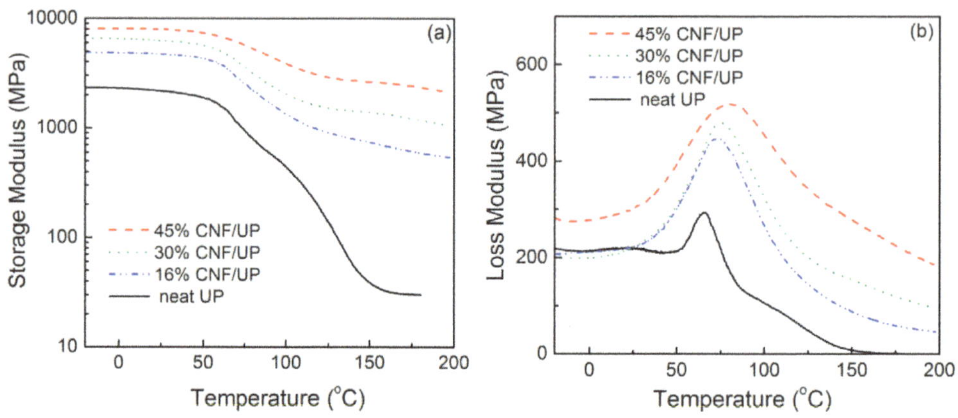

Figure 2. DMTA plots of (a) Storage Modulus vs temperature (b) Loss modulus vs temperature for neat UP and CNF/UP composites.

Table 1. Data for CNF/UP characteristics and mechanical properties.

	CNF vol %	Elastic Modulus (GPa)	Yield strength (MPa)	Strain hardening co-efficient	Ultimate strength (MPa)	Strain to failure (%)	Moisture sorption (wt %)	Toughness K_{IQ}(MPa.m$^{0.5}$)	T_g(°C)
RH 50	0	2.9 ± 0.2	49 ± 2.0	–	54.5 ± 4.8	3.4 ± 0.3	0.9	2.5 ± 0.3	66
	16	5.8 ± 0.4	88 ± 2.1	0.27 ± 0.02	106 ± 3.4	6.4 ± 0.6	2.2	3.5 ± 0.3	73
	30	7.0 ± 0.4	93 ± 6.4	0.40 ± 0.03	156 ± 9.0	7.4 ± 0.8	2.6	6.4 ± 1.0	75
	45	8.5 ± 0.6	103 ± 1.4	0.42 ± 0.02	173 ± 12	7.6 ± 1.3	4.6	6.1 ± 0.7	78
	100	14.9 ± 0.8	111 ± 4.0	0.50 ± 0.02	243 ± 16	6.0 ± 0.2	7.8	9.2 ± 0.4	–

RH 90	0	2.7 ± 0.1	44 ± 1.5	–	52.0 ± 1.7	4.3 ± 0.4	2.3		
	16	4.0 ± 0.4	52 ± 2.5	0.48 ± 0.04	97.7 ± 6.6	9.9 ± 0.5	5.7		
	30	5.0 ± 0.5	52 ± 2.7	0.60 ± 0.02	115 ± 24	7.5 ± 2.2	6.9		
	45	6.0 ± 0.3	54 ± 5.9	0.82 ± 0.04	155 ± 18	8.5 ± 1.7	7.8		
	100	9.4 ± 0.4	65 ± 5.0	0.70 ± 0.07	226 ± 14	8.3 ± 1.2	14		

The glass transition temperature (T_g), calculated from the loss modulus peak is reported in Table 1. T_g of the composites increases strongly with increased CNF content – from 66 °C for the neat UP to 78 °C for the 45% CNF/UP composite. This is apparent in Fig. 2b as a shift in the peak to higher temperatures. The T_g increases by ca. 7 °C when 16% CNF is added to the matrix, and then increases gradually with CNF content. Increased T_g indicates constrained molecular mobility in the UP network due to the proximity of high specific surface area CNF fibrils. In the present system, carbonyl groups from UP may interact with the surface hydroxyls of CNF via hydrogen bonding and other secondary interactions [9]. The CNF/UP interface has an important role in determining the composite properties, as will be discussed in the coming sections.

To further evaluate the mechanical performance, uniaxial tensile tests were carried out. The stress-strain curves are reported in Fig. 3 and mechanical properties as a function of CNF content are presented in Fig. 4. The strong reinforcement effect of the CNF network is obvious in Fig. 3a. Neat UP shows limited plastic yielding, and ultimate fracture takes place just beyond the yield point. The composite curves are similar in terms of characteristic tensile behavior. The composites show linear behavior until a strain of 1–2%, after which they yield and undergo linear strain hardening until fracture. Despite the quantitative differences, this characteristic stress strain behavior is preserved at high humidity (Fig. 3b), indicating that the CNF network characteristics are weakened but not totally lost even at RH 90%.

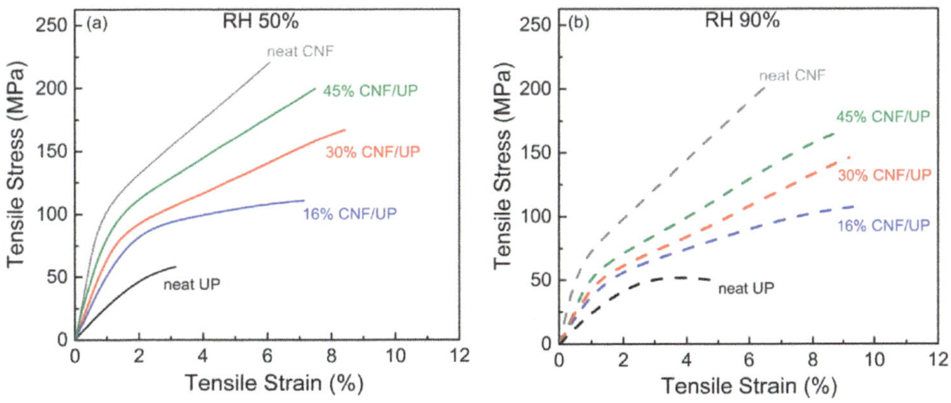

Figure 3. Representative stress strain curves for neat UP and CNF/UP composites conditioned at (a) Relative humidity of 50% and (b) Relative humidity of 90%. The numbers in the legend indicate the volume % of CNF. The curves for neat CNF are obtained from Ref. [23].

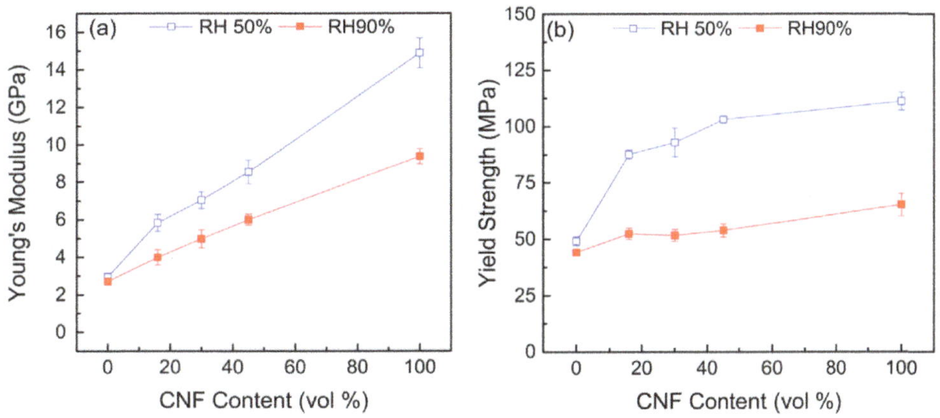

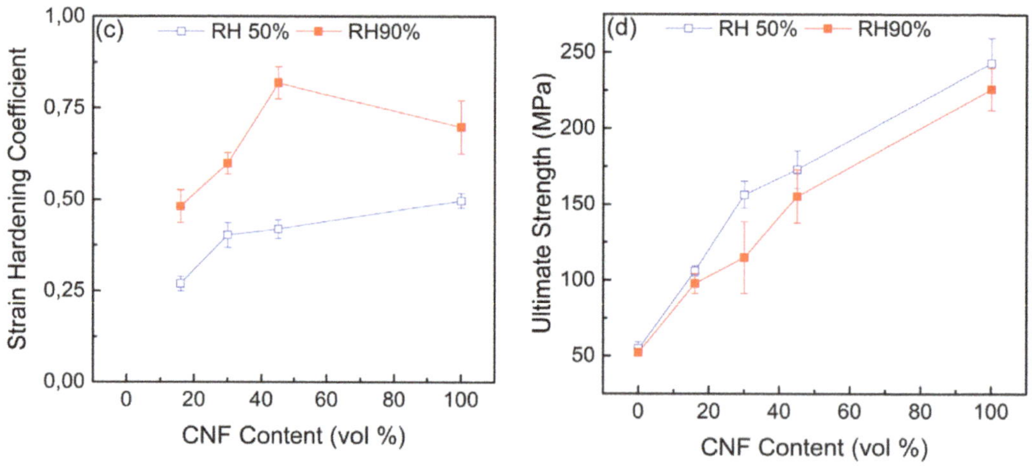

Figure 4. The effect of CNF content on (a) Young's Modulus (b) Yield Strength (c) Strain Hardening Coefficient (see Eq. 1) and (d) Ultimate Strength at a relative humidity of 50% (hollow squares) and 90% (full squares).

The neat UP fails at a strain as low as 3.4%, which increases to 7.6% when 45 vol% CNF is added. This is contrary to typical effects from addition of reinforcing particles or fibers, where the strain to failure tends to decrease. In the present case, the strain to failure for the composites surpassed that of the neat UP as well as the neat CNF network. This synergy between the two components is interesting and unusual. Failure in thermosets typically starts with the formation of shear bands, which then intersect so that nanoscale cavities are formed followed by crack growth to cause macroscopic fracture [34]. The CNF network may limit the formation and growth of nanoscale cavities, thereby delaying biocomposites failure to higher strains.

For better understanding, mechanical properties as a function of CNF content are plotted in Fig. 4 and also reported in Table 1. At a relative humidity of 50%, the reinforcement effects are strong – modulus and ultimate strength increase as the CNF content increases. The mechanical performance of composites based on CNF and UP have been reported to increase up to a certain CNF content (typically a few wt%), after which the dispersion of the CNF becomes a problem – both strength (and occasionally modulus) begin to drop when CNF content increases [9] and [13]. In the present work, the strength and modulus increase continuously with the addition of the CNF network, indicating that CNF agglomeration is limited even at 45 vol% CNF. This is primarily due to the processing strategy adopted, where a preformed CNF network was used so that the CNF agglomeration during mixing/drying was prevented to a large extent. The CNF network increases the yield strength of the matrix, in particular at 15 vol% CNF, and CNF also allows for substantial plastic deformation and delays catastrophic UP failure, seeFig. 3. This is very different from biocomposites based on plant fibers with diameters at the 10 μm scale, where high fiber content correlates with low ductility [35] due to failure initiation at low strain from fiber-matrix debonding.

The composite modulus increases almost linearly with CNF content (Fig. 4a). However, closer scrutiny of modulus and to some extent strength reveals that lower CNF contents provide a relatively stronger reinforcement effect. This can be better understood by back-calculating the effective CNF modulus ($E_{CNF,\ effect}$) in the different composites (see equations (2), (3), (4) and (5)). The effective modulus so calculated, represents the contribution in modulus from the fiber phase which can be interpreted to reveal insights about aggregation effects in the composite. The value of $E_{CNF,\ effect}$ is 42 GPa for 16% CNF/UP and goes down to 29 GPa and 25 GPa for 30% CNF/UP and 45% CNF/UP respectively. This means that the relative contribution from CNF to the composite modulus is higher for 16% CNF/UP than at higher CNF content. One may speculate that an important reason is that there is a larger fraction of CNF agglomerates at high CNF content, and this results in less efficient reinforcement due to the CNF–CNF interfaces. As water enters the interface at elevated humidity, load transfer between fibrils will be compromised. The strain-hardening coefficient (slope in plastic region) also increases with CNF content, see Fig. 4c. Reorientation and stiffening of the CNF network during deformation results in higher stress as the CNF content is increased. Since the coefficient is lower for neat CNF, the UP matrix has a positive contribution to the strain-hardening coefficient by bonding the nanofibers together so that a specific strain results in higher stress. There is very little effect of CNF content on yield strength at 90%RH, and this is interesting. Most likely, the reason is decreased CNF–UP interface adhesion due to interfacial water. It indicates that interfacial adhesion is important for yield strength.

Fracture Surfaces

High resolution FE-SEM images of the composites are presented in Fig. 5. A layered morphology can be discerned even at 16 vol% CNF. However, it is not possible to see any signs of CNF agglomeration. On the contrary, individual fibrils are apparent as white dots and are well dispersed at the scale of 100 nm, especially for the 16 vol% CNF/UP. The higher magnification images (Fig. 5b, d) reveal roughness at the scale of 1 μm, corresponding to local plastic deformation of the UP.

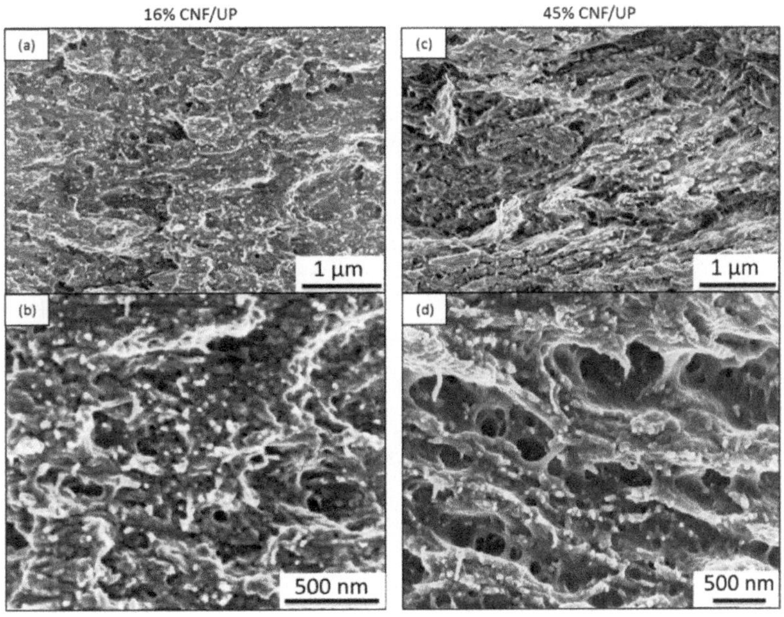

Figure 5. FE-SEM micrographs at different magnifications of cross-sections fractured during tensile tests at RH 50% of (a,b) 16 vol% CNF/UP composite and (c,d) 45 vol% CNF/UP composite.

Role of Moisture

In order to better understand the interfacial characteristics, composites were tested after conditioning at 90% relative humidity. The modulus and yield strength drop considerably. The composites adsorb significant amounts of moisture (see Table 1). At high humidity, water molecules are likely to be preferably located at the high energy CNF/UP interface, reducing the interactions and interfacial adhesion between CNF and UP.

This is especially evident for yield strength data; at RH 90%, the yield strength almost drops down to the value of neat UP. This implies that while load transfer was favorable between UP and CNF at RH 50%, this was not the case at RH 90%. The yield phenomenon is likely to be associated with local shear yielding and CNF/UP debonding. In a composite with high specific interfacial area (small diameter CNF, high volume fraction), yielding may depend strongly on interfacial CNF/UP adhesion.

Although modulus and yield strength drop significantly at RH 90%, the ultimate strength is not affected to the same extent (Fig. 4d). This is primarily because the characteristic strain hardening behavior of the CNF network is retained, and perhaps even improved at high humidity. The slope in the plastic region appear to be unaffected by moisture (Fig. 3). However, a closer analysis using the quantitative values of strain hardening (Table 1) reveals that the strain hardening coefficients become much higher at RH 90% (seeFig. 4c), probably because of the lower stress levels. This implies that the orientation/realignment phenomenon associated with the plastic deformation of CNF network is more dominant since it can operate at lower stress. The orientation/realignment of fibrils combined with the increased matrix ductility at high RH, provide mechanisms so that the final strength of the material is fairly well preserved also at high humidity. In earlier cellulose fiber composites work, the moisture sensitivity of UP composites with cellulosic fibers is significant, and is present also when the fibers are chemically modified with silane compounds [12] and [35]. For silane-treated pure bacterial cellulose nanofibers [12], the silane modification may be incomplete so that hygroscopic hydroxyls are still present at the cellulose nanofiber surface.

Comparison between CNF/EP and CNF/UP Biocomposites

In earlier work [23], we reported on CNF/EP biocomposites. This enables a useful comparison between two of the most commonly used thermosets reinforced by a nanostructured CNF fibril network. Such comparisons are often challenging for nanocomposites, since small differences in degree of nanofiber dispersion have strong effects on properties. An interesting observation is that stress-strain behavior at low humidity (RH 50%) for different CNF contents is fairly similar for composites based on EP and UP, probably reflecting the dominance of CNF network deformation. The composites have similar modulus and strength and both show the characteristic strain-hardening behavior. The UP matrix is somewhat stronger/stiffer than the EP matrix [23] (Elastic Modulus – 2.1 GPa and 2.9 GPa, Yield strength: 26 MPa and 49 MPa, Ultimate Strength: 32 MPa and 54 for neat EP and neat UP resins respectively). This is reflected in the slightly better properties of the CNF/UP nanocomposite.

It is interesting to note that while the CNF/UP composites show somewhat better properties than the CNF/EP composite at RH 50%, the case is reversed at RH 90% (seeFig. 6). The CNF/EP composite is almost unaffected at high humidity and retains its mechanical properties, while CNF/UP shows strongly reduced modulus and yield strength. This emphasizes the role of the CNF/polymer interface – while the CNF/EP interface contains covalent bonds from amine-promoted hydroxyl–epoxide reactions[23], there is no such possibility in CNF/UP. Instead, the interface is dominated by secondary interactions such as hydrogen bonds and van der Waals interactions. Moisture is preferentially localized at the interface and properties are reduced at RH 90%.

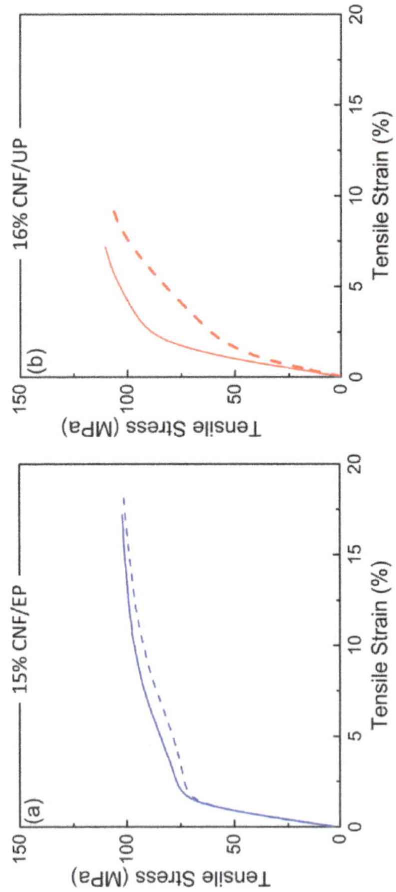

Figure 6. Representative stress strain curve of (a) CNF/EP composite (from Ref. [23]) and (b) CNF/UP composite from the present study. Solid lines represent uniaxial tensile tests at RH 50% and dashed lines at RH 90%.

Apparent Fracture Toughness

In order to further investigate the fracture behavior, single-edge notched specimens were tested and the data is reported in Fig. 7. The addition of CNF has a positive effect on the apparent fracture toughness K_Q of the materials. K_Q increases from 2.5 MPa $m^{0.5}$ for neat UP to 6.4 MPa $m^{0.5}$ for the 30% CNF/UP. Builes et al. studied fracture toughness for CNF/UP composites at 1 vol% of CNF and reported only marginal increase [14]. Note that thin specimens do not meet conditions of plain strain [32]. Most likely, the values reported in Fig. 7 are higher than would be obtained from thick specimens meeting plain strain criteria. The fracture surfaces of the 16% CNF/UP are presented at different magnifications in Fig. 7. Fig. 7a and c shows laminar features at the scale of 5–10 µm, which correspond to the underlying material structure. At higher magnifications, roughness features are more apparent and nanofibers are discerned as white dots (Fig. 7c, d). Interestingly, the nanofiber pull-out lengths are very short (Fig. 7d), note the 500 nm scale bar. Occasional fibrils are 100 nm, but the large majority is much shorter. Short pull-out lengths indicate strong interfacial adhesion, which seems reasonable for well-dispersed nanofibers of 5–7 nm diameter at RH 50%.

The crack growth in the present experiment is immediately catastrophic as the critical stress is reached for a given crack length. The increased fracture toughness with CNF content therefore means that the load-carrying ability of the CNF network provides increased resistance to crack growth initiation. However, toughening by fiber bridging appears to be a negligible mechanism, due to the short nanofiber length.

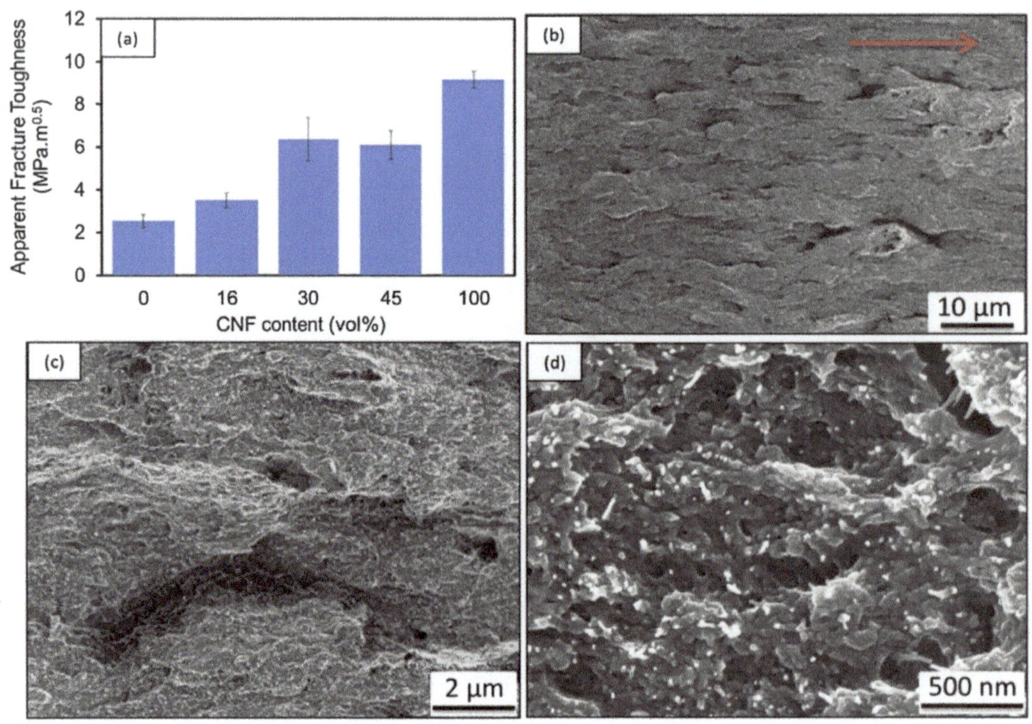

Figure 7. (a) Fracture toughness as a function of CNF content in the CNF/UP composites. (b–d) SEM micrographs at different scales from cross sections in the 16% CNF/UP composite studied by crack growth experiments. Arrow indicates the direction of crack growth (same for all images).

CONCLUSIONS

CNF was successfully used to reinforce an unsaturated polyester resin, without the need of complex surface modifications. A prepreg-based processing concept was used to incorporate as much as 45 vol% of CNF in the resin with preserved nanostructural characteristics. The processing steps used can be envisaged as a continuous process, with some adaptations from the conventional papermaking route. The nanoscale dispersion of the CNF nanofibers was confirmed and is important for the combination of properties in the present materials (increased strength and ductility).

Strong CNF–UP synergy was observed in the composites, reflecting the successful nanostructuring of the materials. Despite the use of a comparably brittle UP thermoset, both modulus and strength increased with increasing CNF content up to the maximum content (45 vol%). The T_g also increased from 66 °C for the neat material to 78 °C for the 45 vol% composite, indicating favorable secondary CNF–UP interactions through hydrogen bonding, Van der Waals interactions and similar mechanisms. The ductility of the composites and microscopy of the fracture surfaces provide support for well-dispersed individual CNFs in a continuous UP matrix. The interfacial adhesion was favorable at RH 50%, since CNF nanofibers on fracture surfaces were very short (<100 nm). The favorable dispersion at the nanoscale combined with strong interactions at the interface leads to remarkable improvements in isotropic in-plane mechanical properties of the UP. Comparable plant fiber (≈30 μm diameter) reinforced biocomposites show lower strength and low ductility and higher moisture sensitivity due to the presence of lignin and hemicellulose and the less favorable nature of the fiber/matrix interface.

Crack growth experiments showed substantially increased apparent fracture toughness with increased CNF content. The main toughening mechanisms were intrinsic in nature, in front of the crack tip, whereas fiber bridging from CNF was negligible. Interestingly, in conventional tensile tests, the strain to failure of the composites was higher than that of neat UP as well as the CNF. Possibly, cavity growth in UP is constrained by the nanoscale fibrous network.

The hydrophilic nature of CNF surfaces and thus the CNF/UP interface reduced mechanical properties at high humidity. The yield strength, for example, decreased from 93 MPa at RH 50% to 52 MPa at RH 90% for the composite with 30 vol% CNF. Although the properties at 50%RH were comparable for CNF/EP and CNF/UP, the latter material was much inferior at RH 90%. Most likely, the CNF–UP interfacial adhesion is reduced since moisture becomes concentrated at the interface at high humidity.

ACKNOWLEDGMENT

Matti Rauva and Tuula Mannermaa at Ashland, Finland are gratefully acknowledged for providing the Polyester resin. Financial support from VINNOVA through the BiMaC Innovation Excellence Centre and the Knut and Alice Wallenberg through the Wallenberg Wood Science Center (WWSC) is gratefully acknowledged.

REFERENCES

1. M. Henriksson, G. Henriksson, L.A. Berglund, T. Lindstrom, An environmentally € friendly method for enzyme-assisted preparation of Microfibrillated Cellulose (MFC) nanofibers, Eur. Polym. J. 43 (2007) 3434e3441.
2. T. Saito, Y. Nishiyama, J.-L. Putaux, M. Vignon, A. Isoga, Homogeneous suspensions of individualized microfibrils from TEMPO-catalyzed oxidation of native cellulose, Biomacromolecules 7 (2006) 1687e1691.
3. T. Saito, S. Kimura, Y. Nishiyama, A. Isogai, Cellulose Nanofibers Prepared by TEMPO-Mediated Oxidation of Native Cellulose, Biomacromolecules 8 (2007) 2485e2491.
4. L.A. Berglund, T. Peijs, Cellulose biocomposites e from bulk moldings to nanostructured systems, MRS Bull. 35 (2010) 201e207.
5. S.J. Eichhorn, A. Dufresne, M. Aranguren, N.E. Marcovich, J.R. Capadona, S.J. Rowan, et al., Review: current international research into cellulose nano- fibres and nanocomposites, J. Mater. Sci. 45 (2009) 1e33.
6. R.J. Moon, A. Martini, J. Nairn, J. Simonsen, J. Youngblood, Cellulose nanomaterials review: structure, properties and nanocomposites, Chem. Soc. Rev. 40 (2011) 3941e3994.
7. H.V. Boenig, Polyesters, unsaturated, in: N. Bikales, J. Conrad, A. Ruks, J. Perlman (Eds.), Encylcopedia of Polymer Science, vol. 11, Wiley-Interscience Publishers, New York, 1971, pp. 129e168.
8. J.-P. Pascault, H. Sautereau, J. Verdu, R.J.J. Williams, Thermosetting Polymers, Marcel Dekker, New York, 2002.
9. C.J. Chirayil, J. Joy, L. Mathew, J. Koetz, S. Thomas, Nanofibril reinforced unsaturated polyester nanocomposites: morphology, mechanical and barrier properties, viscoelastic behavior and polymer chain confinement, Ind. Crops Prod. 56 (2014) 246e254.
10. C.J. Chirayil, L. Mathew, P.A. Hassan, M. Mozetic, S. Thomas, Rheological behaviour of nanocellulose reinforced unsaturated polyester nanocomposites, Int. J. Biol. Macromol. 69C (2014) 274e281.
11. L. Hu, Y. Wan, F. He, H.L. Luo, H. Liang, X. Li, et al., Effect of coupling treatment on mechanical properties of bacterial cellulose nanofibre-reinforced UPR ecocomposites, Mater. Lett. 63 (2009) 1952e1954.
12. C. Gao, Y. Wan, F. He, H. Liang, H. Luo, J. Han, Mechanical, moisture absorption, and photodegradation behaviors of bacterial cellulose nanofiber- reinforced unsaturated polyester composites, Adv. Polym. Technol. 30 (2011) 249e256.
13. H. Kargarzadeh, R.M. Sheltami, I. Ahmad, I. Abdullah, A. Dufresne, Cellulose nanocrystal: a promising toughening agent for unsaturated polyester nanocomposite, Polymer 56 (2015) 346e357.
14. D.H. Builes, J. Labidi, A. Eceiza, I. Mondragon, A. Tercjak, Unsaturated polyester nanocomposites modified with fibrillated cellulose and PEO-b-PPO-b-PEO block copolymer, Compos. Sci. Technol. 89 (2013) 120e126.

15. M. Henriksson, L.A. Berglund, P. Isaksson, T. Lindstrom, T. Nishino, Cellulose nanopaper structures of high toughness, Biomacromolecules 9 (2008) 1579e1585.
16. H. Sehaqui, Q. Zhou, O. Ikkala, L.A. Berglund, Strong and tough cellulose nanopaper with high specific surface area and porosity, Biomacromolecules 12 (2011) 3638e3644.
17. H. Fukuzumi, T. Saito, A. Isogai, Influence of TEMPO-oxidized cellulose nanofibril length on film properties, Carbohydr. Polym. 93 (2013) 172e177.
18. H. Sehaqui, Q. Zhou, L.A. Berglund, High-porosity aerogels of high specific surface area prepared from Nanofibrillated Cellulose (NFC), Compos. Sci. Tech. 71 (2011) 1593e1599.
19. H. Yano, J. Sugiyama, A.N. Nakagaito, M. Nogi, T. Matsuura, M. Hikita, et al., Optically transparent composites reinforced with networks of bacterial nanofibers, Adv. Mater. 17 (2005) 153e155.
20. S. Iwamoto, A.N. Nakagaito, H. Yano, M. Nogi, Optically transparent composites reinforced with plant fiber-based nanofibers, Appl. Phys. A Mater. Sci. Process. 81 (2005) 1109e1112.
21. M. Henriksson, L. Fogelstrom, L.A. Berglund, M. Johansson, A. Hult, Novel € nanocomposite concept based on cross-linking of hyperbranched polymers in reactive cellulose nanopaper templates, Compos. Sci. Technol. 71 (2011) 13e17.
22. M. Henriksson, L.A. Berglund, Structure and properties of cellulose nanocomposite films containing melamine formaldehyde, J. Appl. Polym. Sci. 106 (2007) 2817e2824.
23. F. Ansari, S. Galland, M. Johansson, C.J.G. Plummer, L.A. Berglund, Cellulose nanofiber network for moisture stable, strong and ductile biocomposites and increased epoxy curing rate, Compos. Part A Appl. Sci. Manuf. 63 (2014) 35e44.
24. I. Siro, D. Plackett, M. Hedenqvist, M. Ankerfors, T. Lindstr om, Highly trans- € parent films from carboxymethylated microfibrillated cellulose: the effect of multiple homogenization steps on key properties, J. Appl. Polym. Sci. 119 (2011) 2652e2660.
25. K. Prakobna, S. Galland, L.A. Berglund, High-performance and moisture-stable cellulose-starch nanocomposites based on bioinspired core-shell nanofibers, Biomacromolecules 16 (2015) 904e912.
26. K. Prakobna, C. Terenzi, Q. Zhou, I. Furo, L.A. Berglund, Core-shell cellulose nanofibers for biocomposites e nanostructural effects in hydrated state, Carbohydr. Polym. 125 (2015) 92e102.
27. L.J. Gibson, M.F. Ashby, Cellular Solids: Structure and Properties, second ed., Cambridge University Press, Cambridge, 1997.
28. W.D. Callister, Fundamentals of Materials Science and Engineering, John Wiley and Sons, New York, 2001.
29. B.D. Agarwal, L.J. Broutman, Analysis and Performance of Fiber Composites, second ed., Wiley, New York, 1990.
30. J.C. Halpin, J.L. Kardos, The Halpin-Tsai equations: a review, Polym. Eng. Sci. 16 (1976) 344e352.
31. I. Diddens, B. Murphy, M. Krisch, M. Muller, Anisotropic elastic properties of cellulose measured using inelastic X-ray scattering, Macromolecules 41 (2008) 9755e9759.
32. A.J. Kinloch, R.J. Young, Fracture mechanics, in: Fracture Behaviour of Polymers, Applied Science Publishers, London & New York, 1983, pp. 74e106.
33. S. Houssine, A. Liu, Q. Zhou, L. Berglund, Fast preparation procedure for large, flat cellulose and cellulose/inorganic nanopaper structures, Biomacromolecules 11 (2010) 2195e2198.
34. A.J. Kinloch, R.J. Young, Shear yielding, in: Fracture Behaviour of Polymers, Applied Science Publishers, London & New York, 1983, pp. 107e146.
35. M. Abdelmouleh, S. Boufi, M.N. Belgacem, A. Dufresne, A. Gandini, Modification of cellulose fibers with functionalized silanes: effect of the fiber treatment on the mechanical performances of cellulose-thermoset composites, J. Appl. Polym. Sci. 98 (2005) 974e984.

CHAPTER 20

THE DEVELOPMENT OF A HIGH TEMPERATURE TENSILE TESTING RIG FOR COMPOSITE LAMINATES

Joseph Mills-Brown, Kevin Potter, Steve Foster, and Tom Batho

Keywords: A. Ceramic–matrix composites (CMCs); B. Mechanical properties; B. High-temperature properties; D. Mechanical testing

INTRODUCTION

The recent resurgence and growing interest in high temperature structures to maximise design space and performance in aerospace and motorsport applications, has led to the need for greater understanding of high temperature composite materials. Fibre reinforced composite materials are typically required for their low weight and high mechanical performance, properties which are necessary for structural application. In addition to this, temperatures up to 1000 °C can be encountered, and as such, ceramic matrix composites are often sought [1]. However, the suitability of materials to perform in these environments is difficult to ascertain due to a lack of high temperature design data.

Mechanical property data is needed to understand the behaviour and limitations of composite materials at high temperatures, as well as allow accurate modelling to take place. Several mechanical tests are available to assess the mechanical properties of a material, of which tensile and flexural testing are the most common. Tensile testing is generally preferred to flexural testing, as flexural testing produces maximum stresses only in small regions at the test specimen surface, which can lead to localised effects. In addition to this, several failure mechanisms can be seen in flexural testing; tensile, compression and shear, which can allow more than one mechanism to dominate results. Tensile testing, however, produces maximum stresses throughout the test specimen as a whole, and the data achieved is more established and appropriate regarding the design and modelling of structures. There are several components and aspects to tensile testing; the test coupon, grips and strain measurement are all required [2]. Typically tensile testing of composite materials uses hydraulic grips on flat rectangular samples, equipped with bonded end tabs to reduce stress concentrations and through thickness crushing. This is especially important with composite materials as they tend to be weak in the through thickness direction. However, the adhesives used to bond end tabs to composite coupons are only suitable up to a maximum of 300 °C. In addition to this, at temperatures up to 1000 °C attention is needed to both the gripping mechanism and material that the grips are constructed from. Strain gauges are usually employed to measure tensile strain, although these are typically bonded to the coupon, and the same problems with end tab adhesives arise. Therefore, high temperature tensile testing requires unique gripping, heating and strain measuring devices as well as an optimised specimen geometry, in order to successfully test tensile properties of composite coupons at high temperatures.

Current high temperature tensile test equipment is hard to source and typically impractical for use with thin composite laminates. Most available tensile test facilities with the ability to reach 1000 °C are designed for metallics with the use of cylindrical test coupons. Strain measurements are typically taken through the use of clip gauges, which can require notches in the coupon and can impart thermal stress concentrations. This study aims to provide a high temperature tensile test, suitable for thin, flat fibre reinforced composite laminates. The use of thin coupons, approximately 1 mm, is required as it is representative of those used in structural applications and of the material thicknesses readily available for high temperature composites. In addition to this, there are several requirements that are sought to expand the ease-of-use of the test setup and drive the design. These include a simple, quick and cost effective high temperature tensile test, which makes use of small coupons to reduce cost, and where possible, integration

Citation: Joseph Mills-Brown, Kevin Potter, Steve Foster, Tom Batho, The development of a high temperature tensile testing rig for composite laminates, Composites Part A: Applied Science and Manufacturing, Volume 52, September 2013, Pages 99-105, ISSN 1359-835X, http://dx.doi.org/10.1016/j.compositesa.2013.04.009.

into existing tensile equipment. The ability to add a high temperature capability to existing test equipment will add novelty to the test design as the majority of high temperature testing requires significant investment in both cost and time.

HIGH TEMPERATURE TEST METHOD

The methods discussed for conducting high temperature tensile testing of fibre reinforced composite materials have highlighted the different approaches that can be taken. The requirements for this study stated that a simple, quick and cost effective high temperature tensile test should be chosen. In addition to this, thin, small coupons are needed due to the cost and availability of ceramic matrix composites, which are a focus of this test method. It is also advantageous for time and resource considerations to make use of existing and available test equipment where possible. With these requirements the most appropriate high temperature tensile test method was developed.

Grip Design and Test Setup

After an extensive review of high temperature testing methods [2], [3], [4], [5] and [6], hot passive grips, which grip the coupon through a wedge interference fit and apply edge loading, were selected as the most appropriate gripping technique. The coupon geometry is simplified and less manufacturing is required, as opposed to other methods such as pin grips, which require the drilling of several holes to the test coupon. This grip design can also be easily used with conventional servohydraulic tensile test machines available to this study. The grips will run hot, again to simplify the test procedure but also to remove thermal gradient problems associated with cold grips. In addition to this, the initial cost of the grips is lower than cold grips. This is because less manufacturing is required, as internal coolant networks are no longer necessary within the grip body, which also removes the need for ancillary pump systems to circulate the coolant. The use of hot grips demands that the grips be made from materials which retain sufficient strength at test temperatures. Nickel alloys, specifically Inconel 718 grade, have been selected as the most appropriate grip material for this reason. Although coupon test temperatures of 1000 °C were sought, only localised surface heating of the grips was observed at test temperatures during testing. This was due to the speed of coupon heating and the thermal mass of the grips themselves.

The grip design is similar to that used by Holmes [3] in the testing of SiC reinforced ceramic composites at 1200 °C. The grips comprise of the main grip attachment with an oversized cut-out for the coupon to be inserted, centreing inserts placed either side of the coupon once in the grip, retaining plates either side of the centreing inserts to reduce associated stress concentrations at the root from the coupon cut-out, and finally blinds to prevent the light produced by the rear emitter interfering with the optical strain measurement equipment. Misalignment of the coupon can cause significant bending stresses which can induce failure outside of the gauge section [3]. The centreing inserts reduce coupon misalignment and twist, keeping the coupon aligned between opposing grips, whilst also allowing lateral forces to be imparted on the coupon from the retaining plates. The use of the retaining plates to exert a degree of pressure laterally onto the centreing inserts and coupon face is important when testing thin coupons, as this mitigates any edge crushing effects on the coupon as the inserts are of the same material as the grip body. The grip assembly has been illustrated in Fig. 1, which shows the individual components discussed. Initial finite element analysis (FEA) of the grip setup indicated that significant stresses were developing at the root edges of the coupon cut-out. To reduce these effects the retaining plates were incorporated into the grip setup to reduce splaying of the cut-out forward edges, and as such, stresses at the root edges.

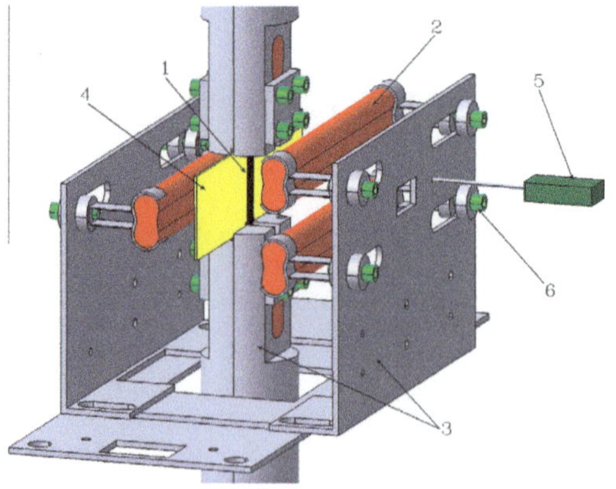

Figure 1. Final grip assembly and test setup; coupon (1), three IR emitters (2), grips and rig (3), blinds (4), thermocouple (5) and bolts (6). (For interpretation of the references to colour in this figure legend, the reader is referred to the web version of this article.)

Heating of the coupon is conducted through an indirect heating approach, specifically infrared (IR) emitters. Several indirect heat sources could be used but short-wave IR emitters produced by Heraeus Noblelight Ltd. were chosen due to their small size and ability to reach temperatures as high as 2500 °C. In addition to this, a gold coating is specified on the back surface of the emitter so heat is typically radiated in one direction, minimising heating of other components and maximising the efficiency of the emitter. Initial testing of the test rig indicated that even at maximum emitter power the temperature of the rig and extended grip ends does not exceed 80 °C. In addition to this, the time required to reach uniform coupon test temperatures is typically between 1 and 2 min depending on the temperature sought. This is significant when comparing against conventional testing and heating methods which can require more than 60 min to achieve uniform temperatures. The thermal conductivity of the material tested also has an influence on the rate of heating, however, the use of thin coupons reduces this significantly, highlighting the effectiveness of the IR emitters and their placement within the setup. As heating of the coupon is through radiation, the high temperature capability of the emitters allows us to vary the power or distance of the emitters from the coupon and still achieve the test temperatures required. This extra flexibility in emitter placement is useful in the positioning of strain and temperature measurement equipment. Three emitters are used, placed either side of the coupon faces, allowing the test coupon to be seen from one side. The emitter setup is demonstrated in Fig. 1 and Fig. 2. The power of the emitters allows the setup to be unenclosed, which again provides easier access for strain and temperature measurement. Control of the emitters is by varying the power supplied to the emitters in order to achieve a specific temperature. Calibration is required to evaluate the appropriate power and the distance of the emitters to the coupon, to achieve the required coupon test temperature.

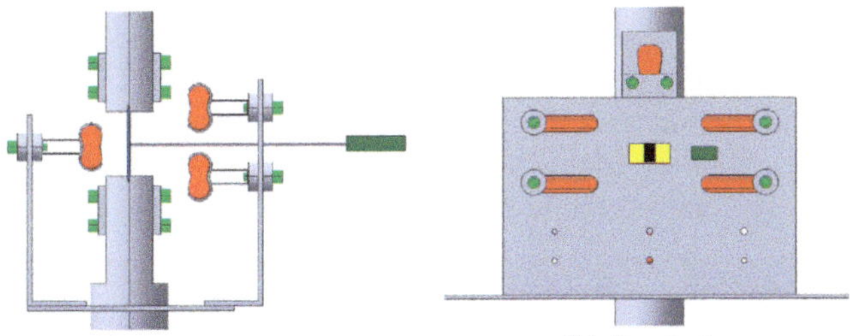

(a) Side view (b) Front view

Figure 2. IR emitter setup in the final grip assembly. (For interpretation of the references to colour in this figure legend, the reader is referred to the web version of this article.)

Several strain measurement techniques are available, however, with consideration given to the test requirements, video extensometery has been chosen. Video extensometers are non-contact strain measurement devices, which reduce external influence on the test coupon [7]. Speed of testing is greatly increased and minimal coupon preparation is required, as opposed to high temperature strain gauges, which require significant sample preparation time and consumables. To conduct video extensometery, a speckle pattern is needed on the coupon to achieve the necessary contrast required for accurate video strain measurements. Coupons were speckled using Dykem®, a white high temperature paint produced by ITW Dymon. Blinds were included on either side of the test coupon to remove issues of excessive light caused by the rear emitter. Light produced when the rear emitter is in operation is otherwise directed down the video extensometer camera, which has the effect of over exposing the video and silhouetting the test coupon, resulting in inaccurate strain measurements. The blinds were made from the same composite to be tested, which also provided additional regions for temperature measurement. In addition to this, neutral density (ND) filters were incorporated into the camera to mitigate the levels of light produced from the front emitters reflecting off the test coupon and entering the camera during testing.

Temperature measurement is important not only to determine the test temperature of the coupon but also as part of a feedback loop to calibrate and eventually control the IR emitters. High temperature type-*K* thermocouples are used as they offer the simplest solution and are easily integrated into a future control setup. To avoid influencing the test coupon, thermocouples can be placed on the blinds either side of the coupon. These are made from the same composite being tested and sit under the exact same thermal conditions. Initially several will be used to determine the temperature distribution, but once calibrated to ensure uniform heating, the number could be reduced. The complete test rig is illustrated as a CAD model in Fig. 1, where the base plate includes support holes for attachment to the test machines. Fig. 2 illustrates the positioning of the camera viewing port, where strain measurements of the gauge section will be made. The grips are oversized for safety reasons, to ensure reduced heating of the load cell and surrounding tensile test machine. Holes included in the rig are required for attachment of cables and other ancillary components required to operate the emitters, as well as provide access for high temperature thermocouples.

Tensile testing was conducted using an Instron 1342 mechanical tester. Tests were performed over a temperature range of 20–800 °C, with a cross-head displacement rate of 1 mm/min used. At least 4 coupons were tested at each temperature increment. ASTM specifications C1366 [8] and C1359 [9] were followed at all times. Composite tensile strength was calculated using Eq. (1), where F and A are the applied force at breaking point and area of the coupon respectively. Failure strain was measured using a video extensometer supplied by Imetrum Ltd. and combined with the composite tensile stress in Eq. (2) to yield the tensile modulus.

$$\sigma = \frac{F}{A} \quad (1)$$

$$E = \frac{\sigma}{\epsilon} \quad (2)$$

Coupon Design

Several high temperature composite tensile testing reviews were consulted and it was found that for passive grips little changed in the fundamental specimen design. Tapering of the specimen width between the tab end and gauge length is a consistent feature across studies and was sought in the final design. Worthem [10] investigated how best to homogeneously stress material volumes to initiate failure in the heated gauge section of composite coupons, using tapered widths. This showed that a gradual taper is necessary in achieving reliable tensile failure of composite materials, a feature which was emulated in the design of the coupon used here. Ultimately a coupon design from ASTM standards[8], [9] and [11], echoed by Grande [12] and Holmes [3], for tensile testing of ceramic composites at elevated temperatures was chosen. This provided a simple flat geometry with a suitably gradual taper, as illustrated in Fig. 3, and has been successfully implemented with SiC–SiC composites at temperatures up to 1200 °C [3]. Thicknesses of 3 mm are suggested by Holmes [3], however, the specimen thickness can be decreased or increased as required. The decision to use 1 mm thick material was taken due to the requirements of the study.

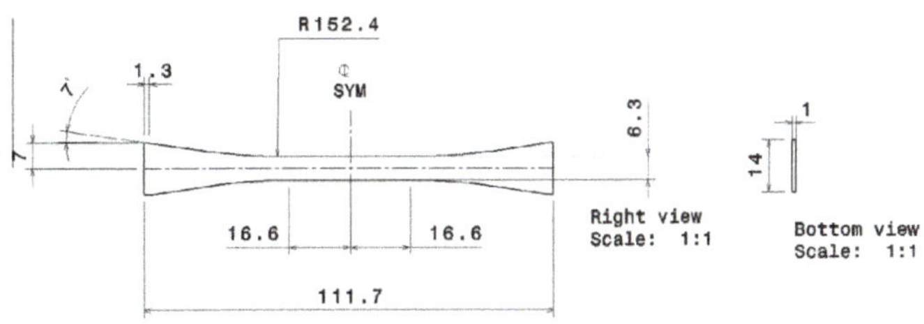

Figure 3. Tensile coupon geometry.

Materials

High temperature composite materials suitable for testing up to 1000 °C needed to be selected in order to verify the test design. In addition to this, the material should be representative of those to be used in high temperature structures in order to maximise the reach of the research. Typically, traditional engineering ceramic matrix composites (CMCs) would be envisioned for these applications, however, one material which has shown potential for application in high temperature structures is polysialate matrix composite.

Polysialates are ceramics derived from inorganic polymers and processed through a polymerisation chemical activation, rather than the extreme temperature processing synonymous with traditional engineering ceramics. This gives them a number of advantages over typical CMC materials such as low production times, environmental friendliness and low density [13]. High temperature resistance, up to 1200 °C, and an incombustible nature, with no harmful smoke or fumes being released [14], [15] and [16]are also retained. Polysialates can be reinforced with several different reinforcements, although the most common are carbon or silicon carbide (SiC) fibres, typically in a woven architecture. Some of the most comprehensive research in this field was conducted by Davidovits [17], who first applied the term geopolymer to these materials, with several terms now being used, such as inorganic polymers, or in this case, polysialates. Very little thermal or mechanical data exists for polysialate composites, with the majority of data available limited to testing conducted at ambient temperatures. This study aims both to develop the testing method and to generate useful high temperature data for this class of composites. The polysialate composite materials used in this study were produced by Pyromeral Systems, as they produce the most developed and utilised polysialate-type materials commercially available. Composites are available with both carbon and silicon carbide reinforcement, but given the temperature range explored in this study, silicon carbide reinforcement was selected. This material is referred to as PyroSic® and is a polysialate matrix composite, reinforced with Tyranno silicon carbide fibres in a plain weave architecture. Due to quality control concerns all materials were produced in-house by Pyromeral Systems. Test coupons made use of 1 mm thick material constructed from 4 woven plies using a 0/90 layup, as this is representative of the material likely to be employed in high temperature structures. Coupons were cut using computer numerically controlled (CNC) equipment with the 0 fibre direction orientated along the coupon length, to achieve the geometry and tolerances required for use in the passive gripping mechanism.

FINITE ELEMENT VALIDATION

The grip design was modelled using FEA to establish the loading distribution on the coupon and determine any problems with the design. Three dimensional (3D) quarter models of the grip setup were created and analysed using Abaqus CAE. The grip and coupon were modelled, with the influence of centreing inserts, retaining plates and bolts represented using relevant boundary conditions and loads.

The geometry of the grip meant that solid elements were needed in order to effectively capture the behaviour of the setup. The grip body made use of C3D10 10-node quadratic tetrahedrons as the complex geometry, including holes and curvatures, proved difficult to mesh with hexagonal elements. The coupon was modelled using C3D20 20-node quadratic hexagonal bricks and incorporated a more refined mesh than that used in the grip, as stress within the coupon is of primary interest. The final mesh used is the outcome of a mesh refinement study, which optimised the mesh against the stresses achieved and the required computational time. The model is illustrated in Fig. 4, where the equivalent mesh densities are visible. Symmetry boundary conditions were used to represent the complete geometry, with surface interactions used to simulate contact and friction between the coupon and relevant components. Load was applied to the end of the coupon in the form of a ramped uniform tensile stress, which ensured uniform loading of the coupon and simulated tensile test conditions. The applied pressure of 275 MPa is representative of the room temperature tensile failure stress of PyroSic, as given in the datasheet [18]. In addition to this, thermal boundary conditions simulating a temperature rise from 20 °C to 1000 °C were implemented on the coupon, in order to study the effect that the coefficient of thermal expansion (CTE) has on stresses within the grip and coupon. Temperature dependant CTE data for both PyroSic and Inconel alloys were taken from previous thermal characterisation and manufacturers datasheets respectively.

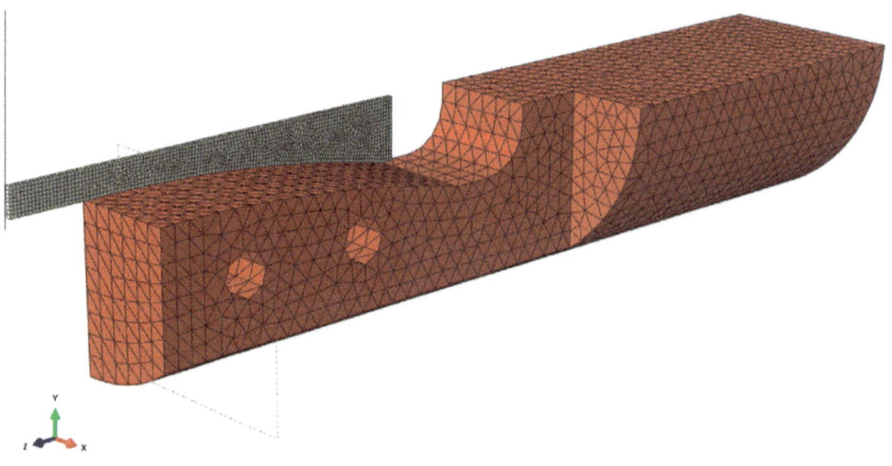

Figure 4. Quarter model of grip and coupon; coupon 1387 elements, grip 22478 elements. (For interpretation of the references to colour in this figure legend, the reader is referred to the web version of this article.)

The temperature conditions are in all probability an over estimation. The coupon is unlikely to achieve 1000 °C over the whole geometry due to contact with the grips, which act as a thermal sink. Also, the grips will most probably achieve more local temperature concentrations, closer to the emitter heat sources, rather than a uniform temperature distribution. However, these conditions produce a worst case scenario which should always be designed for. Boundary conditions and interactions within the model are more complex than first appears, as the coupon interacts against the grip and centreing insert surfaces with a degree of friction. Assumptions are made to simulate these effects, which could induce some error in the results achieved. These conditions and the quarter model design permitted suitable run times, whilst still providing adequate information on the stress state within the grip and coupon.

Results

Fig. 5 shows the longitudinal S11 tensile stresses present in the coupon at 25 °C and 1000 °C. The results show that tensile stress across the coupon simulated at 25 °C reduces from 288 MPa in the gauge section to approximately 0 MPa at the tab end. Similarly, at 1000 °C, the stress reduces in comparative manner, but with slightly higher magnitude, from 290 MPa to becoming slightly compressive at approximately −13 MPa. In both instances the stress is higher in the gauge section, promoting tensile failure in this region as desired. The taper in the coupon produces this effect, where suitable transition in tensile load is observed, and with it, confidence in the accurate high temperature tensile testing of composite coupons.

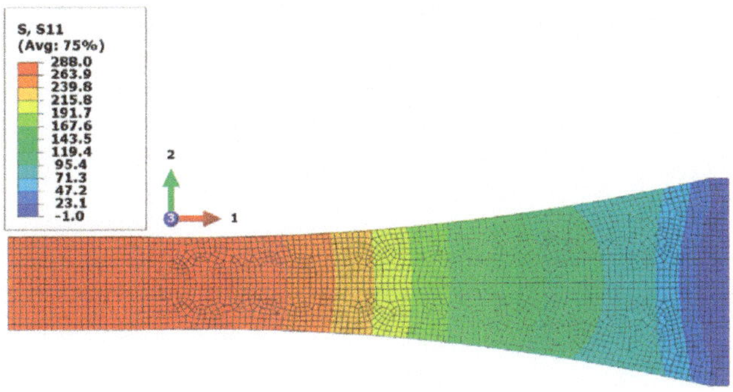

(a) S11 25°C

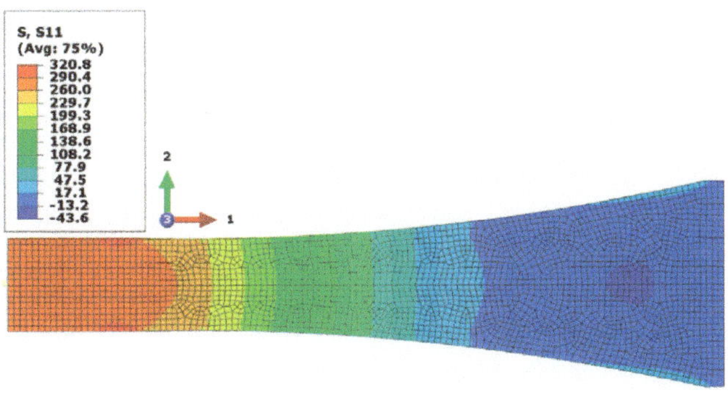

(b) S11 1000°C

Figure 5. Longitudinal S11 coupon tensile stresses at 25 °C and 1000 °C. (For interpretation of the references to colour in this figure legend, the reader is referred to the web version of this article.)

Small stress concentrations are noted on the edges of the gauge section. This is probably caused by 'pinching' of the coupon by the edge of the coupon cut-out, as the grip expands due to CTE effects and is constrained by the retaining plates. In addition to this, greater compressive stress is observed in the 1000 °C simulation due to compression of the grip on the coupon, again in all likeliness due to CTE expansion, although the magnitudes are very low. It is also possible that these effects could be due to limitations in the modelling technique. It is worth noting that at this stage in the design process the tensile strength of PyroSic at elevated temperatures was unknown. In reality the strength could be higher or lower at elevated temperatures, which would have an impact on failure, although FEA results indicate that tensile failure should still dominate, due to the gradual taper in the coupon geometry.

Transverse S22 stresses across the coupon, perpendicular to the tensile stress, and S33 stresses through the thickness, are less significant. Fig. 6 shows that compressive contact stresses are observed near the coupon edge in both the ambient and high temperature simulations, although this is expected due to the nature of the gripping mechanism used. Localised contact stresses of −50 MPa are observed in the 25 °C simulation, which gives rise to maximum stress of −40 MPa across the coupon width, but applied away from the gauge section. A similar trend is observed with the 1000 °C simulation, where contact stress is now higher at −150 MPa, but reduces to a maximum or −100 MPa across the coupon width. Compressive contact stresses are in both cases below the −170 MPa compressive resistance of PyroSic [18]. This increase in stress magnitude over the low temperature simulation, is coupled with a shift in area of influence closer to the gauge region. This is again due to thermal expansion effects, with the influence area shifting as the grip profile in this region is more constrained due to the retaining plate and associated components, encouraging the profile to expand preferentially in the direction of the coupon. S33 stresses through the coupon thickness are negligible with only the clamping force from the centreing insert and retaining plate evident.

(a) S22 25°C

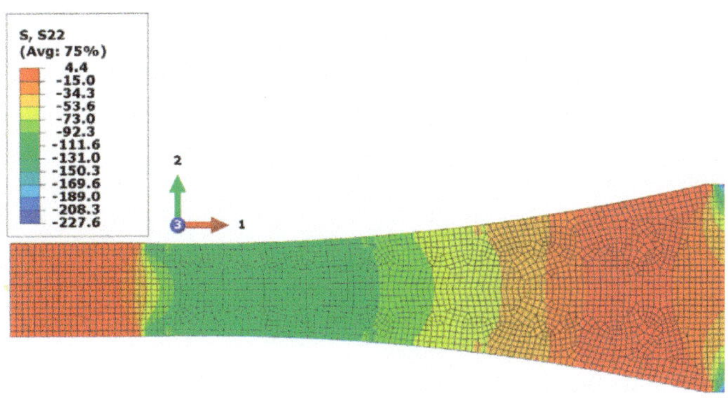

(b) S22 1000°C

Figure 6. Transverse S22 coupon stresses across the coupon at 25 °C and 1000 °C. (For interpretation of the references to colour in this figure legend, the reader is referred to the web version of this article.)

The compressive gripping stresses perpendicular to the tensile load could cause some localised edge crushing.

However, the stress should not effect tensile results as they are away from the gauge region and of significantly lower magnitude. In addition to this, the centreing inserts will act to mitigate any edge crushing effects and stop any out of plane deformation from occurring, whilst ensuring true tensile failure of the coupon. A small but significant compressive stress concentration can be observed in the high temperature simulation, in the rear corner of the coupon tab region. This is thought to be unrepresentative of reality given its location and magnitude, and is probably due to boundary effects, over constraint or singularities. However, it serves to highlight the complexity of the interactions and, in any case, has no effect on the performance of the test in performing accurate tensile loading.

Stresses in the grips never exceed the high temperature strength of Inconel alloys, which retain a yield strength of 275 MPa at 900 °C [19]. The main stress regions are as anticipated, located in contact areas between the coupon and grip in the S22 and S33 directions. Similarly as with the coupon, small regions of stress concentrations are prominent at the edge of the model, however, this is most probably due to over constraint and boundary effects.

Strain distribution in the coupon follows a similar pattern as that for the tensile stress. Maximum elongation of 0.34 mm is confined to the gauge section, with very little change in the strain magnitude or location with the change in temperature. Total strain achieved mimics the 0.9% quoted in the datasheet, which confirms material properties are well simulated by the model.

Modelling was conducted to determine if the grips would be properly loaded and to ascertain if any unexpected effects were occurring. Full confidence in the results is difficult given the complexity of boundary conditions implemented, however, it serves to indicate tensile failure is most probable even at high temperatures, although practical testing of the setup was required to validate the analysis and confirm this.

RESULTS AND DISCUSSION

Test Validation

Testing of the final design and setup was undertaken to verify the choices made and FEA predictions. This included establishing whether the coupons were being correctly loaded, in order to ensure accurate tensile properties can be evaluated from the results. In addition to this, the region of failure was noted in order to determine the primary failure mechanism. The passive grips were validated by testing PyroSic coupons against control tests of coupons tested in conventional hydraulic grips, at room temperature. Instron tensile machines were used in both tests. The results achieved for both the passive grips and control tests are given in Table 1, which includes the standard deviation present in the test data.

Table 1. Room temperature tensile test data.

Grip type	Strain (%)	Strength (MPa)	Modulus (GPa)
Passive	0.99 ± 0.09	288 ± 14.0	32.0 ± 2.5
Hydraulic	0.89 ± 0.08	266 ± 18.5	33.4 ± 3.9

The results show that the passive grips compare extremely well with the control tests, highlighting that the grips can produce data similar to traditional testing methods. In addition to this, all test coupons tested in the passive grips failed in the gauge section, as predicted in FEA and should guarantee a tensile failure mode and accurate tensile data, even at elevated temperatures. The average strength results for the passive grip are similar to those achieved in the control tests at 288 MPa. In addition to this, a standard deviation of 5% highlights the reproducibility of the passive grip and test setup. Modulus results for the passive grips are also very reliable at 32 GPa with a standard deviation of 8%. This is slightly lower than the modulus of 33 GPa achieved in the control tests, but neither the stresses nor strains measured are significantly different from those measured using the conventional test methodology.

Strain results were slightly higher for the passive grip, at 0.99%, than that achieved for the control tests at 0.89%, again with an associated standard deviation of 9%. All test coupons tested in both the passive and hydraulic grips demonstrated a linear stress strain behaviour.

Further confidence in the test is gained when data is compared against data sheet values for PyroSic materials [18]. The strength and strain data from both the passive grip setup and control tests agree well with each other, and data available in the PyroSic data sheet, which gives values of 275 MPa and 0.9% for tensile strength and strain to failure respectively.

High Temperature Tests

Tensile strain, strength and modulus were measured using the high temperature tensile test described. Maximum test temperatures were restricted to 760 °C due to restrictions with the IR emitter placement and material performance. This was due to the available proximity of the IR emitters to the test coupon, which were impeded by direct interference from the grips. An increase in coupon length would remove this interference and allow the front two emitters greater access with regard to heating the test coupon, which in turn would result in higher available test temperatures. All test coupons failed in the gauge section and all tests produced linear stress strain relationships, with uniform heating exhibited in the coupons at each test temperature.

The tensile properties of the polysialate composite tested undergo significant change over the test temperature range. Polysialate composite failure strain increases from approximately 1.0–1.3%, as shown in Fig. 7. Conversely, strength decreases from 288 MPa to 213 MPa, which ultimately reduces the stiffness of the material from 32.0 GPa to 17.3 GPa, as illustrated in Fig. 8.

The error bars shown in the test results highlight the reproducibility of the test method. Error experienced in the higher temperature results is comparable to that in the room temperature and hydraulic control tests. This gives confidence in the data and test design, where the error present is therefore more likely to be present due to bulk material inconsistencies in a brittle material. Standard deviation in the vast majority of results is less than 10%.

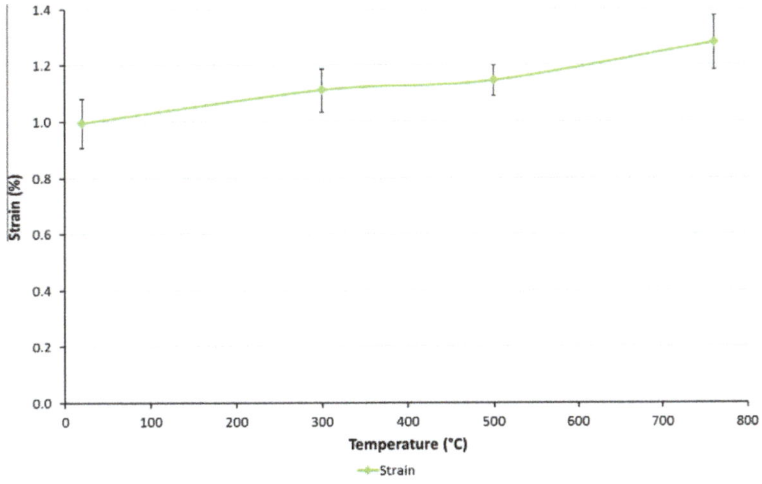

Figure 7. Strain performance of polysialate composites at elevated temperatures, error bars represent ± one standard deviation. (For interpretation of the references to colour in this figure legend, the reader is referred to the web version of this article.)

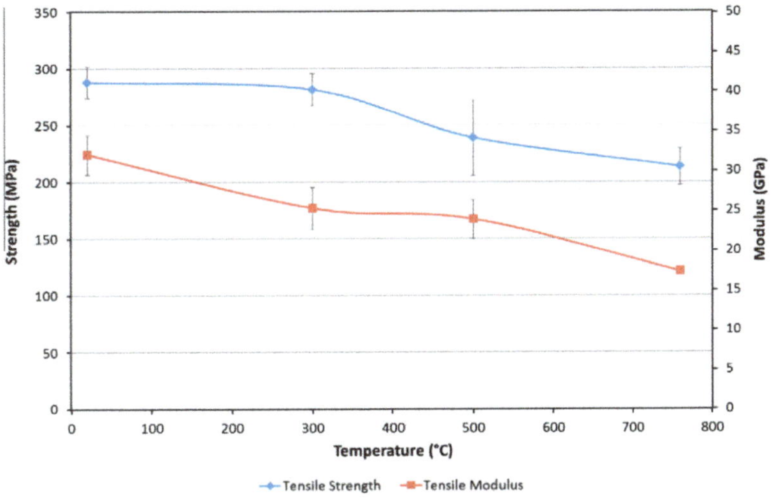

Figure 8. Strength and stiffness performance of polysialate composites at elevated temperatures, error bars represent ± one standard deviation. (For interpretation of the references to colour in this figure legend, the reader is referred to the web version of this article.)

CONCLUSION

The results show the test procedure is capable of giving reliable results in the high temperature testing of thin composite coupons. Grip design and the IR emitters provided uniform tensile loading and rapid heating of coupons respectively, with test run and turn around times significantly faster than conventional high temperature methods. Results strongly suggest that accurate high temperature tensile testing can be performed with this grip design and test setup, with the data achieved accurate and offering a good representation of how composite materials behave under tensile loading at elevated temperatures. Temperatures of 760 °C were achieved during testing but this could easily be increased with an adjustment in coupon size.

The test rig heating design is suitable for both ceramic and polymer matrix composites, and it may be possible to adapt it for compression and bending tests. The inclusion of control mechanism for temperature control would be a useful addition to future test changes and to ensure accurate and reproducible temperature and heating rates. The rapid heating and cooling achievable with this design of test rig allows it to be used to efficiently investigate important engineering properties, such as the impact of thermal cycling and thermal shock under load.

ACKNOWLEDGEMENTS

The author thanks McLaren Racing Ltd. for the supply of materials used and financial contribution throughout the study. The financial contribution of the ESPRC in sponsoring this work as part of the ACCIS DTC (Advanced Composite Centre for Innovation and Science Doctoral Training Centre) at the University of Bristol, Grant Number EP/G036772/1, is also gratefully acknowledged.

REFERENCES

1. Ohnabe H, Masaki S, Onozuka M, Miyahara K, Sasa T. Potential application of ceramic matrix composites to aero-engine components. Compos Part A: Appl Sci Manuf 1999;30(4):489–96.
2. Gyekenyesi JZ, Hemann JH. High temperature tensile testing of ceramic composites. Tech Rep Contractor Report-180888; NASA; 1988.

3. Holmes JW. A technique for tensile fatigue and creep testing of fiber-reinforced ceramics. J Compos Mater 1992;26:916–33.

4. Essam G, Syers G. Novel test methods for mechanical strength characterisation of engineering ceramics. Int J High Technol Ceram 1988;4:275–80 [Mechanical testing of engineering ceramics at high temperatures].

5. Hermansson T, Dunlop G. A high-temperature tensile testing rig for ceramic materials. Int J High Technol Ceram 1988;4:263–8.

6. Ruggles-Wrenn M, Mall S, Eber C, Harlan L. Effects of steam environment on high-temperature mechanical behavior of NextelTM720/alumina (N720/A) continuous fiber ceramic composite. Compos Part A: Appl Sci Manuf 2006;37(11):2029–40.

7. Gyekenyesi JZ, Hemann JH. Optical strain measuring techniques for high temperature tensile testing. Tech Rep Contractor Report-179637; NASA; 1987.

8. C1366. Standard test method for tensile strength of monolithic advanced ceramics at elevated temperatures. ASTM International; 2009.

9. C1359. Standard test method for monotonic tensile strength testing of continuous fiber-reinforced advanced ceramics with solid rectangular crosssection test specimens at elevated temperatures. ASTM International; 2011.

10. Worthem DW. Flat tensile specimen design for advance composites. Tech Rep Contractor Report-185261; NASA; 1990.

11. C1275. Standard test method for monotonic tensile behaviour of continuous fibre-reinforced advanced composites with solid rectangular cross-section test specimens at ambient temperature. ASTM International; 2010.

12. Grande DH. Testing and properties of high temperature glass–ceramic matrix composites. Ph D thesis. Massachusetts Institute of Technology; 1987.

13. Lyon RE, Balaguru PN, Foden A, Sorathia U, Davidovits J, Davidovics M. Fireresistant aluminosilicate composites. Fire Mater 1997;21(2):67–73.

14. Hammell J, Balaguru P, Lyon R. Strength retention of fire resistant aluminosilicate carbon composites under wet dry conditions. Compos Part B: Eng 2000;31(2):107–11.

15. He P, Jia D, Lin T, Wang M, Zhou Y. Effects of high-temperature heat treatment on the mechanical properties of unidirectional carbon fiber reinforced geopolymer composites. Ceram Int 2010;36(4):1447–53.

16. Lyon RE, Sorathia U, Balaguru PN, Foden A, Davidovits J, Davidovits M. Fire response of geopolymer structural composites. First Int Conf Compos Infrastruct 1996;00762524:972–81.

17. Davidovits J. Geopolymers inorganic polymeric new materials. J Mater Educ 1994;16:91–138.

18. PyroSic Datasheet. Pyromeral systems; 2009.

19. Inconel alloy 718 Datasheet. Special metals; 2007.

CHAPTER 21

CARBON NANOTUBE REINFORCED CEMENT COMPOSITES

Tanvir Manzur, Nur Yazdani, and Md. Abul Bashar Emon

INTRODUCTION

In the field of nanotechnology, carbon nanotube (CNT) has the prime focus as one of the most major and significant areas of research. There is a keen interest in developing nanotechnology for cement and concrete. Not only the chemistry of cement hydration products but also the physical behavior of these hydration products can be modified through application of nanotechnology due to their dimensional scale. The mechanical properties of CNT manifest immense potential to be used as reinforcements in composite materials. In addition to their high strength and elasticity, CNTs have extremely high aspect ratios, with values typically higher than 1000 : 1 and reaching as high as 2,500,000 : 1 [1]. CNTs' ultra high aspect ratios [2], extreme high strength [3], high modulus [4], and elasticity [5] make them an idle candidate for outstanding reinforcing material within any composite matrix. The size and aspect ratios of CNT are such that it is possible to distribute them on a much finer scale than commonly used micro reinforcing fibers [1]. Cracks can be interrupted very quickly during propagation in a properly dispersed CNT reinforced matrix. Eventually, bridge coupling effect of CNT ensures relatively lower crack widths and guarantees load-transfer across voids and cracks. Therefore, CNT reinforced composites can have higher flexural and compressive strengths as well as smaller strain and higher density. These mechanical properties of CNT reflect its prospect as reinforcement within the matrix of cementitious composites.

Citation: Tanvir Manzur, Nur Yazdani, and Md. Abul Bashar Emon, "Carbon Nanotube Reinforced Cement Composites" Journal of Nanomaterials, vol. 2016, Article ID 1421959, 10 pages, 2016. doi:10.1155/2016/1421959

A number of researches have already been conducted on the application of CNT for enhancing mechanical properties of cementitious composites. It has been found that research on CNT-cement composites started in the early 1990s. Chen and Chung [6] in 1993 observed an increase of 85%, 205%, and 22% in flexural strength, flexural toughness, and compressive strength, respectively, with the use of short carbon fibers in the cement mortar. However, significant research outcomes have been observed in mid 2000s. Li et al. [7] in 2004 showed that both compressive and flexural strength of the cement mortar with CNTs were higher than plain cement paste. In another study [8], the same researchers in 2005 showed that porosity of cement composite can be reduced by adding CNT. Makar et al. [1] in 2005 found that addition of SWNTs (single-walled nanotubes) can accelerate the hydration of cement at early age by providing additional spaces for hydration products to grow. This behavior of CNT is analogous to their performance as nucleating agent in different applications. For example, growth of Si_3N_4 [9], ZrO_2 [10], $CaCO_3$ [11], and nanodiamonds [12] on MWNT, MWNT, MWNT, and SWNT, respectively, can be mentioned. Cwirzen et al. [13] in 2008 obtained 10% higher flexural strength for MWNT reinforced composites as compared to plain cement mortar. A CNT-to-cement proportion of 0.042 (percent of cement weight) was used in that research. Agullo et al. [14] in 2009 observed an increase in early age compressive strengths by adding low concentration of MWNT within cement composite. On the other hand, 28-day compressive strength was found to be similar for both normal cement mortar and nanotube reinforced composites. In another study in 2009, Makar and Chan [15] observed that nanotubes can influence the morphology of hydration products. Konsta-Gdoutos et al. [16] in 2010 found that cement composites having both short and long MWNT as reinforcement showed higher flexure and modulus than those of plain cement paste. Manzur and Yazdani [17] in 2010 observed increase in compressive strength of cement mortar through addition of two different sizes of untreated commercial MWNT. Hunashyal et al. [18] in 2011 obtained higher tensile strength and better stress-strain relationship in direct tension for cement samples with CNT. Han et al. [19] in 2013 showed that addition of MWNT could decrease both water sorptivity and permeability coefficient of CNT reinforced cement-based materials. In 2014, Manzur et al. [20] found that MWNT with OD (outside diameter) 20 nm or less obtained higher compressive strengths as compared to larger OD-MWNT when used in mortar and the maximum compressive strength was achieved by the smallest size of MWNT considered in the study. Such behavior is due to the fact that smaller size MWNTs can be distributed on much finer scale which in turn ensures efficient filling of nanopore space within the cement matrix. Finer distribution of nanotubes also enhances the performance of nanotubes as nucleating agent during cement hydration and, hence, encourages the formation of higher stiffness C-S-H phase. Lelusz [21] in 2014 found that compressive strength of cement composites decreased with the increase in CNT dosage. This research used different dosage rates of MWNT ranging from 0.00% to 0.12% by weight of cement. A suitable mix proportion was proposed by Manzur and Yazdani [22] in another study in 2015 to produce surface treated MWNT reinforced cement composites based on flexural and compressive strengths. It is, therefore, quite evident that CNT reinforcement could result in robust and improved cementitious composites. The reasons behind higher compressive strength and lesser porosity of CNT reinforced cementitious composites have been pointed out as better hydration and filling of nanospace by nanotubes in previous studies. Better hydration is caused by the role of nanotubes as nucleating agent as already mentioned above. On the other hand, enhanced flexural and tensile behavior of cement mortar reinforced by nanotubes is mainly due to crack bridging and fiber pullout actions of nanotubes. However, performance of CNT-cement composites is observed to be quite variable. Properties of CNT reinforced cementitious composites depend on several factors and mixing procedures and have been well documented by Parveen et al. [23]. The authors of the present study have studied the effect of different important parameters on different mechanical properties of CNT-cement composites and suggested a suitable mixing technique [17] and a tentative optimum mix proportion [22]. In this paper, a possible functional application of CNT-cement mortar is discussed.

A potentially useful application of CNT within construction industry is in producing nanotubes reinforced cement mortar that may be employed for concrete rehabilitation, such as surface restoration and crack repair. Several repair, retrofit, and strengthening techniques are currently available to enhance the integrity and durability of concrete structures with cracks, spalling, and so forth. Current available materials like epoxy for crack repair and cementitious grouts for spalling overlay are good solutions for concrete repair. However, epoxy repair is often not suitable for repair of concrete spalling. In case of rapid hardening cementitious grout, repair depth and volume are usually kept small due to the high cost and rapid heat generation of the rapid hardening materials. Therefore, there is always a need for innovative high performing concrete repair materials with good mechanical, rheological, and durability properties. In this regard, CNT-cement composites appear to have good potential to produce a repair mortar having the desired properties.

RESEARCH OBJECTIVES

The applicability of MWNT reinforced cementitious composites in concrete repair works has been investigated in the present study. A high performance cementitious mortar intended for use in concrete repair should meet several performance criteria. High strength and good bonding with the existing concrete substrate are two important characteristics of a repair material. Setting times are important for fresh cement mortar to be used as repair material. The mortar should not suffer from bleeding since bleeding may adversely affect the durability of repair material. The bond between the repair material and the concrete base often presents a weak link in the repaired structure. This can be evaluated through slant shear test. Since no data is available on the properties of MWNT reinforced cement mortar as concrete repair material, the following tasks were carried out in the current research:

- Effect of MWNT addition on setting time was investigated.
- Effect of MWNT addition on bleeding was investigated.
- Bond strength between the repair materials (MWNT-cement composite, normal cement mortar, and epoxy resin) and concrete substrate was investigated through slant shear tests.

MATERIALS USED

Ordinary Type II Portland cement was used as cementitious material in this study. Special graded sand according to ASTM C109 [24] was utilized as fine aggregates. A commercially available surface treated MWNT was obtained from the manufacturer (Cheap Tubes Inc.) in powder form having outside diameter (OD) of 10–20 nm. Surface treatment of MWNT was done using a mixture of nitric and sulfuric acids to develop hydroxyl or carboxyl end groups on outer surface of nanotubes. This size of nanotube was selected based on results of the previous study by the authors since this size of surface treated MWNT resulted in relative higher strength [20]. Seven different sizes (outside diameters) of MWNT were used. It was found that, for a given mix proportion, smaller size MWNT reinforcement produced relatively higher compressive strength. Figure 1 shows the Transmission Electron Microscope (TEM) image of the MWNT supplied by the manufacturer. Properties of MWNT are provided in Table 1 which was also supplied by the manufacturer (Cheap Tubes Inc.). Polycarboxylate based super plasticizer ADVA Cast 575 (produced by W. R. Grace & Co.) was used as plasticizer in the experiments to improve the workability of the mix and also as a surfactant for the MWNT. Texas Department of Transportation (TXDOT) utilizes Pro Poxy 300 for quick repair of cracks in bridges. For slant shear test of specimens jointed by epoxy, Pro Poxy 300 Fast was used.

Table 1: Properties of used treated MWNT.

	OD (outside diameter)	Length	Purity	Ash	SSA (Specific Surface Area)	EC (Electrical Conductivity)	COOH content (wt%)
MWNT	10–20 nm	10–30 μm	>95 wt%	<1.5 wt%	>233 m^2/g	>10^{-2} s/cm	2.00

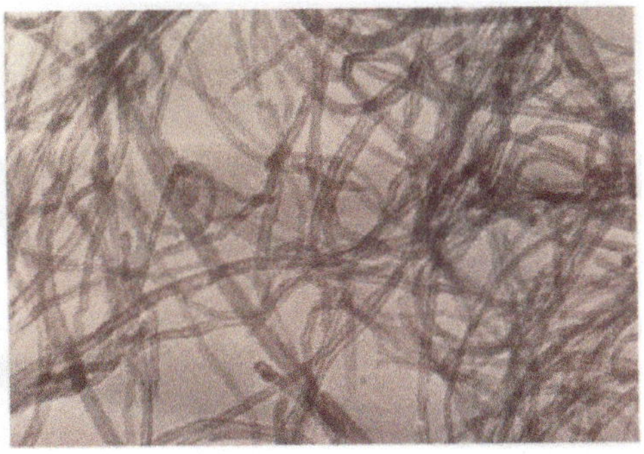

Figure 1: TEM image of used treated MWNT with OD of 10–20 nm.

EXPERIMENTAL INVESTIGATION

Initially, a comprehensive parametric study was conducted in order to obtain a suitable mixing technique for dispersing nanotubes within cement matrix and a tentative optimum mix proportion for producing MWNT reinforced cement composites. Composite samples with different dosages of treated multiwalled carbon nanotubes (MWNTs) and control samples without MWNT were prepared and tested. Samples were made using different mix proportions having different w/c (water to cement) ratios and plasticizer proportions. Table 2shows proportion of different components used in the initial phase of study. The dosage rate of MWNT was also varied to select an optimum concentration. A suitable technique using ultrasonication was employed in dispersing MWNT within the cement matrix [17, 25]. In a typical mixing process, the required amount of plasticizer was first mixed with water and sonicated for few minutes. MWNT was then added in sequence and sonicated in plasticizer mixed water for 5 minutes for each application. After the last addition of MWNT, the whole mixture was sonicated for another 15 minutes. The amount of MWNT for each sequence was selected in such a way that total sonication time remained constant at 40 minutes. Once sonication was completed, cement, sand, and water (with dispersed nanotubes) were mixed in a mixer with flat beater for about 6 minutes as per ASTM C109 [24] requirements. Through SEM images (Figure 2), it has been observed that such sonication can ensure uniform dispersion of nanotubes within cement matrix. The behavior of MWNT reinforced composites in compression and flexure was then evaluated and statistical analysis was carried out. It was found that MWNT reinforced composites performed better both in compression and in flexure. From the initial study, it was observed that composites reinforced with MWNT having OD smaller than 20 nm yielded relatively higher strength as compared to composites with larger MWNT [20]. It is well established that properties of MWNT reinforced composites primarily depend on proper distribution of nanotubes. Usually, higher water content is required to produce MWNT-cement composites for relative stable dispersion of nanotubes [22, 26]. However, higher water content often produces weaker composites due to its negative effect on cementitious materials. It has been observed that utilization of plasticizer as surfactant makes the dispersion of MWNT into aqueous solution more stable and eventually produces stronger composites [27, 28]. Polycarboxylate based super plasticizer was used as surfactant in this study to make the MWNT more soluble than that of pristine nanotubes. The tentative optimum proportion of plasticizer to be used as surfactant was found as 0.008 in terms of weight of cement [22]. This proportion of plasticizer produced the maximum strength both in compression and in flexure. It was also observed that use of plasticizer increased the flow values of the mix which can be used as an indirect measure of dispersion quality of nanotubes in cement matrix [29]. In the absence of plasticizer, higher water content was required to achieve comparable flow values. It was found that the tentative optimum dosage rate of MWNT in terms of weight of cement ranged between 0.1% and 0.3% with lower concentration of MWNT performed better in flexure. Figures 3 and 4 show the compressive and flexural strengths of controls samples and composite samples with different dosage rates of MWNT at different ages, respectively. In each case, twelve samples were prepared and the average strength of twelve samples and corresponding error bands are presented in Figures 3 and 4. It was evident from the initial study that MWNT reinforcement within cement produced composites with significantly higher strengths. MWNT-cement composites exhibited around 26%, 21%, 15%, and 14% higher compressive strength at 3, 7, 28, and 100 days, respectively, as compared to control samples without plasticizer for 0.10% addition. The increment in flexural strength was observed to be about 24%, 25%, and 20% at 3, 7, and 28 days, respectively, for similar amount of MWNT addition. It was evident that composites with nanotubes achieved relatively higher 7-day compressive strengths than 28-day strength, as compared to control samples. The reason behind such behavior of nanotube reinforced cement composites is the presence of nanotubes that affects the morphology of cement hydration products. MWNT influences both the initial C_3A and the C_3S hydration products by accelerating the rate of hydration process particularly at early stages since they act as a matrix for the development of C-S-H and $Ca(OH)_2$ [1, 22]. It was also observed that composites with higher proportion of MWNT (usually higher than 0.3% weight of cement) exhibited lower strength values. Higher amount of MWNT resulted in inadequate dispersion due to agglomeration which eventually produced less workable and unstable mixes. It was found that attainment of homogenous distribution using higher amount of nanotubes was extremely difficult even with sonication process. Water also got entrapped between clumped nanotubes and, consequently, hindered the hydration process. Finally, the applicability of MWNT reinforced cement composites as concrete repair material was evaluated through setting time, bleeding, and slant shear tests since these parameters are important to assess the quality of a cementitious repair material. MWNT dosage rates of 0.1% and 0.3% were utilized for producing composites to be used as repair material.

Table 2: Proportion of different components used in the initial phase of the study and corresponding strengths of the control and composite samples.

Sample ID	MWNT proportion in terms of cement wt	Plasticizer proportion in terms of cement wt	Plasticizer proportion in terms of MWNT wt	Compressive strength (MPa)				Flexural strength (MPa)		
				3 days	7 days	28 days	100 days	3 days	7 days	28 days
CS	—	—	—	23.7	27.7	33.3	37	2.86	3.3	4.67
CS + *P*	—	0.008	—	26.1	28.6	32.8	38.6	—	—	—
0.05% + *P*	0.0005	0.008	16	27.8	30.8	36	40.2	3.1	3.55	4.93
0.10% + *P*	0.0010	0.008	8.0	29.8	33.5	38.3	42.1	3.54	4.11	5.58
0.20% + *P*	0.0020	0.008	4.0	28.9	33.9	37.8	42.9	3.59	3.89	5.45
0.30% + *P*	0.0030	0.008	2.7	28.4	35.3	38.5	42.6	3.46	3.95	5.42
0.50% + *P*	0.0050	0.008	1.6	26.1	29.2	33	35.7	—	—	—

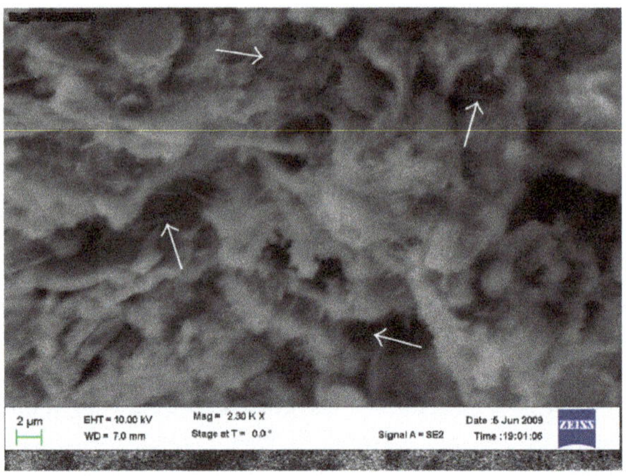

Figure 2: SEM image of crushed mortar sample (MWNTs are fairly uniformly distributed as shown by arrows).

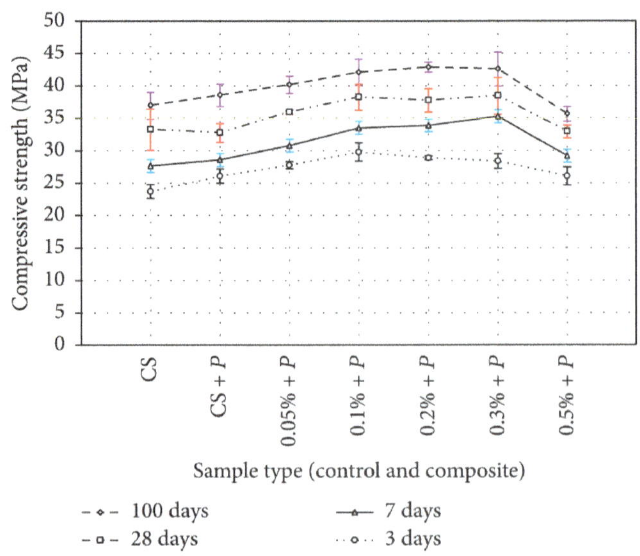

Figure 3: Compressive strengths of control and composite samples with different dosage rate of MWNT at different ages.

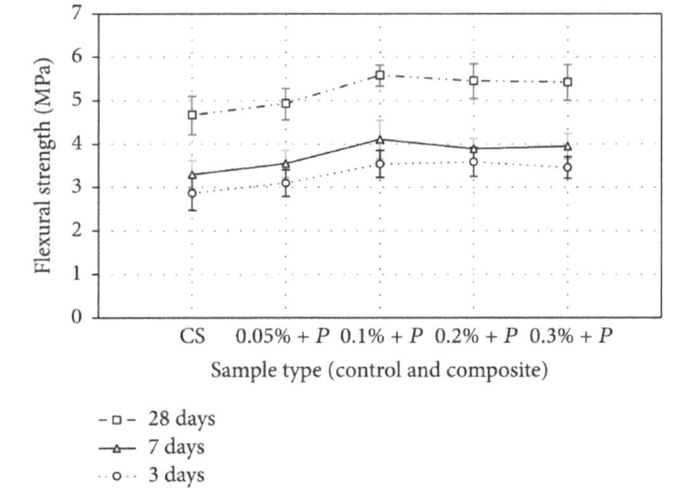

Figure 4: Flexural strengths of control and composite samples with different dosage rate of MWNT at different ages.

SETTING TIME OF COMPOSITES

The time of setting represents the onset of the solidification phase at which fresh mortar can no longer be properly handled or injected. This test is significant for evaluating the effect of MWNT addition on setting time of cementitious mortar. The setting time is particularly important for repair purposes regarding practical use of repair mortar in field. The setting times were determined using the Vicat apparatus described in ASTM C807 [30]. The ASTM C807 test procedure is for determination of the time of setting of hydraulic cement mortar. The modified setting time test utilizes thicker steel needle of 2.0 mm instead of 1.0 mm needle. The 1.0 mm needle is used for determination of setting time of cement paste.

Sample Preparation and Testing

Control and composite samples reinforced with 0.1, 0.2, and 0.3% MWNT were prepared and tested for time of setting. Samples were produced using w/c ratio of 0.50 and 0.60. Control samples with plasticizer addition were also made for comparison. Two different plasticizer proportions of 0.005 and 0.008 as surfactant were used to prepare composites. Samples were prepared using the rotary mixture with flat beater. After preparation, the samples were kept in moisture room for 30 minutes without being disturbed. For testing, samples were then placed on Vicat apparatus and the needle was set in contact with the surface of mortar samples. The needle was then released to penetrate and allowed to set for 30 seconds. The reading of penetration was measured and recorded. The penetration readings were then determined at every 30 minutes till the needle failed to touch the bottom of the mold. After that the reading was taken at 10-minute intervals until a reading of 10 mm or less was obtained. From all recorded penetration readings, the time when penetration was 10 mm was determined through interpolation. Each penetration of needle was done at a distance of 10 mm from any previous penetration and from the inside edge of the mold. The difference in time between the mixing of mortar and a penetration of 10 mm was the setting time of cementitious mortar.

Results and Discussion on Setting Time

It was observed that composite samples with nanotubes settled significantly early as compared to control samples for the same mix proportions. Control samples with w/c ratio of 0.50 had time of setting of around 175 minutes. Composites reinforced with 0.1% to 0.3% MWNT showed setting time ranging between 125 and 135 minutes. Control samples with plasticizer proportion of 0.008 had setting time of about 340 minutes. Setting times for 0.3% and 0.1% MWNT reinforced composites with plasticizer proportion of 0.008 as surfactant were observed as 245 and 260 minutes, respectively. Hence, about 25% less time was required by the MWNT composite to set than that of control samples for mixes with plasticizer proportion of 0.008. Similar pattern was also observed for other mix proportions of control and composite samples. In case of mixes with w/c ratio of 0.60, MWNT composites required about 14% less time for setting as compared to control samples. It is, therefore, evident that addition of nanotubes decreases the initial setting time of cement mortar since presence of nanotubes accelerates the rate of hydration of initial C_3A by acting as nucleating agent. And, the stiffening and setting characteristics of cement paste primarily depend on hydration of aluminates. It was also found that the variation in setting time with the change in concentration of MWNT between 0.1% and 0.3% was little. Hence, it appears that amount of nanotubes within this range has similar effect on cement hydration which is also evident from strength results. Figures 5–7 show the comparison between times of setting of control and composite samples having different mix proportions. Each bar in the graphs represents one sample for the respective mix proportion; that is, the three bars under CS in Figure 5 are representing setting time of three control samples having w/c ratio of 0.50. In Figure 6, *P* means plasticizer and different percent values represent amount of MWNT; that is, 0.3% + *P*-0.005 means composite samples with 0.3% MWNT and plasticizer proportion of 0.005. All samples with plasticizer were made using w/c ratio of 0.50.

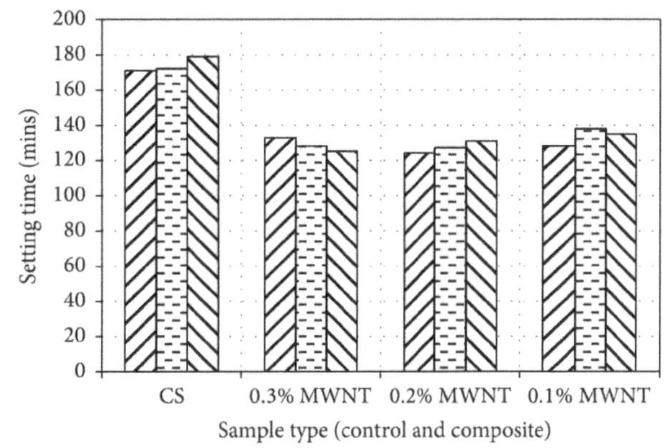

Figure 5: Time of setting of control samples without plasticizer and composites with different dosage rate of MWNT having w/c ratio of 0.50.

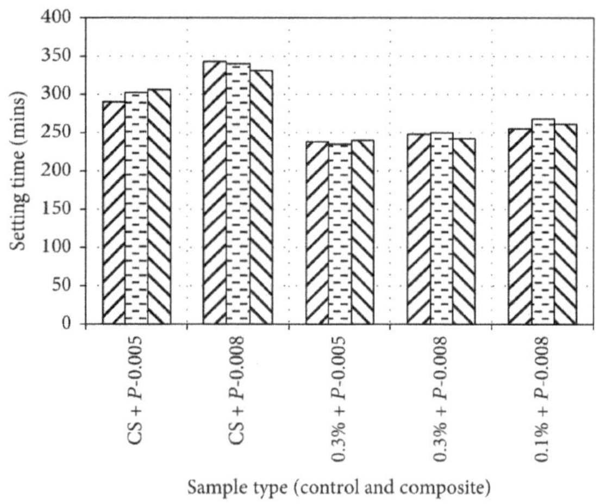

Figure 6: Time of setting of control samples and composites with plasticizer (w/c ratio: 0.50).

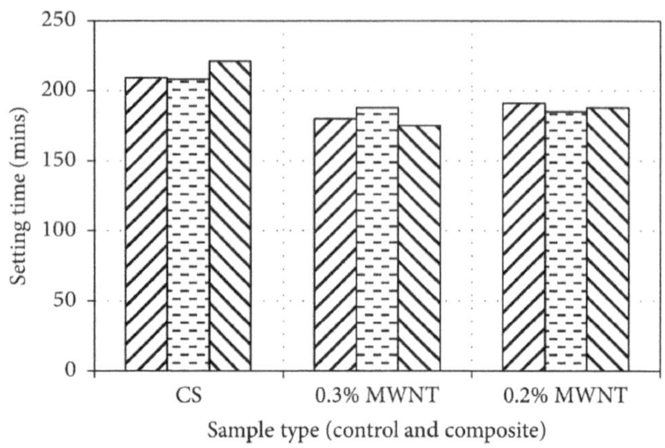

Figure 7: Time of setting of control samples without plasticizer and composites with different dosage rate of MWNT having w/c ratio of 0.60.

BLEEDING OF COMPOSITES

Bleeding is the appearance of free water on the surface of the unset mortar, as the relatively heavy solid particles settle because of gravity, and represents the quality of the mortar mix. Bleeding of the freshly mixed mortar was measured following the procedure given in ASTM C940 [31]. Excessive bleeding may leave several uncontrolled open channels within the mortar mass and eventually results in porous and weak samples. It also adversely affects the durability of the sample. Therefore, composite samples should be free from bleeding to be used as an effective repair material.

Sample Preparation and Testing

800 mL of freshly mixed cement mortar was produced for each test of bleeding. Both control and composite samples were made and tested for bleeding. Composite samples were prepared with 0.1% and 0.3% dosage rate of MWNT having different mix proportions for comparison. After completion of mixing, the temperature of the mix was measured and recorded. The mix was then poured into a 1000 mL glass graduated cylinder and covered. The difference between the upper surface of mortar and bleed water was then recorded if any difference was observed. The process was continued at 15-minute intervals for the first 60 minutes and then at 1-hour intervals. The height of bleed water was then expressed as a percent of the original height of the mortar. This height was referred to as the percent final bleed.

Results and Discussion on Bleeding

Table 3 provides the bleeding results and corresponding temperatures of both control and composites samples. It was observed that none of the samples, either control or composite, exhibited bleeding. Therefore, it is evident that addition of nanotubes does not affect the stability of the cementitious mix. The temperature of MWNT reinforced cement composites has been found significantly higher than that of control samples. Presence of MWNT accelerates the hydration process which eventually increases the temperature of the mix.

Table 3: Bleeding test result of control and composite samples.

Type of sample	Amount of MWNT (%)	w/c ratio	Amount of plasticizer (as part of cement by wt.)	Bleeding (%)	Temp (°C)
Control	NA	0.50	NA	None	19.4
Composite	0.3	0.50	NA	None	27.8
Composite	0.3	0.60	NA	None	31
Composite	0.3	0.50	0.008	None	28.8
Composite	0.1	0.50	NA	None	26.7
Composite	0.1	0.50	0.008	None	27.2

BONDING STRENGTH OF COMPOSITES AND REPAIR MATERIALS

The bond strength between the repair material and concrete substrate plays an important role in the efficiency of repair material used in concrete structures. Slant shear tests are typically employed to measure this bond strength since it is considered as the most effective method of compressive shear testing [32–34]. The bond between the repair material and the old concrete usually presents a weak link in the repaired structure. The MWNT reinforced composites should exhibit good bonding strength with existing concrete substrate to be utilized as repair material. Slant shear test was conducted to evaluate the bonding strength. Slant shear strengths of control and composite mortar were measured following the procedure given in ASTM C882/C882M-05 [35] and DMS 4655 [36].

Epoxy resins are widely used to repair cracks in concrete structures. TXDOT utilizes Pro Poxy 300 for quick repair of cracks in bridges. Slant shear test of specimens jointed by Pro Poxy 300 Fast was therefore carried out for comparison.

Sample Preparation and Testing

The cylinders with 75 mm-by-150 mm dimensions were produced according to ASTM C882/C882M and DMS 4655 procedures. The compressive strength of cylinders should be higher than the slant shear strength of the repair material in order to suppress the failure of precast cylinders. After suitable curing, the slant test cylindrical specimens were saw cut into halves with a diagonal bonding area at an angle of 30° from the vertical, as illustrated in Figure 8. Fresh normal cement and CNT-cement mortar layers with a thickness of 3 mm were then placed on the diagonal bonding area between the two saw cut halves. Also epoxy resin was used on the bonding area of other specimens for comparison purpose. The thickness of epoxy resin was used as 0.5 mm according to ASTM standard. The slant shear strengths of specimens repaired by CNT-cement mortar and epoxy resin were obtained by conducting a series of compression tests.

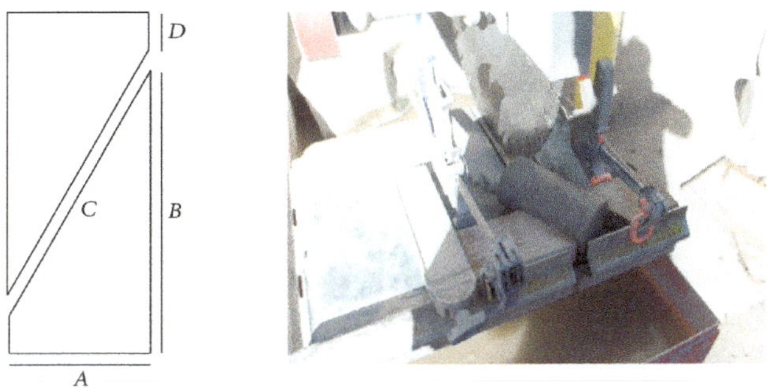

(A) Diameter 75 ± 2 mm
(B) Height 140 ± 2 mm
(C) Slant height 150 ± 2 mm
(D) Base height 10 ± 2 mm

Figure 8: Dimensions of cylindrical specimens used for the standard slant shear test of ASTM C882/C882M and cutting arrangement.

The 28-day compressive strength of concrete cylindrical specimen must be between 34.5 and 41.4 MPa to be used for slant shear testing as per DMS 4655. Concrete cylinders were made following the ACI mix design procedures. The w/c ratio of 0.44 was used for the concrete mix. The gradation of coarse aggregates is shown in Figure 9. The bulk specific gravity of coarse aggregate was obtained as 2.64. The fineness modulus (FM) of fine aggregate was calculated as 2.68. The FM was determined using the formula as shown in the following equation. The standard sieves for FM calculation of fine aggregates are 4.75 mm, 2.36 mm, 1.18 mm, 600 μm, 300 μm, and 150 μm. The bulk specific gravity of coarse aggregate and FM of fine aggregate were required to proportion the concrete mix as per ACI mix design method for the design strength. Consider

$$\text{FM} = \frac{\sum(\text{cumulative percent retained on standard sieves})}{100}. \quad (1)$$

Slant shear tests were carried out at 3, 7, and 28 days. Composites samples were made with 0.1% and 0.3% dosage rate of MWNT with plasticizer proportion of 0.008 as surfactant. The w/c ratio was kept at 0.50. Control samples were made with w/c ratio of 0.50. In each case, a total of 3 cylindrical specimens were jointed together and tested at designated days. Figure 10 shows specimens prepared through bonding by MWNT reinforced composite and epoxy.

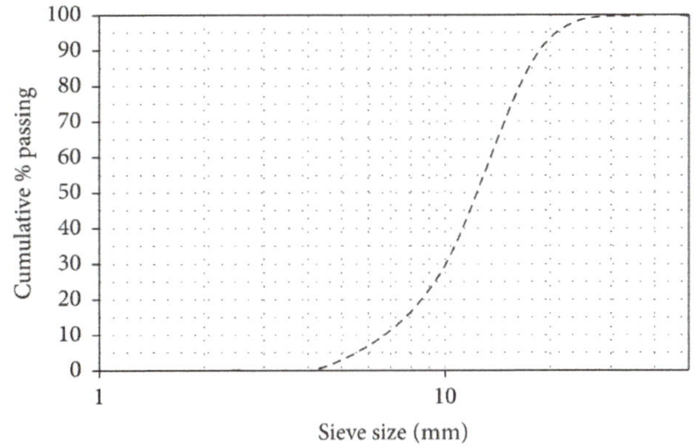

Figure 9: Gradation of coarse aggregates.

(a)

(b)

Figure 10: (a) Cylindrical specimens bonded together by MWNT composite. (b) Cylindrical specimens bonded by epoxy.

Results and Discussion on Slant Shear

Two sets (each containing 3 samples) of cylindrical specimens were tested for compressive strength at 28 days. Average compressive strengths of the two sets were 41.3 and 39.7 MPa, respectively, which fall within the limit given by DMS 4655.

The slant shear test results are shown in Figure 11. It was found that slant shear strength of epoxy at 3 and 7 days was much greater (almost double) than that of both control and composite mortar. This type of behavior was expected as Pro Poxy 300 Fast is a quick hardening epoxy resin and gains strength very rapidly. The slant shear strengths of 0.1% MWNT reinforced composites were about 20% and 23% higher than that of normal cement mortar at 3 and 7 days, respectively. At 28 days, the 0.1% MWNT composites had about 22% higher slant shear strength than control mortar. Such increment in slant shear strength of MWNT composites was due to the enhanced strength properties of MWNT reinforced cementitious composites. Control mortar showed about 53% and 50% less slant shear strengths as compared to Pro Poxy 300 at 3 and 7 days, respectively, whereas the 3- and 7-day slant shear strengths of 0.1% MWNT-cement mortar were about 44% and 40% less than that of Pro Poxy 300, respectively. However, the 28-day slant shear strengths of cylindrical specimens repaired using MWNT reinforced mortar were relatively close to that repaired using epoxy as compared to 3- and 7-day strength. It is evident from the obtained results that cementitious mortar reinforced with surface treated MWNT showed enhanced performance in slant shear strength compared to control mortar at all ages considered in this study. On the other hand, as compared to epoxy resin, MWNT reinforced composites exhibited relatively lesser bonding strength particularly at early ages. However, cement mortar with 0.1% MWNT content achieved almost 80% of the slant shear strength of Pro Poxy at 28 days. Hence, it can be said that MWNT reinforcement within cement composites showed good potential to be used as repair grout. Nevertheless, further research is needed to improve the bonding performance of such nanotubes reinforced mortar. Figure 12 shows the specimens after slant shear test for both MWNT reinforced mortar and epoxy bonded specimens. In both cases, failure occurred at the bonding plane.

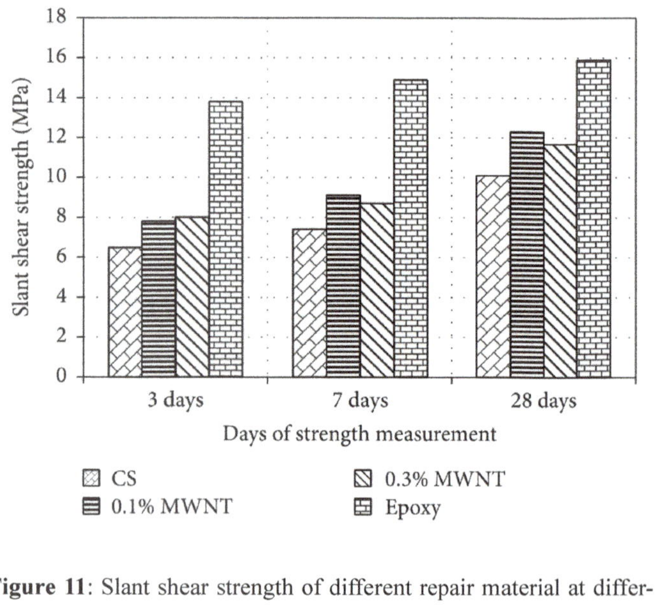

Figure 11: Slant shear strength of different repair material at different ages.

(a)

(b)

Figure 12: (a) Specimens (bonded by MWNT reinforced composite) after testing. (b) Specimens (bonded by epoxy) after testing.

CONCLUSION

MWNT reinforced cementitious composite has already proven its enhanced mechanical properties in terms of both compressive and flexural strength. However, past researches have shown variable increase in strengths of nanotube reinforced cementitious composites ranging from none to more than 80% for different mixing process and concentration of CNT. Hence, it is imperative to develop a suitable mixing technique and recommend a tentative optimum mix ratio for producing nanotube reinforced cement composites. The mixing procedure and mix proportions developed and used in this study have been found to be satisfactory in producing stable MWNT reinforced composites with enhanced mechanical properties. The composites thus produced exhibited about 15% and 20% higher compressive and flexural strength than those of plain cement mortar, respectively. The suitability of the produced MWNT reinforced composites as concrete repair material was then investigated through setting time, bleeding, and slant shear tests. The setting time results show that nanotubes reinforced cement mortar hardened quite more rapidly than normal cement mortar. It required almost 25% less time to set as compared to plain cement mortar. Rapid setting is very important for a material to be used as repair material. It was also observed that composites with nanotubes showed no sign of bleeding. Bleeding can adversely affect the overall strength and durability of a repair material. Slant shear strengths of nanotubes reinforced composites were also found to be higher than that of control plain mortar samples at all ages of 3, 7, and 28 days. The 0.1% dosage rate performed a little better as compared to 0.3% dosage rate in slant shear tests. As compared to epoxy, MWNT reinforced composites exhibited lower slant shear strengths particularly at early ages of 3 and 7 days. The reason behind such behavior was the extremely rapid hardening characteristics of Pro Poxy 300 Fast. The 28-day slant shear strengths of composites achieved relatively higher value which was almost equal to 80 percent of epoxy resin strength. Therefore, it can be said that initial test results on applicability of MWNT reinforced cement composites as repair material demonstrate good potential. However, this type of investigation is presently at budding phase. As a result, more research is required in this field. Several issues like flowability of MWNT-cement mortar, pullout strength, and so forth need to be investigated before recommending practical application of such nanotube reinforced repair material. In addition, any health concern regarding practical use of nanotubes should also be kept in mind. In this study, experiments were conducted following appropriate safety precautions under controlled environment of laboratory. Chemical-resistant gloves, safety goggles, respirators, and protective clothing were used. Comprehensive investigation regarding potential health hazards of using nanotubes in field scale must be conducted simultaneously.

REFERENCES

1. J. Makar, J. Margeson, and J. Luh, "Carbon nanotube/ cement composites-early results and potential application," in Proceedings of the 3rd International Conference on Construction Materials: Performance, Innovations and Structural Implications, pp. 1–10, Vancouver, Canada, August 2005.
2. L. X. Zheng, M. J. O'Connell, S. K. Doorn et al., "Ultralong single-wall carbon nanotubes," Nature Materials, vol. 3, no. 10, pp. 673–676, 2004.
3. M.-F. Yu, B. S. Files, S. Arepalli, and R. S. Ruoff, "Tensile loading of ropes of single wall carbon nanotubes and their mechanical properties," Physical Review Letters, vol. 84, no. 24, pp. 5552–5555, 2000.
4. J.-P. Salvetat, J.-M. Bonard, N. H. Thomson et al., "Mechanical properties of carbon nanotubes," Applied Physics A, vol. 69, no. 3, pp. 255–260, 1999.
5. D. A. Walters, L. M. Ericson, M. J. Casavant et al., "Elastic strain of freely suspended single-wall carbon nanotube ropes," Applied Physics Letters, vol. 74, no. 25, pp. 3803–3805, 1999.
6. P.-W. Chen and D. D. L. Chung, "Concrete reinforced with up to 0.2 vol% of short carbon fibres,"Composites, vol. 24, no. 1, pp. 33–52, 1993.
7. H. Li, H.-G. Xiao, J. Yuan, and J. Ou, "Microstructure of cement mortar with nano-particles," Composites Part B: Engineering, vol. 35, no. 2, pp. 185–189, 2004.
8. G. Y. Li, P. M. Wang, and X. Zhao, "Mechanical behavior and microstructure of cement composites incorporating surface-treated multi-walled carbon nanotubes," Carbon, vol. 43, no. 6, pp. 1239–1245, 2005.
9. C. Balázsi, F. Wéber, Z. Kövér et al., "Application of carbon nanotubes to silicon nitride matrix reinforcements," Current Applied Physics, vol. 6, no. 2, pp. 124–130, 2006.
10. F. Lupo, R. Kamalakaran, C. Scheu, N. Grobert, and M. Rühle, "Microstructural investigations on zirconium oxide-carbon nanotube composites synthesized by hydrothermal crystallization," Carbon, vol. 42, no. 10, pp. 1995–1999, 2004.
11. D. Tasis, S. Pispas, C. Galiotis, and N. Bouropoulos, "Growth of calcium carbonate on non-covalently modified carbon nanotubes," Materials Letters, vol. 61, no. 28, pp. 5044–5046, 2007.
12. M. L. Terranova, S. Orlanducci, A. Fiori et al., "Controlled evolution of carbon nanotubes coated by nanodiamond: the realization of a new class of hybrid nanomaterials," Chemistry of Materials, vol. 17, no. 12, pp. 3214–3220, 2005.
13. A. Cwirzen, K. Habermehl-Cwirzen, and V. Penttala, "Surface decoration of carbon nanotubes and mechanical properties of cement/carbon nanotube composites," Advances in Cement Research, vol. 20, no. 2, pp. 65–73, 2008.
14. J. V. Agullo, V. C. Ligero, D. P. Rico et al., "Mortar and concrete reinforced with nanomaterials,"Nanotechnology in Construction, vol. 3, pp. 383–388, 2009.
15. J. M. Makar and G. W. Chan, "Growth of cement hydration products on single-walled carbon nanotubes," Journal of the American Ceramic Society, vol. 92, no. 6, pp. 1303–1310, 2009.
16. M. S. Konsta-Gdoutos, Z. S. Metaxa, and S. P. Shah, "Highly dispersed carbon nanotube reinforced cement based materials," Cement and Concrete Research, vol. 40, no. 7, pp. 1052–1059, 2010.
17. T. Manzur and N. Yazdani, "Strength enhancement of cement mortar with carbon nanotubes: early results and potential," Journal of the Transportation Research Board, no. 2142, pp. 102–108, 2010.
18. A. M. Hunashyal, S. V. Tippa, S. S. Quadri, and N. R. Banapurmath, "Experimental investigation on effect of carbon nanotubes and carbon fibres on the behavior of plain cement mortar composite round bars under direct tension," ISRN Nanotechnology, vol. 2011, Article ID 856849, 6 pages, 2011.
19. B. Han, Z. Yang, X. Shi, and X. Yu, "Transport properties of carbon-nanotube/cement composites,"Journal of Materials Engineering and Performance, vol. 22, no. 1, pp. 184–189, 2013.
20. T. Manzur, N. Yazdani, and M. A. B. Emon, "Effect of carbon nanotube size on compressive strengths of nanotube reinforced cementitious composites," Journal of Materials, vol. 2014, Article ID 960984, 8 pages, 2014.
21. M. Lelusz, "Carbon nanotubes influence on the compressive strength of cement composites," Technical Transactions, Civil Engineering 1-B/2014, 2014.
22. T. Manzur and N. Yazdani, "Optimum mix ratio for

carbon nanotubes in cement mortar," KSCE Journal of Civil Engineering, vol. 19, no. 5, pp. 1405–1412, 2015.

23. S. Parveen, S. Rana, and R. Fangueiro, "A review on nanomaterial dispersion, microstructure, and mechanical properties of carbon nanotube and nanofiber reinforced cementitious composites," Journal of Nanomaterials, vol. 2013, Article ID 710175, 19 pages, 2013.

24. American Society for Standard Testing and Materials (ASTM), Standard Test Method for Compressive Strength of Hydraulic-Cement Mortars, ASTM C109-07, American Society for Standard Testing and Materials, West Conshohocken, Pa, USA, 2008.

25. T. Manzur, Nano-modified cement composites and its applicability as concrete repair material [Ph.D. thesis], University of Texas at Arlington, Arlington, Tex, USA, 2011.

26. S. Musso, J.-M. Tulliani, G. Ferro, and A. Tagliaferro, "Influence of carbon nanotubes structure on the mechanical behavior of cement composites," Composites Science and Technology, vol. 69, no. 11-12, pp. 1985–1990, 2009.

27. C. Gay and F. Sanchez, "Performance of carbon nanofibercement composites with a high-range water reducer," Journal of the Transportation Research Board, vol. 2142, pp. 109–113, 2010.

28. A. Yazdanbakhsh, Z. Grasley, B. Tyson, and R. K. Abu Al-Rub, "Distribution of carbon nanofibers and nanotubes in cementitious composites," Journal of the Transportation Research Board, vol. 2142, pp. 89–95, 2010.

29. T. Manzur and N. Yazdani, "Importance of flow values in qualitative evaluation of carbon nanotube reinforced cementitous matrix," Malaysian Journal of Civil Engineering, vol. 25, no. 1, pp. 71–80, 2013.

30. American Society for Standard Testing and Materials (ASTM), "Standard test method time for of setting of hydraulic cement mortar by modified vicat needle," ASTM C 807-08, West Conshohocken, Pa, USA, 2009.

31. American Society for Standard Testing and Materials (ASTM), Standard Test Method for Expansion and Bleeding of Freshly Mixed Grouts for Preplaced-Aggregate Concrete in the Laboratory, ASTM C940-98, American Society for Standard Testing and Materials, West Conshohocken, Pa, USA, 2008.

32. K. N. Bakshs, Evaluation of bond strength between overlay and substrate in concrete repairs [M.S. thesis], School of Architecture and the Built Environment, Royal Institute of Technology (KTH), Stockholm, Sweden, 2010.

33. R. Abbasnia, M. Khanzadi, and J. Ahmadi, "Mortar mix proportions and free shrinkage effect on bond strength between substrate and repair concrete," in Concrete Repair, Rehabilitation and Retrofitting II, M. G. Alexander, H. Beushausen, and F. Dehn, Eds., Taylor & Francis Group, London, UK, 2009.

34. E. N. B. S. Júlio, F. A. B. Branco, and V. D. Silva, "Concrete-to-concrete bond strength. Influence of the roughness of the substrate surface," Construction and Building Materials, vol. 18, no. 9, pp. 675–681, 2004.

35. American Society for Standard Testing and Materials (ASTM), "Standard test method for bond strength of epoxy-resin systems used with concrete by slant shear," ASTM C 882-05, ASTM, West Conshohocken, Pa, USA, 2008.

36. Texas Department of Transportation (TXDOT), "Concrete repair materials," Departmental Material Specification DMS-4655, TXDOT, 2009.

CHAPTER 22

NON-DESTRUCTIVE TESTING METHODS OF COMPOSITE MATERIALS

S. Gholizadeh

Keywords: NDT; Composite materials; Condition monitoring

INTRODUCTION

There are varieties of methods to evaluate materials or components and non-destructive methods are an important category of them with many applications. The field of Non-Destructive Evaluation (NDE) or Non-Destructive Testing (NDT) involves the identification and characterization of damages on the surface and interior of materials without cutting apart or otherwise altering the material (Lockard, 2015). In other words, NDT refers to the evaluation and inspection process of materials or components for characterization or finding defects and flaws in comparison with some standards without altering the original attributes or harming the object being tested. NDT techniques provide a cost effective means of testing of a sample for individual investigation or may be applied on the whole material for checking in a production quality control system (Newswire, 2013).

A wide variety of NDT methods plays major roles in testing of composite materials (Scott & Scala, 1982). The applications of composite NDT may include manufacturing (Venkataraman, 2001), pipe and tube manufacturing (Hufenbach et al. 2011; Schneider, 1984), storage tanks (Castaings & Hosten, 2008), aerospace (Liew et al. 2011; Yekani Fard et al. 2014) military and defense (Bennett et al. 2013), nuclear industry (Vavilov et al. 2015), and composite defects characterization (Fotsing et al. 2014). Numerous techniques are used in the composite NDT field, including ultrasonic testing (Peng et al. 2012), thermographic testing (Kroeger, 2014), infrared thermography testing (Vavilov et al. 2015), radiographic testing (Tan et al. 2011), visual testing (VT) or visual inspection (VI) (Bossi & Giurgiutiu, 2015), acoustic emission testing (AE) (Sarasini & Santulli, 2014), acousto-ultrasonic (Su et al., 2014), shearography testing (Hung et al. 2013), optical testing (Liu et al. 2014), electromagnetic testing (Yang et al. 2013), liquid penetrant testing (Kalinichenko et al. 2013), and magnetic particle testing (Lu et al. 2013).

This paper reviews NDT methods for composite evaluation followed by categorizing them and discussion about their advantages and disadvantages and describes non-destructive testing methods of composite material, in order to have a comprehensive review of NDT of composites.

NON-DESTRUCTIVE TESTING METHODS CATEGORIES FOR COMPOSITES

Composite materials have become a major component in the construction of aircraft and spacecraft due to their properties of high strength and low weight (Xu & Li, 2012). Composites are materials that composed of more than one base material, with the base materials maintaining their own structures and properties rather than forming a combined alloy. Based on the reviews of the NDT methods, they can categorize in different ways according to the applications and situations of the testing. The following sections describe these categories.

Contact Methods VS Non-Contact Methods

The basic types of NDT methods include contact and non-contact methods and both of them have their specific applications in testing and evaluating the composites. Most NDT techniques require good contact between the sensor and tested composite surface to obtain reliable data. Contact methods are traditional ultrasonic testing, eddy current testing, magnetic testing, electromagnetic testing, and penetrant testing. Another approach to speed up the data collection process is to

Citation: S. Gholizadeh. Non-destructive testing methods of composite materials, Procedia Structural Integrity, Volume 1, 2016, Pages 50-57, ISSN 2452-3216, http://dx.doi.org/10.1016/j.prostr.2016.02.008.

eliminate the need for physical contact between the sensor and tested structure. Noncontact methods are through transmission ultrasonic, radiography testing, thermography, shearography, and visual inspection (Newswire, 2013). Optical methods (e.g. thermography, holography or shearography) are mostly noncontact. Table 1 categorized NDT methods to contact methods and non-contact methods.

Table 1: Contact & Non-Contact NDT Methods

Contact Methods	Non-Contact Methods
Traditional ultrasonic testing	Through transmission Ultrasonic
Eddy current testing	Radiography testing
Magnetic testing	Thermography
Electromagnetic	Infrared Testing
Penetrant testing	Holography
Liquid penetrant	Shearography
-	Visual inspection

Physical Properties & Structural Integrity

There are diverse ways of testing composites. It is important to pay enough attention to factors such as efficiency and safety in analyzing the best method to use. Furthermore, the method chosen should minimize the costs incurred in the operation.

Material defefcts are the major sources of composite failures. These damages as a result of failure in the material composites can manifest as matrix cracking, fiber fracture fibre debonding and fiber pull-out (Mckuur, 2006). Structural integrity is a formalized process which utilize advanced non-destructive testing (NDT) methods in order to detect, localize and determine a size of damage (Andrzej Katunin et al. 2015). Non-destructive tests can categorize based on the factors that they evaluated. Categories of non-destructive tests are mentioned in Table 2.

Table 2: Category of NDT Methods Based on the Detecting Factors

Category	Applications
The estimation of the physical and mechanical properties, and the material defects detection, in composite	Measurement of: dynamic mechanical analysis (DMA) (Šturm et al. 2015) fiber amount of portion (El-Sabbagh et al. 2013) mechanical strength and stiffness (Ray, 2006) elastic constants (Rojek et al. 2005) material content (El-Sabbagh et al. 2013) damage failing initiation and subsequent damage evolution (Talreja, 2008) delamination (Ghadermazi e tal. 2015) construction connected with laminate (Scarselli et al. 2015) condition of resin cure (Aggelis & Paipetis, 2012) condition regarding to fiber/ matrix interface (Kersemans et al., 2014)
To determine the integrity of structural components which are manufactured from composite	Detection of: cracks and debonding (Giurgiutiu, 2016) mechanical rubbing (Gostautas et al. 2005) fibre pull out (Short et al. 2002) fibre breakage (Narita et al. 2014)

The mechanical properties of a material determine its manufacturability, performance, and longevity. This means that knowledge of the mechanical properties is essential for physical and mechanical characterization of a composite material (Šturm et al. 2015). Mayer & Council (1993) showed that the mechanical performance of GFRP composites depends on the fibers' strength and modulus, the strength of the matrix and the chemical stability. Ray (2006) has demonstrated that the higher the temperature of the environment and the longer is the exposure time, the larger will be the decrease in the strength and modulus of the GFRP. Rojek et al. (2005) found that the best methods for elasticproperties determinations are the ultrasound methods. El-Sabbagh et al. (2013) studied the feasibility of using an ultrasonic longitudinal sound wave in the definition of the fiber content and the distribution in natural fiber. Talreja (2008) investigated about the mechanical characterization and damage behavior of composite materials. Delamination type defects are detected experimentally on Glass/epoxy composites by employing step heating thermography by Ghadermazi et al. (2015).

Inspection Type versus NDT Method

According to the literature review, there are many types of inspections to evaluate composites and there are many proposed methods by researchers for each of which. For damage identification in aircraft composite structures, aircraft composites assessment, and health monitoring of aerospace composite structures the suggested methods in the literature is ultrasonic testing (Andrzej Katunin et al. 2015; Staszewski et al. 2009), thermographic testing (Andrzej Katunin et al. 2015; Maierhofer et al. 2014), vibration methods (Andrzej Katunin et al. 2015; Loutas et al. 2012; Rizos et al. 2008), infrared thermography (Meola & Carlomagno, 2014), shearography (Růžek et al. 2006), and XCT (Bull et al. 2013).

Ultrasonic testing is the most applied method in health monitoring of a composite wing-box structure (Grondel et al. 2004), damage identification in aircraft composite structures (Andrzej Katunin et al. 2015; Polimeno & Meo, 2009), aircraft composites assessment (Růžek et al. 2006), health monitoring of aerospace composite structures (Loutas et al., 2012), and structural health monitoring (SHM) (Staszewski et al. 2009). Table 3 presents the inspection type and the methods that have used for each one. This table provides a quick suggestion for the readers to make a decision when needs to evaluate a specific type of composite.

Table 3: Inspection Type and NDT Methods

Inspection Type	NDT Method
Damage identification in aircraft composite structures (Andrzej Katunin et al. 2015; Polimeno & Meo, 2009) Aircraft composites assessment (Růžek et al. 2006) Health monitoring of aerospace composite structures (Loutas et al. 2012)	Ultrasonic Testing (Andrzej Katunin et al. 2015) Thermographic Testing (Maierhofer et al. 2014; Usamentiaga et al. 2013) Vibration Methods (Loutas et al. 2012; Rizos et al. 2008) Infrared Thermography (Meola & Carlomagno, 2014) Shearography (Růžek et al. 2006) XCT (Bull et al. 2013)
Health monitoring of a composite wing-box structure (Grondel et al. 2004)	Ultrasonic Testing (Grondel et al. 2004; Andrzej Katunin et al. 2015)
Structural Health Monitoring (SHM) (Staszewski et al. 2009)	Ultrasonic Testing (Andrzej Katunin et al. 2015; Staszewski et al. 2009)
Damage in GFRP (Meola & Carlomagno, 2010; Rique et al. 2015)	Thermographic Testing (Maierhofer et al. 2014; Usamentiaga et al. 2013) Radiography (Ataş & Soutis, 2013; Rique et al. 2015)
Auto-detection of impact damage in carbon fiber composites (Usamentiaga et al. 2013) Characterizing damage in CFRP structures (Ataş & Soutis, 2013; Maierhofer et al. 2014)	Thermographic Testing (Maierhofer et al. 2014; Usamentiaga et al. 2013) Radiography (Ataş & Soutis, 2013; Rique et al. 2015)
Impact damage in glass/epoxy with manufacturing defects (Meola & Carlomagno, 2014)	Infrared Thermography (Meola & Carlomagno, 2014)
damage assessment in sandwich structures (Meo et al. 2005) parameters influencing the damping of a structure (Adams et al. 1978) the structures behavior (Cawley; & Adams, 1979) dynamic characteristics for damage detection of structures (Cawley & Adams, 1978) skin damage statistical detection and restoration assessment (Rizos et al. 2008)	Vibration Methods (Loutas et al. 2012; Rizos et al. 2008; Trendafilova et al. 2008)
Multiple Cracks Detection (Zhang et al. 2010)	Neutron Radiography (Zhang et al. 2010)

Visual Testing (VT) - (VI - Visual Inspection)

Visual testing (VT) - (VI - visual inspection) should be the most basic type of NDT that many instances use because it can save both time and money by reducing the amount of other testing, or in some cases reducing the need for other types of testing all together. The most important advantage of the visual inspection is its quick process. The other advantage of visual inspection is the relative affordability of the process. The visual inspection needs no equipment but this method has its intrinsic disadvantage

Ultrasonic Testing (UT)

Ultrasonic testing (UT) evaluation system consists of a transmitter and receiver circuit, transducer tool, and display devices. Based on the information carried by the signal, crack location, flaw size, its orientation and other characteristics could be achieved (Lu, 2010). Advantages of ultrasonic testing include speed of scan, good resolution and flaw detecting capabilities, and ability of use in the field. Disadvantages include difficulty of set up, needed skill to scan a part accurately, and the need of test sample to insure accurate testing. This type of testing is excellent for use in an assembly line where the same part design must test repeatedly. There are two approaches of ultrasonic NDT generally used in different applications; pulse echo and through transmission approaches. Both of these approaches use high frequency sound waves on the order of 1-50 MHz to detect internal flaws in a material (Garney, 2006). Ultrasonic testing conducted in three modes, transmission, reflection, and back scattering. Each of which uses a range of transducers, coupling agents, and frequencies (Stonawski et al. 2008). Pulse echo ultrasonic method can readily locate defects in homogeneous materials. In this method, the operator more concerns about the transit time of the wave and the energy loss due to attenuation and wave scattering on flaws. It helps to locate inconsistency in a material whether it is homogeneous or heterogeneous (Warnemuende, 2006). For large defect detection, location, and imaging purposes, and quality control, ultrasonic pulse velocity measurements are quite suitable (Oguma et al. 2012). The through transmission ultrasonic method is different from conventional ultrasonic methods. This method keeps the transducer and receiver off the surface and at a fixed distance away from the sample. This is particularly advantageous when complex geometries do not allow for the contact of a traditional transducer and receiver to the surface of the part.

The most commonly used indicators of properties are wave propagation velocity and amplitude (or energy) loss. Some of the testing methods described herein only address one property, while others, more versatile, may measure two or three (Ducharne et al. 2015). Most applications consider only the pulse velocity and relate it to different parameters. Considering energy loss can discover a few additional characteristics of a material (Karabutov & Podymova, 2014a). A number of authors have studied the method of pulse attenuation analysis (El-Sabbagh et al. 2013; Genovés et al. 2015). Scattering, absorption and geometric are three parameters that affect the attenuation. Small discontinuities like grain boundaries are the source of scattering.

Thermography Testing

Thermography testing also called thermal imaging. The thermal conductivity of a material may change by the presence of defects, thermography inspection used for thin parts because when defects moved deeper under the surface of a part, they tend to produce less heat fluctuation than defects seen closer to the surface of the part. As a generally rule, defects that have a diameter smaller than their depth in the part, cannot be picked up by this type of inspection. A flaw, such as a delamination or impact damage causes a change in the thermal radiation of the area (Meyendorf et al., 2013). There are many advantages and disadvantages to this type of inspection. One advantage is it can inspect a large surface of a part. The second advantage is that unlike many other types of inspection it does not have to couple. This allow for the inspection of parts where only one side of the part is accessible to inspection. Disadvantages of this type of inspection include the need for sensitive and expensive instrumentation, the need for highly skilled inspectors to run the instruments, and the lack of clarity of defects if they fall too deeply under the surface of the part. Infrared Thermography Testing (IRT) is based on the recording of the thermal radiation emitted by a surface of a specimen by means of an infrared camera (Mulaveesalaa & Tuli, 2008).

Radiographic Testing (RT)

Radiographic testing (RT) is the most commonly used testing method (Lockard, 2015). The most common type of damage to composites is a delamination resulting in an air pocket; a delamination can only be seen in RT if its orientation is not perpendicular to the x-ray beam. There are many types of radiography and each has specific applications. Conventional radiography is the most useful when the parts are neither too thick nor too thin. For thin parts, 1 to 5 mm, low voltage radiography is used and γ-rays radiography is good for thick parts. These types of radiography are useful in detecting large voids, inclusions, trans-laminar cracks, non-uniform fiber distribution, and fiber misorientation such as fiber wrinkles or weld lines (Garney, 2006). Another type of radiography uses γ-rays to penetrate the composite. Gamma rays radiography is good for thick parts because the gamma rays have shorter wavelengths. Penetrant-enhanced is another type of radiography employed specifically to detect small matrix cracks, and delaminations in a sample (Ataş & Soutis, 2013). There are varieties of radiographic testing methods for different applications. These methods are film radiography (Burkle & Lemle, 1993), computed radiography (Tan et al. 2011), computed tomography (Katunin et al. 2015), and digital radiography (Aidi et al. 2015). X-ray Computed Tomography (XCT) is a nondestructive technique for visualizing interior features within solid objects, and for obtaining digital information on their 3-D geometries and properties. The great advantage of XCT in comparison with the projection radiology is the 3-D visualized image of the structure while in projection radiology the image is only 2-D. Therefore, the XCT data is readable quickly and simply. XCT will modify the scale of observation from macroscopic to microscopic scale so the results of the XCT method are very reliable (Bayraktar et al. 2008).

Electromagnetic Testing (ET)

Electromagnetic testing (ET) methods use magnetism and electricity to detect and evaluate fractures, faults, corrosion or other conditions of materials. ET induces electric currents, magnetic fields, or both inside a test object and observes the electromagnetic response. Electromagnetic (EM) methods include Eddy Current Testing (EC) (Koyama et al. 2013), Remote Field Testing (RFT), Magnetic Flux Leakage (MFL) and Alternating Current Field Measurement (ACFM). In each of these techniques, the underlying physics is fundamentally different as the fields described by different classes of partial differential equations (PDEs).

Acoustic Emission (AE)

Acoustic emission (AE) is an effective method of imperfection analysis. This mechanical vibration generated by material defects such as matrix micro cracking, fiber-matrix debonding, localized delamination, or fiber pullout and breakage (Arumugam et al. 2011; S. Gholizadeh et al. 2015). The stress waves that result from these types of defects spread out concentrically from their origin and are detected by an array of highly sensitive peizoelectrics.

Acoustic emission technique is different from most other NDE techniques in two aspects. The first difference is the origin of the signal. Instead of supplying energy to the object, this method listens to the "sound" generated by energy released in the object. The second difference is the method that AE deals with dynamic processes in a material. The ability to discern between developing and stagnant defects is significant. Other advantages of AE method include high sensitivity, fast and global inspection using multiple sensors, permanent sensor mounting for process control, and no need to disassemble and clean a specimen (Lu, 2010). The second advantage of the AE is that it is very useful in detecting many different types caused by fatigue loading. Fatigue damage types which acoustic emission testing can detect include fatigue cracks, fiber fractures, matrix micro-cracks, fiber-matrix debonding, and delamination. The drawback to this type of testing is the great skill that is required to correlate acoustic emission data to specific types of damage mechanisms.

Acousto-Ultrasonic

Acousto-ultrasonic is a combination method of acoustic and ultrasonic testing that used specifically to determine the severity of internal imperfections and inhomogeneity in a composite. In nondestructive testing, the acoustic/ultrasonic class of testing has great potential based on optimal economy, flexibility and sensitivity. However, no available method is sensitive or reliable enough to effectively detect. It is useful method because it allows non-critical flaws to see and assess. The second advantage is that it is a good indicator of accumulated damage in a structure due to fatigue loading or impact damage. The disadvantage of this type of inspection is the setup and pre-calculations that is mandatory before any testing. The second disadvantage is that this type of testing is not useful to detect individual large flaws such as delamination or voids (Schroeder et al. 2002).

Shearography Testing

Shearography testing is a laser optical method. The failure of composites usually happens by stress concentrations and the criticality of defects will easily deduct by the degree of strain concentrations around a particular defect, this is an advantage of shearography (Hung et al. 2013). A second advantage of shearography is that it is less susceptible to noise than many other types of nondestructive testing. This is good because it allow less skilled users to be able to inspect and determine the usability of a part without extensive training. A major disadvantage of shearography is that characterization of defect types other than delamination is extremely difficult. Therefore it is sometimes paired with other types of non-destructive evaluation techniques that can help to identify certain defects.

CONCLUSION

This paper reviewed NDT methods for composite evaluation by categorizing their advantages and disadvantages as well as describing NDT methods of composite materials in order to have a comprehensive review of NDT of composites. Due to the fact that composite tools are mostly used in critical-safety applications for example in aircraft primary constructions, the non-destructive testing of composite materials has become more crucial and demanding. Factors such as efficiency and safety should be used in analyzing the best method to be used. Furthermore, the method chosen should minimize the costs incurred in the operation. It is based on methods that depend on the use of physical values to determine the characteristics of materials. In addition, non-destructive tests use physical principles to identify and evaluate faults or destructive defects.

REFERENCES

1. Adams, R., Cawley, P., Pye, C., & Stone, B. (1978). A vibration technique for non-destructively assessing the integrity of structures. Journal of Mechanical Engineering Science, 20(2), 93-100.
2. Aggelis, D. G., & Paipetis, A. S. (2012). Monitoring of resin curing and hardening by ultrasound. Construction and Building Materials, 26(1), 755-760. doi: http://dx.doi.org/10.1016/j.conbuildmat.2011.06.084
3. Aidi, B., Philen, M. K., & Case, S. W. (2015). Progressive damage assessment of centrally notched composite specimens in fatigue. Composites Part A: Applied Science and Manufacturing, 74(0), 47-59. doi: http://dx.doi.org/10.1016/j.compositesa.2015.03.022
4. Arumugam, V., Kumar, C. S., Santulli, C., Sarasini, F., & Stanley, A. J. (2011). A Global Method for the Identification of Failure Modes in Fiberglass Using Acoustic Emission. Journal of Testing and Evaluation, 39(5).
5. Ataş, A., & Soutis, C. (2013). Subcritical damage mechanisms of bolted joints in CFRP composite laminates. Composites Part B: Engineering, 54(0), 20-27. doi: http://dx.doi.org/10.1016/j.compositesb.2013.04.071
6. Bayraktar, E., Antolovich, S. D., & Bathias, C. (2008). New developments in non-destructive controls of the composite materials and applications in manufacturing engineering. Journal of Materials Processing Technology, 206(1–3), 30-44.
7. Bennett, L. G. I., Lewis, W. J., & Hungler, P. C. (2013). The Development of Neutron Radiography and Tomography on a SLOWPOKE-2 Reactor. Physics Procedia, 43, 21-33. doi: http://dx.doi.org/10.1016/j.phpro.2013.03.003
8. Bossi, R. H., & Giurgiutiu, V. (2015). 15 - Nondestructive testing of damage in aerospace composites. In P. E. Irving & C. Soutis (Eds.), Polymer Composites in the Aerospace Industry (pp. 413-448): Woodhead Publishing.
9. Bull, D. J., Spearing, S. M., Sinclair, I., & Helfen, L. (2013). Three-dimensional assessment of low velocity impact damage in particle toughened composite laminates using micro-focus X-ray computed tomography and synchrotron radiation laminography. Composites Part A: Applied Science and Manufacturing, 52(0), 62-69. doi: http://dx.doi.org/10.1016/j.compositesa.2013.05.003
10. Burkle, W. S., & Lemle, L. P.(1993).The Effect of Order-of-Film Placement in Composite Film Radiography. Materials Evaluation,51(3), 327-29
11. Castaings, M., & Hosten, B. (2008). Ultrasonic guided waves for health monitoring of high-pressure composite tanks. NDT & E International, 41(8), 648-655. doi: http://dx.doi.org/10.1016/j.ndteint.2008.03.010
12. Cawley, P., & Adams, R. D. (1978). The predicted and experimental natural modes of free-free CFRP plates. Journal of Composite Materials, 12(4), 336-347.
13. Cawley;, P., & Adams, R. D. (1979). The location of defects in structures from measurements of natural frequencies. The Journal of Strain Analysis for Engineering Design, 14(2), 49-57.
14. Ducharne, B., Guyomar, D., Sébald, G., & Zhang, B. (2015). 10 - Modeling energy losses in power ultrasound transducers. In J. A. GallegoJuárez & K. F. Graff (Eds.), Power Ultrasonics (pp. 241-256). Oxford: Woodhead Publishing.
15. El-Sabbagh, A., Steuernagel, L., & Ziegmann, G. (2013). Characterisation of flax polypropylene composites using ultrasonic longitudinal sound wave technique. Composites Part B: Engineering, 45(1), 1164-1172. doi: http://dx.doi.org/10.1016/j.compositesb.2012.06.010
16. Fotsing, E. R., Ross, A., & Ruiz, E. (2014). Characterization of surface defects on composite sandwich materials based on deflectrometry. NDT & E International, 62, 29-39. doi: http://dx.doi.org/10.1016/j.ndteint.2013.11.004
17. Garney, G. (2006). Defects found through non-destructive testing methods of fiber reinforced polymeric composites. (1438996 M.S.), California State University, Fullerton, Ann Arbor. Retrieved from http://search.proquest.com/docview/304910161?accountid=28930 ProQuest Dissertations & Theses Global database.
18. Genovés, V., Gosálbez, J., Miralles, R., Bonilla, M., & Payá, J. (2015). Ultrasonic characterization of GRC with high percentage of fly ash substitution. Ultrasonics, 60, 88-95. doi: http://dx.doi.org/10.1016/j.ultras.2015.02.016
19. Ghadermazi, K., Khozeimeh, M. A., Taheri-Behrooz, F., & Safizadeh, M. S. (2015). Delamination detection in glass–epoxy composites using step-phase thermography (SPT). Infrared Physics & Technology, 72, 204-209. doi: http://dx.doi.org/10.1016/j.infrared.2015.08.006
20. Giurgiutiu, V. (2016). Chapter 5 - Damage and Failure of Aerospace Composites. In V. Giurgiutiu (Ed.), Structural Health Monitoring of Aerospace Composites (pp. 125-175). Oxford: Academic Press.

21. Gostautas, R., Ramirez, G., Peterman, R., & Meggers, D. (2005). Acoustic Emission Monitoring and Analysis of Glass Fiber-Reinforced Composites Bridge Decks. Journal of Bridge Engineering, 10(6), 713-721.
22. Grondel, S., Assaad, J., Delebarre, C., & Moulin, E. (2004). Health monitoring of a composite wingbox structure. Ultrasonics, 42(1–9), 819-824.
23. Hufenbach, W. B., R. Thieme, M. Tyczynski, T. (2011). Damage monitoring in pressure vessels and pipelines based on wireless sensor networks. Procedia Engineering, 10, 340-345. doi: http://dx.doi.org/10.1016/j.proeng.2011.04.058
24. Hung, Y. Y., Yang, L. X., & Huang, Y. H. (2013). 5 - Non-destructive evaluation (NDE) of composites: digital shearography. In V. M. Karbhari (Ed.), Non-Destructive Evaluation (NDE) of Polymer Matrix Composites (pp. 84-115): Woodhead Publishing.
25. Kalinichenko, N. P., Kalinichenko, A. N., Lobanova, I. S., & Borisov, S. S. (2013). Methods for the manufacture of nonmetallic reference specimens for liquid-penetrant inspection. Russian Journal of Nondestructive Testing, 49(11), 668-672.
26. Karabutov, A. A., & Podymova, N. B. (2014a). Quantitative analysis of the influence of voids and delaminations on acoustic attenuation in CFRP composites by the laser-ultrasonic spectroscopy method. Composites Part B: Engineering, 56(0), 238-244.
27. Katunin, A., Danczak, M., & Kostka, P. (2015). Automated identification and classification of internal defects in composite structures using computed tomography and 3D wavelet analysis. Archives of Civil and Mechanical Engineering, 15(2), 436-448.
28. Katunin, A., Dragan, K., & Dziendzikowski, M. (2015). Damage identification in aircraft composite structures: A case study using various nondestructive testing techniques. Composite structures, 127(0), 1-9. doi: http://dx.doi.org/10.1016/j.compstruct.2015.02.080
29. Kersemans, M., De Baere, I., Degrieck, J., Van Den Abeele, K., Pyl, L., Zastavnik, F., . . . Van Paepegem, W. (2014). Nondestructive damage assessment in fiber reinforced composites with the pulsed ultrasonic polar scan. Polymer Testing, 34, 85-96.
30. Koyama, K., Hoshikawa, H., & Kojima, G. (2013). Eddy Current Nondestructive Testing for Carbon Fiber- Reinforced Composites. Journal of Pressure Vessel Technology, 135(4), 041501-041501. doi: 10.1115/1.4023253
31. Kroeger, T. (2014). Thermographic inspection of composites. Reinforced Plastics, 58(4), 42-43.
32. Liew, C. K., Veidt, M., Rajic, N., Tsoi, K., Rowlands, D., & Morton, H. (2011). Inspections of Helicopter Composite Airframe Structures using Conventional and Emerging Nondestructive Testing Methods. Journal of Testing and Evaluation, 39(6).
33. Liu, P., Groves, R. M., & Benedictus, R. (2014). 3D monitoring of delamination growth in a wind turbine blade composite using optical coherence tomography. NDT & E International, 64(0), 52-58. doi: http://dx.doi.org/10.1016/j.ndteint.2014.03.003
34. Lockard, C. D. (2015). Anomaly detection in radiographic images of composite materials via crosshatch regression. (1586713 M.A.), Mills College, Ann Arbor. Retrieved from ProQuest Dissertations & Theses Global database.
35. Loutas, T. H., Panopoulou, A., Roulias, D., & Kostopoulos, V. (2012). Intelligent health monitoring of aerospace composite structures based on dynamic strain measurements. Expert Systems with Applications, 39(9), 8412-8422. doi: http://dx.doi.org/10.1016/j.eswa.2012.01.179
36. Lu, Y. (2010). Non-destructive Evaluation on Concrete Materials and Structures using Cement-based Piezoelectric Sensor. (3457480 Ph.D.), Hong Kong University of Science and Technology (Hong Kong), Ann Arbor. Retrieved from http://search.proquest.com/docview/880396184?accountid=28930 ProQuest Dissertations & Theses Global database.
37. Lu, Z. Y., Zhang, Q. L., & Liu, X. (2013). New Magnetic Particle Cassette NDT Intelligent Detection Device. 2013 Fourth International Conference on Intelligent Systems Design and Engineering Applications, 403-406. doi: Doi 10.1109/Isdea.2013.496
38. Maierhofer, C., Myrach, P., Reischel, M., Steinfurth, H., Röllig, M., & Kunert, M. (2014). Characterizing damage in CFRP structures using flash thermography in reflection and transmission configurations. Composites Part B: Engineering, 57(0), 35-46.
39. Mayer, R. M., & Council, D. (1993). Design with Reinforced Plastics: A Guide for Engineers and Designers: Springer Netherlands.
40. Mckuur, G. (2006). Defect and Fracture Detection Using Acoustic Emission Monitoring: University of Cape Town.
41. Meo, M., Zumpano, G., Piggott, M., & Marengo, G. (2005). Impact identification on a sandwich plate from wave propagation responses. Composite structures, 71(3–4), 302-306. doi: http://dx.doi.org/10.1016/j.compstruct.2005.09.028
42. Meola, C., & Carlomagno, G. M. (2014). Infrared thermography to evaluate impact damage in glass/epoxy with manufacturing defects. International Journal of Impact Engineering, 67(0), 1-11. doi: http://dx.doi.org/10.1016/j.ijimpeng.2013.12.010
43. Meyendorf, N.G.H., P.B. Nagy, and S. Rokhlin (2013), Nondestructive Materials Characterization: With Applications to Aerospace Materials: Springer Berlin Heidelberg.
44. Mulaveesalaa, R., & Tuli, S. (2008). Applications of Frequency Modulated Thermal Wave Imaging For Non-destructive Characterization, Kollam, Kerala.
45. Narita, F., Shindo, Y., Takeda, T., Kuronuma, Y., & Sanada, K. (2014). Loading Rate-Dependent Fracture Properties and Electrical ResistanceBased Crack Growth Monitoring of Polycarbonate Reinforced with Carbon Nanotubes Under Tension. Journal of Testing and Evaluation, 43(1).
46. Newswire, P. (2013, 2013 Feb 07). Global Non-Destructive Testing (NDT) Equipment Market - By Technology (Ultrasonic, Eddy Current, Electromagnetic,

Radiography, Thermography), Verticals (Manufacturing, Petrochemical, Aerospace, Automotive, Power Generation) & Geography (2013 - 2018), PR Newswire.

47. Oguma, I., Goto, R., & Sugiura, T. (2012). Ultrasonic inspection of an internal flaw in a ferromagnetic specimen using angle beam EMATs. Przeglad Elektrotechniczny, 88(7B), 78-81.

48. Peng, W., Zhang, Y., Qiu, B., & Xue, H. (2012). A Brief Review of the Application and Problems in Ultrasonic Fatigue Testing. AASRI Procedia, 2(0), 127-133. doi: http://dx.doi.org/10.1016/j.aasri.2012.09.024

49. Polimeno, U., & Meo, M. (2009). Detecting barely visible impact damage detection on aircraft composites structures. Composite structures, 91(4), 398-402. doi: http://dx.doi.org/10.1016/j.compstruct.2009.04.014

50. Ray, B. C. (2006). Temperature effect during humid ageing on interfaces of glass and carbon fibers reinforced epoxy composites. Journal of Colloid and Interface Science, 298(1), 111-117. doi: http://dx.doi.org/10.1016/j.jcis.2005.12.023

51. Rique, A. M., Machado, A. C., Oliveira, D. F., Lopes, R. T., & Lima, I. (2015). X-ray imaging inspection of fiberglass reinforced by epoxy composite. Nuclear Instruments and Methods in Physics Research Section B: Beam Interactions with Materials and Atoms, 349(0), 184-191. doi: http://dx.doi.org/10.1016/j.nimb.2015.03.003

52. Rizos, D. D., Fassois, S. D., Marioli-Riga, Z. P., & Karanika, A. N. (2008). Vibration-based skin damage statistical detection and restoration assessment in a stiffened aircraft panel. Mechanical Systems and Signal Processing, 22(2), 315-337. doi: http://dx.doi.org/10.1016/j.ymssp.2007.07.012

53. Rojek, M., Stabik, J., & Wróbel, G. (2005). Ultrasonic methods in diagnostics of epoxy–glass composites. Journal of Materials Processing Technology, 162, 121-126.

54. Růžek, R., Lohonka, R., & Jironč, J. (2006). Ultrasonic C-Scan and shearography NDI techniques evaluation of impact defects identification. NDT & E International, 39(2), 132-142. doi: http://dx.doi.org/10.1016/j.ndteint.2005.07.012

55. S. Gholizadeh, Leman, Z., & Baharudin, B. T. H. T. (2015). A review of the application of acoustic emission technique in engineering. Structural Engineering and Mechanics, An International Journal, 54(6), 1075-1095.

56. Sarasini, F., & Santulli, C. (2014). 10 - Non-destructive testing (NDT) of natural fibre composites: acoustic emission technique. In A. Hodzic & R. Shanks (Eds.), Natural Fibre Composites (pp. 273-302): Woodhead Publishing.

57. Schneider, H. (1984). The nondestructive testing of tubes and pipes for nuclear application. Nuclear Engineering and Design, 81(1), 69-76.

58. Schroeder, J. A., Ahmed, T., Chaudhry, B., & Shepard, S. (2002). Non-destructive testing of structural composites and adhesively bonded composite joints: pulsed thermography. Composites Part A: Applied Science and Manufacturing, 33(11), 1511-1517.

59. Scott, I. G., & Scala, C. M. (1982). A review of non-destructive testing of composite materials. NDT International, 15(2), 75-86.

60. Short, G., Guild, F., & Pavier, M. (2002). Delaminations in flat and curved composite laminates subjected to compressive load. Composite structures, 58(2), 249-258.

61. Staszewski, W. J., Mahzan, S., & Traynor, R. (2009). Health monitoring of aerospace composite structures – Active and passive approach. Composites Science and Technology, 69(11–12), 1678-1685. doi: http://dx.doi.org/10.1016/j.compscitech.2008.09.034

62. Stonawski, O., & Engineering, S. I. U. a. C. M. (2008). Non-destructive Evaluation of Carbon/carbon Brakes Using Air-coupled Ultrasonic Inspection Systems: Southern Illinois University at Carbondale.

63. Šturm, R., Grimberg, R., Savin, A., & Grum, J. (2015). Destructive and nondestructive evaluations of the effect of moisture absorption on the mechanical properties of polyester-based composites. Composites Part B: Engineering, 71, 10-16.

64. Su, Z., Zhou, C., Hong, M., Cheng, L., Wang, Q., & Qing, X. (2014). Acousto-ultrasonics-based fatigue damage characterization: Linear versus nonlinear signal features. Mechanical Systems and Signal Processing, 45(1), 225-239.

65. Talreja, R. (2008). Damage and fatigue in composites–a personal account. Composites Science and Technology, 68(13), 2585-2591.

66. Tan, K. T., Watanabe, N., & Iwahori, Y. (2011). X-ray radiography and micro-computed tomography examination of damage characteristics in stitched composites subjected to impact loading. Composites Part B: Engineering, 42(4), 874-884.

67. Usamentiaga, R., Venegas, P., Guerediaga, J., Vega, L., & López, I. (2013). Automatic detection of impact damage in carbon fiber composites using active thermography. Infrared Physics & Technology, 58(0), 36-46.

68. Vavilov, V. P., Budadin, O. N., & Kulkov, A. A. (2015). Infrared thermographic evaluation of large composite grid parts subjected to axial loading. Polymer Testing, 41(0), 55-62.

69. Vavilov, V. P., Plesovskikh, A. V., Chulkov, A. O., & Nesteruk, D. A. (2015). A complex approach to the development of the method and equipment for thermal nondestructive testing of CFRP cylindrical parts. Composites Part B: Engineering, 68, 375-384.

70. Venkataraman, B. (2001). NDT as a Support for Materials Processing and Manufacturing, Encyclopedia of Materials: Science and Technology (Second Edition) (pp. 5959-5963). Oxford: Elsevier.

71. Warnemuende, K. (2006). Amplitude modulated acousto-ultrasonic non-destructive testing: Damage evaluation in concrete. (3243085 Ph.D.), Wayne State University, Ann Arbor. Retrieved from http://search.proquest.com/docview/304968621?accountid=28930 ProQuest Dissertations & Theses Global database.

72. Xu, B., & Li, H. Y. (2012). Advanced composite materials and manufacturing engineering : selected, peer

reviewed papers from the 2012 international conference on advanced composite materials and manufacturing engineering (CMME2012) October 13-14, 2012, Beijing, China. Durnten-Zurich, Switzerland ; Enfield, NH: Trans Tech Publications.

73. Yang, S.-H., Kim, K.-B., Oh, H. G., & Kang, J.-S. (2013). Non-contact detection of impact damage in CFRP composites using millimeter-wave reflection and considering carbon fiber direction. NDT & E International, 57(0), 45-51.

74. Yekani Fard, M. S., Seid Mohammadali. Raji, Brian B.. Chattopadhyay, Aditi. (2014). Damage characterization of surface and sub-surface defects in stitch-bonded biaxial carbon/epoxy composites. Composites Part B: Engineering, 56, 821-829.

75. Zhang, P., Wittmann, F. H., Zhao, T. J., Lehmann, E. H., Tian, L., & Vontobel, P (2010). Observation and quantification of water penetration into Strain Hardening Cement-based Composites (SHCC) with multiple cracks by means of neutron radiography. Nuclear Instruments and Methods in Physics Research Section A: Accelerators, Spectrometers, Detectors and Associated Equipment, 620(2–3), 414-420.

CHAPTER 23

IRON OXIDE-PEROVSKITE COMPOSITE MATERIALS

Cristina Dueso, Claire Thompson, and Ian Metcalfe

Keywords: Hydrogen production; Chemical looping; Perovskite; Iron oxide; Composite; Oxygen carrier material

INTRODUCTION

Chemical looping hydrogen production typically utilises iron oxide as an oxygen carrier as this material has several thermodynamically favourable phase changes that can be exploited [1], [2], [3] and [4]. In this process a carbonaceous fuel, here carbon monoxide, is used to reduce the iron oxide (producing carbon dioxide). Then water is used to re-oxidise the iron and produce hydrogen. This is performed in a cyclic manner. In order to have good utilisation of the redox capacity of the material, we need to use an oxygen carrier material (OCM) particle size, l, that is similar to or less than the effective oxygen diffusional lengthscale (effective diffusion coefficient, D, divided by an equivalent first order surface rate constant, k) such that the surface rate is controlling, e.g. l is similar to or less than D/k. The diffusional lengthscale of oxygen within iron oxide is expected to be quite short as loss of surface area in iron oxide OCMs significantly affects effective OCM capacity [5] and [6]. Therefore a small particle size is required, which in turn means that a large surface area is required. Iron oxide, however, suffers from thermal sintering at high temperatures (>800 °C) and as a result loses surface area while particle size increases.

These problems could be overcome if the iron oxide was embedded and well dispersed in a stable matrix. This matrix should have a diffusional lengthscale for oxygen comparable to or larger than its particle size. Doped lanthanum ferrites are known to have a relatively long effective diffusional lengthscale for oxygen with, e.g., $La_{0.9}Sr_{0.1}FeO_{3-\delta}$ (LSF911) having a diffusional lengthscale of around 60 μm at 900 °C [7]. Here we use $La_{0.7}Sr_{0.3}FeO_{3-\delta}$ (LSF731) as the matrix with particle sizes of 80–160 μm albeit at the slightly lower temperature of 850 °C in the isothermal work. Thus we may expect, if the properties of LSF911 and LSF731 are not dissimilar and the temperature difference has limited impact, that the diffusional lengthscale of LSF731 is similar to its particle size. This matrix material does not in itself require a high oxygen capacity on redox cycling (this capacity being provided by the iron oxide), rather it provides a rapid oxygen transfer pathway to access a distributed reserve of iron oxide and prevents excessive sintering of iron oxide. It is expected that the composite material would effectively combine the benefits of both materials individually, i.e. the high oxygen capacity of the iron oxide and the superior stability and oxygen transport properties of LSF731 [8].

The idea of embedding the oxygen carrier within a stable matrix was first suggested by Thursfield et al. [1]. Galinsky et al. [9] prepared an OCM containing 60 wt.% iron oxide supported on $La_{0.8}Sr_{0.2}FeO_3$ by solid-state synthesis (SSR). Reactivity tests performed in a TGA showed that the presence of LSF enhanced the reactivity of the OCM with H_2, CO and CH_4 5-70-fold compared with inert supports, such as TiO_2, and reduced carbon formation. The authors proposed that the use of a mixed ionic-electronic conductive (MIEC) support facilitated O^{2-} transport from the iron oxide and improved the resistance of the material. The same research group investigated the influence of the preparation method on the performance of $Fe_2O_3@La_xSr_{1-x}FeO_3$ OCMs for partial oxidation of methane [10]. A sol–gel method obtained an OCM which was more reactive and exhibited less carbon formation than the one prepared by SSR. An OCM combining iron oxide and $CaTi_{0.85}Fe_{0.15}O_{3-\delta}$ (CTF) showed better properties for syngas conversion than composite materials with other perovskites, such as LSF and BCF ($BaCe_{0.7}Fe_{0.3}O_{3-\delta}$), due to a higher

Citation: Cristina Dueso, Claire Thompson, Ian Metcalfe, "Iron oxide-perovskite composite materials" http://dx.doi.org/10.1016/j.apenergy.2015.05.062.

stability [11] and indeed redox properties in general appear to be affected by support-iron oxide interactions [12]. Iron oxide supported on $Ca_{0.8}Sr_{0.2}Ti_{0.8}Ni_{0.2}O_3$ presented better stability and less carbon formation with CH_4 as fuel than when CeO_2 and $MgAl_2O_4$ were used as supports. Substitution of Ni^{2+} on the B-site with Fe^{3+} improved the perovskite structure stability.

This work aims to definitively prove the idea that well-dispersed clusters of iron oxide embedded in an LSF731 matrix can improve the stability and maintain the reactivity of the OCM on repeated cycling.

EXPERIMENTAL

OCM Preparation

Several materials with a composite structure made of iron oxide clusters embedded in a matrix of LSF731 were prepared in this work to be used as OCMs for hydrogen production by chemical looping. Fig. 1 shows a schematic diagram of the OCM particles. Iron oxide is defined to be in the form of a cluster when located within the LSF731 matrix. Otherwise the iron oxide may be in the form of a particle (>40 μm) or powder (<40 μm). Two different preparation methods, mechanical mixing and a modified Pechini method, are described below.

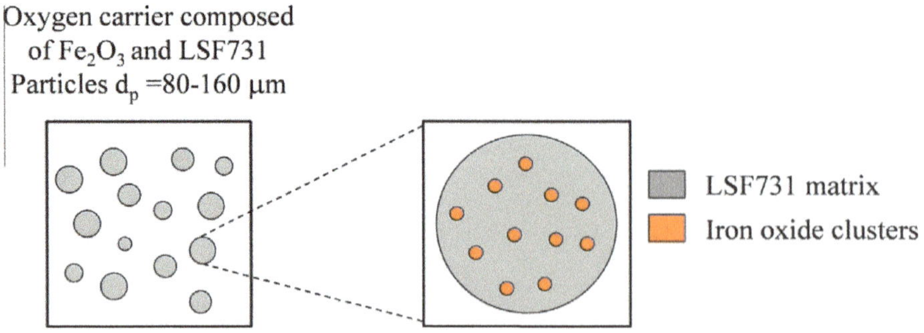

Figure 1. Schematic diagram of the OCM composite particles.

Mechanical Mixing

To prepare the composite materials by the mechanical mixing method, commercial sub-micron powders of LSF731 and iron oxide were mixed in two different mass ratios: 7:3 and 8.9:1.1. Pellets of 2 g were formed from these mixtures by applying 1500 kgf/cm^2 for 30 s with a hydraulic press, followed by sintering at 1250 °C in air for 4 h in a tubular furnace. The desired particle size was obtained by crushing and sieving the pellets (80–160 μm). This resulted in iron oxide clusters embedded in an LSF731 matrix.

The effect of the iron oxide cluster size was also studied. For this purpose, OCM particles were prepared by mixing commercial LSF731 sub-micron powder and iron oxide powder with a size range of 20–40 μm following the method described above. These 20–40 μm powders were obtained from pellet formation from the iron oxide sub-micron powders, calcination at 1250 °C for 4 h, crushing and sieving to 20–40 μm.

OCM particles containing only either iron oxide or LSF731 were also prepared as reference materials. Particles in the range 80–160 μm were made from commercially available iron oxide sub-micron powders. 2 g of the powders were pelletized with the hydraulic press, sintered at 1250 °C in air for 4 h in the tubular furnace, crushed and sieved to 40–80 μm and 80–160 μm particle sizes. The same process was also employed to fabricate LSF731 particles of specific particle sizes (80–160 μm).

Pechini Method

OCMs with a composite structure containing LSF731 and iron oxide were prepared from the metallic nitrates using a modified Pechini method. Reaction (1) shows the stoichiometry for the LSF731 preparation which was chosen such that the amount of ammonium nitrate was minimized and the by-products were gaseous water, carbon dioxide and nitrogen.

$$0.7[La(NO_3)_3 \cdot 6H_2O] + 0.3\ Sr(NO_3)_2 + Fe(NO_3)_3 \cdot 9H_2O + 1.684C_6H_8O_7 + 0.858NH_4NO_3 \rightarrow La_{0.7}Sr_{0.3}FeO_3 + 10.105\ CO_2 + 21.653H_2O + 3.708\ N_2 \quad (1)$$

Stoichiometric amounts of the metal nitrates, $La(NO_3)_3{\cdot}6H_2O$, $Sr(NO_3)_2$ and $Fe(NO_3)_3{\cdot}9H_2O$, citric acid and ammonium nitrate were dissolved in de-ionized water separately and stirred for 30 min to get clear solutions. Citric acid was used as a complexation agent while ammonium nitrate acted as an igniter, accelerating the combustion process. The ammonium nitrate solution was added after the citric acid to the nitrate solution. To ensure good mixing of all the constituents, the final solution was stirred for one hour.

Either iron oxide sub-micron powders or iron oxide 20–40 μm powders were added later to this mixture in the right proportion to get a mass ratio in the final material of 7:3 or 8.9:1.1. The solution was then transferred to a 2 L beaker and heated gradually on a hot plate up to ~200 °C inside a fume cupboard. As the water evaporation in the mixture neared completion, the solution turned brownish and viscous. Finally, in the form of a gel, it combusted with visible flames releasing gases according to reaction (1). The fluffy powder obtained was homogenised in a pestle and mortar and packed into ceramic crucibles to be dried overnight at 80–100 °C. After that, the solid was calcined in air at 800 °C for 2 h to convert the amorphous gel into crystalline LSF731. To obtain a suitable particle size for the chemical looping reactors, pellets were prepared from the obtained powders by uniaxially pressing approximately 2 g of the material and sintering at 1200 °C in air for 6 h. These pellets were subsequently crushed and sieved (80–160 μm).

Table 1 shows all the OCMs prepared for this work by different methods. The composite materials are designated with LSF followed by the weight fraction content of iron oxide and the first letters of the preparation method, MM for mechanical mixing and Pec for the modified Pechini method. The letter P indicates that sub-micron iron oxide powders were used while 20–40 indicates that 20–40 μm iron oxide powders were utilized in the preparation of the OCMs.

Table 1. Properties of the OCMs prepared in this work.

Sample	Preparation method	Iron oxide content (%)	Particle size (μm)	Cluster size (μm)	BET (m^2/g)
LSF731	MM	N/A	80–160	N/A	0.10
FeP	As received	100	Sub-micron	N/A	25.5
Fe40-80	MM	100	40–80	N/A	Unavailable
Fe80-160	MM	100	80–160	N/A	0.125
LSF-30Fe-Pec-P	Pechini	30	80–160	Sub-micron	4.97
LSF-30Fe-Pec-20-40	Pechini	30	80–160	20-40	1.10
LSF-11Fe-Pec-P	Pechini	11	80–160	Sub-micron	0.47
LSF-11Fe-Pec-20-40	Pechini	11	80–160	20–40	0.29
LSF-30Fe-MM-P	MM	30	80–160	Sub-micron	0.79
LSF-30Fe-MM-20-40	MM	30	80–160	20–40	<0.01
LSF-11Fe-MM-P	MM	11	80–160	Sub-micron	<0.01
LSF-11Fe-MM-20-40	MM	11	80–160	20–40	Unavailable

Thermogravimetric Analysis

The reactivity of the materials for hydrogen production from water splitting by chemical looping was studied in a thermogravimetric analyser Mettler-Toledo TGA/DSC 1 at atmospheric pressure and 850 °C. The particles were exposed alternatively to reducing and oxidising conditions. Three redox cycles were performed in all cases. A permanent flow of nitrogen of 50 ml STP/min (all flowrates are measured at STP) was used as protective gas for the balance. In addition to this flow, the reacting gas (50 ml/min) was fed to the TGA reactor during the operation. Prior to reacting with the solid sample, the reactive gas was mixed with the protective gas. The initial sample mass used in the TGA experiments was always 20–21 mg.

During the reduction period, 10% of carbon monoxide in nitrogen was introduced in the TGA for 30 min. This time was selected as a compromise to perform as many tests as possible while the duration was sufficient to study the kinetics of the reaction. After this, nitrogen was fed for 15 min to avoid the mixing of the reducing and the oxidising gases. The oxidation took place in two different stages. First, the particles were oxidized by 10% carbon dioxide in nitrogen for 20 min. Following a 15 min nitrogen purge, samples were fully oxidized with air for 10 min in order to return to the initial state.

In chemical looping, hydrogen is produced from water splitting during the oxidation with water. Nevertheless, in these experiments water was substituted by carbon dioxide, thermodynamically equivalent at 850 °C. The values of ΔG for reactions (2) and (3) at 850 °C are very similar, as well as the corresponding equilibrium constants. The validity of the substitution of steam by carbon dioxide was experimentally verified in a previous work by Liu et al. [13] reproducing the same cyclic conversion in a fluidized bed at 850 °C with both gases.

$$3Fe + 4H_2O(g) \rightarrow Fe_3O_4 + 4H_2(g) \quad \Delta G_{850°C} = -12.234\ kJ/mol \tag{2}$$

$$3Fe + 4CO_2(g) \rightarrow Fe_3O_4 + 4CO(g) \quad \Delta G_{850°C} = -16.391\ kJ/mol \tag{3}$$

Cycling Operation in a Microreactor

In order to determine the amount of hydrogen produced and the stability of the materials with time, 25 redox cycles were performed in a fully automated and programmable gas supply microreactor (CATLAB, Hiden Analytical, UK) at 1 atm and 850 °C. After the fresh sample (~50 mg) was placed inside, the microreactor was flushed with a helium flow of 50 ml/min prior to commencing testing. Each sample was subjected to multi-cycle isothermal reduction under 5% carbon monoxide in helium and isothermal oxidation under 5% water in helium with inlet flow rates of 50 ml/min. Reduction and oxidation steps were performed for 30 min. Helium was flushed for 15 min after reduction and oxidation in order to evacuate the reactor of any reactive gas or any product and to avoid mixing of the reducing and oxidising gases. The outlet gases of the microreactor, i.e. carbon monoxide, carbon dioxide, hydrogen, water and helium, were fed into a soft ionisation quadrupole mass spectrometer (QMS) through a heated capillary line for continuous online analysis. Temperature of the sample was monitored and controlled by an internal K-type thermocouple inserted into an inert alumina sleeve. Water was delivered into the microreactor by a saturator system (Grant, UK) using helium as a carrier. The set point of the water bath was 30.1 °C which corresponds to a water mole fraction of 4.2% in the carrier gas at atmospheric pressure. Nevertheless, the water fraction measured by a cold mirror dew point hygrometer (CMH-1, Alpha Moisture Systems, UK) used to calibrate the water signal for the QMS, indicated that this value oscillated between 5.0 and 5.5%, depending on the lab temperature. These differences are taken into account later.

Characterisation

Fresh OCM particles were physically and chemically characterised by several techniques. BET surface area was determined by adsorption/desorption of nitrogen at 77 K in a Beckman Coulter SA3100 instrument. The microstructure of the particles and elemental distribution in the solids were observed by a scanning electron microscope (SEM) XL30 ESEM-FEG equipped with a Rontec Quantax EDX analyser. The identification of crystalline chemical species was carried out by powder X-ray diffraction performed in a PANalytical X'Pert Pro Multipurpose Diffractometer (MPD).

RESULTS AND DISCUSSION

Characterisation

SEM analysis was used to determine the morphology of the fresh OCMs while the elemental composition was studied by EDX. The obtained particles had an irregular shape due to the kind of preparation methods used, including a pellet crushing step. The morphology of the surface of several OCM particles, LSF-30Fe-Pec-P, LSF-30Fe-MM-P, LSF-30Fe-Pec-20-40 and LSF-30Fe-Pec-MM-20-40, is shown in Fig. 2. The samples prepared by the Pechini method (a, c) were more porous than those obtained by mechanical mixing (b, d). Similar trends were obtained in the BET analyses with the samples prepared by the Pechini method having higher surface areas.

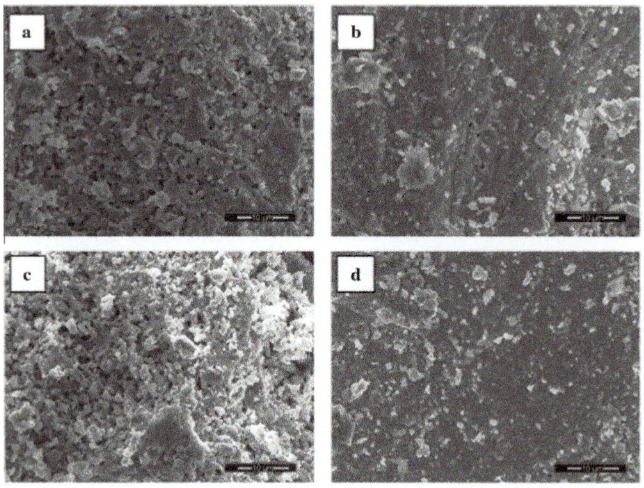

Figure 2. SEM pictures of (a) LSF-30Fe-Pec-P, (b) LSF-30Fe-MM-P, (c) LSF-30Fe-Pec-20-40 and (d) LSF-30Fe-MM-20-40.

The elemental composition of the OCMs was studied by EDX analyses which proved that the desired composite structure, encapsulating iron oxide in an LSF731 matrix, was obtained using both preparation methods. Areas with a higher concentration of iron oxide were found in the particles when either sub-micron iron oxide powders or 20–40 µm iron oxide powders were used during the preparation. As an example, Fig. 3 shows a SEM and EDX of the LSF-30Fe-MM-20-40 OCM. Iron oxide is in the form of layers, typical of the structure of haematite, while LSF731 is in the form of small granules. Grain boundaries can be distinguished. Fig. 3b shows the EDX analysis of the same OCM which confirms the presence of iron oxide in specific areas of the particle surrounded by the LSF731.

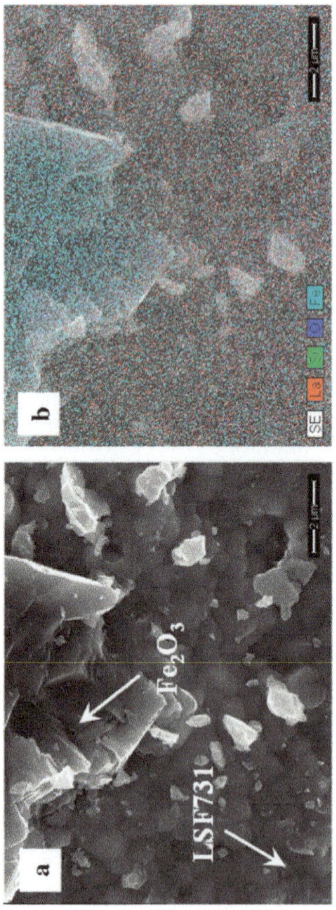

Figure 3. (a) SEM picture and (b) EDX analysis of LSF-30Fe-MM-20-40.

Formation of the desired phases, iron oxide and LSF731, was confirmed by XRD analyses (not shown here) of the composite materials prepared by either mechanical mixing or a modified Pechini method. Reaction between iron and strontium to form strontium ferrites at high temperature could take place to a small extent during the preparation of the composite materials. Nevertheless, this interfacial phase was not detected in the XRD analyses performed. Galinsky et al. [9] found that strontium ferrite phase coexisted with the perovskite and iron oxide due to an incomplete solid-state reaction during the preparation of a composite material. After the first redox cycle in a TGA, the ferrite was incorporated to the LSF structure.

Effect of the Cluster Size

In order to determine if the size of the iron oxide clusters in the composite structure could have an effect on the reactivity of the OCMs, a preliminary study was carried out in the TGA. Iron oxide particles of different sizes (40–80 μm, 80–160 μm) and sub-micron powder were tested feeding carbon monoxide during the reduction and carbon dioxide during the oxidation at 850 °C and atmospheric pressure. Fig. 4 shows the normalised weight gain during the reduction and the oxidation, respectively, during the first and the third redox cycle. The weight variation was normalised by dividing by the expected available oxygen in the iron oxide, assuming that it starts in the form of Fe_2O_3. During the reduction period of the first cycle, the sub-micron powder, FeP, was reduced faster and to a significantly higher extent than the iron oxide particles (Fe40-80 and Fe80-160). This could be related to the surface area of the particles shown in Table 1. FeP had a surface area two orders of magnitude higher than Fe80-160. Therefore the diffusional lengthscales required to access the oxygen are shorter. In the third cycle, a significant decrease in weight loss was observed for the iron powders, FeP, while the larger particles Fe40-80 and Fe80-160 showed little difference in weight loss. It is believed that FeP experienced sintering and agglomeration increasing the particle size to lengthscales that are now in excess of the diffusional lengthscale. During the oxidation, a similar reduction in weight gain was observed for the iron oxide powders due to sintering or agglomeration. Values for the 80–160 μm and 40–80 μm particles were similar in both cycles, although the larger particles presented a higher oxidation rate. Thermal treatment of the particles during the preparation could cause sintering and grain growth. As grain boundary diffusion could be the rate determining step, during reaction the grain structure would be expected to influence kinetics, explaining the results shown in Fig. 4. It must be highlighted that FeP was not subjected to sintering at 1250 °C as were Fe40-80 and Fe80-160. Operation at high temperature favoured the sintering of the sub-micron iron oxide powders.

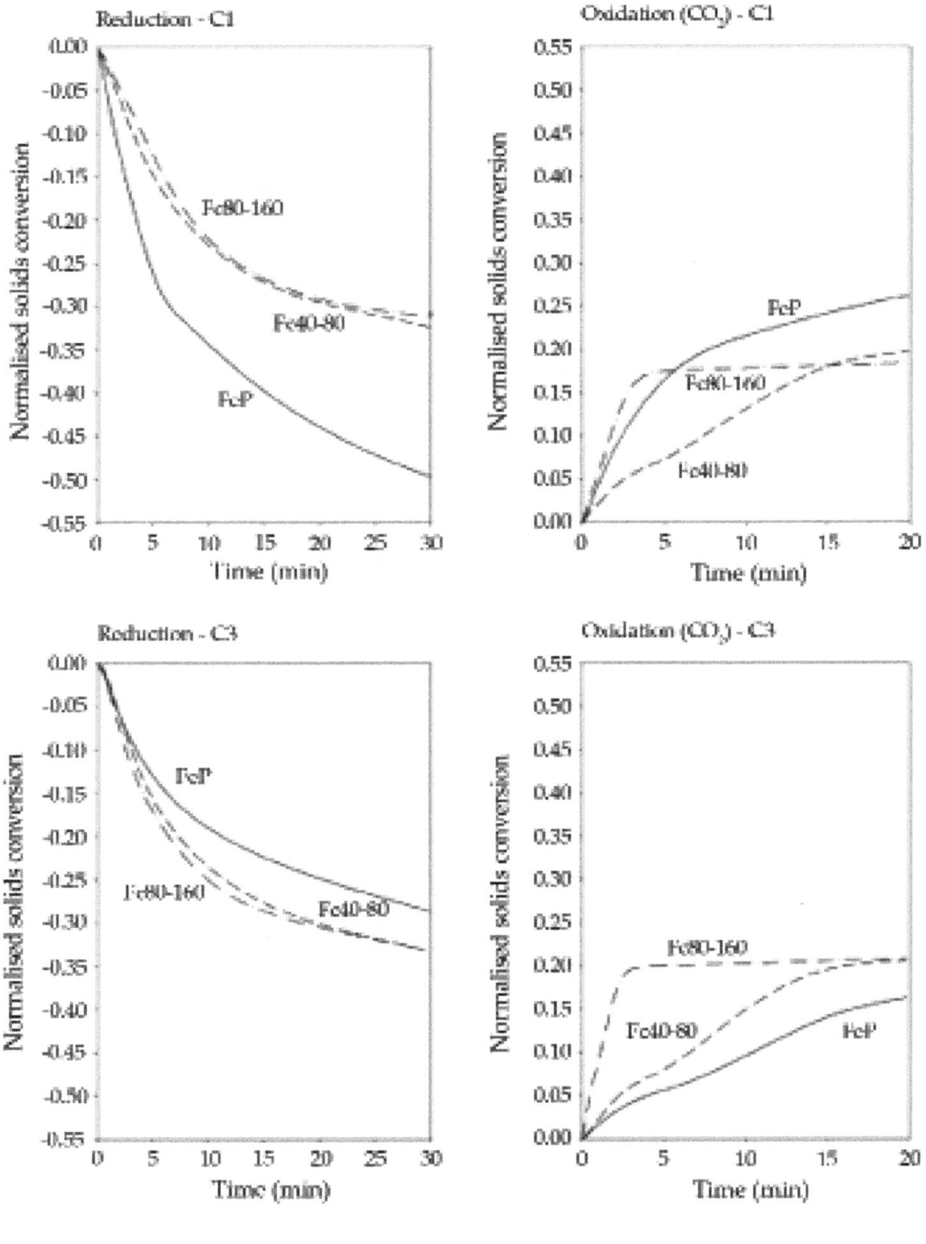

Figure 4. Weight loss and weight gain during the reduction and the oxidation, respectively, of the first and third redox cycles for particles of iron oxide with different sizes (80–160 μm and 40–80 μm) and sub-micron powders. Reduction: 10% carbon monoxide in nitrogen; Oxidation: 10% carbon dioxide in nitrogen; Temperature: 850 °C.

The oxygen consumed, summarised in Table 2, was calculated as the number of oxygen micromoles gained by the sample during the oxidation normalized with respect to the total amount of oxygen micromoles present in the iron oxide (as Fe_2O_3). Only between 16% and 26% of this oxygen was involved in the oxidation reaction suggesting that oxidation kinetics with CO_2 were slow.

Table 2. Oxygen consumption during the oxidation period of the first and the third redox cycle in the TGA for iron oxide particles with different particle size.

Iron oxide sample	Oxygen consumed (µmol O/µmol O in iron oxide)	
	Cycle 1	Cycle 3
FeP	0.26	0.16
Fe40-80	0.19	0.21
Fe80-160	0.18	0.21

Results obtained with the iron oxide particles of different sizes (sub-micron, FeP, 40–80 µm and 80–160 µm) in Table 2 and Fig. 4, illustrated that the highest reactivity corresponded to the commercial sub-micron powders during the first redox cycle. This suggests that if these iron oxide powders could be well dispersed into a LSF731 matrix, avoiding agglomeration and sintering, a significant increase in the reactivity of the composite material should be observed. To analyze the effect of the nominal cluster size on the reactivity several tests were performed in the TGA under the same operating conditions as experiments in Fig. 4. No clear effect of the cluster size was observed. SEM-EDX analyses of the composite OCMs demonstrated that the size of the iron oxide clusters was on the scale of 20 µm despite the initial cluster size used during the preparation (Fig. 3). Agglomeration of the sub-micron powders took place then during the OCM preparation. This could explain the similarities in the reactivity tests when sub-micron iron oxide and 20–40 µm powders were used.

Reactivity of the Composite Materials

The reactivity of the composite materials was studied in a TGA performing CO/CO_2 redox cycles at 850 °C. Fig. 5 shows the normalised solids conversion obtained during the third redox cycle. The normalised solids conversion was defined as the ratio between the weight gain during the reduction or oxidation with respect to the expected available oxygen.

$$\text{Normalised solids conversion} = X = \frac{\Delta m}{\text{Expected available oxygen}} = \frac{\Delta m}{(\alpha \cdot R_{O,Fe_2O_3} + (1-\alpha) \cdot R_{O,LSF731} \cdot \beta) \cdot m} \quad (4)$$

where Δm is the weight gain of the sample, m is the actual sample weight, α is the weight fraction of iron oxide in the OCM (30% or 11%) and $R_{O,Fe2O3}$ and $R_{O,LSF731}$ are the oxygen capacities of iron oxide and LSF731, respectively. The oxygen capacity of an OCM, R_O, can be calculated as:

$$R_O = \frac{m_{ox} - m_{red}}{m_{ox}} = \frac{3 \cdot M_O}{M_i} \quad (5)$$

where m_{ox} is the weight of the fully oxidised form of the OCM and m_{red} is the weight of the fully reduced form. M_O is the molecular weight of oxygen and M_i is the molecular weight of either LSF731 or Fe_2O_3. For iron oxide, $R_{O,Fe2O3}$ would be 30% when Fe_2O_3 is reduced completely to Fe, and $R_{O,LSF731}$ is equal to 21.1%. As not all the oxygen in the LSF731 is available to react, an additional coefficient, β, is required. β is the weight fraction of available oxygen from the LSF731 which was obtained from the first redox cycle in the TGA, equal to 6.1 wt.%.

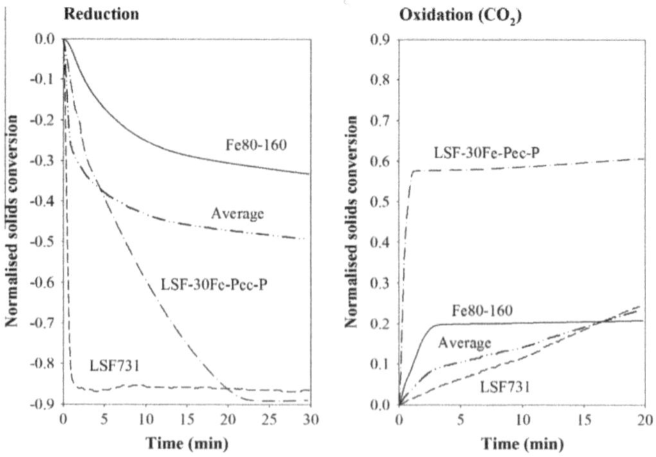

Figure 5. Normalised solids conversion for the reduction and oxidation of (___) Fe80-160, (_ __) LSF731 and (_._.) LSF-30Fe-Pec-P during the third redox cycle in the TGA. (••••••••••••) Average normalised solids conversion considering the weight fraction of oxygen from LSF731 and iron oxide is included in the figure. Reduction: 10% carbon monoxide in nitrogen; Oxidation: 10% carbon dioxide in nitrogen; Temperature: 850 °C.

The normalised solids conversion and reaction rate for LSF731 appear high during reduction. The slope of the LSF731 can be deceptive, however, as significantly less oxygen is available due to the beta coefficient associated with the perovskite. The iron oxide, on the other hand, has very high oxygen availability yet the normalised solids conversion appears much lower. This is due the slow kinetics of iron oxide reduction.

The average normalised solids conversion at any time shown in Fig. 5, was modelled as the addition of the individual normalised solid conversion of LSF731 and iron oxide multiplied by the respective weight fraction of available oxygen in the OCM coming from the iron oxide, γ_{Fe2O3}, or the LSF731, γ_{LSF731}.

$$\gamma_{Fe_2O_3} = \frac{\alpha \cdot R_{O,Fe_2O_3}}{\alpha \cdot R_{O,Fe_2O_3} + (1-\alpha) \cdot \beta \cdot R_{O,LSF731}} \quad (6)$$

$$\gamma_{LSF731} = \frac{(1-\alpha) \cdot \beta \cdot R_{O,LSF731}}{\alpha \cdot R_{O,Fe_2O_3} + (1-\alpha) \cdot \beta \cdot R_{O,LSF731}} \quad (7)$$

$$Average_t = \gamma_{Fe_2O_3} \cdot X_{Fe_2O_3} + \gamma_{LSF731} \cdot X_{LSF731} = \gamma_{Fe_2O_3} \cdot \frac{\Delta m_{Fe_2O_3,t}}{R_{O,Fe_2O_3} \cdot m_{Fe_2O_3,0}} + \gamma_{LSF731} \cdot \frac{\Delta m_{LSF731,t}}{R_{O,LSF731} \cdot \beta \cdot m_{LSF731,0}} \quad (8)$$

where $m_{i,0}$ is the initial weight of the solid i, and $\Delta m_{i,t}$ is the weight gain at any given time, t. Sub-script i is either iron oxide or LSF731. The absolute value of the LSF-30Fe-Pec-P normalised solids conversion was higher than the average during both the reduction and oxidation. This seems to indicate that the reactivity of this material did not correspond just to the simple additive model proposed (Eq. (8)). This improvement in reactivity could be due to the inhibition of sintering and agglomeration of the iron oxide when well dispersed within the LSF731 matrix. It would be possible to construct an average normalised solids conversion using the results from the first reduction (shown in Fig. 4) with sub-micron iron oxide powder; sintering being yet to occur to a high extent. Regardless of if the average is calculated from first or third cycle conversions, or if sub-micron powder, FeP, or particles, Fe80-160, of iron oxide are used, the composite material, LSF-30Fe-Pec-P, always performs better than any average model. This improvement must be due to a synergetic effect. It is suggested that the structure of the composite material enhanced the availability of oxygen and, therefore, the hydrogen production, as shown below.

The rate of oxygen uptake was calculated as the first derivate of the moles of oxygen reacted during the oxidation with time. The composite material LSF-30Fe-Pec-P showed an enhanced rate of oxygen uptake during the oxidation with carbon dioxide compared to iron oxide and LSF731 alone, as shown in Fig. 6.

In the case of Fe80-160, a decrease in the carbon dioxide and hydrogen production was observed during the reduction and the oxidation, respectively, with the number of cycles. Evidence for phase changes could not be distinguished in terms of any kinetic impact on the outlet gas phase mole fractions. When LSF731 was used, hydrogen was formed only for a very short period of time. A very narrow peak can be observed in Fig. 7. Nevertheless, the shape of the curves was the same during the 1st and the 25th cycle, indicating that the material behaviour was stable with time as demonstrated by Murugan et al. [8] in a previous work. Carbon dioxide and hydrogen production were much higher using the composite material LSF-30Fe-Pec-P than with either the LSF731 or iron oxide alone. A larger peak was obtained with clear kinetic evidence of what may be phase transitions between Fe–FeO and FeO–Fe_3O_4. Mass balances showed that, within measurement uncertainty, carbon formation did not take place with any of the OCMs.

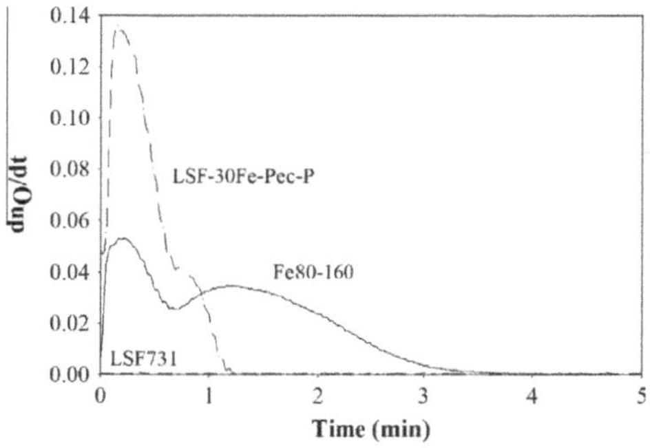

Figure 6. Rate of oxygen uptake during the oxidation period of the third redox cycle working with (___) Fe80-160, (___) LSF731, and (_._.) LSF-30Fe-Pec-P as OCMs. Oxidising agent: 10% carbon dioxide in nitrogen; Temperature: 850 °C.

Cyclic Operation in the Microreactor

Several experiments were carried out in the CATLAB in order to determine the hydrogen produced with different LSF731- and iron oxide-containing OCMs and the stability of these materials over a larger number of cycles. The sample (50 mg) was initially pre-oxidised with 5% oxygen in helium. After that, 25 cycles using 5% carbon monoxide in helium during the reduction and 5% water in helium during the oxidation were performed at 850 °C and 1 atm.

Fig. 7 shows the gas molar fractions measured with the QMS during the 1st and 25th cycle for Fe80-160, LSF731, and LSF-30Fe-Pec-P OCMs. Sharp initial carbon monoxide peaks at the beginning of the reduction were due to valve changes and backpressure in the CATLAB system and should be ignored. A mixture of hydrogen and water was always obtained when the oxidation period started and pure hydrogen was not detected in any case.

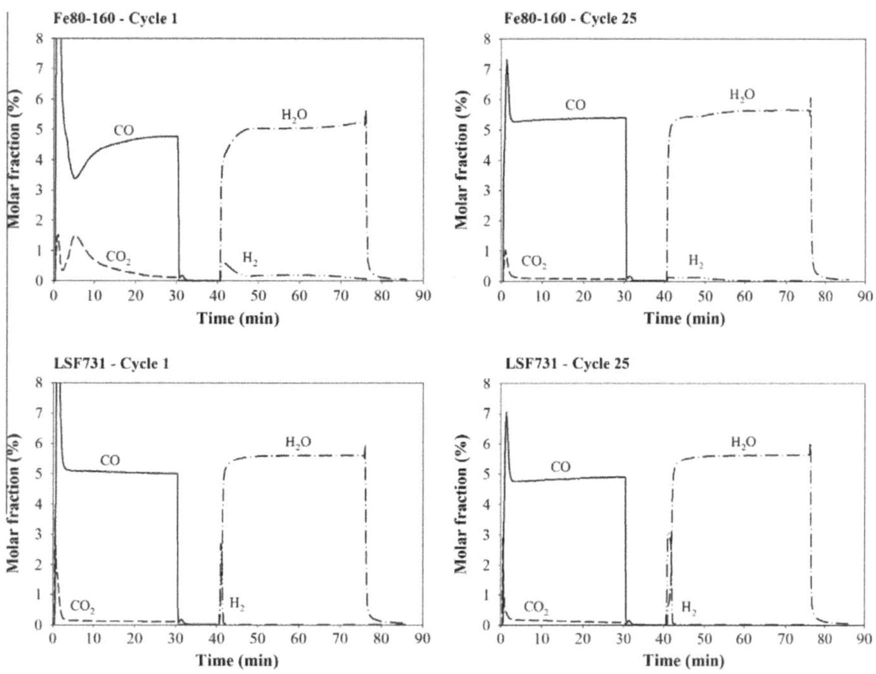

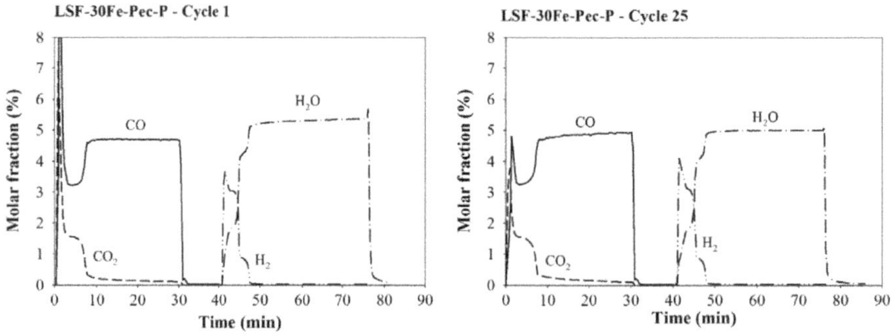

Figure 7. Mole fraction during the 1st and 25th reduction and oxidation using Fe80-160, LSF731 and LSF-30Fe-Pec-P as OCMs. Reduction: 5% carbon monoxide in helium; Oxidation: 5% water in helium; Temperature: 850 °C. Approximate sample mass: 50 mg.

The hydrogen production in micromoles, calculated from the integral of the hydrogen mole fraction versus time data, is depicted in Fig. 8. Although the Fe80-160 OCM was more stable than the iron oxide sub-micron powder, FeP, during the experiments performed in the TGA and was capable of maintaining reactivity over three cycles, a significant decrease in hydrogen production was observed over 25 cycles in the CATLAB. With LSF731 or LSF-30Fe-Pec-P, the hydrogen production remained constant after the few cycles that were needed for stabilization. The hydrogen production with LSF-30Fe-Pec-P in micromoles (367 μmol) during the oxidation period of the 25th cycle was 8 times higher than the production with Fe80-160 (46 μmol) and 4 times higher than the production obtained with the LSF731 (93 μmol). The expected available oxygen (Eq.(4)) for each material is equivalent to the expected amount of hydrogen production on a weight basis. In micromoles, the expected amount of hydrogen that could be obtained using the LSF-30Fe-Pec-P is 299 μmoles. LSF-30Fe-Pec-P produced 23% more hydrogen than expected. This indicates that the perovskite is more reducible in the composite form. This may be related to the synergetic behaviour observed in the TGA data.

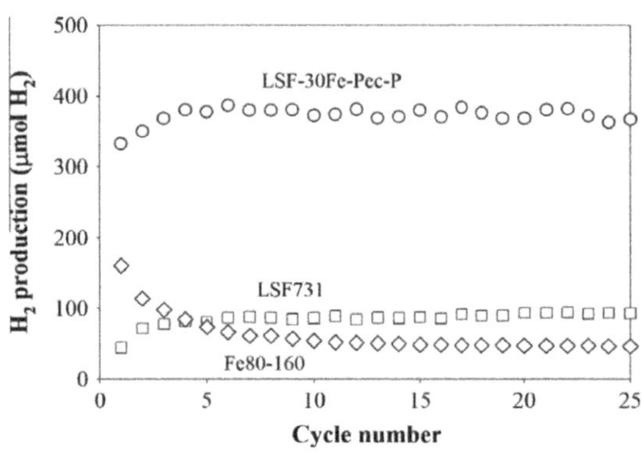

Figure 8. Hydrogen production in μmol during isothermal cycling in the CATLAB with (◇) Fe80-160, (□) LSF731 and (○) LSF-30Fe-Pec-P as OCMs. Reduction: 5% carbon monoxide in helium; Oxidation: 5% water in helium; Temperature: 850 °C. Approximate sample mass: 50 mg.

Effect of the Preparation Method and the Iron Oxide Content

Fig. 9 shows the hydrogen produced during the oxidation period with 5% water in helium using OCMs with 30 and 11 wt.% iron oxide fractions. The micromoles of hydrogen produced were lower when the material was prepared by mechanical mixing. It is suggested that the higher porosity of the OCMs prepared by Pechini method facilitated redox processes, improving the water splitting and increasing the amount of hydrogen produced. The micromoles of hydrogen obtained with the OCMs containing 11 wt.% of iron oxide were lower than with 30 wt.% of iron oxide as expected. The expected available oxygen of the OCMs is dominated by the iron oxide since the fraction of reducible oxygen in the LSF731 was very low. When the hydrogen production was normalized with respect to the expected available oxygen in the samples, there was no significant influence of the iron oxide content on the amount of hydrogen produced. As in the previous section, all the composite materials formed more hydrogen than expected considering that only 6.1 wt.% of the oxygen in the LSF731 was active for the process (β). The synergetic effect in the composite materials was observed again.

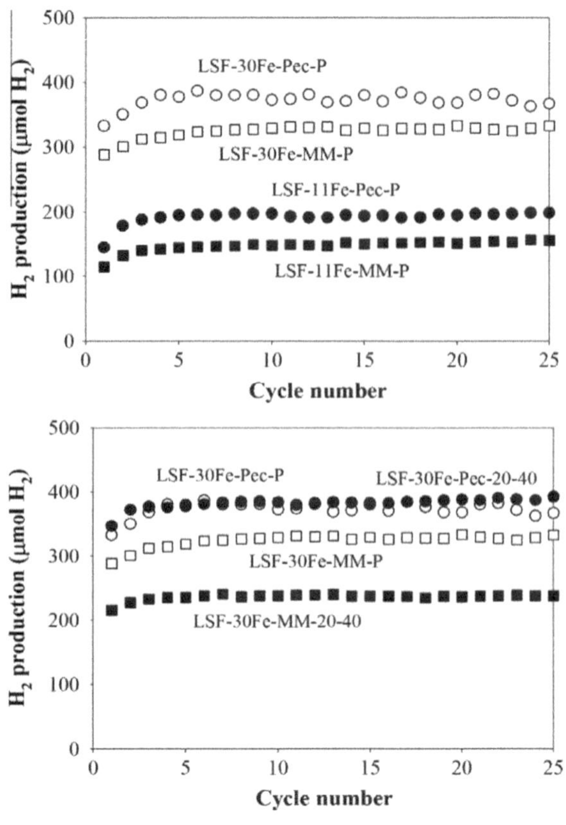

Figure 9. Hydrogen production during isothermal cycling in the

CATLAB (a) Effect of the preparation method and (b) effect of the cluster size. Reduction: 5% carbon monoxide in helium; Oxidation: 5% water in helium; Temperature: 850 °C. Approximate sample mass: 50 mg.

From an industrial point of view, a material with a higher content of iron oxide would be more suitable as it would allow a higher hydrogen production at a lower cost due to the lower LSF731 fraction. To prepare the large quantities of OCM needed in an industrial installation, a preparation method as simple as possible should be used. This would reduce the steps and time needed to get the final material and, as a consequence, the cost of the manufacture. Results obtained with the OCM prepared by Pechini method or MM were very similar in terms of hydrogen production. This suggests that the more involved Pechini method adds no extra value and the simple mechanical mixing method would be sufficient to prepare a composite OCM with better performance than the individual components.

Values of hydrogen production were similar using either Fe_2O_3 sub-micron powders during the preparation or 20–40 μm powders as can be seen in Fig. 9b. Differences observed between the OCMs LSF-30Fe-MM-P and LSF-30Fe-MM-20-40, both prepared by mechanical mixing, are due to a different mass of the sample used in the CATLAB. Data shown in the graph are not normalized.

CONCLUSIONS

In this work, the behaviour of several OCMs composed of LSF731 and iron oxide for hydrogen production by chemical looping was investigated. The objective was to obtain a composite structure with iron oxide clusters well dispersed and embedded in an LSF731 matrix. The OCMs were prepared by two different methods, i.e. mechanical mixing and a modified Pechini method, with different iron oxide weight fractions, 11 and 30 wt.%. SEM pictures and EDX analysis confirmed that the desired structure could be obtained. Although sub-micron powder and 20–40 μm powder of iron oxide were used to study the effect of the iron oxide nominal cluster size on reactivity, a clear conclusion could not be made since some agglomeration of the sub-micron powder took place during the preparation resulting in little difference between actual cluster sizes.

CO/CO_2 redox cycles performed in a TGA with LSF-30Fe-Pec-P showed that the reactivity of the composite OCMs did not merely correspond to the proportional addition of the oxygen capacities of both individual components, LSF731 and iron oxide. There was a synergetic effect resulting in a higher capacity of the composite material.

The stability of the composite materials for hydrogen production was studied in a microreactor, reducing with 5% carbon monoxide in helium and oxidising with 5% water in helium at 850 °C over 25 redox cycles. Hydrogen production was stable for LSF731 and LSF-30Fe-Pec-P but iron oxide particles (80–160 μm) experienced a decrease in hydrogen production due to agglomeration. The hydrogen production in micromoles during the oxidation of the 25th cycle was eight times higher for the composite material than for iron oxide and four times higher than for LSF731 alone. A 23% increase in the expected hydrogen production, equivalent to the expected available oxygen, was observed for LSF-30Fe-Pec-P. This suggests that LSF731 is more reducible when part of the composite material.

The results demonstrate that composite OCMs for chemical looping can be rationally designed. Dispersion of small iron oxide clusters within an LSF731 matrix avoids iron oxide agglomeration while the perovskite facilitates oxygen transport to and from the iron oxide. A combination of both compounds in the OCM particles allows an increase in the amount of hydrogen obtained in a chemical looping water–gas shift process.

ACKNOWLEDGEMENTS

Authors wish to thank Dr Stuart Scott, Wen Liu and Mohammad Ismail from Cambridge University for their help with the TGA work. CD thanks EPSRC for funding via a PLATFORM Grant, EP/G012865/1. CT thanks EPSRC for funding via a doctoral training award.

REFERENCES

1. Thursfield A, Murugan A, Franca R, Metcalfe IS. Chemical looping and oxygen permeable ceramic membranes for hydrogen production – a review. Energy Environ Sci 2012;5:7421–59.
2. Hacker V, Fankhauser R, Faleschini G, Fuchs H, Friedrich K, Muhr M, et al. Hydrogen production by steam-iron process. J Power Sources 2000;86: 531–5.
3. Bohn CD, Müller CR, Cleeton JP, Hayhurst AN, Davidson JF, Scott SA, et al. Production of very pure hydrogen with simultaneous capture of carbon dioxide using the redox reactions of iron oxides in packed beds. Ind Eng Chem Res 2008;47:7623–30.
4. Rydén M, Arjmand M. Continuous hydrogen production via the steam-iron reaction by chemical looping in a circulating fluidized-bed reactor. Int J Hydrogen Energy 2012;37:4843–54.
5. Dennis JS, Scott SA. In situ gasification of a lignite coal and CO2 separation using chemical looping with a Cu-based oxygen carrier. Fuel 2010;89:1623–40.
6. Bohn CD, Cleeton JP, Müller CR, Chuang SY, Scott SA, Dennis JS. Stabilizing iron oxide used in cycles of reduction and oxidation for hydrogen production. Energy Fuel 2010;24:4025–33.
7. Bouwmeester HJM, Kruidhof H, Burggraaf AJ. Importance of the surface exchange kinetics as rate limiting step in oxygen permeation through mixedconducting oxides. Solid State Ionics 1994;72(2):185–94.
8. Murugan A, Thursfield A, Metcalfe IS. A chemical looping process for hydrogen production using iron-containing perovskites. Energy Environ Sci 2011;4:4639–49.
9. Galinsky NL, Huang Y, Shafiefarhood A, Li F. Iron oxide with facilitated O2 transport for facile fuel oxidation and CO2 capture in a chemical looping scheme. ACS Sustain Chem Eng 2013;1:364–73.
10. Shafiefarhood A, Galinsky N, Huang Y, Chen Y, Li F. Fe2O3@LaxSr1xFeO3 coreshell redox catalyst for methane partial oxidation. ChemCatChem 2014;6:790–9.
11. Chen Y, Galinsky N, Wang Z, Li F. Investigation of perovskite supported composite oxides for chemical looping conversion of syngas. Fuel 2014;134:521–30.

12. Galinsky NL, Shafiefarhood A, Chen Y, Neal L, Li F. Effect of support on redox stability of iron oxide for chemical looping conversion of methane. App Catal B-Environ 2015;164:371–9.

13. Liu W, Dennis JS, Scott SA. The effect of addition of ZrO2 to Fe2O3 for hydrogen production by chemical looping. Ind Eng ChemRes 2012;51:16597–609.

INDEX